CRC Handbook of Electrophoresis

of

Volume II
Lipoproteins in Disease

Editors

Lena A. Lewis

Emeritus Consultant, Division of Research and of Laboratory Medicine
The Cleveland Clinic Foundation
Clinical Professor of Chemistry
Cleveland, Ohio

Jan J. Opplt

Associate Professor of Pathology
Clinical Professor of Chemistry
Cleveland State University
Director, Division of Clinical Chemistry Pathology
Cleveland Metropolitan General Hospital
Cleveland, Ohio

CRC Press, Inc.
Boca Raton, Florida

Library of Congress Cataloging in Publication Data

Lewis, Lena Armstrong, 1910-
    CRC handbook of electrophoresis.

    Bibliography: p.
    Includes index.
    1. Contents: v.1. Lipoproteins. — v.2. Lipoproteins in disease.
Electrophoresis. I. Opplt, Jan J., joint author.
II. Title.
QD117.E45L48   543´.087   78-10651
ISBN 0-8493-0572-1.

    Direct all inquiries to CRC Press, 2000 N.W. 24th Street, Boca Raton, Florida, 33431.

© 1980 by CRC Press, Inc.

International Standard Book Number 0-8493-0572-1

Library of Congress Card Number 78-10651
Printed in the United States

PREFACE

The concept of a handbook of electrophoresis, comparable to the *Handbook of Chromatography,* appeared to be a sound and plausible idea, when first suggested. Some modification of this idea, however, resulted after consideration of the vastness of the electrophoretic literature which has accumulated during the 40 years since the monumental work of Arne Tiselius. Thus, it was decided to give in-depth coverage to electrophoresis as applied to the study of lipoproteins in the human being. This aspect was selected because of the many different electrophoretic techniques which have been employed in elucidating physiological variations and biochemical and physical properties of lipoproteins and of apolipoprotein molecules. Knowledge in the lipoprotein field has developed as in electrophoresis, rapidly, and has in some aspects been significantly affected by the application of developments in electrophoretic techniques.

The *Handbook of Electrophoresis,* Volumes I and II, provides basic information concerning lipoprotein molecules, their electrophoretic properties, basic principles of electrophoresis, and the techniques employed in studying lipoproteins. Alterations in lipoprotein electrophoretic patterns in many diseases states are presented, and their significance, discussed. A detailed bibliography of the literature dealing with lipoproteins is included in the appendix, which provides invaluable information for all who want to pursue in-depth investigation of particular aspects of the lipoprotein literature.

EDITORS

Dr. Lena A. Lewis is Emeritus Consultant at The Cleveland Clinic Foundation in the Division of Laboratory Medicine and in the Division of Research. She is a Clinical Professor of Chemistry at Cleveland State University. Dr. Lewis has an AB from Lindenwood College (1931), an MA from Ohio State University (1938), and Ph.D. from Ohio State University. She served for 10 years on the editorial board of Clinical Chemistry, is author of Electrophoresis in Physiology (1950), 2nd edition, 1960, Charles C Thomas, Springfield, Illinois. Dr. Lewis received an honorary LLD degree in 1952 from Lindenwood College, was elected a fellow in the New York Academy of Science in 1977 and received the Boehringer-Mannheim Award from the American Association of Clinical Chemists for outstanding contributions to Clinical Chemistry in the field of lipids and lipoproteins.

She is a contributor to *Endocrinology, Clinical Endocrinology and Metabolism,* the *American Journal of Medicine, Clinical Chemistry,* and the *American Journal of Physiology, Science.*

Dr. Lewis was the President (1970-1971) of the Northeast Ohio chapter of the American Association for Clinical Chemists. She also is a member of the American Physiological Society, the Endocrine Society, The AAAS; Fellow Atherosclerosis Council of the American Heart Association, the New York Academy of Science (Fellow), and the Ieota Club (International Women's Service Club.

She is listed the *American Men of Science, Who's Who in America* (Midwest section), the *Who's Who in American Education,* the *American Women in Science, Personalities of the West & Midwest,* and the *Who's Who in the World of Women.*

Dr. Jan J. Opplt received his education at the Charles University in Prague, Czechoslovakia. He was awarded the following degrees: M.S. (1942), M.D. (1949), Ph.D. (1952), Scientific Degree in Medical Science (1966), and Scientific-Educational Degree-Docent (1969).

He specialized in Clinical Pathology (1956) and in Clinical Chemical Pathology (1968).

He served as Assistant Professor in the Department of Medical and Clinical Chemistry School of Medicine, Charles University (1948—1950). Thereafter, he was appointed as Chairman of the Department of Clinical Chemistry and later as Associate Professor (Docent) at the School of Medicine and Hygiene, Charles University (1951—1969).

From 1970 to the present he has been serving as Clinical Professor in the Department of Chemistry, Cleveland State University. Since 1971, he has been acting as Director of the Division of Clinical Chemistry, Cleveland Metropolitan General Hospital and as Associate Professor of Pathology, Case Western Reserve University.

He acted as a Member of the Czechoslovak Medical Society J.E. Purkyne's (1948—1969) and on its Board of Clinical Biochemistry (1962—1969). He was also a member of the European Diabetes Society (1962—1969). He is presently active in the following professional societies: Cleveland Academy of Science (member since 1969); American Heart Association (Fellow since 1970, Scientific Committee, Ohio Chapter, 1976—1978; American Association of Clinical Chemists (member since 1970, Chairman of the Cleveland Section, 1975) and National Academy of Clinical Biochemistry (Fellow since 1979).

His awards and honors include: Prize of J.E. Purkyne, ACPS (CPSP, 1952, Prague, CSR; Fellowship Award, WHO, 1959, (Professors Tiselius and Svedberg) Uppsala, Sweden; Fellowship Award, CCF, 1970, (Dr. L.A. Lewis), USA; Gold Award, American Society of Clinical Pathologists (ASCP and CAP), 1971, USA; Honorable Men-tion Award, Ohio State Medical Association, 1973, USA; Certificate of Recognition, CSU, 1977, USA and Service Award, American Heart Association, 1978, USA.

His main professional interests are in the fields of Internal Medicine, Clinical Chemical Pathology, Biochemistry, and Physical Chemistry. He developed courses for medical students, residents and graduate students in differential diagnostic procedures and clinico-pathologic correlations as well as clinical chemical pathology, used in therapeutic and preventive medicine.

He has devoted 27 years, investigating the physiological and pathological metabolism of plasma proteins and lipoproteins. He developed specific techniques for separation and physical analyses, which opened his study of plasma lipoproteins and apolipoproteins on a molecular basis. His publications include 87 scientific papers, 22 abstracts and contributions in seven books.

ADVISORY BOARD

CONTRIBUTORS

Paul S. Bachorik
Assistant Professor
Department of Pediatrics
School of Medicine
Johns Hopkins University
Baltimore, Maryland

Robert C. Bahler
Associate Professor of Medicine
School of Medicine
Case Western Reserve University
Cleveland, Ohio

M. Barclay
Woodland Clearing
Greenwich, Connecticut

R. K. Barclay
Woodland Clearing
Greenwich, Connecticut

K. K. Carroll
Professor of Biochemistry
Department of Biochemistry
University of Western Ontario
London, Ontario
Canada

Charles J. Glueck
Professor of Medicine
Associate Professor of Pediatrics
University of Cincinnati Medical School
Cincinnati, Ohio

Manjula S. Kumar
Head, Radioimmunoassay Section
The Cleveland Clinic Foundation
Cleveland, Ohio

Lena A. Lewis
Emeritus Consultant, Divisions of
  Research and of Laboratory Medicine
The Cleveland Clinic Foundation
Clinical Professor of Chemistry
Cleveland State University
Cleveland, Ohio

Allen H. Mackenzie
Head, Sections of Clinical and Pediatric
  Rheumatology
Department of Rheumatic and
  Immunologic Disease
The Cleveland Clinic Foundation
Cleveland, Ohio

Betty H. Masket
Chemist
Review Branch
Division of Extramural Affairs
National Heart, Lung and Blood Institute
National Institutes of Health
Bethesda, Maryland

Herbert K. Naito
Head, Lipid-Lipoprotein Laboratories
Division of Laboratory Medicine
The Cleveland Clinic Foundation
Senior Scientist
Department of Atherosclerosis and
  Thrombosis Research
Division of Research
Clinical Associate Professor
Department of Chemistry
Cleveland State University
Cleveland, Ohio

Maryanne S. Olynyk
Research Technologist
Department of Atherosclerosis
  and Thrombosis Research
Division of Research
The Cleveland Clinic Foundation
Cleveland, Ohio

Jan J. Opplt
Associate Professor of Pathology
Clinical Professor of Chemistry
Cleveland State University
Director, Division of Clinical Chemistry
  Pathology
Cleveland Metropolitan General Hospital
Cleveland, Ohio

Marie A. Opplt
Senior Research Assistant
Department of Medicine
Case Western Reserve University
Cleveland, Ohio

D. C. K. Roberts
Lecturer
Human Nutrition Unit
Department of Biochemistry
University of Sydney
Sydney, New South Wales

Dietrich Seidel
Professor of Clinical Chemistry
Head, Department of Clinical Chemistry
Institute of Clinical Chemistry
Medical School
University of Göttingen
Göttingen, Germany

Paula M. Steiner
Assistant Director, Clinical Chemistry
Cincinnati General Hospital
Cincinnati, Ohio

Heinrich Wieland
Doctor
Institute of Clinical Chemistry
Medical School
Institute of Göttingen
Göttingen, Germany

Charles E. Willis
Staff-Senior Consultant
Department of Biochemistry
The Cleveland Clinic Foundation
Cleveland, Ohio

NOMENCLATURE OF LIPOPROTEINS

The nomenclature of lipoproteins has developed over the years as knowledge of their physical, chemical, and immunological properties has been elucidated. The plasma lipoproteins have been known for many years to have different electrophoretic properties. The major plasma lipoproteins of normal, healthy human beings have the electrophoretic mobility of α- and of β-globulins, respectively. In addition to the major components, additional fractions of characteristic mobility and lower concentration may be demonstrated in some plasma.

The various nomenclatures of lipoproteins are based chiefly upon the method of isolation or characterization of the fraction, e.g., (1) chemical procedures, involving precipitation, electrophoretic, and chromatographic procedures; (2) ultracentrifugation using density gradients, and (3) immunologic techniques. The following table includes *most* of the names which are currently (1977) being used to identify lipoproteins and which have been used by various authors of this volume. Elucidation of the complexity of the lipoprotein structure has resulted in a much more involved nomenclature than in the early days, and extensive discussions of preferred terminology have taken place.

Table 1
VARIOUS NOMENCLATURES OF LIPOPROTEINS

Method of study on which nomenclature based	Name of fraction
	α-Lipoproteins
Electrophoresis	α-lipoproteins, α-Lp, lipoproteins with electrophoretic mobility of α_1-globulins; they may be resolved as single or multiple bands depending on type of support media
Ultracentrifugation	HDL, high-density lipoprotein (d 1.063—1.21 g/ml); HDL$_2$, subclass of HDL (d 1.063—1.125 g/ml); HDL$_3$, subclass of HDL (d 1.125—1.21 g/ml); flotation rate at d 1.21 of $-$S 0—10.
Apolipoprotein composition — chemical, immunologic	Apo A, apolipoprotein A consisting of two nonidentical polypeptides, A-I and A-II: A-I contains glutamic acid as C-terminal and aspartic acid as N-terminal amino acid A-II contains glutamic acid as C-terminal and pyrrolidine carboxylic acid as N-terminal amino acid apo D, apolipoprotein D present in HDL$_3$; apo E$_{1-3}$, apolipoprotein E$_{1-3}$, present in HDL; apo C-II, apolipoprotein C-II present in HDL
	Pre-β-Lipoproteins
Electrophoresis	Pre-β-lipoproteins, pre-β-Lp, lipoproteins with electrophoretic mobility of α_2-globulins; they have this mobility when agarose, paper, or starch powder is used as support medium; they migrate slower than β-lipoprotein when gels with sieving effect, such as acrylamide or starch gel, are used; pre-β-Lp may be resolved as single or multiple bands.
Ultracentrifugation	VLDL, very low-density lipoproteins (d < 1.006 g/ml); isolated from serum after previous removal of chylomicron; flotation rate at d 1.21 of $-$S 70—400

Table 1 (continued)
VARIOUS NOMENCLATURES OF LIPOPROTEINS

Method of study on which nomenclature based	Name of fraction
Apolipoprotein composition — chemical, immunologic	Apolipoprotein of VLDL contains apo C-II, B, and E_{1-3} polypeptides; apo C is an apolipoprotein consisting of 3 nonidentical polypeptides: C-I is characterized by N-threonine and C-serine, C-II by N-threonine and C-glutamic acid, C-III by N-serine and C-alanine

β-Lipoproteins

Electrophoresis	β-lipoproteins, β-Lp, lipoproteins with electrophoretic mobility of β-globulins; β-Lp may be resolved as single or double bands, depending on buffer and support medium used
Ultracentrifugation	LDL, low-density lipoproteins (d 1.006—1.063 g/ml); LDL_1, subclass of LDL (d 1.006—1.019 g/ml); LDL_2, subclass of LDL (d 1.019—1.063 g/ml); flotation rate at d 1.21 of −S 25—40; at d 1.063, S_f0—20 may be divided into −S 40—70, i.e., S_f 12—20, intermediate density fraction and −S 25—40, S_f 0—12 LDL
Apolipoprotein composition — chemical, immunologic	Apo B, apolipoprotein B is major apoprotein of β-lipoprotein; β-Lp also contains C-II and E_{1-3}

Other fractions

Electrophoresis on agar	Lp-X, lipoprotein X, a lipoprotein of β- or slow β-globulin mobility and low density, characteristically found in obstructive jaundice patients' sera; best identified by electrophoresis on agar, where it has unusual property of migrating to γ-globulin position
Immunologic and genetic studies	Lp(a) is a polymorphic form of β-Lp which is of importance in genetic studies, and in those patients who have received multiple transfusions

TABLE OF CONTENTS

TABLE OF CONTENTS
CRC HANDBOOK OF ELECTROPHORESIS
VOLUME I

Lipoprotein Changes Induced by Physiological and Disease Processes

Genetically Directed Hyperlipoproteinemias

FAMILIAL HYPER-α-LIPOPROTEINEMIA

C. J. Glueck and P. M. Steiner

More than 25 years ago in the early investigations of the interaction of lipoproteins with ischemic heart disease (IHD), several groups reported that α-lipoprotein cholesterol (C-HDL) correlated negatively and β-lipoprotein cholesterol (C-LDL) positively with development of IHD.[1-4] The "anti-risk" nature of HDL and its hypothesized role in mobilizing cholesterol from the artery wall has been recently reviewed by Miller and Miller.[5] In population studies, C-HDL is "inversely related to CHD prevalence . . . this relationship is essentially independent of total and LDL cholesterol,"[6] CHD being coronary heart disease. In population groups having low IHD event rates, C-HDL levels are considerably higher than in males of comparable age who have higher IHD mortality.[7] While controlling for serum cholesterol, an inverse relationship between IHD prevalence and HDL cholesterol was reported in Hawaiian Japanese men.[8] In a very dissimilar population group from Evans County, Georgia,[9] Tyroler et al. observed that black males had lower frequencies of coronary heart disease, "controlling for the standard risk factors in univariate and in multivariate logistic risk function analyses." Tyroler et al. observed significantly higher LDL cholesterol and total triglycerides in whites and higher HDL cholesterol in blacks, in comparisons matched for age, sex, and total serum cholesterol.[9] Tyroler et al. concluded:[9] "the black-white lipoprotein fraction differences in Evans County are consistent with a negative coronary risk factor role of elevated HDL cholesterol ..."

C-HDL is inversely correlated with plasma low- and very low-density lipoprotein cholesterols (C-LDL and C-VLDL, respectively)[3,10,11] In subjects having familial hypercholesterolemia,[12-15] C-LDL elevations are accompanied by low levels of C-HDL. Accelerated IHD in the familial hypercholesterolemias and hypertriglyceridemias is associated with both elevated C-LDL and C-VLDL and subnormal C-HDL.[12,16] This provides a much higher than normal ratio of atherogenic (C-LDL, C-VLDL) to "anti-atherogenic" lipoproteins (C-HDL).[17-20]

The authors recently described a new type of familial dyslipoproteinemia, familial hyper-α-lipoproteinemia, in 18 kindreds.[18-20] In extensive prevalence studies of lipids and lipoproteins in family units in the Cincinnati area (5000 kindreds), qualitative increments in α-lipoproteins which appeared to aggregate in certain families were first observed by paper lipoprotein electrophoresis. Putative normal limits for C-HDL were then established for 168 free-living control subjects, with the 90th percentile 70 mg/dl and the mean ± 1 SD C-HDL being 55 ± 12 mg/dl. As an upper normal limit for C-HDL, 70 mg/dl was arbitrarily chosen.[18] Affected probands and relatives had distinctive elevations of C-HDL, slight elevations of total cholesterol, no elevation of C-LDL and C-VLDL, and normal to low triglyceride levels. Simple segregation analysis involving 84 offspring of 22 hyper-α × normal α matings revealed a ratio of hyper-α to normal of 37:47, a ratio not significantly different from 1:1 ($X^2_1 = 1.2$), that ratio which is consistent with autosomal dominant transmission.[18] Males and females from kindreds with familial hyper-α-lipoproteinemia had life expectancies which were 5 and 7 years longer, respectively, than others in U.S. (life table) populations (P < 0.002).[18-20] Combined morbidity and mortality from myocardial infarction was threefold greater in normal control kindreds than in kindreds with familial hyper-α-lipoproteinemia.[18-20] The (mean ± SE) ratio of C-LDL to C-HDL in kindreds with familial hyper-α-lipoproteinemia was 1.21 ± 0.06, twofold lower than the ratio, 2.4 ± 0.12, in a control population, P < 0.01.[20] This relatively low ratio of C-LDL to C-HDL, which is qualitatively similar to the "protective" cholesterol/phospholipid ratio reported 25 years ago,[1,2] probably relates to prolonged longevity and reduced morbidity and mortality from myocardial infarction in familial hyper-α-lipoproteinemia.[18-20]

Recognition of hyper-α-lipoproteinemia, until C-HDL determinations are done routinely as part of lipid and lipoprotein sampling, will inevitably be accomplished by electrophoresis, coupled with appropriate subsequent quantitative C-HDL determinations. Qualitative and quantitative aspects of C-HDL determination are summarized below.

QUALITATIVE AND QUANTITATIVE ASPECTS OF C-HDL DETERMINATION IN NORMAL AND HYPERALPHALIPOPROTEINEMIC SUBJECTS

Hyper-α-lipoproteinemia is often detected initially through routine paper[21] or agarose-gel[22] lipoprotein electrophoresis by increased staining of the band with α_1-globulin mobility. Quantitation of C-HDL requires separation of HDL and subsequent measurement of C-HDL. HDL isolation may be accomplished either by precipitation of particles, VLDL and LDL,[23,24] or by sedimentation of HDL in the preparative ultracentrifuge.[25]

Plasma VLDL and LDL are precipitated by sodium heparin and manganese chloride according to Burstein and Samaille.[23] Plasma samples are obtained from fasting subjects in vacuum tubes containing crystalline disodium EDTA (1mg/ml).[26] Plasma in 2-ml aliquots is placed in 15-ml conical, glass-stoppered centrifuge tubes. Samples are maintained in an ice bath or at 4°C throughout the procedure. With a microliter syringe, 0.08 ml of sodium heparin (5000 U.S.P. units per milliter and 0.10 ml of $1.0M$ MnCl$_2$ are added by vortex mixing after each addition. Tubes are allowed to stand for 30 min and then centrifuged for 30 min at 1600 G. The high-density lipoprotein (HDL)-containing supernatant is removed with a Pasteur pipette, and analyzed for cholesterol.[26] Frequently, when samples have increased concentrations of VLDL and triglyceride, sedimentation of the precipitate will not be complete. If this occurs, HDL must be separated by preparative ultracentrifugation.[26]

It is advantageous to monitor supernatants by performing agarose-gel lipoprotein electrophoresis to ensure complete precipitation of LDL. Immunodiffusion against antiserum to β-lipoprotein[25] is an even more sensitive method for checking supernatants for LDL. Because of differences in the preparation of heparin from different sources and the considerable molecular weight variation of commercial heparin,[27] it is necessary to check each individual lot for its ability to completely precipitate VLDL and LDL by agarose-gel electrophoresis.[26] Some HDL has been reported to precipitate with heparin and manganese chloride;[28] however, this has not been found to be a cause of underestimation of HDL in this procedure.[29]

HDL is separated in the preparative ultracentrifuge by adjusting plasma density to 1.063, ultracentrifugation, and recovery of the infranatant. Plasma density is adjusted by addition of potassium bromide according to the following formula:

$$X = \frac{V_i\,(d_f - d_i)}{1 - \bar{V}\,d_f}$$

where X is grams of solid KBr added, V_i is initial volume of the solution, d_f is the final density desired, d_i is the initial density, and $\bar{V}$ is the partial specific volume of KBr.[25,30] Densities of the solutions should be confirmed by pycnometry. A convenient procedure is to add 5 ml of plasma to a cellulose nitrate ultracentrifuge tube into which 0.417 g KBr has previously been weighed. The KBr is dissolved, and the tube, filled with KBr solution, d 1.063, and capped. Samples are centrifuged at 105,000 G for 18 hr at 10°C. Lipoprotein fractions are recovered using a tube slicer to cut the cellulose nitrate tube. The HDL-containing infranatant is quantitatively recovered with saline washes to a final volume of 5 ml and then assayed for cholesterol.

Agarose-gel lipoprotein electrophoresis performed on the two fractions d < 1.063 and d > 1.063 is done to verify the separation of HDL. The staining sensitivity of a diluted control of pure β-lipoprotein, which has been analyzed for cholesterol, is used to determine the lower limit of detection. If electrophoresis is not sufficiently sensitive to detect 5 to 10 mg percent or less of LDL cholesterol, immunodiffusion may be employed to detect any LDL.[25]

In some plasma samples, "sinking pre-β-lipoprotein," the Lp(a) antigen, which is an antigenic variant of LDL with a density range overlapping that of HDL, will be recovered in the ultracentrifugally fractionated plasma at d > 1.063.[31] Although Lp(a) contains some cholesterol, the amount of Lp(a) recovered is generally not sufficient to render C-HDL inaccurate.[29,32]

ACKNOWLEDGMENT

This study was supported in part by the General Clinical Research Center (RR 00068-14) and NHLI contract NO1-HV-2-2914-L. A portion of this work was done during Dr. Glueck's tenure as an Established Investigator of the American Heart Association from 1971 to 1976.

REFERENCES

1. Gofman, J. W., Jones, H. B., Lindgren, F. T., Lyon, T. P., Elliot, H. A., and Strisower, B., Blood lipids and human atherosclerosis, *Circulation,* 2, 161, 1950.
2. Barr, D. P., Russ, E. M., and Eder, H. A., Protein lipid relationships in human plasma. II. In atherosclerosis and related conditions, *Am. J. Med.,* 11, 480, 1951.
3. Nikkila, E., Studies on lipid-protein relationships in normal and pathological sera and effect of heparin on serum lipoprotein, *Scand. J. Clin. Lab. Invest. Suppl.,* 5, 1, 1953.
4. Jencks, W. P., Hyatt, M. R., Jetton, M. R., Mattingly, T. W., and Durrum, E. L., A study of serum lipoproteins in normal and atherosclerotic patients by paper electrophoretic techniques, *J. Clin. Invest.,* 35, 980, 1956.
5. Miller, G. J. and Miller, N. E., Plasma-high-density lipoprotein concentration and development of ischemic heart disease (technical note), *Lancet,* 1(7897), 16, 1975.
6. Castelli, W. P., Doyle, J. T., Gordon, T., Hames, C., Hulley, S. B., Kagan, A., McGee, D., Vicic, W. J., and Zukel, W. J., HDL cholesterol levels (HDLC) in coronary heart disease (CHD): A cooperative lipoprotein phenotyping study, *Circulation Suppl.,* 2(96), 51, 1975.
7. Bang, H. W., Dyerberg, J., and Nielson, A., Plasma lipid and lipoprotein pattern in Greenland west-coast Eskimos, *Lancet,* 1, 1143, 1971.
8. Gulbrandsen, C. L., Rhoades, G. G., and Kagan, A., Cholesterol fractions and coronary heart disease in Hawaii Japanese men, *Circulation Suppl.,* 50, 100, 1974.
9. Tyroler, H. A., Hames, C. G., Krishan, I., Heyden, S., Cooper, G., and Cassel, J. C., Black-white differences in serum lipids and lipoproteins in Evans County, *Prev. Med.,* 4, 541, 1975.
10. Nichols, A. V., Human serum lipoproteins and their relationships, *Adv. Biol. Med. Phys.,* 11, 109, 1967.
11. Carlson, L. A., Lipoprotein fractionation, *J. Clin. Path.,* Suppl. 5, 32, 1973.
12. Fredrickson, D. S., Levy, R. I., and Lees, R. S., Fat transport in lipoproteins—an integrated approach to mechanisms and disorders, *N. Engl. J. Med.,* 276, 32, 94, 215, 273, 1967.
13. Fredrickson, D. S., Levy, R. I., and Lindgren, F. T., A comparison of heritable abnormal lipoprotein patterns as defined by two different techniques, *J. Clin. Invest.,* 47, 2446, 1968.

14. **Fredrickson, D. S. and Levy, R. I.,** Familial hyperlipoproteinemia, in *The Metabolic Basis of Inherited Disease,* 3rd ed., Stanbury, J. B., Wyngaarden, J. B., and Fredrickson, D. S., Ed., McGraw-Hill, New York, 1972, 546.

15. **Stone, N. H., Levy, R. I., Fredrickson, D. S., and Verter, J.,** Coronary artery disease in 116 kindred with familial type II hyperlipoproteinemia, *Circulation,* 49, 476, 1974.

16. **Carlson, L. A., and Bottiger, L. E.,** Ischaemic heart-disease in relation to fasting values of plasma triglycerides and cholesterol, *Lancet,* 1, 865, 1972.

17. **Glueck, C. J.,** Alpha-lipoprotein cholesterol, beta-lipoprotein cholesterol, and longevity, *Artery,* 2(3), 196, 1977.

18. **Glueck, C. J., Fallat, R. W., Millett, F., Gartside, P., Elston, R. C., and Go, R. C. P.,** Familial hyperalpha-lipoproteinemia: studies in 18 kindreds, *Metabolism,* 24, 1243, 1975.

19. **Glueck, C. J., Fallat, R. W., Spadafora, M., and Gartside, P.,** Longevity syndromes, *Circulation Suppl.,* 2(272), 51, 1975.

20. **Glueck, C. J., Gartside, P., Fallat, R. W., Sielski, J., and Steiner, P. M.,** Longevity syndromes: familial hypobeta and familial hyperalphalipoproteinemia, *J. Lab. Clin. Med.,* 88(6), 941, 1976.

21. **Lees, R. S. and Hatch, F. T.,** Sharper separation of lipoprotein species by paper electrophoresis in albumin-containing buffer, *J. Lab. Clin. Med.,* 61, 518, 1963.

22. **Noble, R. P.,** Electrophoretic separation of plasma lipoproteins in agarose gel, *J. Lipid Res.,* 9, 693, 1968.

23. **Burstein, M. and Samaille, J.,** Sur un dosage rapide du cholesterol lie aux alpha et aux beta-lipoproteines du serum, *Clin. Chim. Acta,* 5, 609, 1960.

24. **Burstein, M. and Scholnick, H. R.,** Lipoprotein-polyanion-metal interactions, in *Advances in Lipid Research,* Paoletti, R. and Kritchevsky, D., Eds., Academic Press, 1973, 67.

25. **Hatch, F. T. and Lees, R. S.,** Practical methods for plasma lipoprotein analysis, in *Advances in Lipid Research,* Paoletti, R. and Kritchevsky, D., Eds., Academic Press, 1968, 33.

26. Manual of Laboratory Operations, Lipid Research Clinics Program, Vol. 1, Lipid and Lipoprotein Analysis, DHEW, Pub. No. 75, National Heart, Lung, and Blood Institute, Bethesda, Maryland, 1974, 628.

27. **Nader, H. B., McDuffie, N. M., and Dietrich, C. P.,** Heparin fractionation by electrofocusing: presence of 21 components of different molecular weights, *Biochem. Biophys. Res. Commun.,* 57, 2, 1974.

28. **Srinivasan, S. R., Radhakrishnamurthy, B., and Berenson, G. S.,** Studies on the interaction of heparin with serum lipoproteins in the presence of Ca^{2+}, Mg^{2+}, and Mn^{2+}, *Arch. Biochem. Biophys.,* 170, 334, 1975.

29. **Ishikawa, T. T., Brazier, J. B., Steiner, P. M., Stewart, L. E., Gartside, P. S., and Glueck, C. J.,** A study of the heparin-manganese chloride method for determination of plasma alpha-lipoprotein cholesterol concentration, *Lipids,* 11(8), 628, 1976.

30. **Radding, C. M. and Steinberg, D.,** Studies on the synthesis and secretion of serum lipoproteins by rat liver slices, *J. Clin. Invest.,* 39, 1560, 1960.

31. **Ehnholm, C. H., Garoff, K., Simons, K., and Aro, H.,** Purification and quantitation of the human plasma lipoprotein carrying the Lp(a) antigen, *Biochim. Biophys. Acta,* 236, 431, 1971.

32. **Wilson, D. E. and Spiger, M. J.,** Method for quantitative plasma lipoprotein measurement without ultracentrifugation, *J. Lab. Clin. Med.,* 82, 473, 1973.

ELECTROPHORESIS IN THE DETERMINATION OF PLASMA LIPOPROTEIN PATTERNS

P. S. Bachorik

INTRODUCTION

General

The association between elevated concentrations of plasma cholesterol and increased cardiovascular risk has focused attention on the circulating lipoproteins as the major, indeed almost exclusive, carriers of circulating lipids. Electrophoresis has been used extensively both to investigate lipoprotein metabolism and to clinically evaluate hyperlipoproteinemia. The present discussion focuses on the clinical use of the techniques.

Lipoproteins are complex molecules containing characteristic protein and lipid components and are currently classified on the basis of (1) their hydrated densities,[1,2] (2) their electrophoretic mobilities,[3,4] or (3) their apoprotein compositions.[5] The first two systems of nomenclature are probably the most widely used clinically and describe the lipoproteins in terms of properties that depend on the whole lipoprotein molecule. They will be used in the present discussion insofar as possible. There is, however, a growing appreciation of the need to consider individual **apoprotein** components of the lipoproteins as they relate to atherogenesis and lipoprotein metabolism since they may reflect one or another lipoprotein disorder. Apoproteins are mentioned here as necessary. A summary of some of the chemical and physical properties of the normal plasma lipoproteins is presented in Table 1.

Normally Occurring Plasma Lipoproteins

The major classes of lipoproteins found in normal human plasma are chylomicrons, very low-density lipoproteins (VLDL), low-density lipoproteins (LDL), and high-density lipoproteins (HDL). Chylomicrons appear in plasma after meals and contain about 85% triglycerides by weight. They are synthesized in the intestine from dietary fat and are not detected in normal fasting plasma. Chylomicrons have densities less than 0.95 g/ml, and significant amounts are evident as a white, floating "cream" layer after plasma stands in the refrigerator overnight. Chylomicrons remain at the origin when plasma is subjected to electrophoresis in agarose gel or on paper. VLDL contain about 50% triglycerides by weight and are synthesized primarily in the liver from triglycerides of endogenous origin. They have densities between 0.95 to 1.006 g/ml and can be separated from plasma as a floating layer by ultracentrifugation of plasma without density adjustment. VLDL migrate with α_2-globulins on electrophoresis and are also called pre-β-lipoproteins. LDL are the major cholesterol-bearing lipoproteins in man and are thought to be breakdown products of VLDL metabolism.[6] They occupy the density range from 1.006 to 1.063 g/ml, move electrophoretically with the β-globulins, and are referred to as β-lipoproteins. HDL are rich in protein and phospholipids. They have densities between 1.063 to 1.21 g/ml, migrate electrophoretically with α-globulins, and are called α-lipoproteins. Their significance in lipoprotein metabolism is not entirely clear, but they may be involved in the removal of cholesterol from,[7,8] or in the limiting of uptake of LDL by[9] peripheral tissues.

In addition to the major lipoproteins, a minor lipoprotein class Lp(a)[10-14] may also be observed. This lipoprotein, which contains apo B, the major apoprotein of LDL, overlaps the LDL-HDL density region and sediments when plasma is ultracentrifuged at d 1.006 g/ml. It has an electrophoretic mobility similar to the pre-β-lipoproteins and has been commonly referred to as "sinking pre-β" lipoprotein. Lp(a) is probably

Table 1

PHYSICAL AND CHEMICAL PROPERTIES OF THE PLASMA LIPOPROTEINS

Lipoprotein class	Density range (g/ml)	Electrophoretic migration (paper or agarose)	Chemical Composition[102] (Percent by wt)				Apoproteins	
			Protein	Cholesterol	Triglycerides	Phospholipids	Major	Minor
Chylomicrons	<0.95 (S$_f$ >400)	Origin	2	10	81	7	Apo C-I, -II, -III	Apo B (Apo A-I, -II)
VLDL	0.95—1.006 (S$_f$ 20—400)	Pre-β (α$_2$)	10	17	60	13	Apo B, Apo C-I, -II, -III	Apo E (Apo A-I, -II)
LDL	1.006—1.063 (S$_f$ 0—20)	β	20	45	12	23	Apo B	Apo C-I, -II, -III
HDL	1.063—1.21	α	53	14	5	28	Apo A-I, -II	Apo C-I, -II, -III, Apo D.

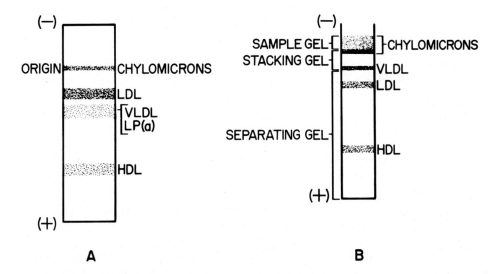

FIGURE 1.   Diagrammatic representation of the relative electrophoretic mobilities of the lipoproteins. A. In paper, agarose gel, or cellulose acetate. B. In 3% polyacrylamide gel. Note the reverse order of β-pre-β migration in the two illustrations.

present in low concentrations in most individuals,[13,15] although the frequency with which it is observed in the laboratory depends on the sensitivity of methods used to detect it.[15,16] The significance of the lipoprotein is not understood but may be associated with increased risk.[15,17] Its presence is usually simply noted. A representation of the electrophoretic mobilities of the major lipoprotein classes is illustrated in Figure 1.

The metabolism of the plasma lipoproteins is quite complex and is incompletely understood at present. Hydrolysis of chylomicron and VLDL triglycerides is catalyzed at the capillary endothelial surface by lipoprotein lipase. One of the apoproteins present in these lipoproteins, apo C-II (Table 1), is a cofactor for this reaction. Hydrolysis of VLDL is accompanied by the transfer of C apoproteins to HDL. Phosphoryl choline and unesterified cholesterol are also transferred from VLDL to HDL, where they are acted on by the enzyme lecithinolesterol acyltransferase. This enzyme catalyzes the transfer of one molecule of fatty acid from phosphatidyl choline to the 3-OH group of cholesterol to form cholesteryl ester and lysolecithin, thereby promoting further removal of these lipids from VLDL. Apo A-I, the major apoprotein of HDL, is a cofactor for the esterification reaction. The "remnant" lipoprotein particles formed consist mostly of cholesteryl esters, phospholipids, and apo B; they are presumably taken up by the liver for further catabolism. VLDL is thought to be ultimately converted to LDL.

Hyperlipoproteinemia

The abnormal increase in plasma cholesterol, triglycerides, or both results from changes in the concentrations of one or more of the plasma lipoproteins and may be primary, in that it reflects aberrant lipid or lipoprotein metabolism, or secondary, resulting from underlying disorders other than those of lipoprotein metabolism. Secondary hyperlipidemias are usually resolved when the underlying primary disorder is treated.Primary hyperlipoproteinemias are usually genetically based, and their genetics can be monogenic or polygenic. The clinical diagnosis and management of primary hyperlipoproteinemia requires physical examination, determination of which of the lipoproteins are elevated, tests to rule out secondary causes, and family screening.

About 10 years ago, Fredrickson et al.[18] proposed a system to classify the primary hyperlipidemias into five (later modified to six)[19] phenotypes, depending on which of the lipoproteins are elevated. The system was primarily concerned with clarifying the relationship between elevated cholesterol and triglycerides, their underlying lipoprotein elevations, and certain clinically recognizable disorders. Since then, it has become increasingly apparent that lipoprotein phenotypes should not be thought of as representing discrete lipoprotein disorders, because more than one biochemical lesion can be associated with a particular plasma lipoprotein pattern.[20-23] Furthermore, whereas increased concentrations of the cholesterol and triglyceride-rich lipoproteins are usually associated with increased cardiovascular risk, this is not true for HDL. Elevated HDL-cholesterol concentrations, in fact, appear to be related to reduced incidence of coronary heart disease.[24,25] These observations have stimulated interest in determination of HDL as an additional factor to be considered in assessing risk.

The shift in emphasis from plasma lipids to lipoproteins has created the need for readily available methods of accurately determining plasma lipoprotein patterns; electrophoresis, because of its relative ease, speed, and low cost, has been widely used for this purpose. Unfortunately, attempts to accurately classify lipoprotein patterns, based only on total cholesterol and triglyceride concentrations and lipoprotein electropherograms of unfractionated plasma or serum, have had limited success. In some cases, they have led to disillusionment with the routine use of electrophoresis for this purpose.[20,26] On the other hand, the technique has been more successfully employed in conjunction with other fractionation techniques, and certain recent advances promise to further increase its usefulness in lipoprotein screening.

SUPPORTING MEDIA

General

Electrophoresis of plasma lipoproteins has been performed in a number of supporting media, including starch block,[27] cellulose acetate,[28-30] polyacrylamide gel,[31-33] paper,[4,34-37] and agarose gel[34,35,38-40] among others. General reviews of lipoprotein electrophoretic methods are available.[2,41,42]

Paper and agarose-gel electrophoresis have enjoyed the widest popularity. Of the two, agarose gel is preferred because of its greater sensitivity and better resolution of the lipoproteins. Although cellulose acetate is sensitive and has a higher resolution than paper, agarose gel seems to be somewhat more sensitive, can better resolve chylomicron, β-, and pre-β-bands,[29,43] and has been most intensively standardized. The relative migrations of the lipoproteins are the same in all three systems (Figure 1). Polyacrylamide-gel electrophoresis is more technically demanding. It has a limited sample capacity and has usually been used only to identify or confirm the type III pattern (see below).

Paper

Paper electrophoresis is performed in albumin-containing buffer. Addition of albumin improves the separation of lipoproteins and gives better defined bands.[4] Samples of approximately 20 μl are applied to paper strips that have been equilibrated with the buffer and subjected to electrophoresis for about 16 hr.[2,4,41,44] The electrophoretic strips are dried and are then usually stained with oil red O for 1½ hr. The lipoproteins are visible as red bands; the background is also stained. The β- and α-bands are usually observed, and the prebeta band, visible when plasma triglycerides exceed about 150 mg/dl or when an appreciable concentration of Lp(a) is present. The stain is prepared with heating and must be stored and used at 37°C because of its limited solubility. The electrophoretic strips form a permanent record of the analysis, and the procedure does not require much time. Paper electropherograms can be difficult to interpret, however. The stained background reduces sensitivity and the resolution of β- and pre-β-bands is not always complete.[45] Quantitation of lipoproteins in paper electropherograms has not been too successful.

Agarose Gel

The agarose gel procedures used are based on the method of Noble.[38] Approximately 10-μl samples are applied to 0.6% agarose gels that have been cast on a sheet of plastic and that contain 0.05% human or bovine albumin. Albumin is not usually added to the buffer. A number of samples can be run at the same time. Electrophoresis proceeds for about 90 min, after which the sheet is fixed in ethanol-acetic acid, dried, and stained, usually with Sudan black B or fat red 7B. The electropherograms have a clear, almost colorless background particularly suitable for densitometry. β- and pre-β-lipoproteins are usually well resolved; a pre-β-band is observed when plasma triglycerides exceed about 50 mg/dl. Two pre-β-bands are sometimes observed.[16,34,40,46] The slow component moves near, but faster than, the β-band and is more easily observed in lipoproteins of d < 1.006 g/ml. It is probably related to particles intermediate between VLDL and LDL and can be difficult to differentiate from β-VLDL,[47] the β-migrating lipoprotein of < 1.006 g/ml which appears in type III hyperlipoproteinemia (see below). Carlson and Ericsson[16] reported the presence of the slow pre-β-band in 35% of the normal males and 25% of the normal females they examined. Vessby et al.[47] have observed it in 47% of the non-type III relatives of type III subjects. The α-band can be similarly resolved into two bands under appropriate conditions.[40,46,48,49]

Precast agarose gels are commercially available in kit form from several sources. While their use may sacrifice some resolution, they are relatively inexpensive, use smaller samples, have shorter running times (about 20 min), do not require cooling, and perform quite well for general purposes. They are amenable to quantitative use, with proper calibration.[50,51] Correlation coefficients of 0.94 (β vs. S_f0—20), 0.99 (pre-β vs. S_f20—400), and 0.88 (α vs. HDL) have been reported between electrophoresis in one such kit and ultracentrifugation.[50]

Polyacrylamide Gel

Polyacrylamide gel procedures are based on the method of Davis[52] and have been adapted for use with lipoproteins. Several modifications have been used; these differ in various details as well as whether uniform concentrations[32,53-56] or concentration gradient[57-60] gels are used. Preparation of the gels is time consuming, and the number of samples that can be processed at one time is limited; polyacrylamide-gel electrophoresis is therefore less widely used than other methods. It has been used with selected samples to confirm the presence of the type III pattern (see below). The methods use low-concentration separating gels and incorporate stacking and sample gels.

In one popular procedure,[32] a 3% separating gel is cast and overlaid with a 2.5% stacking gel. When polymerization is complete, samples are prestained by mixing them with a sample-gel solution containing Sudan black B. The mixture is applied to the gel column and allowed to polymerize. Electrophoresis proceeds for about 25 min at a constant current of 4 to 5 mA per gel, and gels are immediately photographed to provide a permanent record of the analysis. It is necessary to use both stacking and sample gels in order to obtain reproducible results and adequate separation of the lipoproteins.[55] The separation obtained with 3% gels is schematically illustrated in Figure 1.

When 3.5 or 3.75% gels are used, the lipoproteins are resolved further and additional bands appear.[32,55] A summary of some of the various electrophoretic procedures have been used for lipoprotein phenotyping is found in Table 2.

LIPOPROTEIN PATTERNS

General

The lipoprotein pattern exhibited by an individual is, from the laboratory standpoint, a description of his lipoprotein status at the time of sampling. It includes estimates of the concentrations of individual lipoproteins and an assessment of the presence of unusual lipoproteins. Consideration of the use of electrophoresis in the determination of lipoprotein patterns must be preceded by a description of the patterns observed in hyperlipidemic patients.

Although, as mentioned above, a particular lipoprotein pattern may be exhibited in more than one disorder, laboratory determination of lipoprotein patterns is useful in narrowing the range of possibilities and, in fact, in deciding whether the patient with mild hypercholesterolemia should be considered abnormal at all, as in the case of the individual with a normal LDL-cholesterol concentration who has a higher than normal total cholesterol concentration because of a high HDL-cholesterol concentration. Lipoprotein patterns observed in hyperlipidemia can be classified into types corresponding to the six original phenotypes.

The present brief survey describes lipoprotein patterns observed by laboratory examination of hyperlipidemic plasma, without regard to underlying primary or secondary causes or to changes induced by therapy. The reader who is interested in the hyperlipidemias and their reationship to atherosclerosis is referred to a recent comprehensive discussion on the subject.[61] Examination of lipoprotein patterns is performed on plasma following a suitable fast, to ensure clearance of postprandial chylomicrons. A 12-hr fast is recommended.

Normal Lipoprotein Pattern

"Normal" lipoprotein patterns can be simply defined as those of individuals in whom concentrations of the circulating lipoproteins are within the normal range and who exhibit no unusual lipoproteins. Unfortunately, the decision as to whether a particular lipoprotein pattern is normal is more complex, because as pointed out by Kwiterovich,[62] concentrations of plasma lipids and lipoproteins in a given population are influenced by a number of genetic and environmental factors including age, sex, obesity, degree of physical activity, diet, seasonal variation, and social surroundings. Furthermore, current ideas of what constitute unusual lipoproteins are changing as understanding of lipoprotein metabolism expands and methods are improved (see below). One further consideration is that lipid and lipoprotein concentrations, prevailing within a given population, are probably not fixed but may change over a period of time, as dietary patterns and other environmental conditions change. The most widely used normal ranges for plasma lipid and lipoprotein concentrations at the moment are probably those from the Molecular Disease Branch of the National Heart, Lung, and Blood Institute.[18] Results of the recent large-scale screening efforts of the Lipid Research Clinics program are beginning to emerge and suggest that normal ranges will have to be revised. For this reason, the present discussion does not include references to currently used normal ranges. Furthermore, it is being increasingly appreciated that a distinction will probably have to be made between **normal** lipid and lipoprotein ranges, defined as those exhibited by most of the population, and **healthy** ranges, defined as those outside of which atherosclerotic risk is unacceptably high.

Table 2
SOME ELECTROPHORETIC METHODS USED TO DETERMINE PLASMA LIPOPROTEIN PATTERNS[a]

Author and ref no.	Year	Medium, special conditions	Buffer	Cell and special equipment	Electrical conditions	Time of electrophoresis	Temp	Special handling	Staining	Evaluation	Densitometry	Methodological Error
Hatch, F. T. and Lees, R. S.[41]	1968	Paper, 20-μl sample	1100 ml barbital, pH 8.6, ionic strength 0.1, containing 0.001 M EDTA and 1% bovine albumin; buffer can be used for 25 to 30 runs	Durrum® cell 8 samples per cell	Constant voltage, 110 V (7—10 mA); reverse polarity each run	16 hr	Room temp[ ]	Equilibrate paper strips for 2 hr before applying sample; then dry strips at 80 to 100°C before staining	Oil red O in 60% ethanol at 37°C for 6 to 8 hr	Qualitative or semiquantitative	Red bands on a stained background	±30%
Naito, H. K. and Lewis, L. A.[100]	1973	Paper, 20-μl sample	Barbital, 0.05 M pH 8.6	Durrum cell	Constant voltage, 110 V	18 hr	Room temp	Presoak paper strips in buffer containing 1% bovine albumin; then dry strips at 100 to 110°C before staining	Oil red O in 50% acetone at 37°C for 2.5 hr	Qualitative	Clear, colorless background	
Noble, R. P.[8]	1968	Agarose, 0.4% stabilized with 0.12% agar, cast on polyester film strips (35×150 mm) or sheets (13×35 cm); gels contain 0.5% bovine albumin; 20- or 33-μl serum samples, electrophoresis can be performed in 0.6% agarose without agar	Barbital, 0.05 M pH 8.4	Horizontal, water-cooled cell	Constant current, 10 mA per strip or 80 mA per sheet	2 hr	25°C	Fix gels in 75% ethanol, containing 5% acetic acid, for 30 min; dry at 80 to 85°C for 20 min	Sudan black B in 60% ethanol for 1 hr with agitation or 6 hr without agitation: oil red O in 60% ethanol at 37°C for 2 hr with agitation or 6 hr without agitation	Qualitative	Clear, colorless background; Scan at 600 nm (Sudan black B)	
Noble, R. P. et al.[7]	1969	Agarose, 0.5% cast on polyester film strips (35×85mm) 33 μl serum sample	Barbital, 0.05 M, pH 8.6	Horizontal, water-cooled cell	Constant current 10 mA per strip (ca. 250 V)	2 hr	25°C	Fix gels in 85% ethanol containing 5% acetic acid for 45 min; dry at 85°C for 15 min	Oil red O in 60% ethanol at 37°C for 18 hr	Semiquantitative	Red bands on clear colorless background	±30%; correlation coefficients, 0.94 (pre-β vs. S, 20—400), 0.83 (β vs. S, 0—20), 0.76 (α vs. HDL)
Hatch, F. T. et al.[56] and Lindgren, F. T. et al.[57]	1973, 1975	Agarose, cast on plastic slides (BioGram A lipoprotein kit, BioRad® Laboratories, Richmond, Cal.)approximately 1-μl sample applied	Barbital, 0.025 M. pH 8.6	Durrum cell without racks for paper strips; slides run face down	Constant voltage 150 V (3.6 to 3.8 mA per slide); reverse polarity each run	15 to 25 min	Room temp	Slides are soaked in buffer containing 0.5% albumin for 24 hr before use and drained in the open for 20 to 25 min before use to remove excess surface moisture; then gels, fixed in 60% ethanol, containing 5% acetic acid and dried with stream of warm air	Fat red 7B in 83% methanol containing 0.017 N NaOH and Triton® X-100, 3 drops per 100 ml stain for 15 min	Quantitative, calibrated against analytical ultracentrifuge; uses internal or external standardization	Red bands on clear colorless background	Correlation coefficients, 0.98 to 0.99 (pre-β vs. VLDL, S, 20—400), 0.92—0.94 (β vs. LDL, S, 0—20), 0.77—0.88 (α vs. HDL)
Van Melsen, A. et al.[102]	1974	Agarose, 0.83% containing 0.3% bovine albumin, cast on a glass plate (10×17.5cm); 20 μl sample applied	Barbital, ionic strength 0.025 pH 8.6		Constant voltage, 220 V (20 mA per plate)	1 hr	<30°C	Gels, fixed in 47.5% methanol, containing 3% ethanol and 2% acetic acid, for 2 hr; dry at 50°C	Sudan black B in 60% ethanol, containing 1% zinc acetate for 1 hr; stain changed every 15 days	Quantitative; calibration based on separate determination of serum and HDL cholesterol	Black bands on clear, colorless background; scanned at 590 mm	Reproducibility of 28 analyses of one sample (percent of densitometric scan): α, 17.8±0.9%; β, 55.8±1.2%; pre-β, 26.3±1.0%)
Papadopoulos, N. M. and Kintzios, J. A.[49,48,111]	1971, 1969, 1967	Agarose, 0.5%, cast on glass slides; 3μl serum applied	Horizontal, petroleum ether-cooled cell		Constant voltage, 150 V	10 min	Room temp	Gels, fixed in 2% acetic acid	Sudan black B in 60% ethanol for 1 hr	Qualitative		
Papadopoulos, N. M. and Herbert, P. N.[48]	1977							Similar method used to detect type III; see Reference 95	Lipoprotein, precipitated in gel with 0.2m MgCaCl, and 0.6% dextran sulfate	Quantitative; calibrated with pure isolated lipoproteins	Precipitated bands on a clear, colorless background; scanned at 500 nm	Coefficient of variance <5% for α-, β-, and pre-β-lipoproteins; standard curves, linear up to at least 10 g/l of each of lipoproteins

Table 2 (continued)
SOME ELECTROPHORETIC METHODS USED TO DETERMINE PLASMA LIPOPROTEIN PATTERNS[a]

Author and ref no.	Year	Medium, special conditions	Buffer	Cell and special equipment	Electrical conditions	Time of electrophoresis	Temp	Special handling	Staining	Evaluation	Densitometry	Methodological Error
Neubeck, W. et al.	1977	Agarose, 0.8%, containing 0.2% human albumin, cast on glass plates (75×25mm);5 μl sample applied	Barbital, 0.05 M pH 8.6	LBK = electrophoresis system (LKB, Bromma, Sweden)	10 V per cm	90 min	Room temp	Strips impregnated with buffer before use	Oil red O in 60% ethanol for 8 to 16 hr at 37°C	Qualitative		Relative peak areas of lipoprotein bands of two samples, each analyzed on two separate days, agreed within 0.6 to 3.8%
Winkelman, J. et al.	1969	Cellulose acetate strips, 25×120mm; 10-μl sample applied	Barbital, ionic strength 0.05, pH 8.6; 750 ml buffer used	Gelman® cell (Gelman Instrument Co., Ann Arbor, Mich.) or Photovolt cell (Photovolt Corp. New York, N.Y.)	Constant voltage. 150 V	90 min	Room temp	Strips impregnated with buffer before use; lipoprotein lipids can be extracted with organic solvents and quantitated	Oil red O in 70% methanol at 37°C overnight	Qualitative or semiquantitative in terms of relative peak area	Red bands on a translucent background	
Chin, H. P. and Blankenhorn, O. H.	1968	Cellulose acetate strips, 1×6 ¼"; 10-μl sample applied	Barbital, ionic strength 0.075, pH 8.6	Horizontal cell	Constant voltage, 150 V (2.5 mA per strip)	90 min	22±2°C	Strips soaked in buffer before use	Ozone oxidation followed by staining with Schiff's reagent	Qualitative or quantitative, based on densitometric scans and separate serum lipid determinations	Lipoprotein bands on a clear background; scan at 550 nm	
Postma, T. and Stroes, J. A. P.	1968	Cellulose acetate strips, 2.5×17 cm; 6-7 μl samples applied	Barbital, 0.07 M, pH 8.6	Gelman Deluxe electrophoresis chamber, Gelman Instruments, Ann Arbor, Mich.)	Constant voltage, 175 V	90 min	Room temp	Strips soaked in buffer at 4°C before use	Oil red O 70% methanol at 37 to 39°C for 6 to 7 hr	Qualitative	Red bands on clear background	
Charman, R. C. and Landowne, R. A.	1967	Cellulose acetate strips, 1×6" 5-μl samples applied	Barbital, ionic strength 0.05, pH 8.6, containing 1×10^{-4} M EDTA and 1% albumin	Gelman Deluxe cell (Gelman Instruments, Ann Arbor, Mich.)	Constant voltage, 240 V (2.5 mA per strip)	90 min	Room temp	Procedure requires stacking gel and sample gel for adequate separation and reproducibility	Sample prestained with Sudan black B; photographed to provide permanent record of analysis	Qualitative		
Frings, C. S. et al.	1971	Polyacrylamide gel, 3.75% cast in glass tube (7×75 mm); 20-μl sample applied	TRIS-glycine, pH 8.3	Canalco® system (Canalco Inc., Rockville, Md.)	5mA per gel	35 min	Room temp	Stacking and sample gels employed	Sample prestained with Sudan black B	Quantitative analysis of LDL; calibrated with sera of known LDL cholesterol concentrations; standard curve linear to about 200 mg/dl LDL cholesterol	Scan at 623 nm	Coefficient of variation 3.4 and 5.4% for two serum samples analyzed 24 times each; correlation coefficient, 0.90 (β vs. ultracentrifugally determined LDL cholesterol)
Moran, R. F. et al.	1972	Polyacrylamide gel (concentration not specified),40-mm gel height; 5-μl sample applied	Not specified	Canalco system (Canalco, Inc., Rockville, Md.); gels and buffers prepared with manufacturer-supplied lipoprotein reagent kit, Canalco QDL, modified	5mA per gel	50 min	Room temp	Stacking and sample gels employed	Sample prestained with acetylated Sudan black B; use of acetylated Sudan black B led to better resolution of chylomicrons	Qualitative		
Hall, F. F. et al.	1972	Polyacrylamide gel, 3%, cast in glass tubes (5×80 mm)	TRIS-glycine, pH 8.6	Canalco system (Canalco, Inc. Rockville, Md.)	5 mA per gel	35 min	4 to 6°C	Stacking and sample gels employed	Sample prestained with Sudan black B	Qualitative		
Masket, B. H. et al.	1973	Polyacrylamide gel, 3%, cast in glass tubes (5×70mm)	TRIS-glycine, pH 8.6	Laboratory-fabricated apparatus	4 to 5 mA per gel	25 min	Room temp		Sample prestained with Sudan black B	Qualitative		

[a] The following abbreviations are used in the table: HDL = high-density lipoproteins; VLDL = very low-density lipoproteins; LDL = low-density lipoprotein.

Fasting plasma which has been freshly drawn from the normal individual is clear and does not contain detectable levels of chylomicrons. Total cholesterol and triglyceride concentrations are within the normal range, as are the concentrations of the three major classes of lipoproteins. The minor lipoprotein class Lp(a) may also be present.

Types I and V Patterns

Primary type I hyperlipoproteinemia is an extremely rare disorder characterized in part by the genetic absence of lipoprotein lipase, the enzyme responsible for clearance of postprandial chylomicrons. The type I pattern is also occasionally displayed in several other disease states, such as diabetes and hypothyroidism,[63,64] and has also been observed during use of oral contraceptives.[65] Freshly drawn plasma from a patient who has been on a diet containing a normal amount of fat is extremely turbid or milky in appearance. Triglycerides are markedly elevated, usually between 1000 to 4000 mg/dl, and the increase resides almost exclusively in chylomicrons. When the plasma is allowed to stand in the refrigerator ("standing plasma" test), chylomicrons are visible as a pronounced supernatant cream layer. VLDL are normal or slightly elevated, and the infranatant plasma is clear. Concentrations of HDL and LDL are usually depressed. From a practical standpoint, it is well to bear in mind that chylomicrons over a clear infranatant may also be observed in normal nonfasting plasma and may also be reflected as an increase in plasma triglycerides, although the increase is not nearly as extreme as in type I. The degree of postprandial hypertriglyceridemia is variable and appears to be related to fasting triglyceride levels.[66,67]

Primary type V hyperlipoproteinemia, while not as rare as type I, is nonetheless relatively uncommon. The type V pattern is also obtained in a number of secondary disorders, including diabetes mellitus, hypothyroidism, nephrosis, alcoholism, and others.[68] Freshly drawn plasma is turbid, and triglycerides are elevated due to the presence of both fasting chylomicrons and increased VLDL concentrations. The plasma on standing has a floating chylomicron cream layer and a turbid infranatant reflecting increased VLDL concentration. LDL and HDL concentrations are decreased.[44,69,70]

Separation of type I and type V patterns requires additional information, because as mentioned above, VLDL can be slightly elevated in type I and mild VLDL elevations may not produce an unequivocally turbid infranatant in standing plasma. Intravenous injection of heparin causes immediate release of lipoprotein lipase to the circulation. The enzyme hydrolyzes the plasma triglycerides to free fatty acids and causes a marked transient decrease in the plasma triglyceride concentration of the hypertriglyceridemic patient (Figure 2). Several methods have been developed to quantitate heparin-induced release of lipoprotein lipase.[71-74] These methods all incorporate manipulations designed to separate lipoprotein lipase activity from that of a hepatic triglyceride lipase which is also released upon introduction of heparin. Negligible lipoprotein lipase activity is released in type I hyperlipoproteinemia; whereas, normal or low, but significant, levels are observed in type V, depending on the methods used.[68] Separation of type I from type V patterns, therefore requires some assessment of the presence of heparin-released lipoprotein lipase.[18]

Release of post-heparin lipolytic activity in the type V subject can be observed qualitatively as a change in the post-heparin lipoprotein electropherogram. A fasting blood sample is obtained first. Heparin (10 units/kg body weight) is then injected intravenously. This dose of heparin does not cause maximal release of the enzyme but is sufficient for qualitative use. A second blood sample is obtained 15 min after injection, and plasma from both samples is subjected to lipoprotein electrophoresis. The electropherogram of post-heparin plasma is strikingly different from that of the control (Figure 2). Chylomicron and pre-β-bands are greatly decreased or entirely eliminated. The

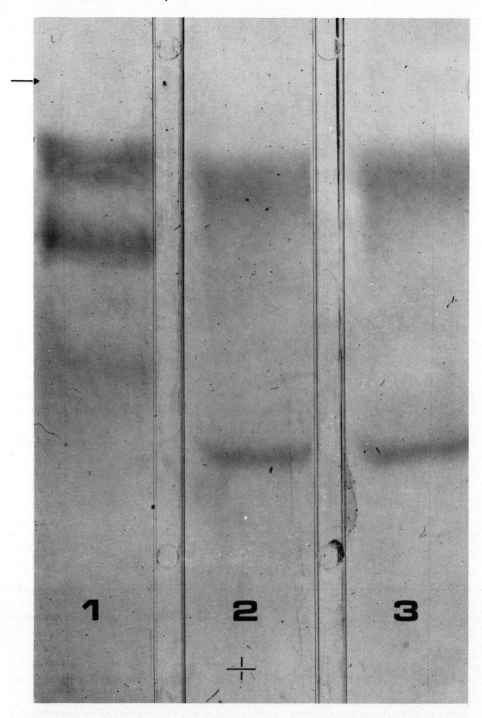

FIGURE 2.   Qualitative agarose electrophoretic test for detection of post-heparin lipolytic activity in unfractionated hyperlipidemic plasma. Sample was drawn from a fasting patient (1) immediately before intravenous administration of heparin; (2) 5 min after; and (3) 15 min after. Lipid values (mg/dl) are pre-heparin, cholesterol (chol), 184, triglycerides (TG), 208; 5 min post-heparin, chol, 173, TG, 76; 15 min post-heparin, chol, 171, TG, 64. Electropherograms were stained with Sudan black B.

β-band becomes more diffuse and its mobility increases; the alpha band becomes sharper. These changes occur in part because charged free fatty acids that are released by the enzyme associate with lipoprotcins and alter the electrophoretic pattern. The post-heparin electropherogram of the type I patient is unchanged.

Type II Pattern

The type II pattern is relatively common and is defined as an increase in the concentration of LDL alone (type IIA). It may be accompanied by an increased concentration of VLDL (type IIB). Secondary causes include hypothyroidism, nephrosis, dysgammaglobulinemia, myxedema, and others.[75] Increased LDL concentration is usually reflected as an elevation of total plasma cholesterol and is accompanied by increased plasma triglyceride concentrations in type IIB. Fasting plasma is free of chylomicrons. While it is true that hypercholesterolemic plasma with normal triglyceride levels can usually be confidently assigned the type IIA pattern without measurement of LDL cholesterol, it is well to bear in mind that high HDL concentrations can elevate total cholesterol in the presence of normal LDL concentrations and the assignment can be more confidently made if the HDL cholesterol level is also known.

Type IV Pattern

The type IV pattern is also relatively common and is defined as increased VLDL concentration in the presence of normal LDL levels. Plasma triglycerides are elevated, and cholesterol is normal or somewhat elevated if VLDL concentrations are very high. Freshly drawn plasma may be turbid, when the triglyceride concentration exceeds about 250 mg/dl and chylomicrons are absent. HDL levels are normal or somewhat low.

Type III Pattern

Type III hyperlipoproteinemia is an uncommon disorder that presents rather difficult laboratory and diagnostic problems. Plasma cholesterol and triglycerides are both elevated in the untreated patient, and chylomicrons may be present. The lipoprotein pattern in patients originally described as representing this disorder includes what was considered to be an abnormal lipoprotein species in the VLDL density range. This lipoprotein species, commonly called β-VLDL or "floating β" lipoproteins, migrates electrophoretically with or slightly faster[22,76] than normal β-lipoproteins. Paper electropherograms of unfractionated plasma usually show a characteristic "broad-β" band with little or no resolution between β- and pre-β-lipoproteins. B-VLDL is richer in cholesterol than VLDL, and its presence is reflected as an increase in the cholesterol:triglyceride ratio of lipoproteins of d < 1.006 g/ml.* Assignment of the type III pattern, therefore, has rested on the observation of (1) β-VLDL in plasma and (2) a higher than normal VLDL cholesterol:triglyceride ratio.

As the study of type III hyperlipoproteinemia has progressed and as more sensitive laboratory techniques have been employed, the situation has become more complicated. First, recent work suggests that β-VLDL may not be an abnormal lipoprotein species at all but may represent or be closely related to an intermediate in the conversion of VLDL to LDL, which is normally very low in concentration but which accumulates because of an interference with normal VLDL metabolism.[77,78] Given sufficiently sensitive techniques then, the intermediate may be observed in normolipidemic samples. The second point, which is undoubtedly related to the first, is that β-VLDL

---

* Since most circulating triglyceride is associated with this density fraction, it is common practice to determine the ratio of VLDL cholesterol to plasma triglyceride. A ratio > 0.3 has been taken as indicative of type III.[76]

has been observed on occasion in patients with other forms of hyperlipoproteinemia.[76] In a recent retrospective study of over 3000 quantitative determinations of lipoprotein patterns in 182 hypertriglyceridemic patients, Fredrickson et al.[76] concluded that quantitative estimation of the cholesterol and triglyceride composition of lipoproteins of d < 1.006 may better distinguish the type III patient than the presence of β-VLDL by electrophoresis. Third, it is difficult to know how to classify borderline or atypical samples, in which either the β-VLDL or the VLDL cholesterol:triglyceride ratio criterion, but not both, is satisfied or in which β-VLDL is observed in one analytical system but not in another.

Albers et al.[79] have recently examined current laboratory methods used to assign the type III pattern and have suggested procedures to minimize misclassification. They use agarose electrophoresis to detect β-VLDL and determine the ratio of cholesterol:triglyceride in lipoproteins of density < 1.006 g/ml by more than one method. Furthermore, rather than using a fixed ratio cut off, they found that discrimination is improved by use of a variable cutoff that depends on triglyceride concentration. This approach attempts to address the variability of the ratio that results from the imprecision of lipid analyses at low lipid concentrations and account for differences in the ratio that depend on the relative amounts of VLDL and β-VLDL present. This approach may be an improvement but as noted by the authors, absence of a known specific genetic marker for type III continues to make diagnosis of the disorder uncertain.

Recently, several studies have focused on apoproteins in type III. Apoprotein E, the so-called arginine-rich apoprotein, is one of the normal protein components of VLDL and is observed as a protein band with an apparent molecular weight of 39,000, by polyacrylamide-gel electrophoresis of VLDL proteins in the presence of sodium dodecyl sulfate.[80] The concentration of this apoprotein, which normally makes up about 10% of the total VLDL protein is considerably higher in type III patients.[81,82] Utermann et al.[82] have separated apolipoprotein E into three components, E-I, E-II, and E-III, by polyacrylamide-gel isoelectric focusing of VLDL proteins. They have found that the third component apo E-III is deficient in type III subjects. Warnick et al.[83] reported similar findings. They determined the ratio of apo E-IIIo E-II and reported values <0.2 in type III subjects they examined; whereas, unaffected subjects had ratios > 1.0. They also noted a ratio of < 0.2 in one adult with type IIA hyperlipoproteinemia and in a normal child. Intermediate ratios (0.2 to 1.0) were found in normal subjects and subjects with type IIA, IIB, and IV hyperlipoproteinemias.

Assessment of apo E-III may prove extremely helpful, but satisfactory resolution of the problems involved in recognizing type III awaits a better understanding of the significance of β-VLDL as it relates to normal lipoprotein metabolism. It will probably require a clearer definition of what is meant by type III hyperlipoproteinemia.

LIPOPROTEIN ELECTROPHORESIS IN THE ASSIGNMENT OF LIPOPROTEIN PATTERNS

Qualitative Interpretation of Lipoprotein Patterns

Three general approaches have been taken in the use of lipoprotein electrophoresis to assign lipoprotein patterns. First, electropherograms of unfractionated plasma or serum have been qualitatively assessed to identify elevated lipoprotein classes evidenced by abnormally intense staining bands or presence of the characteristic "broad β" band in type III. Known normal and abnormal samples are included in each run for comparison. This procedure has met with limited success for a number of reasons. The phenotyping system was described on the basis of a selected population of patients with rather definitive paper electrophoretic patterns. As has been pointed out,[84] patterns encountered in general medical practice are usually not as clear cut. In addition, methodological problems intervene to confuse the interpretation to an extent that depends on the specific method employed. These include the ability of the particular procedure to clearly separate lipoproteins, the comigration of more than one lipoprotein class, the sensitivity of the staining procedures used, and differential dye uptake by the different lipoproteins.

Winkelman and Ibbott,[85] several years ago, examined lipoprotein patterns determined qualitatively by paper electrophoresis alone. The study was conducted on samples submitted by physicians for determination of lipoprotein patterns and were therefore from a relatively unselected population. The authors were able to assign definitive patterns to only 34% of the samples they examined; the rest were assigned to several categories of nondefinitive patterns and included samples from normal subjects, those with mild type II, III, and IV patterns determined by other criteria, and treated subjects with these forms of hyperlipoproteinemia. Furthermore, electrophoretic interpretations were consistent with lipid determinations in only 58% of the samples, for which quantitative lipid analyses were available. Problems arose in large part from the insensitivity of paper electrophoresis, the incomplete separation of β- from pre-β-lipoproteins resulting in misidentification of bands, and differing interpretations of the same electropherograms by different experienced individuals.

Electrophoresis on cellulose acetate is more sensitive and more clearly separates β from pre-β components than paper.[28-30] Winkelman et al.[86] compared qualitative use of paper and cellulose acetate electrophoresis and found that cellulose acetate electrophoretic interpretations were consistent with lipid analyses in 76% of the examined samples. Disagreement between two experienced interpreters occurred only half as often as with paper electrophoresis. Of 183 samples reported, 52% were assigned definitive patterns compared with 32% for paper electrophoresis.[85]

The same group of workers[87] attempted to further examine the usefulness of qualitative cellulose acetate electrophoresis by asking the submitting physicians whether lipoprotein patterns determined in the laboratory were consistent with clinical diagnoses. They received responses to 564 of 866 inquiries. Of these, approximately 89% were consistent. Consistency was greater when the lipoprotein pattern was assigned on the basis of electrophoresis **and** quantitative lipid determinations on the same samples than when only electropherograms were considered. Several factors must be considered in interpreting these three studies. First, the method of assigning the patterns to which the electrophoretic results were compared was not based on quantitative determination of the lipoproteins, but on total cholesterol, triglyceride, and phospholipid determinations. Second, as pointed out by the authors,[87] uniform clinical diagnostic criteria were probably not used by all their respondents. It was impossible to judge the extent to which the laboratory report itself may have influenced the judgment of consistency or nonconsistency. Nonetheless, the studies were performed under conditions and with samples, which obtain in routine laboratory and medical practice and clearly demonstrate some of the problems associated with the qualitative use of lipoprotein electrophoresis. Several specific kinds of problems might be mentioned. A normotriglyceridemic sample with a relatively high concentration of Lp(a), which has pre-β electrophoretic mobility, could be interpreted as having elevated VLDL concentration if only an electropherogram of unfractionated plasma is used. If β- and pre-β-lipoproteins are both elevated and not sufficiently resolved from one another, a type IIB pattern could be misinterpreted as a type III pattern and total cholesterol and triglyceride concentrations might not be of much use in differentiating the two. Furthermore, it is possible to make the opposite mistake.

Noble,[38] during development of agarose-gel system for the electrophoretic separation of lipoproteins, observed that some type III subjects exhibited characteristic "broad β" bands, while others exhibited a pattern in which β and pre-β regions stained more intensely than the region between them, even though lipoproteins with both β- and pre-β- mobility had densities < 1.006 g/ml when separated ultracentrifugally.* It is evident that the latter pattern could be mistaken for a type IIB or IV pattern, if one relied only on electrophoresis of unfractionated plasma, and that, again, knowledge of the total cholesterol and triglyceride concentration might not necessarily aid in interpretation.

Electrophoretic Lipoprotein Quantitation

The second general approach has been to use electrophoresis to quantitate various lipoprotein classes by densitometric scanning of electropherograms of unfractionated plasma or serum. This procedure has been used with various kinds of supporting media including paper,[34,39] cellulose acetate,[88] agarose gel,[38,50,89-91] and polyacrylamide gel.[31] A variation of this technique has been described in which extinction values of stained lipoprotein bands are determined after elution from the supporting medium.[92] The success of this refinement is necessarily limited by the same factors that affect qualitative interpretation of electrophoretic patterns. Its use presents the user with the additional requirement of calibration, in order to relate the integrated densitometric scans to individual lipoprotein concentrations. In most cases, densitometric scanning has been performed on electropherograms that have been stained with an appropriate lipid stain such as Sudan black B, oil red O, or fat red 7B. The stains are taken up to a different extent by different lipids.[41,92,93] For example, Sudan black B and oil red O are taken up predominantly by cholesteryl esters and triglycerides and to a much lesser extent by unesterified cholesterol and phospholipids. Since lipid compositions of the various lipoprotein classes differ, intensity of staining varies not only with lipoprotein concentration but also with lipoprotein class.

The calibration problem has been approached in several ways. The most direct method is that in which integrated areas under each lipoprotein peak, expressed as a percent of the total area under all peaks, are related directly to lipoprotein concentrations determined by some quantitative method, without regard to differential lipoprotein staining. Noble et al.[34] using this approach, found sufficient correlation between paper or agarose-gel electrophoresis and analytical ultracentrifugation to conclude that the electrophoretic procedure was a potentially useful semiquantitative procedure for clinical purposes.

Similar procedures have been used by others. Van Melsen et al.[90] converted the area under each lipoprotein peak to a numerical lipoprotein-cholesterol value by making a calculation which used percent of total area occupied by the lipoprotein peak, independent measurements of total and HDL cholesterol, and a factor representing average cholesterol content of each lipoprotein as a percent of its total lipid content. Another approach has been an attempt, in hyperlipidemic samples, to relate percent of total area occupied by one or another lipoprotein directly to a particular lipoprotein pattern rather than to lipoprotein concentration.[89].

The most intensive efforts to quantitate lipoproteins by electrophoresis are those of Hatch et al.,[50] who devised a procedure for converting integrated densitometric scans of agarose electropherograms to numerical values for α-, β- and pre-β-lipoproteins. To accomplish this, they used external calibration against the β component of a frozen serum standard, for which the β-lipoprotein concentration had been established with the analytical ultracentrifuge. These authors also presented a method of internal cali-

---

*   An example of this behavior in samples encountered by the author is illustrated in Figure 3.

bration based on independent measurements of total cholesterol, triglyceride, or total lipids and factors representing the average weight fractions of lipids in each lipoprotein class. This procedure was designed to overcome variation in the application to the gel of small amounts of sample (1 μl), which would affect external calibration. Both of these calibration procedures incorporate factors which normalize the differential staining of different lipoproteins to β-lipoproteins.

Lindgren et al.[51] have further refined the internal calibration procedure by accounting for the possible presence of β-VLDL and Lp(a) lipoproteins. The technique incorporates electrophoresis of the ultracentrifugally prepared plasma fractions of density <1.006 and >1.006 g/ml. LDL and VLDL results obtained from electropherograms of unfractionated plasma are then corrected by amounts representing the contributions of β-VLDL and Lp(a), respectively. The method is relatively rapid and gives results in terms of analytical ultracentrifugal values that reflect total lipoprotein mass. Dyerberg and Hjorne,[92] taking still another approach, calculated lipoprotein concentrations from the extinction of electrophoretic bands after elution. They accomplished this by using independent measurements of plasma lipids, factors representing relative amount of stain taken up by each class, and average lipid and protein compositions of major lipoprotein classes.

Electrophoretically separated lipoproteins can be visualized in agarose gels by polyanion precipitation.[94] Neubeck et al.[91] have applied quantitative densitometric scanning to polyanion precipitates. Calibration is accomplished with pure isolated lipoproteins. This procedure may more closely reflect lipoprotein concentrations than lipid-staining procedures, because the precipitated band represents the whole lipoprotein molecule rather than just the lipid portion of the complex; however, since precipitability is affected by lipoprotein composition, conditions must be carefully controlled.

The foregoing, of course, is not an exhaustive review of electrophoretically based lipoprotein quantitation procedures but rather serves to illustrate attempts to take advantage of the ease and availability of electrophoresis in the routine laboratory. It is probably fair to say that lipoprotein electrophoresis has been most widely used qualitatively and that semiquantitative procedures based on total cholesterol or triglyceride concentrations would more likely be used in the routine laboratory. Quantitative procedures based directly on calibration with lipoproteins require standard materials that currently are not commercially available and require special equipment and expertise to prepare.

Qualitative Determination of Unusual Lipoproteins

The third general approach has been to use electrophoresis to qualitatively assess the presence of unusual lipoproteins. When used in this way, the procedure is relied upon only to **supplement** other methods of lipoprotein quantitation rather than to assign lipoprotein patterns directly. One major aim has been to assess the presence of β-VLDL in type III samples. Lipoproteins of <1.006 g/ml are separated from those of d > 1.006 g/ml by preparative ultracentrifugation. Aliquots of both ultracentrifugal fractions and the unfractionated sample are concomitantly subjected to electrophoresis in paper or agarose gel and examined for the presence of lipoproteins with β- mobility in the d <1.006 g/ml fraction.[35,44] This fraction usually contains only pre-β-lipoproteins (and chylomicrons, if present) in non-type III samples; however, when β-VLDL is present, it is observed either as a broad band, overlapping the β-pre-β region, or a defined band whose mobility is either identical with or very slightly greater than that of β-lipoproteins (Figure 3). The comparison is made with the β components of the unfractionated sample and the heavy ultracentrifugal fraction; therefore, each sample serves as its own control. Identification of β-VLDL can be difficult, even when the electrophoretic analysis is performed on the ultracentrifugal fraction. For example,

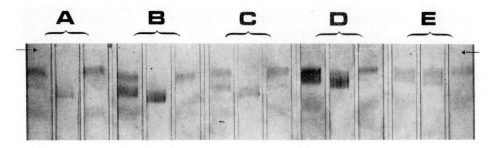

FIGURE 3.   Agarose-gel electropherograms of, from left to right, unfractionated plasma, lipoproteins of d <1.006 g/ml and lipoproteins of d >1.006 g/ml. Plasma fractions were obtained by ultracentrifugation without density adjustment at 105,000× g for 18 hr at 10°C. Arrows denote sample application. Sample A had normal concentrations (mg/dℓ) of total cholesterol and triglycerides (chol, 231; TG, 148); sample B, type IV lipid values of chol, 244, TG, 340; sample C, type IIB lipid values of chol, 286, TG, 205 (note presence of "sinking pre-β" lipoproteins in density >1.006 g/ml fraction); sample D, type III lipid values of chol, 265, TG, 250 (note "broad β" band in unfractionated plasma and density <1.006 g/ml fraction); sample E, type III lipid values of chol, 226, TG, 219 (note defined β and pre-β bands in unfractionated plasma and in density <1.006 g/ml fraction).

some lipemic samples show trailing from the pre-β through the β region toward the origin or from the pre-β to the β region, without a defined band in the latter region. Recognition of β-VLDL in these cases requires judgment and experience.[79] Lp(a), if present in sufficient concentration, is observed as a band with pre-β- mobility in the heavy ultracentrifugal fraction (Figure 3). This procedure requires an ultracentrifuge, and its use has been restricted primarily to lipoprotein research laboratories. Recent development of an inexpensive air-driven table top ultracentrifuge* may increase the general availability of the test.

Polyacrylamide-gel electrophoresis has also been used to detect the type III pattern in unfractionated plasma.[32] Chylomicrons and VLDL do not enter a 3.0% polyacrylamide gel, LDL normally migrates slowly as a well-defined band, and HDL moves rapidly near the electrophoretic front (Figure 1). Type III samples, however, show a markedly diminished or completely absent LDL band. The polyacrylamide-gel procedure has been used in conjunction with electrophoresis of unfractionated samples in other supporting media to detect the type III pattern.[32] β-migrating lipoproteins are observed on paper, cellulose acetate, or agarose electropherograms, but not on polyacrylamide-gel electropherograms. It is well to note that this procedure can also be difficult to interpret, because samples are encountered which show a diminished β band, hence a type III pattern, in polyacrylamide gels; however, they do not have β-VLDL by agarose electrophoresis of lipoproteins of d < 1.006 g/ml and vice versa. The test was originally applied by combining polyacrylamide-gel with **paper** electrophoresis.[32] It may not be as successful when used with the more sensitive agarose-gel procedure.[79]

Several recent developments potentially increase the usefulness of electrophoresis in assigning the type III pattern. First, VLDL and β-VLDL can be precipitated with heparin in the presence of Mg^{+2} under conditions in which β-lipoproteins remain soluble. Wieland and Seidel[95] have taken advantage of this and have described a procedure in which lipoproteins of d <1.006 g/ml are selectively precipitated in agarose electropherograms of unfractionated samples. β-VLDL is observed as a precipitate in the β region. The procedure is more sensitive than those which employ lipid staining because β-lipoproteins are not visualized. It can be difficult, however, to distinguish β-VLDL from trailing pre-β-lipoproteins in some hyperlipidemic samples.[79]

* This ultracentrifuge was developed by Beckman Instruments, Inc., Palo Alto, Cal.

Papadopoulos and Herbert[48] have recently described the presence of a double β-lipoprotein band in lipid-stained agarose electropherograms of type III plasma. The slower of the two bands was found to sediment with lipoproteins of d >1.006 g/ml and was felt to correspond to LDL. The other band was associated with lipoproteins of density <1.006 g/ml and was believed to represent β-VLDL. Successful application of this technique demands use of fresh samples, low gel concentration, and close attention to the details of gel preparation, buffer concentration, sample application and time and temperature of electrophoresis. Nonetheless, it does not require use of the ultracentrifuge, appears to be quite sensitive, and may be useful in detecting the type III pattern even in mild or treated patients.

Use of isoelectric focusing to assess apo E-III as an aid to establishing the presence of the type III pattern has been mentioned above. Recognition of the deficiency of apo E-III in type III is important, because it may represent a qualitative difference in apoprotein composition of the type III patient. It could also conceivably lead to identification of a specific biochemical, e.g., enzyme, lesion in this disorder. Although isoelectric focusing is technically demanding in that it requires the isolation and delipidation of lipoproteins of d <1.006 g/ml before analysis, it may prove extremely valuable for the study and diagnosis of type III hyperlipoproteinemia. Electrophoresis has also been applied to the detection of Lp-X,[94,96,97] the abnormal lipoprotein characteristic of obstructive liver disease,[98,99] as discussed elsewhere in this volume.

CONCLUSION

The attractive simplicity of electrophoresis has elicited its widespread, routine use in lipoprotein phenotyping and has led to many serious attempts to improve its reliability. It is not ideally suited to screening, when the aim is to detect hyperlipidemia; quantitative plasma lipid determinations serve this purpose better. Electrophoretic techniques have been generally applied qualitatively or semiquantitatively because of the technical difficulties of quantitative calibration and general lack of available, standard materials. Availability of such materials would probably stimulate greater application of quantitative methods and undoubtedly increase its usefulness; however, whether employed qualitatively or quantitatively, the inherent limitations of electrophoresis of unfractionated samples work to circumscribe its usefulness as the sole or major method on which assignment of lipoprotein patterns is based. It is more successfully applied when its object is limited, that is, when it is used as a supplement to other methods of fractionation and quantitation to detect unusual lipoproteins. Even when used in this way, it can be difficult to interpret. This, of course, is not to suggest that lipoprotein electrophoresis is useless or that efforts to improve it should not continue, but rather that it should be applied with an appreciation of its limitations and should not be asked to provide more information than it can.

ACKNOWLEDGMENT

The author wishes to express his appreciation to Drs.Peter O. Kwiterovich, Peter D. S. Wood, and John J. Albers for their helpful suggestions during preparation of this manuscript. This work was supported in part by NIH Contract NO1-HV1-2158L and NIH Grant No. 5R01 HL 17898-03

REFERENCES

1. **Gofman, J. W., Glazier, F., Tamplin, A., Strisower, B., and DeLalla, O.,** Lipoproteins and coronary heart disease and atherosclerosis, *Physiol. Rev.,* 34, 589, 1954.
2. **Lindgren, F. T., Jensen, L. C., and Hatch, F. T.,** The Isolation and quantitative analysis of serum lipoproteins, in *Blood Lipids and Lipoproteins: Quantitation, Composition, and Metabolism,* Nelson, G., Ed., Interscience, New York, 1972, 181.
3. **Noble, R. P., Hatch, F. T., Mazrimas, J. A., Lindgren, F. T., Jensen, L. C., and Adamson, G.L.,** Comparison of lipoprotein analysis by agarose gel and paper electrophoresis with analytic ultracentrifugation, *Lipids,* 4, 55, 1969.
4. **Lees, R. S. and Hatch, F. T.,** Sharper separation of lipoprotein species by paper electrophoresis in albumin-containing buffer, *J. Lab. Clin. Med.,* 61, 518, 1963.
5. **Alaupovic, P.,** Apolipoproteins and lipoproteins, *Atherosclerosis,* 13, 141, 1971.
6. **Bilheimer, D. W., Eisenberg, S., and Levy, R. I.,** The metabolism of very low density lipoprotein proteins.I. Preliminary in vitro and in vivo observations, *Biochim.Biophys. Acta,* 260, 212, 1972.
7. **Glomset, J. A.,** The plasma lecithinolesterol acyltransferase reaction, *J. Lipid Res.,* 9, 155, 1968.
8. **Stein, Y., Glangeaud, M.C., Fainaru, M., and Stein, O.,** The removal of cholesterol from aortic smooth muscle cells and landschutz ascites cells by fractionation of human high density apoprotein, *Biochim. Biophys. Acta,* 380, 160, 1975.
9. **Carew, T. E., Koschinsky,T., Hayes, S. B., and Steinberg, D.,** The mechanism by which high-density lipoproteins may slow the atherogenic process, *Lancet,* 1, 1315, 1976.
10. **Sodhi, H. S.,** New lipoprotein differing in charge and density from known plasma lipoproteins, *Metabolism,* 18, 852, 1969.
11. **Enholm, C., Garoff, H., Simons, K., and Aro, H.,** Purification and quantitation of the human plasma lipoprotein containing the Lp(a) antigen, *Biochim. Biophys. Acta,* 236, 431, 1971.
12. **Rider, A. K., Levy, R. I., and Fredrickson, D. S.,** Sinking prebeta lipoprotein and the Lp antigen, *Circulation,* 42, 10, 1970.
13. **Albers, J. J., Cabana, V. G., Warnick, G. R., and Hazzard, W. R.,** Lp(a) lipoprotein: relationship to sinking prebeta lipoprotein, hyperlipoproteinemia and apoprotein B, *Metabolism,* 24, 1047, 1975.
14. **Albers, J. J. and Hazzard, W. R.,** Immunochemical quantification of human plasma Lp(a) lipoprotein, *Lipids,* 9, 15, 1974.
15. **Albers, J. J., Adolphson, J. L., and Hazzard, W. R.,** Radioimmunoassay of human plasma Lp(a) lipoprotein, *J. Lipid Res.,* 18, 331, 1977.
16. **Carlson, L. A. and Ericsson, M.,** Quantitative and qualitative serum lipoprotein analysis. 1. Studies in healthy men and women, *Atherosclerosis,* 21, 417, 1975.
17. **Dahlen, G., Berg, K., Gillnas, T., and Ericson, C.,** Lp(a) lipoprotein/prebeta lipoprotein in Swedish middle-aged males and in patients with coronary heart disease, *Clin. Genet.,* 7, 334, 1975.
18. **Fredrickson, D. S., Levy, R. I., and Lees, R. S.,** Fat transport in lipoproteins: an integrated approach to mechanisms and disorders, *N. Engl. J. Med.,* 276, 34, 94, 148, 215, 273, 1967.
19. **Beaumont, J. L., Carlson, L. A., Cooper, G. R., Fetpas, Z., Fredrickson, D. S., and Strasser, T.,** Classification of hyperlipidemias and hyperlipoproteinemias, *Bull. WHO,* 43, 891, 1970.
20. **Fredrickson, D. S.,** It's time to be practical, *Circulation,* 51, 209, 1975.
21. **Goldstein, J. L., Hazzard, W. K., Schrott, H. G., Bierman, E. L., and Motulsky, A. G.,** Hyperlipidemia in coronary heart disease, *J. Clin. Invest.,* 52, 1533, 1973.
22. **Hazzard, W. R., Goldstein, J. L., Schrott, H. G., Motulsky, A. G., and Bierman, E. L.,** Hyperlipidemia in coronary heart disease. III. Evaluation of lipoprotein phenotypes of 156 genetically defined survivors of myocardial infarction, *J. Clin. Invest.,* 52, 1569, 1973.
23. **Hazzard, W. R., O'Donell, T. F., and Lee, Y. L.,** Broad-beta disease (type III hyperlipoproteinemia) in a large kindred. Evidence for a monogenic mechanism, *Ann. Intern. Med.,* 82, 141, 1975.
24. **Miller, G. J. and Miller, N. E.,** Plasma high density lipoprotein concentration and the development of ischemic heart disease, *Lancet,* 1, 16, 1975.
25. **Gordon, T., Castelli, W. P., Hjortland, M. C., and Kannel, W. B.,** The prediction of coronary heart disease by high density and other lipoproteins. An historical perspective, in *Hyperlipidemia, Diagnosis and Therapy,* Rifkind, B. M. and Levy, R. I.,Eds., Grune & Stratton, New York, 1977, 71.
26. **Immarino, R. M.,** Lipoprotein electrophoresis should be discontinued as a routine procedure, *Clin. Chem.,* 21, 300, 1975.
27. **Bierman, E. L., Porte, D., O'Hara, D. D., Schwartz, M., and Wood, F. C.,** Characterization of fat particles in plasma of hyperlipidemia subjects maintained on fat-free, high carbohydrate diets, *J. Clin. Invest.,* 44, 261, 1965.
28. **Charman, R. C. and Landowne, R. A.,** Separation of human plasma lipoprotein by electrophoresis on cellulose acetate, *Anal. Biochem.,* 19, 177, 1967.

29. Chin, H. P. and Blankenhorn, D. H., Separation and quantitative analysis of serum lipoproteins by means of electrophoresis on cellulose acetate, *Clin. Chim. Acta,* 20, 305, 1968.

30. Postma, T. and Stroes, J. A. P., Lipid screening in clinical chemistry, *Clin. Chim. Acta,* 22, 569, 1968.

31. Moran, R. F., Castelli, W. P., and Moran, M. V., Quantitation of beta lipoprotein (LDL) cholesterol by densitometric evaluation of disc electropherograms, *Clin. Chem.,* 18, 217, 1972.

32. Masket, B. H., Levy, R. I., and Fredrickson, D. S., The use of polyacrylamide gel electrophoresis in differentiating type III hyperlipoproteinemia, *J. Lab. Clin. Med.,* 81, 794, 1973.

33. Raymond, S., Miles, J. L., and Lee, J. C. J., Lipoprotein patterns in acrylamide gel electrophoresis, *Science,* 151, 346, 1966.

34. Noble, R. P., Hatch, F. T., Mazrimas, J. A., Lindgren, F. T., Jensen, L. C., and Adamson, G. L., Comparison of lipoprotein analysis by agarose gel and paper electrophoresis with analytical ultracentrifugation, *Lipids,* 4, 55, 1969.

35. *Lipid Research Clinics Program Manual of Laboratory Operations. Lipid and Lipoprotein Analysis,* Vol. 1, Department of Health, Education, and Welfare Publication, DHEW Publ No. (NIH) 75-628, National Institutes of Health, Bethesda, Md., 1974.

36. Dangerfield, W. G. and Smith, E. B., Investigation of serum lipids and lipoproteins by paper electrophoresis, *J. Clin. Pathol.,* 18, 132, 1955.

37. Masket, B. H., Lipoprotein fractionation by paper electrophoresis, in *Manual of Procedures for the Applied Seminar on the Clinical Pathology of the Lipids,* Sunderman, F. W., Ed., Philadelphia Association of Clinical Science, Philadelphia, 1971.

38. Noble, R. P., Electrophoretic separation of plasma lipoproteins in agarose gel, *J. Lipid Res.,* 9, 693, 1968.

39. Hatch, F. T., Moore, J. L., Lindgren, F. T., Jensen, L. C., Freeman, N. K., and Willis, R. D., Semi-quantitative paper electrophoresis of serum lipoproteins, *Circulation,* 36, 16, 1967.

40. Papadopoulos, N. M. and Kintzios, J. A., Varieties of human serum lipoprotein pattern. Evaluation by agarose gel electrophoresis, *Clin. Chem.* 17, 427, 1971.

41. Hatch, F. T. and Lees, R. S., Practical methods for plasma lipoprotein analysis, *Adv. Lipid Res.,* 6, 1, 1968.

42. Lindgren, F. T., Preparative ultracentrifugal laboratory procedures and suggestions for lipoprotein analysis, in *Procedures for Lipid and Lipoprotein Analysis,* Perkins, E., Ed., American Oil Chemists Society, 1975.

43. Heiberg, A., A comparative study of different electrophoretic techniques for classification of hereditary hyperlipoproteinemias, *Clin. Genet.,* 4, 450, 1973.

44. Fredrickson, D. S. and Levy, R. I., Familial hyperlipoproteinemia, in *The Metabolic Basis of Inherited Disease,* Stanbury, J. B., Wyngaarden, J. B., and Fredrickson, D. S., Eds., McGraw-Hill, New York, 1972, 531.

45. Hatch, F. T., Moore, J. L., Lindgren, F. T., Jensen, L. C., Freeman, N. K., and Willis, R. D., Semiquantitative paper electrophoresis of serum lipoproteins, *Circulation,* 36, 16, 1967.

46. Papadopoulos, N. M. and Kintzios, J. A., Determination of human serum lipoprotein patterns by agarose gel electrophoresis, *Anal. Biochem.,* 30, 421, 1969.

47. Vessby, B., Hedstrand, H., Lundin, L. G., and Olsson, V., Inheritance of type III hyperlipoproteinemia. Lipoprotein patterns in first degree relatives, *Metabolism,* 26, 225, 1977.

48. Papadopoulos, N. M. and Herbert, P. N., The β-lipoprotein doublet in type III hyperlipoproteinemia, *Clin. Chem.,* 23, 978, 1977.

49. Papadopoulos, N. M. and Bedynek, J. L., Serum lipoprotein patterns in patients with coronary atherosclerosis, *Clin. Chim. Acta,* 44, 153, 1973.

50. Hatch, F. T., Lindgren, F. T., Adamson, G. L., Jensen, L. C., Wong, A. W., and Levy, R. I., Quantitative agarose gel electrophoresis of plasma lipoproteins: a simple technique and two methods of standardization, *J. Lab. Clin. Med.,* 81, 946, 1973.

51. Lindgren, F. T., Adamson, G. L., Jenson, L. C., and Wood, P. D., Lipid and lipoprotein measurements in a normal adult American population, *Lipids,* 10, 750, 1975.

52. Davis, B. J., Disc electrophoresis. II. Method and application to human serum proteins, *Ann. N. Y. Acad. Sci.,* 121, 404, 1964.

53. Narayan, K. A., Narayan, S., and Kummerow, F. A., Disc electrophoresis of human serum lipoproteins, *Nature,* 205, 246, 1965.

54. Narayan, K. A., Human and rat serum proteins, lipoproteins, ammonium persulfate, gel concentration and disc electrophoresis, *Microchem. J.,* 14, 235, 1969.

55. Frings, C. S., Foster, L. B., and Cohen, P. S., Electrophoretic separation of serum lipoproteins in polyacrylamide gel, *Clin. Chem.,* 17, 111, 1971.

56. Hall, F. F., Ratliff, C. R., Westfall, C. L., and Culp, T. W., Serum lipoprotein electrophoresis. An improved polyacrylamide procedure, *Biochem. Med.,* 6, 464, 1972.

57. **Melish, J. S. and Waterhouse, C.,** Concentration gradient electrophoresis of plasma from patients with hyperbetalipoproteinemia, *J. Lipid Res.,* 13, 193, 1972.
58. **Borrie, P.,** Type III hyperlipoproteinemia, *Br. Med. J.,* 2, 665, 1969.
59. **Pratt, J.J. and Dangerfield, W. G.,** Polyacrylamide gels of increasing concentration gradient for the electrophoresis lipoproteins, *Clin. Chim. Acta,* 23, 189, 1969.
60. **Dangerfield, W. G. and Pratt, J. J.,** An investigation of plasma lipoproteins by polyacrylamide electrophoresis, *Clin. Chim. Acta,* 30, 273, 1970.
61. *Hyperlipidemia: Diagnosis and Therapy,* Rifkind, B. M. and Levy, R. I., Eds., Grune & Stratton, New York, 1977.
62. **Kwiterovich, P. O.,** Pediatric aspects of hyperlipoproteinemia, in *Hyperlipidemia: Diagnosis and Therapy,* Rifkind, B. M. and Levy, R. I., Eds., Grune & Stratton, New York, 1977, 249.
63. **Porte, D., O'Hara, D. D., and Williams, R. H.,** Relation between post heparin lipolytic activity and plasma triglyceride in myxedema, *Metabolism,* 15, 107, 1966.
64. **Bagdade, J. D., Porte, D., and Bierman, E. L.,** Diabetic lipemia, *N. Engl. J. Med.,* 276, 427, 1967.
65. **Hazzard, W. R., Speiger, M. J., Bagdade, J. D., and Bierman, E. L.,** Studies on the mechanism of increased plasma triglyceride levels by oral contraceptives, *N. Engl. J. Med.,* 280, 471, 1969.
66. **Mann, J. A. and Truswell, A. S.,** Effect of controlled breakfast on serum cholesterol and triglycerides, *Am. J. Clin. Nutr.,* 24, 1300, 1971.
67. **Denborough, M. A.,** Alimentary lipemia in ischemic heart disease, *Clin. Sci,* 25, 115, 1963.
68. **Brown, W. V., Baginsky, M. L., and Enholm, C.,** Primary type I and type V hyperlipoproteinemia, in *Hyperlipidemia: Diagnosis and Therapy,* Rifkind, B. M. and Levy, R. I., Eds., Grune & Stratton, New York, 1977, 93.
69. **Simons, L. A. and Williams, P. F.,** The biochemical composition and metabolism of lipoproteins in type V hyperlipoproteinemia, *Clin. Chim. Acta,* 61, 341, 1975.
70. **Kwiterovich, P. O., Farah, J. R., Brown, W. V., Bachorik, P. S., Baylin, S. B., and Neill, C. A.,** The clinical, biochemical and familial presentation of type V hyperlipoproteinemia in childhood, *Pediatrics,* 59, 513, 1977.
71. **Krauss, R. M., Levy, R. I., and Fredrickson, D. S.,** Selective measurement of two lipase activities in post heparin plasma from normal subjects and patients with hyperlipoproteinemia, *J. Clin. Invest.,* 54, 1107, 1974.
72. **Enholm, C., Shaw, W., Greten, H., Langfelder, W., and Brown, M. V.,** Separation and characterization of two triglyceride lipase activities from human post heparin plasma, in *Atherosclerosis III. Proceedings of the Third International Symposium,* Shettler, G. and Weizel, A., Eds., Springer-Verlag, Berlin, 1974, 557.
73. **Bobery, J., Augustin, J., Baginsky, M. L., Tejada, P., and Brown, W. V.,** Quantitative determination of hepatic and lipoprotein lipase activities from human postheparin plasma, *J. Lipid Res.,* 18, 544, 1977.
74. **Huttunen, J. K., Enholm, C., Kinnunen, P. K., and Nikkila, E. A.,** An immunochemical method for the selective measurement of two triglyceride lipases in human post-heparin plasma, *Clin. Chim. Acta,* 63, 335, 1975.
75. **LaRosa, J. C.,** Secondary hyperlipoproteinemia, in *Hyperlipidemia: Diagnosis and Therapy,* Rifkind, B. M. and Levy, R. I., Eds., Grune & Stratton, New York, 1977, 205.
76. **Fredrickson, D. S., Morganroth, J., and Levy, R. I.,** Type III hyperlipoproteinemia: an analysis of two contemporary definitions, *Ann. Intern. Med.,* 82, 150, 1975.
77. **Hazzard, W. R.,** Primary type III hyperlipoproteinemia, in *Hyperlipidemia: Diagnosis and Therapy,* Rifkind, B. M. and Levy, R. I., Eds., Grune & Stratton, New York, 1977, 137.
78. **Hazzard, W. R. and Bierman, E. L.,** The spectrum of electrophoretic mobility of very low density lipoproteins: role of slower migrating species in endogenous lipemia and broad-β disease, *J. Lab. Clin. Med.,* 86, 239, 1975.
79. **Albers, J. J., Warnick, G. R., and Hazzard, W. R.,** Type III hyperlipoproteinemia: a comparative study of current diagnostic techniques, *Clin. Chim. Acta,* 75, 193, 1977.
80. **Utermann, G.,** Isolation and partial characterization of an arginine-rich apolipoprotein from human plasma very-low density lipoproteins, *Hoppe Seylers Z. Physiol. Chem.,* 36, 1113, 1975.
81. **Havel, R. J. and Kane, J. P.,** Primary dysbetalipoproteinemia: predominance of a specific apoprotein species in triglyceride rich lipoproteins, *Proc. Natl. Acad. Sci. U.S.A.,* 70, 2015, 1973.
82. **Utermann, G. Jaeschke, M., and Menzel, J.,** Familial hyperlipoproteinemia type III: deficiency of a specific apolipoprotein (ApoE-III) in the very low density lipoproteins, *FEBS Lett.,* 56, 352, 1975.
83. **Warnick, G. R., Albers, J. J., and Hazzard, W. R.,** Genetics of type III hyperlipoproteinemia; "pseudodominant" transmission in a large kindred, *Circulation,* 56 (III), 21, 1977.
84. **Lees, R. S.,** A progress report on lipoprotein phenotyping, *J. Lab. Clin. Med.,* 82, 529, 1973.
85. **Winkelman, J. and Ibbot, F. A.,** Studies on the phenotyping of hyperlipoproteinemias. Evaluation of paper electrophoresis technique, *Clin. Chim. Acta,* 26, 25, 1969.

86. **Winkelman, J., Ibbot, F. A., Sobel, C., and Wynbenga, D. R.,** Studies on the phenotyping of hyperlipoproteinemias. Evaluation of cellulose acetate technique and comparison with paper electrophoresis, *Clin. Chim. Acta,* 26, 33, 1969.

87. **Winkelman, J., Wynbenga, D. R., and Ibbot, F. A.,** Correlation of laboratory tests and clinical evaluation in phenotyping of lipoproteinemias, *Clin. Chem.* 16, 594, 1970.

88. **Winkelman, J., Wynbenga, D. R., and Ibbot, F.,** Quantitation of lipoprotein components in the phenotyping of hyperlipoproteinemias, *Clin. Chim. Acta,* 27, 181, 1970.

89. **Sirtori, C., Hussanein, K. M., Hussanein, R., and Boulos, B. M.,** Phenotyping of type II and IV hyperlipoproteinemias by a simple quantitative agarose gel lipoprotein electrophoresis, *Clin. Chim. Acta,* 31, 305, 1971.

90. **Van Melson, A., DeGreve, Y., Van Der Weiken, F., Vastesaeger, M., Blaton, V., and Peeters, H.,** A modified method for phenotyping of hyperlipoproteinemia on agarose electrophoresis, *Clin. Chim. Acta,* 55, 225, 1974.

91. **Neubeck, W., Wieland, H., Habenicht, A., Mueller, P., Baggio, G., and Seidel, D.,** Improved assessment of plasma lipoprotein patterns. III. Direct measurement of lipoproteins after gel electrophoresis, *Clin. Chem.* 23, 1296, 1977.

92. **Dyerberg, J. and Hjørne, N.,** Quantitative plasma lipoprotein electrophoresis. Correction for the difference in dye uptake by the lipoprotein fractions, *Clin. Chim. Acta,* 30, 407, 1970.

93. **Jencks, W. P. and Durram, E. L.,** Paper electrophoresis as a quantitative method: the staining of serum lipoproteins, *J. Clin. Invest.,* 34, 1437, 1955.

94. **Seidel, D., Wieland, H., and Ruppert, C.,** Improved techniques for assessment of plasma lipoprotein patterns. I. Precipitation in gels after electrophoresis with polyanionic compounds, *Clin. Chem.* 19, 737, 1973.

95. **Wieland, H. and Seidel, D.,** Improved techniques for assessment of serum lipoprotein patterns. II. Rapid method for diagnosis of type III hyperlipoproteinemia without ultracentrifugation, *Clin. Chem.,* 19, 1139, 1973.

96. **Seidel, D.,** A new immunochemical technique for rapid, semiquantitative determination of the abnormal lipoprotein (Lp-X) characterizing cholestasis, *Clin. Chim. Acta,* 31, 225, 1971.

97. **Neubeck, W. and Seidel, D.,** Direct method for measuring lipoprotein-X in serum, *Clin. Chem.,* 21, 853, 1975.

98. **Seidel, D., Gretz, H., and Ruppert, C.,** Significance of the LP-X test in differential diagnosis of jaundice, *Clin. Chem.,* 19, 86, 1973.

99. **Ritland, S., Blomhoff, J. P., Elgjo, K., and Gjone, E.,** Lipoprotein X (LP-X) in liver disease, *Scand. J. Gastroenterol,* 8, 155, 1973.

100. **Naito, H. K. and Lewis, L. A.,** Rapid lipid staining procedure for paper electropherograms, *Clin. Chem.,* 19, 106, 1973.

101. **Papadoupolos, N. M. and Kintzios, J. A.,** Differentiation of pathological conditions by visual evaluation of serum protein electrophoretic patterns, *Proc. Soc Exp. Biol. Med.,* 125, 927, 1967.

102. **Nichols, A. V.,** Human serum lipoproteins and their interrelationships, *Adv. Biol. Med. Phys.,* 11, 109, 1967.

LIPOPROTEIN ELECTROPHORESIS IN DIFFERENTIATING TYPE III HYPERLIPOPROTEINEMIA

B. H. Masket

INTRODUCTION

A simple, specific diagnostic test to identify patients with type III hyperlipoprotein-emia[1] has been sought since its identification as "xanthoma tuberosum" in 1954.[2] Recognition of primary type III hyperlipoproteinemia is important, because it is not rare and produces early peripheral vascular disease as well as xanthomas (xanthoma striata palmaris in the creases of the hands and tuberous or tendon xanthomas elsewhere) in many of these patients. Type III hyperlipoproteinemia responds rapidly to caloric control and clofibrate, and there is evidence that the course of the disease may be slowed by treatment.[3]

A number of reports have shown that type III lipoproteins of density less than 1.006, very low-density lipoproteins, (VLDL) have abnormal lipid compositions.[4-8] It is on this basis that a tentative diagnostic test can be made using two types of lipoprotein electrophoresis. The electrophoretic lipoprotein pattern in type III is characterized by the presence of "floating β-lipoproteins," β very low-density lipoproteins, β-VLDL) in plasma. These lipoproteins have β mobility on paper electrophoresis and have a density of 1.006. Normally, in this density range, VLDL migrate to a pre-β position, and chylomicrons remain at the origin.[9,10]

Electrophoresis on agarose[11,12] and paper provide comparable separation of the four major lipoprotein families, although agarose more frequently separates several of the families into more than one band.[12] Some preparations of cellulose acetate provide poorer separation of chylomicrons and very low-density lipoproteins.[13,14]

Polyacrylamide gel electrophoresis (PAGE) has also been adapted to electrophoresis of plasma lipoproteins using both uniform gel[15-21] and continuous gradient[22-24] systems. PAGE separates particles on the basis of their size and shape in addition to their charge.[25] It, thus, offers some theoretical advantages for distinguishing among lipoprotein patterns. The plasma lipoprotein patterns obtained with polyacrylamide gel electrophoresis were compared with those obtained by paper electrophoresis in multiple samples from 138 patients previously classified into five different types of primary hyperlipoproteinemia. Special emphasis was placed on exploration of a possible diagnosis of type III hyperlipoproteinemia by electrophoresis alone. In 95% of 118 samples from 27 patients with type III hyperlipoproteinemia, β-migrating lipoproteins were produced on paper, while there was an absence of a discernible β-lipoprotein band in polyacrylamide. This combination occurred in only three of 228 samples from other types of hyperlipoproteinemia. It was concluded that conjoint electrophoresis offers a simple and reasonably accurate means of diagnosis of type III hyperlipoproteinemia.

This report will provide a detailed description of techniques for paper electrophoresis and PAGE that have been found to be useful for lipoprotein separation. Special emphasis is placed on the exploration of a possible diagnosis of type III hyperlipoproteinemia by electrophoresis alone. Particular attention is focused on the use of conjoint electrophoresis on paper and polyacrylamide gel to establish a diagnostic technique for type III without resorting to the ultracentrifuge.

Materials and Methods

Plasma Samples

All plasma samples utilized for this study were obtained from blood collected in disodium EDTA, 1 mg/ml, and kept at 4°C until used. Nearly all samples were electrophoresed within 8 hr of collection, although some were stored for as long as 10 days before use. All of the samples were collected in the morning after a 10- to 14-hr fast and were classified as to lipoprotein type according to analytical techniques previously described.[4] This included quantification of low-density-lipoprotein (LDL) cholesterol by combined use of the preparative ultracentrifuge and precipitation of all lipoproteins except high-density lipoproteins (HDL).[26] Donors included 27 patients classified as having type III hyperlipoproteinemia after demonstration of the presence of β-migrating lipoproteins having a density of less than 1.006.[4] Samples from 111 patients with the other four major types of hyperlipoproteinemia were also examined for comparisons with type III.

Lipoprotein Fractions

A crude preparation of chylomicrons was obtained from patients with type I hyperlipoproteinemia by centrifugation of plasma without denisty adjustment for 10^5 g-min. VLDL, lipoproteins of <1.006 prepared from chylomicron-free plasma; low-density lipoproteins (LDL), d 1.019 to 1.063; and high-density lipoproteins (HDL) d 1.063 to 1.21 were prepared using the ultracentrifuge under conditions previously described.[27]

POLYACRYLAMIDE GEL ELECTROPHORESIS

Equipment and Reagents

Acrylamide (Eastman® 5521)* is recrystallized from acetone as suggested by Rodbard and Chrambach.[28] Acrylamide (Eastman X5521) can be used directly without further recrystallization. *N, N'*-methylenebisacrylamide is also recrystallized from acetone. Both reagents should be stored in the refrigerator. *N, N, N', N'*-tetramethylethylenediamine (TEMED), Eastman 8178, is a clear to slightly yellow liquid which should be kept in a brown bottle in the refrigerator. A concentrated solution of riboflavin (Eastman 5181) can be made up and used indefinitely if kept refrigerated. Tween® 20 can be obtained from the Atlas® Chemical Company, Wilmington, Delaware. Sudan black B can be obtained from National Aniline Division of Allied Chemical® and Dye Corporation, New York, N.Y. Reagent grades of the following are used: 2-amino-2-hydroxymethyl- 1,3-propanediol (TRIS) $1 N$ hydrochloric acid (HCl), ammonium persulfate peroxydisulfate, $((NH_4)_2S_2O_8)$, glycine (ammonia free), and diethylene glycol.

Reagent Solutions

Stock solutions are prepared as shown in Table 1. Stock solutions are kept at 4°C and can be used for as long as 3 months with one exception. The persulfate solution (solution F) should be prepared on the day it is used. Working solutions are prepared from different volumes of the stock solutions as shown in Table 2. The 3.0% gel working solution and the dilute TRIS buffer are freshly prepared before use. The two large-pore gel working solutions can be used for 2 weeks after preparation provided they are stored at 4°C.

A rectangular Plexiglas® chamber can be specially constructed for PAGE which allows the gels to be poured, polymerized, and electrophoresed without removing them from the cell. Electrophoresis is carried out in cylindrical Pyrex® glass tubes (7 mm O.D. × 5 mm I.D. × 70 mm long). The tubes are acid-washed to minimize adherence of water to the glass surface. The bottom of each tube is sealed with parafilm held in

---

* Eastman 5521, Eastman X5521, *N,N'*-methylenebisacrylamide, and Eastman 5181 is produced by Distillation Products Industries, Division of Eastment Kodak Company, Rochester, N.Y.

Table 1
STOCK SOLUTIONS FOR PREPARATIONS OF
POLYACRYLAMIDE GELS

Solution A	$1N$ HCl	48.0 ml
	TRIS	36.3 g
	TEMED	0.23 ml
	Water to make	100 ml
Solution B	$1N$ HCl	48.0 ml
	TRIS	5.98 g
	TEMED	0.46 ml
	Water to make	100 ml
Solution C	Acrylamide	12.0 g
	Bis	1.0 g
	Water to make	100 ml
Solution D	Acrylamide	10.0 g
	Bis	2.5 g
	Water to make	100 ml
Solution E	Riboflavin	4.0 mg
	Water to make	100 ml
Solution F	Ammonium persulfate	0.14 mg
	Water to make	100 ml
Solution G	Tween® 20 (1:100) in water	10 ml
	Saturated solution of Sudan black B in diethylene glycol	5 ml
	Water to make	100 ml
Buffer (to be diluted	TRIS	3.0 g
1:10 with water before	Glycine	14.4 g
using)	Water to make	1 l

Modified slightly from Davis, B. J., *Ann. N.Y. Acad. Sci.*, 121, 404, 1964. With permission.

Table 2
WORKING SOLUTIONS FOR PREPARATION
OF POLYACRYLAMIDE GELS

Separating gel, 3.0% small-pore[a]	Spacer gel, 2.5% large pore[b]	Loading gel, 2.5% large pore[b]
1 part A	1 part B	1 part B
2 parts C	2 parts D	2 parts D
4 parts F	1 part E	1 part E
1 part water	4 parts water	1 part G
		3 parts water

[a] Separating gel should be freshly prepared.
[b] Spacer and loading gels may be used for 2 weeks if kept refrigerated at 4°C.

From Masket, B. H., Levy, R. I., and Frederickson, D. S., *J. Lab. Clin. Med.*, 81, 794, 1973. With permission.

place by 1/4-in. ordinary rubber grommets obtained from an electrical supply store. Pyrex, parafilm, or polypropylene have no inhibitory effect on polymerization except at very acid pH, while soft glass, Plexiglas, and Tygon® may be inhibitory under conditions employed in this procedure.[28]

Preparation of Gels

The technique employed for PAGE is modified from that of Davis[29] by altering solutions C and D (Table 1) to obtain the most satisfactory separation of lipoproteins. The fat stain, Sudan black B, is added to the loading gel to prestain the plasma lipoproteins. The low concentrations of polyacrylamide used do not produce gels that are solid enough to be removed from the tubes and stained after electrophoresis.

Gels are added to the tubes through a 3-in. length of polyethylene tubing[29] attached by a 19-gauge needle to a 1-ml syringe in a Cornwall pipetter. A 1-ml syringe with a slightly bent 25-gauge needle is also useful for layering water to a height of 3 to 4 mm over each gel to flatten the gel interface.

Separating gel (0.7 ml) is added to the 12 tubes in the electrophoresis chamber and layered with water. The gels are allowed to sit for 30 to 45 min, at the end of which polymerization is evident from the slight light scattering visible through the tube. The water is removed from the top of the tube containing the gels by inverting and wiping the inside of the top of each tube with a tissue. Next, 0.1 ml of the spacer gel solution is added to each tube and layered with water. These gels are placed adjacent to a 15-W fluorescent lamp for 25 min. After removing the water from the top of the polymerized spacer gels, 25λ of plasma is applied with a disposable micropipette. Loading gel (0.2 ml) containing Sudan black B stain is added immediately, the top of each tube is covered with parafilm, and the sample is thoroughly mixed with the loading gel by inverting the cell containing the tubes five times. The 1:10 TRIS-glycine electrophoresis buffer is next layered over the loading gels, which are then polymerized by exposure to a fluorescent lamp for 25 min.

Electrophoresis

Sufficient 1:10 TRIS-glycine buffer is added to the top and bottom compartments of the cell such that both ends of the tubes are covered by buffer for a distance of approximately 1 cm. The cell is connected to a constant current power supply and electrophoresis is carried out at a constant current of 4 to 5 mA per tube until the HDL has migrated to a position about 1 cm from the bottom of the tube (about 25 min).

A permanent record of the polyacrylamide gel pattern, similar to that shown in Figure 1,[30] is made immediately after completion of electrophoresis by color photography with a model 180 Polaroid® camera fitted with a +8 close-up lens and a cc20m Kodak® filter. A fluorescent light source is used.

Comparison of Gels and Staining Techniques

In electrophoresis of lipoproteins, resolution of LDL to a single sharp band of minimum width is employed as the primary criterion for optimum conditions.

A 3.0% gel is used for lipoprotein separation. If 2.5% large-pore separating gels are used, diffuse and poorly separated lipoprotein bands are obtained, while 3.5% gels provide a distinct separation but result in the appearance of multiple, lipid-staining bands. Sharper bands are obtained by increasing the usual concentration of Bis to 0.25%,[28] a change which gives better wall adherence and improved mechanical stability of the gel. On this basis, the most satisfactory separating gel is one that contains 3.0% polyacrylamide and 0.25% Bis.

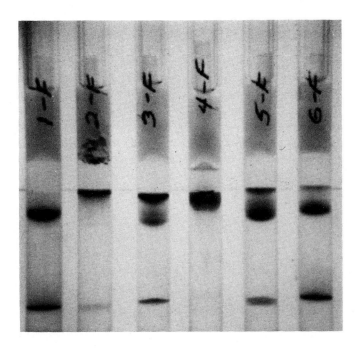

FIGURE 1. Polaroid® photograph of representative polyacrylamide gel lipoprotein patterns. (From Masket, B. H., Levy, R.I., and Frederickson, D. S., *J. Lab. Clin. Med.,* 81, 794, 1973. With permission.)

It is necessary to modify slightly the staining method of McDonald and Ribeiro.[31] Prestaining the plasma according to their technique leads to precipitation of excess dye in the loading gel. This gives a "false chylomicron band," even in some normal plasma samples. The use of pure diethylene glycol or pure alcohol as the solvent for Sudan black B also appears to prevent polymerization of the gel and fails to retain the dye in solution upon dilution of the stock. The nonionic detergent, Tween 20, is added to the solvent. The concentration of Tween 20, 1:100, is the minimum necessary to keep the dye in solution. Higher concentrations greatly decrease migration of lipoproteins.

Migration of Lipoproteins

The migration of lipoproteins in whole plasma or as isolated fractions is depicted schematically in Figure 2. The relative positions correspond to those previously described for PAGE-separated plasma lipoproteins by Narayan.[20]

Isolated chylomicrons remain diffusely distributed in the loading gel, the intensity of staining being greater in that region proximal to the stacking gel. The intensity of staining of the loading gel obtained with plasma samples is roughly proportional to the amount of "cream layer" (chylomicrons) present when plasma is allowed to sit at 4°C for 18 hr. Some staining of the lower portion of the loading gel frequently occurs when high concentrations of VLDL are electrophoresed. (This was observed with pure preparations of VLDL obtained from patients with type IV hyperlipoproteinemia who were on fat-free diets at the time of sampling and whose standing plasma contained no visible chylomicrons.)

VLDL of the usual composition found in controls or subjects with hyperlipoproteinemia other than type III always migrate to the region adjacent to or slightly beyond the interface of spacer and separating gels. The appearance of this region in the initial portion of the separating gel is variable. Sometimes two bands are visible; whereas,

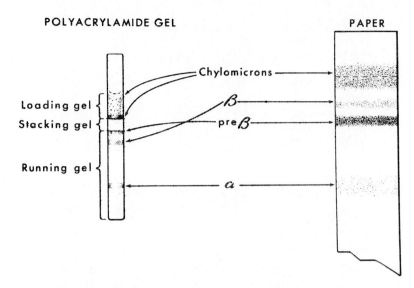

FIGURE 2.   Schematic comparison of the migration of plasma lipoproteins on paper and polyacrylamide gel. (From Masket, B. H., Levy, R. I., and Frederickson, D. S., *J. Lab. Clin. Med.,* 81, 794, 1973. With permission.)

only one is apparent on paper electrophoresis of the same sample. VLDL are often separated into several bands by agarose.[12]

LDL migrates to form a sharp band 0.5 to 0.75 cm from the top of the separating gel. A clear separation of LDL and VLDL is usually obtained. HDL migrates to the farthest position (Figure 2). HDL in isolated fractions or in plasma usually migrates as a single band.

Pattern in Type III

Type III is defined by the presence of lipoproteins that have β mobility on paper, agarose, or similar media but have a density (less than 1.006) lower than conventional β-lipoproteins (LDL).[4,6] Patients with this pattern have two major populations of plasma VLDL particles, which can most easily be separated by starch-block electrophoresis.[6] On this medium, α_2-migrating VLDL have the usual composition present in normal subjects and in other types of hyperlipoproteinemia. The β-migrating VLDL component contains an unusually high content of cholesterol relative to triglyceride.[6]

PAPER ELECTROPHORESIS

Equipment and Reagents

Paper electrophoresis is carried out at room temperature using the Durrum® type electrophoresis cell with a constant voltage power supply.

Electrophoresis is performed in 1% albumin in 0.12 M barbital buffer (pH 8.6) containing 0.001 M disodium ethylenediamine tetraacetate. Whatman® #1 filter paper strips are the supporting media. After electrophoresis, strips are dried and stained with oil red O dissolved in 60% ethanol. Oil red O stains cholesterol esters and triglyceride present in the separated lipoproteins. (See Table 3.)

Preparation and Electrophoresis

Paper strips are prepared for electrophoresis by marking the origin with a medium lead pencil and moistening with buffer from the cell. Eight strips are placed in the cell,

Table 3
WORKING SOLUTIONS FOR PAPER
ELECTROPHORESIS

Buffer (for 1 l at pH 8.6)

Sodium barbital (mol wt = 206.18)	20.6 g
Barbital (mol wt = 184.20)	3.68 g
Disodium ethylenediamine tetraacetate (mol wt = 372.254)	0.372 g
Bovine serum albumin, fraction V (to be added just before using)	10.0 g

Oil red O stain

Oil red O dye	1 g
Ethanol	1.5 l
Water	1.0 l

Note: Reflux the stain mixture for at least 1 hr. Store at 37°C.

even when less than eight samples are to be run. After equilibrating with no current flowing through the cell for 3 or 4 hr (never less than 1 hr), the samples are applied. Twenty μl of unfractionated plasma, 20 μl of ultracentrifugal bottom fraction (d > 1.006), and 40 μl of ultracentrifugal top fraction (20 μl, if lipemic) are used. Both ultracentrifugal fractions should be run concurrently to facilitate identification of floating β-lipoprotein. Electrophoresis is performed at room temperature for 16 hr at constant voltage (120 V), which results in a current flow of 6 to 8 mA per cell at the beginning of the analysis and 7 to 11 mA per cell at the end. Strips are subsequently dried at 90°C for 20 to 30 min, stained with oil red O for approximately 1 to 1½ hr, rinsed with tap water for 1 min, blotted, air-dried, and examined visually.

PRECAUTIONS

Albumin is added to the cell buffer to improve separation and definition of lipoprotein bands. Bovine albumin gives results perfectly comparable to human albumin. If albumin is omitted, lipoproteins are not sharply defined but appear as a smear from the origin to a point beyond the β-lipoprotein band.

The buffer may be used for 2 months or longer provided the polarity of the cell is changed for each run. Cells are kept at room temperature at all times. When freshly prepared buffer is used, the lipoprotein bands are not sharp. This difficulty may be overcome by using buffer that has already been used to moisten the strips before placing them in the cell containing the freshly prepared buffer.

The plasma sample may be kept at 4°C for several days before electrophoresis. It should never be frozen, since freezing irreversibly alters some lipoprotein patterns. After a single freezing and thawing, plasma, rich in pre-β-lipoprotein, displays a large, artifactual "chylomicron" band at the origin. If the plasma, especially when lipemic, is allowed to stand at room temperature for several hours, denatured material accumulates at the origin and the mobility of the lipoprotein bands is increased.

The electrophoretic patterns of lipoproteins are subject to alteration from a number of variables. One must observe the following precautions in order to minimize this variation:

1. Good contact must be made between strips and wicks.

2. End wicks on the cell cover should be quite moist. Also, filter paper strips are placed on the side of the cell cover at the ends next to the end wicks. This minimizes the distortion which usually occurs on the end strip.
3. The fluid level in the cell must be the same on both sides and within the maximum and minimum fluid level lines inscribed on the sides of the cell.
4. The opening on the top of the cell cover is covered with short, overlapping pieces of 3/4-in. waterproof tape so that only one paper strip at a time is exposed when loading.
5. Electrophoresis should be performed in a draft-free area at a relatively constant temperature.
6. Salt accumulates on the baffle adjacent to the positive electrode. This salt should be carefully removed from the baffle with a spatula and returned to the same electrode chamber. Salt is not permitted to accumulate on the center baffle which separates the positive from the negative electrode chamber. Such an accumulation would produce a salt bridge which would result in poor migration and damage to the power supply unit.
7. After each use, the dye mixture is heated to incipient boiling and allowed to cool on standing to 37 to 40°C. Fresh dye mixture is periodically added to maintain the level of the dye bath. The mixture is discarded after 2 months, and a new mixture prepared. When fresh dye mixture is used, strips are stained from 45 to 60 min. After about 1 week, staining time is increased to 1½ hr. The electrophoretic pattern of a normal subject is prepared and stained with each run as a measure of the variation of the staining procedure.
8. Ultracentrifugal top (d <1.006) and bottom (>1.006) fractions should be run side by side in order to facilitate identification of floating β-lipoprotein in the top fraction, if present. If samples are not lipemic, 40 μl of the top fraction are applied to the paper. Twenty μl of the bottom fraction and 20 μl of whole plasma are used.

EXPERIMENTS

Two types of preliminary experiments were performed to compare the mobility of VLDL from type III plasma on paper and polyacrylamide gel. In one experiment, samples from five primary type III patients were ultracentrifuged at d 1.006 and the supernatant and infranatant fractions electrophoresed on polyacrylamide gel. The mobility of the lipoproteins in PAGE was not affected by ultracentrifugation. Upon gel electrophoresis of the supernatant fractions, some lipoproteins migrated to the interface between the stacking and separating gels. The remainder either stayed at the interface of the separating gel like ordinary VLDL or migrated downward through the gel almost as far as the LDL zone. In paper electrophoresis this supernatant fraction always contained lipoproteins having β mobility. Sometimes this was accompanied by a discrete pre-β band; in other samples, a "broad β" band extended unbroken into the pre-β region. On polyacrylamide, the lipoproteins in the infranatant fraction migrated to the usual VLDL and HDL positions. None of the samples displayed a distinct LDL band in the gel. On paper, bands migrating to the β and α positions were present.

In another set of experiments, 11 samples of primary type III plasma which contained a broad β zone were electrophoresed on paper and the position of the β band identified on the wet strip without staining. The zone corresponding to the broad β band of each was eluted with barbital buffer (ionic strength 0.12, pH 8.6), stained, and reelectrophoresed on gel. Sudan black-stained material was spread from the interface to varying depths in the separating gel. In no instance did staining appear in the LDL zone. In both of these experiments, therefore, the LDL in type III plasma did

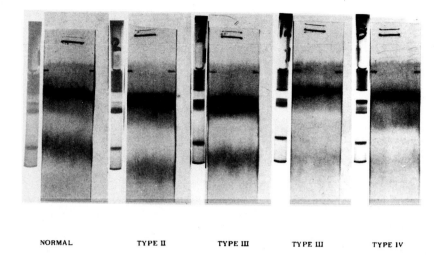

NORMAL   TYPE II   TYPE III   TYPE III   TYPE IV

FIGURE 3.   Comparison of lipoprotein patterns on paper and polyacrylamide gel. From Masket, B. H., Levy, R. I., and Frederickson, D. S., *J. Lab. Clin. Med.*, 81, 794, 1973. With permission.)

not migrate in polyacrylamide gel to the usual position occupied by LDL from normal individuals.

A total of 118 plasma samples from 27 primary type III donors were then electrophoresed on both paper and polyacrylamide, and the patterns compared. Figure 3 shows a comparison of type III patterns on paper and PAGE. In all of the 118 samples, β-migrating lipoproteins were present in paper electrophoresis of the supernatant fraction (d <1.006) obtained by preparative ultracentrifugation.[4] Varying degrees of hyperlipidemia were present in these samples. Results in Table 4 are arranged according to plasma concentrations of cholesterol and triglyceride.

In paper electrophoresis of whole plasma, a broad β-lipoprotein band, previously described as an inconstant feature of type III,[4] was present in 50 of the 118 samples (42%). Patterns on the polyacrylamide gel varied. The vast majority were characterized by a definite VLDL band and HDL band but no LDL band. A discrete band in the usual LDL position, compared to controls run in each set, was present in only five of the 118 samples, and these samples were obtained from treated patients whose cholesterol and triglyceride concentrations were both less than 250 mg/100 ml.

Thus, 113 of 118 type III samples (96 %) were characterized by a combination of the presence of obvious β-migrating lipoproteins in paper electrophoresis and the absence of a band migrating like LDL on polyacrylamide gel. This combination occurred in only three of 228 samples of plasma from all other types of hyperlipoproteinemia which were similarly examined. "False positives" were restricted entirely to plasma obtained from patients with very high triglyceride concentrations (untreated patients with either type I or type V hyperlipoproteinemia).

DISCUSSION

The primary intent of the present study is examination of lipoprotein patterns obtained in a large group of patients with type III hyperlipoproteinemia characterized by a combination of paper electrophoresis, preparative ultracentrifugation, and other tests as described elsewhere.[4,5] Previous reports of lipoprotein analyses on uniform gel

Table 4

RESULTS OF ELECTROPHORESIS OF PLASMA IN TYPE III
HYPERLIPOPROTEINEMIA

No. of Samples[a]	Plasma		Broad β band present on paper electrophoresis	Positive for type III by combination test, paper + PAGE[b]
	Cholesterol (mg/100 ml)	Triglycerides (mg/100 ml)		
74	<250	<250	25	69
23	>250	>250	13	23
6	>250	<250	3	6
15	<250	>250	9	15
118			50	113

[a]   Samples are from 27 patients.

[b]   Presence of β-migrating lipoproteins in paper electrophoresis and absence of a β-migrating band in polyacrylamide gel electrophoresis (PAGE).

From Masket, B. H., Levy, R. I., and Frederickson, D. S., *J. Lab. Clin. Med.*, 81, 794, 1973. With permission.

polyacrylamide have contained only a few examples of some types of hyperlipoproteinemia; none have included patients with type III hyperlipoproteinemia as ascertained by ultracentrifugation, heretofore the only reliable technique for identifying this abnormal pattern.

Polyacrylamide, in combination with electrophoresis on paper and probably agarose, appears to have special usefulness in the diagnosis of type III hyperlipoproteinemia. One hundred and eighteen samples were obtained from 27 patients with proved type III whose plasma lipid concentrations varied widely because of dietary manipulation or drug therapy. In 95% of the samples, the combination of visible β-migrating lipoproteins on paper and absence of a discrete β-lipoprotein band in polyacrylamide occurred and was nearly unique for type III. It was observed in only three of 228 samples from patients with all other types of hyperlipoproteinemia. These latter were severely hyperglyceridemic subjects whose gel patterns contained no discernible LDL or HDL bands, although a faint β-lipoprotein band could be seen on paper. The 5% of "false negative" tests occurred in treated type III patients who were no longer hyperlipidemic.

These experiments do not provide an explanation for the failure of LDL in all type III samples to concentrate into a visible band in the position where these lipoproteins migrate in normal subjects and other types of hyperlipoproteinemia. It is known, however, that in type III, concentration of the normally predominant S_f 0—12 class of LDL is greatly diminished.[5] Lipoproteins of the S_f 12—20 class are greatly increased. Narayan et al. have shown that these two subclasses of LDL migrate to slightly different positions in 3% polyacrylamide gels.[20] The S_f 12—20 class migrates less far and to a position which would correspond to the leading edge of some of the diffuse VLDL bands that have been observed in type III plasma. Distinction of the position of such "VLDL particles" from that of the "normal" (predominantly S_f 0—12) LDL band requires simultaneous electrophoresis of control samples in each batch of polyacrylamide gels.

The only other published polyacrylamide patterns in type III hyperlipoproteinemia were obtained in three patients by Borrie,[24] who stated that the patterns were characteristic of 12 such patients in whom diagnosis was made on clinical grounds. The three patterns appear to have absent or diminished LDL bands. They were obtained, however, by the use of continuous gradient polyacrylamide electrophoresis;[14] thus, the results cannot be properly compared with those obtained using the single density system. Melish and Waterhouse have recently reported further studies of hyperlipoproteinemic plasma with the continuous gradient method and suggest that it may reveal microheterogeneity among certain types of hyperlipoproteinemia.[23]

SUMMARY

On the basis of results obtained with conjoint electrophoresis on type III, it appears worthy of exploration by laboratories engaged in lipoprotein typing, where the unavailability of a diagnostic method for type III, not requiring the ultracentrifuge, has been a source of frustration. In laboratories using paper or agarose electrophoresis routinely, certain criteria exist to help select those samples to be electrophoresed on polyacrylamide as well. The conjoint test should, at the least, be performed when: 1. a "broad β" band is obtained on paper or agarose; 2. the cholesterol and triglyceride concentrations are both elevated and the ratio of the two is between 0.5 and 1.5; or 3. the patient has planar or tuberoeruptive xanthomas. Until the conjoint test has been validated by more experience, however, results should be confirmed by preparative ultracentrifugation whenever possible.[32]

REFERENCES

1. Frederickson, D. S. and Levy, R. I., Familial hyperlipoproteinemia, in *The Metabolic Basis of Inherited Disease,* 3rd ed., Stanbury, J. B., Wyngaarden, J.B., and Fredrickson, D. S., Eds., McGraw-Hill, New York, 1972, 545.
2. Gofman, J. W., DeLalla, O., Glazier, F., Freeman, N. K., Lindgren, F. T., Nichols, A. V., Strisower, B., and Tamplin, A. R., The serum lipoprotein transport system in health, metabolic disorders, atherosclerosis, and coronary heart disease, *Plasma,* 2, 413, 1954.
3. Zelis, R., Mason, D. T., Braunwald, E., and Levy, R. I., Effects of hyperlipoproteinemias and their treatment on the peripheral circulation, *J. Clin. Invest.,* 49, 1007, 1970.
4. Fredrickson, D. S., Levy, R. I.,and Lees, R. S., Fat transport in lipoproteins — an integrated approach to mechanisms and disorders, *N. Eng. J. Med.,* 276, 32, 1967.
5. Fredrickson, D. S., Levy, R. I., and Lindgren, F. T., A comparison of heritable abnormal lipoprotein patterns as defined by two different techniques, *J. Clin. Invest.,* 47, 2446, 1968.
6. Quarfordt, S. H., Levy, R. I., and Fredrickson, D. S., On the lipoprotein abnormality in type III hyperlipoproteinemia, *J. Clin. Invest.,* 50, 754, 1971.
7. Hazzard, W. R., Porte, D., Jr., and Bierman, E. L., Abnormal lipid composition of chylomicrons in broad-beta disease (type III hyperlipoproteinemia), *J. Clin. Invest.,* 49, 1853, 1970.
8. Hazzard, W. R., Porte, D., Jr., and Bierman, E. L., Abnormal lipid composition of very low density lipoproteins in diagnosis of broad beta disease (type III hyperlipoproteinemia), *Metabolism,* 21, 1009, 1972.
9. Lees, R. S. and Hatch, F. T., Sharper separation of lipoprotein species by paper electrophoresis in albumin-containing buffer, *J. Lab. Clin. Med.,* 61, 518, 1963.
10. Masket, B. H., Lipoprotein fractionation by paper electrophoresis, in *Manual of Procedures for the Applied Seminar on the Clinical Pathology of the Lipids,* Sunderman, F. W., Ed., Association of Clinical Scientists, Philadelphia, 1971, VII-1.
11. Noble, R. P., Electrophoretic separation of plasma lipoproteins in agarose gel, *J. Lipid Res.,* 9, 693, 1968.
12. Papadopoulas, N. M. and Kintzios, J. A., Determination of human serum lipoprotein patterns by agarose gel electrophoresis, *Anal. Biochem.,* 30, 421, 1969.
13. Fletcher, M. J. and Styliou, M. H., A simple method of separating serum lipoproteins by electrophoresis on cellulose acetate, *Clin. Chem.,* 16, 362, 1970.
14. Chin, H. P. and Blankenhorn, D. H., Separation and quantitative analysis of serum lipoproteins by means of electrophoresis on cellulose acetate, *Clin. Chim. Acta,* 20, 305, 1968.

15. **Narayan, K. A., Narayan, S., and Kummerow, E. A.,** Disc electrophoresis of human serum lipoproteins, *Nature (London),* 205, 246, 1965.
16. **Raymond, S., Miles, J. L., and Lee, J. C.,** Lipoprotein patterns in acrylamide gel electrophoresis, *Science,*151, 346, 1966.
17. **Narayan, K. A., Dudacek, W. E., and Kummerow, F. A.,**Disc electrophoresis of isolated rat serum lipoproteins, *Biochem. Biophys. Acta,* 125, 581, 1966.
18. **Narayan, K. A.,Creinin, H. L., and Kummerow, F. A.,** Disc electrophoresis of rat lipoproteins, *J. Lipid Res.,* 7, 150, 1966.
19. **Narayan, K. A.,** Human and rat serum proteins, lipoproteins, ammonium persulfate, gel concentration, and disc electrophoresis,*Microchem. J.,* 14, 335, 1969.
20. **Narayan, K. A., Mary, G. E. S., and Friedman, H. P.,** Disc electrophoresis of subclasses of human serum low density lipoproteins, *Microchem. J.,*14, 235, 1969.
21. **Frings, C. S., Foster, L. B., and Cohen, P. S.,** Electrophoretic separation of serum lipoproteins in polyacrylamide gel, *Clin. Chem.,* 17, 111, 1971.
22. **Pratt, J. J. and Dangerfield, W. G.,** Polyacrylamide gels of increasing concentration gradient for the electrophoresis of lipoproteins, *Clin. Chim. Acta,* 23, 189,1969.
23. **Melish, J. S., and Waterhouse, C.,** Concentration gradient electrophoresis of plasma from patients with hyperbetalipoproteinemia, *J. Lipid Res.,* 13, 193, 1972.
24. **Borrie, P.,** Type III hyperlipoproteinemia, *Br. Med. J.,* 2, 665,1969.
25. **Chrambach, A. and Rodbard, D.,** Polyacrylamide gel electrophoresis, *Science,* 172, 440, 1971.
26. **Burstein, M., and Samaille, J.,** Sur un dosage rapide du cholesterol lie aux α-et aux β-lipoproteines du serum, *Clin. Chim. Acta,* 5, 609, 1960.
27. **Havel, R. J., Eder, H. A., and Bragdon, J. H.,** The distribution and chemical composition of ultracentrifugally separated lipoproteins in human serum, *J. Clin. Invest.,* 34, 1345, 1955.
28. **Rodbard, D. and Chrambach, A.,** Estimation of molecular radius, free mobility, and valence using polyacrylamide gel electrophoresis, *Anal. Biochem.,* 40, 95, 1971.
29. **Davis, B. J.,** Disc electrophoresis. II. Method and application to human serum proteins, *Ann. N. Y. Acad. Sci.,* 121, 404, 1964.
30. **Bull, W. H. O.,** Classification of hyperlipidaemias and hyperlipoproteinaemias, 43, 891, 1970.
31. **McDonald, H. J. and Ribeiro, L. P.,** Ethylene and propylene glycol in the prestaining of lipoproteins for electrophoresis, *Clin. Chim. Acta,* 4, 458, 1959.
32. **Masket, B. H., Levy, R. I., and Fredrickson, D. S.,** The use of polyacrylamide gel electrophoresis in differentiating type III hyperlipoproteinemia, *J. Clin. Med.,* 81, 794, 1973.

Lipoprotein Changes as Affected by Nutrition

LIPOPROTEIN CHANGES IN UNDERNUTRITION AND OVERNUTRITION

D. C. K. Roberts and K. K. Carroll

Malnutrition encompasses both under- and overnutrition, although the term is more often associated with the former. This section will be concerned with reviewing changes in serum lipoproteins associated with the two ends of the malnutrition spectrum, protein-calorie malnutrition and obesity. The primary hyperlipidemias associated with atherosclerosis are discussed elsewhere in this book [19] and therefore will not be included in the section on overnutrition and obesity.

PROTEIN-CALORIE MALNUTRITION

Protein-calorie malnutrition occurs most frequently in children under the age of five living in the disadvantaged nations of the world. [20,24,40] Marasmus presents the typical picture of starvation (severely underweight for age, little or no subcutaneous fat, and stunted growth) and results from deficiency of total calories. Kwashiorkor, on the other hand, is associated with edema and usually a fatty liver and is the result of protein deficiency in the presence of adequate caloric intake. Between these two extremes are many varying degrees of caloric and/or protein deficiency states, generally referred to as marasmic kwashiorkor. A scoring system, which combines clinical and biochemical features of these nutritional disorders as a means of differentiating between marasmus, marasmic kwashiorkor, and kwashiorkor, has been proposed by McLaren et al. [23]

Paper electrophoresis has been the principal method of studying lipoprotein changes in protein-calorie malnutrition. In most cases, quantitation has been by densitometric measurement of the lipophilic stain. However, one group of workers extracted the stained regions and assayed for the lipids contained therein. [38,39] It should be mentioned that Dyerberg and Hjørne[6] (using Sudan Black in agarose gels) reported that analysis of dye uptake can underestimate the concentration of α-lipoprotein relative to β-lipoprotein by as much as 20%.

All workers agree that β-lipoprotein is considerably reduced in kwashiorkor (Table 1). Reduced levels of α-lipoprotein have been reported by authors using densitometry as an assessment of the relative quantity of lipoproteins.[4,5,7,21,23] However, Truswell and co-workers[37-39] reported normal levels of α-lipoprotein in kwashiorkor patients based on cholesterol analysis of extracted lipids. On recovery from kwashiorkor, there is a rapid increase of β-lipoprotein to normal levels[4,5,21,22,37] and a transient appearance of pre-β-lipoprotein during the first weeks of treatment.[4,37]

The mechanism underlying these lipoprotein changes is thought to be a deficiency of apoproteins for very low-density lipoprotein (VLDL) synthesis in the liver, and since low-density lipoprotein (LDL) is thought to be derived from VLDL, plasma β-lipoprotein (LDL) is therefore reduced.[8,37] The appearance of pre-β-lipoprotein (VLDL) shortly after treatment is initiated, concomitant with the decrease in liver fat content to normal levels,[8,9,21,37-39] suggests that VLDL production is impaired and the fatty liver of kwashiorkor may also be the result of deficient VLDL production. Other work, however, does not support this idea concerning the origin of the fatty liver. No reduction was found in plasma triglycerides before treatment in patients with severe kwashiorkor.[30,36] Other factors, such as increased hepatic fat synthesis, decreased hepatic fatty acid oxidation, and increased adipose tissue fatty acid mobilization, have been suggested as being equally important.[29]

Table 1
APPLICATIONS OF ELECTROPHORESIS

Author and ref. no.	Year	Pattern	Type or characteristics	Normal values	± SD*	Pathological values	± SD*	Characteristics of subject, age, sex, race	Summary of results
				Kwashiorkor					
Cravioto, J. et al.[5]	1959	Reduced α Reduced β		64[b,c,d] 388[b,c,d]	22(8) 61(8)	21[b,d] 274[b,d]	11(13) 107(13)	1—5 years, male and female, Mexico	Rapid rise to normal levels on treatment
El-Ridi, M. S. and Shahata, A. H.[7]	1966	Reduced α Slightly reduced β		132[b] 245[b]	— (20) — (20)	79[b] 209[b]	— (46) — (46)	Children, male and female, Egypt	Rise in serum protein bound polysaccharides noted
McLaren, D. S. et al.[23]	1967	Reduced α Reduced β		94[b] 403[b]		30[b] 312[b]		1—3 years, male and female, Jordan	Children selected for absence of xeropthalmia; α remained low on treatment
McLaren, D. S. et al.[23]	1967	Reduced α Reduced β		96[b] 413[b]	—(35) —(35)	12[b] 214[b]		1—5 years, male and female, Jordan	
McLaren, D. S. et al.[22]	1968	Reduced β						<2 years, male and female, Jordan	
Coward, W. A. and Whitehead, R. G.[4]	1972	Reduced or absent α Reduced β		1.8[e]	0.1[e](17)	1.3—1.5[e]	0.1—0.2[f]	0.5—3, years, male and female, Uganda	On treatment α present after week, pre-β present after 3 days
Truswell, A. S.[37]	1975	Normal α Reduced β		39[e] 113[e]		35[e] 60[e]	— (19) — (19)	Children, male and female, South Africa	Appearance of pre-β during treatment; gone by 18 days
				Marasmus					
McLaren, D. S., et al.[23]	1967	Normal α Increased β		96[b] 414[b]	— (35) — (35)	93[b] 699[b]		1—5 years, male and female, Jordan	
Gurson, C. T., et al.[13]	1973	Reduced α Normal β		32[b] 68[b]	9(10) 9(10)	24[b] 76[b]	8(30) 8(30)	Mean age, 10.5 months male and female, Turkey	On treatment, α increased to normal by 30 days
Truswell, A. S.[37]	1975	Normal α, Slightly reduced β Pre-β may be present		39[e] 113[e]		31[e] 80[e]	— (18) —(18)	Children, male and female, South Africa	
				Marasmic kwashiorkor					
McLaren, D. S. et al.[23]	1967	Normal α Normal β		96[b] 414[b]	— (35) — (35)	84[b] 481[b]		1—5 years, male and female, Jordan	
				Overnutrition and Obesity					
Rifkind, B. M. et al.[31]	1966	Chylomicrons Pre-β	Type V					Adult, female, U.K.	
Sims, E. A. H. et al.[34,35]	1968, 1974	α and β ratio unchanged, pre-β <20% of total lipoproteins			— (5)			21—33 years, male, U.S.	Experimentally induced obesity over 40-week period; values remained within normal limits
Miettinen, T. A.[25]	1971	No chylomicrons, normal pre-β, normal β Increased pre-β	Normal Type IV				— (10)	24—63 years, male and female, Finland	Of the normals, 6/10 were diabetic; of the type IV, 6/10 had ischemic heart disease.
Glueck, C. J. et al.[12]	1973	Increased pre-β	Type IV					1 — 20 years, male and female with familial hypertriglyceridemia, U.S.	8/23 were obese; weight reduction produced normal pattern.
Balta, N. and Messinger, V.[1]	1974	Chylomicrons Increased pre-β Increased β	Type V				— (7)	Adult, Romania	Qualitative assessment of polyacrylamide gels
Borrie, P. and Slack, J.[3]	1974	Variable chylomicrons Increased pre-β	Type IV—V	22[f] 73[f]	3—60[f] 39—151[f]	732[d,i] 311[d,i]	123—1660[i](6) 46—460[i](6)	34—59 years, male, U.K.	One patient had floating β
Jourdan, M. et al.[17]	1974	Increased pre-β Increased β	Type IV	25[d,h] 163[d,h]	— (2) — (2)	87[d,h] 246[d,h]	43(6) 27(6)	Adult, female, U.S.	2/8 had normal patterns, 6/8 were type IV, 63 days after weight loss, β had decreased but still type IV
Leelarthaepin, B. et al.[18]	1974		Type IIa and IIb					37—65 years, male with coronary heart disease, Australia	12 type IIa, 10 type IIb, and 4 normal; after weight reduction,5 IIa, 2 IIb, 2 mild IV, and 17 normal
Salel, A. F. et al.[32]	1974		Type IV					Adults, male and female with coronary artery disease, U.S.	Of 105 patients with CAD, 56 were obese and 48 had type IV patterns

Table 1 (continued)
APPLICATIONS OF ELECTROPHORESIS

Author and ref. no.	Year	Pattern	Type or character-istics	Normal values	± SD[a]	Patholog-ical values	± SD[a]	Characteristics of sub-ject, age, sex, race	Summary of results
					Overnutrition and Obesity				
Scott, H. W., Jr. et al.[33]	1974	Increased β	Type II, IV, and V				— (34)	Massively obese adults, U.S.	Underwent jejunoileal bypass surgery, lipids became normal post-operatively
Ho, W. K. K. and Chan, W. Y.[16]	1975	Increased pre-β	Type IV	86[b]	34[f](35)	128[b]	87[f](6)	Early 20's, male and fe-male Chinese, Hong Kong	

[a] Number of observations, upon which SD is based, is in parentheses.
[b] Total lipid in mg/dl.
[c] Obtained 26 — 34 days after start of treatment.
[d] Calculated from individual values.
[e] Measurement by immunocrit method in mm of β-lipoprotein.
[f] Represents standard error of the mean (SEM).
[g] Cholesterol in lipoprotein in mg/dl.
[h] Percent area under peaks.
[i] Total lipid in mg/dl by analytical centrifuge.
[j] Represents range of Sd values.
[k] Values in arbitrary scanning units.

Lipoprotein abnormalities are less pronounced in marasmus and marasmic kwashiorkor (Table 1). In general, both α- and β-lipoproteins are little changed, although Truswell and Hansen[37,38] have reported the presence of some pre-β-lipoprotein in marasmus. Gürson et al.[13] found decreased α-lipoprotein in marasmus and McLaren et al.[23] increased β-lipoprotein. As pointed out by Truswell,[37] there is a negative correlation between β-lipoprotein and the degree of fat in the liver, i.e., in marasmic cases with no fatty liver, β-lipoprotein levels are normal; in indiviuals with some degree of fatty liver (marasmic kwashiorkor), the β-lipoprotein level is reduced; and in those with frank kwashiorkor and very fatty livers, the β-lipoprotein level is only about 50% of normal.

OVERNUTRITION AND OBESITY

Techniques for the study of lipoprotein changes in obesity have included most of the available media, such as paper, agarose, cellulose acetate, and polyacrylamide gel.[14] In a recent survey of techniques, it was concluded that agarose-gel electrophoresis was the best method for phenotypic identification of the types of hyperlipidemias.[15] Descriptions of the various abnormalities have usually followed the scheme of Fredrickson and co-workers;[2,10,11] in general, the changes are associated with β- and pre-β-lipoproteins.

In obese subjects, pre-β-lipoprotein is often present in increased quantities either with[1,31] or without[16,17,25,33] chylomicrons, corresponding to type V or IV patterns, respectively (Table 1). However, normal patterns have been found in some obese individuals,[17,18,25] and a composite type IV-V pattern has been reported[3] with increased pre-β-lipoprotein, the variable presence of chylomicrons, and in one instance, floating β-lipoprotein. Experimentally induced obesity in normal men did not result in an abnormal pattern during a 40-week period on diet.[34,35] Both type IV[25,32] and the type IIA and IIB patterns[18] have been found in obese patients with heart disease.

In a study of the incidence of familial hypertriglyceridemia in children up to age 20,[12] a total of 23 out of 113 were found to be type IV. Of these, 8 were obese. Weight reduction and its maintenance resulted in normal lipid patterns. In severely obese adults, jejunoileal bypass surgery resulted in rapid weight loss and the appearance of normal lipoprotein patterns, a reversion from the type II, IV, and V patterns.[33] Weight loss in a group of obese Australian males with coronary heart disease led to establishment of a normal lipoprotein pattern in 60% of the type IIa patients and 80% of the type IIb patients.[18]

The underlying mechanisms behind the lipoprotein changes in obesity are complicated. Plasma lipoprotein levels are the result of input, primarily from the liver, and removal of triglycerides by adipose tissue and cholesterol by the liver. Obese individuals synthesize more cholesterol[25,27] and triglyceride[28] than normal individuals and produce more VLDL. This is sometimes accompanied by enhanced removal of cholesterol,[26] resulting in normal lipoprotein patterns; however, the removal process may be defective[25] or simply overwhelmed by the input so that lipid accumulates in the plasma and the lipoprotein pattern changes. Weight reduction may lessen VLDL production and thus reduce pressure on the removal processes and normalize lipoproteins. Obviously, not all obese individuals respond in this way, and in some, the pattern remains abnormal despite weight loss.[17,18] In these cases a primary hyperlipidemic defect should be suspected.

REFERENCES

1. **Balta, N. and Messinger, V.,** Serum lipoprotein changes in certain internal diseases. Disc electrophoretic study, *Rev. Roum. Med. Intern.,* 12, 91, 1974.
2. **Beaumont, J. L., Carlson, L. A., Cooper, G. R., Fejfar, Z., Fredrickson, D. S., and Strasser, T.,** Classification of hyperlipidaemias and hyperlipoproteinaemias, *Bull. W. H. O.,* 43, 891, 1970.
3. **Borrie, P. and Slack, J.,** A clinical syndrome characteristic of primary Type IV-V hyperlipoproteinaemia, *Br. J. Dermatol.,* 90, 245, 1974.
4. **Coward, W. A. and Whitehead, R. G.,** Changes in serum β-lipoprotein concentration during the development of kwashiorkor and in recovery, *Br. J. Nutr.,* 27, 383, 1972.
5. **Cravioto, J., De La Pena, C. L., and Burgos, G.,** Fat metabolism in chronic severe malnutrition: lipoprotein in children with kwashiorkor, *Metabolism,* 8, 722, 1959.
6. **Dyerberg, J. and Hjørne, N.** Quantitative plasma lipoprotein electrophoresis. Correction for the difference in dye uptake by the lipoprotein fractions, *Clin. Chim. Acta,* 30, 407, 1970.
7. **El-Ridi, M. S. and Shahata, A. H.,** Metabolic disturbances in kwashiorkor, *Proc. Seventh Int. Cong. Nutr.,* Vol. 1, Kühnau, J., Ed., Pergamon Press, Oxford, 1967, 92.
8. **Flores, H., Pak, N., Maccioni, A., and Monckeberg, F.,** Lipid transport in kwashiorkor, *Br. J. Nutr.,* 24, 1005, 1970.
9. **Flores, H., Seakins, A., Brooke, O. G., and Waterlow, J. C.,** Serum and liver triglycerides in malnourished Jamaican children with fatty liver, *Am. J. Clin. Nutr.,* 27, 610, 1974.
10. **Fredrickson, D. S. and Levy, R. I.,** Familial hyperlipoproteinaemia, in *Metabolic Basis of Inherited Disease,* 3rd ed., Stanbury, J. B., Wyngaarden, J. B., and Fredrickson, D. S., Eds., McGraw-Hill, New York, 1972, 545.
11. **Fredrickson, D. S., Levy, R. I., and Lees, R. S.,** Fat transport in lipoproteins — an integrated approach to mechanisms and disorders, *N. Engl. J. Med.,* 276, 34, 94, 148, 215, 273, 1967.
12. **Glueck, C. J., Tsang, R., Fallat, R., Buncher, C. R., Evans, G., and Steiner, P.,** Familial hypertriglyceridemia: studies in 130 children and 45 siblings of 36 index cases, *Metabolism,* 22, 1287, 1973.
13. **Gürson, C. T., Kurdoglu, G., and Saner, G.,** Serum total lipids and lipid fractions in marasmus, *Nutr. Metab.,* 15, 181, 1973.
14. **Hatch, F. T. and Lees, R. S.,** Practical methods for plasma lipoprotein analysis, in *Advances in Lipid Research,* Vol. 6, Eds. Paoletti, R. and Kritchevsky, D., Eds., Academic Press, New York, 1968, 1.
15. **Heiberg, A.,** A comparative study of different electrophoretic techniques for classification of hereditary hyperlipoproteinaemias, *Clin. Genet.* 4, 450, 1973.
16. **Ho, W.K.K. and Chan, W.Y.,** Evaluation of serum lipid and lipoprotein levels in normal Chinese. The influence of dietary habit, body weight, exercise and a familial record of coronary heart disease, *Clin. Chim. Acta,* 61, 19, 1975.
17. **Jourdan, M., Margen, S., and Bradfield, R. B.,** The turnover rate of serum glycerides in the lipoproteins of fasting obese women during weight loss, *Am. J. Clin. Nutr.,* 27, 850, 1974.
18. **Leelarthaepin, B., Woodhill, J. M., Palmer, A. J., and Blacket, R. B.,** Obesity, diet and type II hyperlipidaemia, *Lancet,* 2, 1217, 1974.

19. Lees, R. S., Screening of serum lipoproteins as predictors of atherosclerosis, in *Handbook Series in Electrophoresis*, Vol. 1, Lewis, L. A. and Opplt, J. J., Eds., CRC Press, Boca Raton, 1979.

20. McLaren, D. S., *Nutrition and Its Disorders*, Churchill, Livingstone, Edinburgh and London, 1972, 103.

21. McLaren, D. S., Asfar, P., and Zekian, B., Liver pathology and vitamin A storage in children recovering from protein-calorie malnutrition, *Proc. Seventh Int. Cong. Nutr.*, Vol. 1, Kühnau, J., Ed., Pergamon Press, Oxford, 1967, 108.

22. McLaren, D. S., Faris, R., and Zekian, B., The liver during recovery from protein-calorie malnutrition, *J. Trop. Med. Hyg.*, 71, 271, 1968.

23. McLaren, D. S, Pellett, P. L., and Read, W. W. C., A simple scoring system for classifying the severe forms of protein-calorie malnutrition of early childhood, *Lancet*, 1, 533, 1967.

24. Metcoff, J., Biochemical effects of protein-calorie malnutrition in man, *Annu. Rev. Med.*, 18, 377, 1967.

25. Meittinen, T. A., Cholesterol production in obesity, *Circulation*, 44, 842, 1971.

26. Miettinen, T. A., Mechanisms of hyperlipidaemias in different clinical conditions, *Adv. Cardiol.*, 8, 85, 1973.

27. Nestel, P. J., Schreibman, P. H., and Ahrens, E H., Jr., Cholesterol metabolism in human obesity, *J. Clin. Invest.*, 52, 2389, 1973.

28. Nestel, P. J. and Whyte, H. M., Plasma free fatty acid and triglyceride turnover in obesity, *Metabolism*, 17, 1122, 1968.

29. Prabhu, K. M., Kanagasabapathi, A. S., and Varma, T. N. S., Serum triglycerides and cholesterol in kwashiorkor, *Indian J. Biochem. Biophys.*, 10, 34, 1973.

30. Rao, K. S. J. and Prasad, P. S. K., Serum triglycerides and nonesterified fatty acids in kwashiorkor, *Am. J. Clin. Nutr.*, 19, 205, 1966.

31. Rifkind, B. M., Begg, T., and Jackson, I. D., Relationship of plasma lipids and lipoproteins to obesity, *Proc. R. Soc. Med.*, 59, 1277, 1966.

32. Salel, A. F., Riggs, K., Mason, D. T. Amsterdam, E. A., and Zelis, R., The importance of type IV hyperlipoproteinemia as a predisposing factor in coronary artery disease, *Am. J. Med.*, 57, 897, 1974.

33. Scott, H.W., Jr., Dean, R. H., Younger, R. K., and Butts, W. H., Changes in hyperlipidemia and hyperlipoproteinemia in morbidly obese patients treated by jejunoileal bypass, *Surg. Gynecol. Obstet.*, 138, 353, 1974.

34. Sims, E. A. H., Bray, G. A., Danforth, E., Jr., Glennon, J. A., Horton, E. S. Salans, L. B., and O'Connell, M., Experimental obesity in man. VI. The effect of variations in intake of carbohydrate on carbohydrate, lipid and cortisol metabolism, *Horm. Metab. Res. Suppl. Ser.*, 4, 70, 1974.

35. Sims, E. A. H., Goldman, R. F., Gluck, C. M., Horton, E. S., Kelleher, P. C., and Rowe, D. W., Experimental obesity in man, *Trans. Assoc. Am. Physicians*, 81, 153, 1968.

36. Taylor, G. O., Serum triglycerides and fatty acids in kwashiorkor, *Am. J. Clin. Nutr.*, 24, 1212, 1971.

37. Truswell, A. S., Carbohydrate and lipid metabolism in protein-calorie malnutrition, in *Protein Calorie Malnutrition*, Olson, R. E., Ed., Academic Press, New York, 1975, 119.

38. Truswell, A. S. and Hansen, J. D. L., Fatty liver in protein-calorie malnutrition, *S. Afr. Med. J.*, 43, 280, 1969.

39. Truswell, A. S., Hansen, J. D. L., Watson, C. E., and Wannenberg, P., Relation of serum lipids and lipoproteins to fatty liver in kwashiorkor, *Am. J. Clin. Nutr.*, 22, 568, 1969.

40. Whitehead, R. G. and Alleyne, G. A. O., Pathophysiological factors of importance in protein calorie malnutrition, *Br. Med. Bull.*, 28, 72, 1972.

HORMONAL EFFECTS ON SERUM LIPOPROTEINS

Jan J. Opplt

INTRODUCTION

It is a longstanding clinical observation that women in the postmenopausal age group have a significantly greater incidence of atherosclerotic disease and its complications than in any other period of their life. Generally, loss of hormonal balance, occurring relatively suddenly, is believed to be the main cause of atherosclerotic development. At the same time that lack of ovarian estrogen develops, postmenopausal women show a dramatic change in their serum lipoprotein levels and classes. It is logically possible, although not yet proven, that endocrine changes, lipoprotein abnormalities, and atheroma are a sequence of causes and effects.

According to the infiltration theory, widely accepted today as an explanation for the pathogenesis of atheroma, some lipoprotein classes and/or lipoprotein groups, called "atherogenic," possess the ability to injure the endothelium of arteries, and penetrate the endothelium to the subendothelium or possibly to the media.[98] There is a further possibility that they may not be metabolized in the usual way.[99] Lack of a high-density lipoprotein, which can serve as an acceptor of split lipid components, may cause an accumulation of cholesterol esters in smooth muscle cells of the media and finally their degradation and death.[100] This has been suggested as the first step in the atherogenic process.[101]

ATHEROSCLEROSIS IN POSTMENOPAUSAL WOMEN

Metabolic changes in postmenopausal women have attracted the attention of many investigators, recent[37] as well as past.[6-8,10,12,60-67] It has been established that in women bilaterally oophorectomized before the menopause, there is a fourfold increase of atherosclerotic cardiovascular disease. In one group study over a two-year period, more new cardiovascular events developed in the castrated group than in controls. In castrated women, those treated with estrogen developed significantly less cardiovascular disease than those who were untreated. It has been suggested that if estrogen is to be given as a preventive therapy for atherosclerosis, it should be used primarily in females.[13]

Women castrated below the age of 42 years have significantly more severe coronary disease than normal. According to Parrish et al.[15] it takes an average of about 14.4 ± 2.5 years after castration before excessive coronary atherosclerosis becomes apparent. Opplt and co-workers[88] studied the influence of the removal of ovary on the occurrence of vascular diseases in women. In an analysis of 5638 autopsies, there was a rapid increase in the number of cases of myocardial infarction in women after menopause as compared to the incidence in premenopausal women (Figure 1). Dyslipoproteinemia and the two phenomena were thought to be causally related. There are a number of other studies on cardiovascular pathology in groups of castrated premenopausal women[39] and on women after physiological menopause.[12,40]

Estrogen secretion before and after ovariectomy in the pre- and postmenopausal periods has been studied frequently in the last decade.[41-43,46-49,51] Other hormones and their metabolic fate have been investigated: gonadotropins,[44,45,52-54] tissue cortisol,[50] FSH and LH,[68,69] and progesterone.[55]

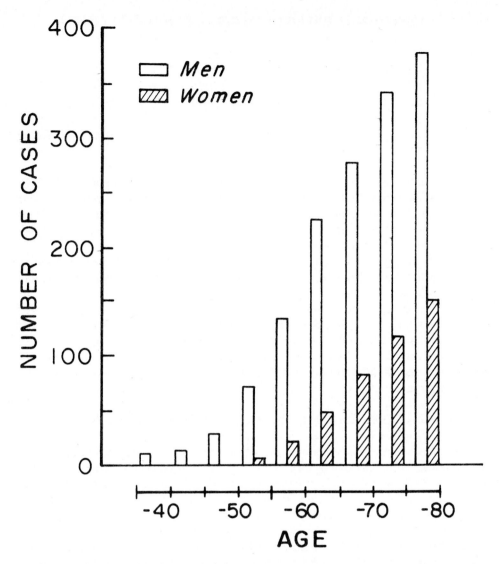

FIGURE 1.   Frequency of fresh myocardial infarction according to age in the group of 5638 autopsies (1955—1965) Faculty of Medical Hygiene (LFH), Prague.

Other metabolic aspects of the climacteric have been described:[56] renin activity in the plasma of women during the menstrual cycle and after menopause has been measured.[57] The elimination of tryptamine[58] in vasomotor disorders during spontaneous and surgical menopause and the elimination of vanyl-mandelic acid[59] in postmenopausal hypertension have been studied.

Considerable evidence is available to indicate that endogenous ovarian estrogen secretion may play a key role in protecting women against clinical symptoms of atherosclerotic coronary heart disease.[8] This has been one of the main reasons for some attempts in treatment of the climacteric syndrome with conjugated or free estrogens.[60-67]

THE INFLUENCE OF ESTROGENS ON LIPIDS AND LIPOPROTEINS IN WOMEN, CHIEFLY POSTMENOPAUSAL

Serum lipids and lipoproteins have been studied mostly in mutual relationship. Although most detailed studies in menopausal women deal with the effect of estrogen on serum lipids and lipoproteins, some concern changes in lipids only.[72-75] A rise of serum cholesterol levels with advancing age has been well established in this group,[5-7,18] although Hamman and co-workers[71] in 1975 found no changes in cholesterol levels in Prima Indian women after the menopause.

Russ et al.[76] and Barr et al.[2] found higher α-lipoprotein (fraction IV + V + VI) and β-lipoprotein (fraction I + II + III) in women than in men of comparable age. Lewis et al.[77] also noted a higher concentration of α-lipoprotein ($-S_{1,21}$ 0—12) but a lower concentration of β-lipoprotein ($-S_{1,21}$ 25—40) in women than in men. Gofman et al.[78] as well as DeLalla et al.[79] reported a high concentration of high-density lipoprotein (HDL) and a lower concentration of low-density lipoprotein (LDL) in women than in men in the 20-to-60-year age group.

Havel et al.[80] as well as Nikkila[81] and Barclay and co-workers[82] confirmed that a higher proportion of cholesterol and phospholipids is present in the high-density lipoprotein of women.

Using chemical fractionation, ultracentrifugation, and paper electrophoresis, it has also been established that men in age groups below 40 have a significantly larger amount of their total blood lipid in the form of β-lipoprotein (LDL and very low-density lipoprotein VLDL) and a correspondingly smaller amount in α-lipoprotein (HDL). The premenopausal female has a higher concentration of blood lipid in α-lipoprotein (HDL) with a relatively lower β-lipoprotein (LDL and VLDL) level. After the menopause the female pattern changes to that of the male with increased β-lipoprotein and decreased α-lipoprotein.[30,32-34]

The effect of estrogen on serum lipids in women has been carefully studied, chiefly in the postmenopausal state.[4,9-12] One of the most significant of previous reports is Eilert's 1949 demonstration of the lipid-shifting effects of estrogens[1]: cholesterol is lowered after the natural menopause as well as after castration, but triglycerides are not changed. Progesterone does not influence plasma lipid levels.[14]

In 1960, Robinson et al. successfully attempted to lower the cholesterol level in 34 women after spontaneous or surgically induced menopause using equine estrogens at doses of either 1.25 or 2.5 mg daily. They suggested that estrogen treatment should be given to women with abnormal serum lipids who had been subjected to hysterectomy.[11] In 1964, Robinson and Lebeau observed the usual changes in lipids with a low dosage of an estrogen (1.25 mg and 2.5 mg of conjugated equine estrogens daily) in groups of postmenopausal women between the ages of 40 and 62 selected on the basis of cholesterol levels higher than 260 mg%. The decrease in cholesterol, β-lipoprotein cholesterol, and β-lipoprotein/α-lipoprotein cholesterol ratio and the increase in α-lipoprotein cholesterol, lipid phosphorus, and triglycerides were modest, but definite.[70]

Large doses of estradiol (160 mg i.m. over 203 weeks) caused a distinct fall in cholesterol levels, but no change in triglycerides or free fatty acids. Total phospholipids tend to fall, but less than cholesterol.[14]

In 1967, Furman et al.[83] published important information concerning dyslipoproteinemia in women after menopause, particularly after treatment with estriols. These women differed from premenopausal women who have less lipid circulating in the form of VLDL. In all lipoprotein fractions, lipids increased with age, resulting in higher serum concentrations of cholesterol, phospholipid, and triglyceride in the older, postmenopausal female. In normolipemic subjects, estrogens increased the serum concentration of α-lipoproteins (HDL), which contained less cholesterol than that HDL isolated prior to estrogen administration. A significant diminution in cholesterol content of the β-lipoprotein fraction (LDL) was also observed. Estrogens regularly increased VLDL as evidenced by increased serum triglyceride levels. In subjects with hypercholesterolemia, estrogens lowered β-lipoproteins (LDL) and increased α-lipoproteins (HDL) and VLDL. In subjects with hyperlipemia, estrogens lowered serum cholesterol levels, mainly in VLDL. Although these observations were not statistically examined, this work of Furman and co-workers provides the best available information until the present. Indeed, in the decade since this publication, only one paper was devoted to the study of lipoproteins and lipids in the menopause and how they are affected by conjugated estrogens.

Notelowitz and Southwood[84] in 1974 published a study concerning the analysis of serum lipoproteins, employing the Thorp® micronephelometer for estimation of the β_1- and pre-β-lipoproteins and chylomicrons. This methodology has so many theoretical and practical weaknesses that it is unacceptable. It is almost unbelievable that in the last 8 years no experienced lipoproteinologist has devoted his attention to the basic problem of the enhancement of atherosclerosis after menopause.

LIPOPROTEINS AND OTHER PROTEINS AFTER OOPHORECTOMY

In 1964 and again in 1968, Opplt and co-workers published original studies concerning metabolic changes in serum proteins and lipoproteins in 50 women who had undergone gynecological operations; 20 of them were surgical castrations.[86,89,90,93,96,97] Total lipoproteinemia declined after an operation for a fibrous uterus with oophorectomy; the difference between the first and sixth day was highly significant statistically. The percentage of α_1 lipoproteins between day 0 and day 1 rose; then it dropped again and did not change substantially; β-lipoprotein values rose markedly from the beginning, the rise being statistically significant. Nonmigrating lipoproteins after varied generally declining. (See group I in Figure 2.)

Within the first 24 hr after abdominal operations, a significant drop of the total lipoprotein levels occurred in all patients, parallel to a fall in hematocrit and total protein levels.[87,94] In patients whose operations included bilateral adnexotomy, i.e., castration, the first drop was followed on the second day by a rise in lipoprotein level which continued to the end of the investigated period. This increase is accounted for solely by the rise of β-lipoprotein concentration. In noncastrated patients (after vaginal plastic operations), changes in the plasma lipoproteins were irregular. After castration the changes of serum lipoproteins thus take a course different from that followed by other types of gynecological surgeries. In addition to the nonspecific humoral response to trauma and possible compensatory hemodilution, other factors seem to participate and to be direct consequences of surgical castration.

The described, arrested decline of total lipoproteinemia and the return to values approaching the initial level can be considered the beginning of changes in plasma lipoproteins which were observed by other authors after prolonged periods following castration.[86,87,94] A remarkable and very important fact is the prompt and early character of this metabolic response to ovariectomy.

In a corresponding clinical study[86,87,94] concerning protein levels and their fractions after gynecological operations, Opplt and co-workers investigated changes in proteinemia and fibrinogen concentration in women operated upon for benign gynecological diseases. In the first 6 days following the operation, a drop in hematocrit, proteinemia, and concentration of albumin and γ-globulins could be observed. Furthermore, an increase in the α_2-globulin fraction and in some patients an increase in the values of

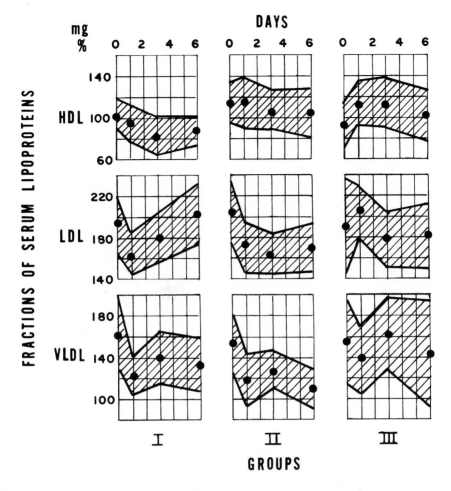

FIGURE 2.   Arithmetic means ($\bar{x}$) and standard deviation(s) of individual lipoprotein fractions (in mg%); I = group of castrated patients, II = group with benign gynecological disorders not requiring oophorectomy, and III = group of vaginal plastic operations.

α_1- and β-globulins were established. Levels of fibrinogen markedly increased till the third day after operation and then dropped. Changes in all cases were substantially of the same character and differed only quantitatively with respect to the kind and extent of the particular type of operation (see Figure 3).

In a group of 30 women, having abdominal hysterectomy and bilateral oophorectomy, and in another group of 30 women having noncastrative operations, levels of macroergic phosphates: APT (adenosine triphosphate) ADP (adenosine diphosphate) and AMP (adenosine monophosphate) were studied before surgery and 2 and 24 hr later.[92] The analytical method of Hoetzl and Laudahn was performed with statistical evaluation by use of the T test at 5% and 1% significance (see Figure 4). It was found that only the ADP concentration significantly changed, dropping in the course of major operations and rising again in the following 24 hr. The preliminary conclusion was that lipoprotein changes in gynecological operations could not be attributed to imbalance in blood levels of the energetically richest (ATP) and poorest (AMP) components.

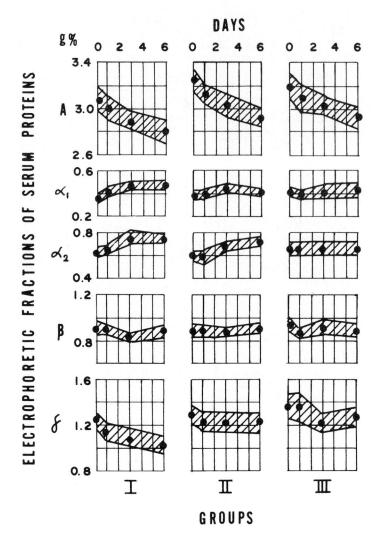

FIGURE 3.   Levels of the main fractions of serum proteins expressed in x̄ (arithmetic means) and SD (standard deviations) before and 1 to 6 days after surgery; I = abdominal hysterectomy, and bilateral oophorectomy, II = minor abdominal surgery, and III = vaginal plastic operations.

THE INFLUENCE OF ESTROGENS ON LIPIDS AND LIPOPROTEINS IN MEN WITH CLINICAL ATHEROSCLEROSIS

The relationship between serum lipid levels, estrogen therapy, and coronary atherosclerosis has been studied chiefly on men for the simple reason that males are most affected by coronary heart disease.[2,3,7,16,17,31,34]

Barr et al. in 1952 was the first to study the influence of estrogen on lipoprotein in men with clinical atherosclerosis. He observed an increase of cholesterol in fractions IV + V + VI (α-lipoproteins), isolated by alcohol fractionation, and a decrease in fractions I + III (β-lipoproteins).[2] Furman and co-workers in 1958, using the analytical ultracentrifuge, reported an increase of HDL ($-S_{1,21}$ 0—12 and sometimes a decrease of LDL ($-_{1,21}$ 25—70) in patients without clinically evident atherosclerosis who had been treated with estrogens.[17]

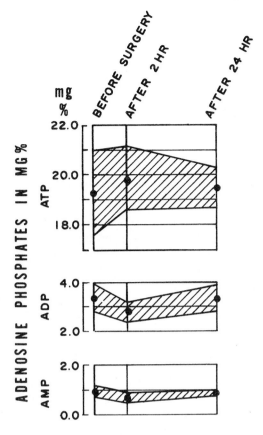

FIGURE 4.   Changes in concentrations ($\bar{x}$ and SD) of
ATP, ADP, and AMP after major gynecological surgeries (hysterectomy and bilateral oophorectomy).

Paper electrophoresis was used as the method for evaluating the effect of long-term
estrogen therapy on serum lipoproteins in men by Danemann et al. in 1960.[28] The α-
lipoprotein:β-lipoprotein ratio was elevated after the oral administration of estrogen.

A most valuable study on the long-term effects of estrogen therapy on serum lipoproteins in men after myocardial infarction was reported in 1961 by Stamler et al.[30] A
typical estrogen effect on cholesterol, cholesterol/phospholipid (C/P) ratio, and α-li-
poprotein (HDL) was achieved after high doses (10 mg/day) of Premarin®. Under
such therapy, a sustained reduction in cholesterol levels was not consistently observed.
Phospholipids rose uniformly and tended to remain at higher than pretreatment levels
with consequent lowering of the C/P ratio. α-lipoprotein levels rose markedly, persisted at "feminine" levels during treatment, and proved to be the most sensitive pa-
rameter.

Using ultracentrifugal analysis, estrogen has been shown to produce a significant
increase in HDL ($-S_{1,21}$ 1—10) and a decrease in the class $-S_{1,21}$ 20—25. Other classes,
representing LDL and VLDL, remained unchanged.[32] This effect has also been described by Marmorston and co-workers[29] as the "interlipid relationship". Changes in
serum lipids under estrogen therapy occur to a maximal degree, i.e., cholesterol levels
decrease and phospholipid levels increase, before the patient develops any clinical manifestation, such as breast tenderness, provided the initial dose is small and dosage is
increased gradually. Estrogen doses, causing no clinical manifestation, may still cause
maximal changes in the serum lipids.

In elderly males, aged 65 to 85, a modification of plasma lipoproteins after estrogen therapy has been observed by Voyles and Evans.[33] Before administration of estrogen, each of the patients was found to have the usual atherogenic pattern with a large β-lipoprotein electrophoretic fraction. After estrogen therapy (5 to 10 mg of Premarin per day) for three or four 3-month periods, each patient demonstrated significant changes in serum lipoproteins to the pattern of a premenstrual female, that is, a decrease in β-lipoprotein and an increase in α-lipoprotein. Approximately 2 weeks after cessation of therapy, there was a reversion to the original pattern of predominantly β-lipoprotein in patients checked.

SUMMARY AND CONCLUSIONS

Millions of women are entering the state of menopause yearly, and other millions use conjugated estrogens daily. These facts create a handful of acute problems.

The incidence of coronary heart disease is estimated to be 20 times higher in men than in women between the ages of 32 to 39 years; however, after the climacteric, the incidence in both sexes is the same. Similarly, the incidence of cardiovascular disease is estimated to be seven times greater in women who have stopped menstruating before the age of 40 years than in a healthy, normal female population of comparable age. This situation worsens after bilateral oophorectomy.

The usual form of investigation of single lipids and their components (cholesterol, triglycerides, and phospholipids) allow only very weak and approximate information about quantitative changes in serum lipoproteins. For semiquantitative or quantitative estimation of serum lipoproteins, paper electrophoresis has been mainly utilized, seldom the ultracentrifuge. Both methods only have a limited value in the evaluation of qualitative changes in serum lipoproteins, although they have made a precious contribution to the explanation of the relationship between lipids and lipoproteins.

In the last several decades, many authors have shown a close connection between changes in the serum lipoprotein spectrum and the clinical manifestations of atherosclerosis. In fact, it has been demonstrated how quickly (within 6 days) lipoprotein changes occur in women after surgical castration. Changes known to occur in the lipoprotein patterns of postmenopausal women are a rapid and persistent increase of cholesterol-rich LDL and a corresponding increase in cholesterol levels.

It has been hypothesized that some subclasses of lipoproteins probably are atherogenic. Wissler's studies were the first to show that certain LDL are specifically atherogenic, meaning that they act directly on cell-cultured intimal and medial cells of human arteries to produce a pathological reaction. In accordance with L. J. Lewis,[98] the main atherogenic action is expected in the LDL fraction. According to the present "infiltration" theory of atherogenesis, only certain lipoprotein subclasses (LDL and IDL) which may then be called "atherogenic", penetrate to any significant extent.

It is established that conjugated estrogens can decrease enhanced development of LDL, which contain most of the serum cholesterol, despite the fact that VLDL are simultaneously increasing as evidenced by increased serum triglyceride levels. Because of these two different types of change, the benefit of estriols in preventing development of atherosclerosis in postmenopausal women is not yet conclusive. It is also not known if certain subclasses of VLDL or IDL may not be particularly atherogenic. Little serious research on serum lipoprotein structure and biological activity has been done in relation to the influence of conjugated estrogens. Nothing has been deduced concerning changes in the quality or conformation of serum lipoproteins; possibly because of this, previous research could not offer a more decisive answer.

REFERENCES

1. **Eilert, M. L.**, The effect of estrogens upon the partition of the serum lipids in female patients, *Am. Heart J.*, 38, 472, 1949.
2. **Barr, D. P., Russ, E. M., and Eder, H. A.**, Influence of estrogens on lipoproteins in atherosclerosis, *J. Assoc. Am. Physicians*, 65, 102, 1952.
3. **Marett, W. C. and Viras, J. R.**, The effect of oral estrogens on serum cholesterol and total lipids, *U.S. Armed Forces Med. J.*, 4, 1439, 1953.
4. **Robinson, R. W., Higano, N., and Cohen, W. D.**, The effects of estrogens on serum lipids in women, *Arch. Intern. Med.*, 100, 739, 1957.
5. **Oliver, M. F. and Boyd, G. S.**, Changes in the plasma lipids during the menstrual cycle, *Clin. Sci.*, 12, 217, 1953.
6. **Wuest, J. H., Dry, T. J., and Edwards, J. E.**, The degree of coronary atherosclerosis in bilaterally oophorectomized women, *Circulation*, 7, 801, 1953.
7. **Rivin, A. U. and Dimitroff, S. P.**, The incidence and severity of atherosclerosis in estrogen-treated males, and in females with a hypoestrogenic or a hyperestrogenic state, *Circulation*, 9, 533, 1954.
8. **Berkson, D. M., Stamler, J., and Cohen, D. B.**, Ovarian function and coronary atherosclerosis, *Clin. Obstet. Gynecol.*, 7, 504, 1964.
9. **Robinson, W. R., Cohen, W. D., and Higano, N.**, Estrogen replacement therapy in women with coronary atherosclerosis, *Ann. Int. Med.*, 48, 95, 1958.
10. **Robinson, W. R., Higano, N., and Cohen, W. D.**, Increased incidence of coronary heart disease in women castrated prior to the menopause, *Arch. Intern. Med.*, 104, 908, 1959.
11. **Robinson, W. R., Higano, N., and Cohen, W. D.**, Effects of long-term administration of estrogens on serum lipids of post-menopausal women, *N. Engl. J. Med.*, 263, 828, 1960.
12. **Higano, N., Robinson, W. R., and Cohen, W. D.**, Increased incidence of cardiovascular disease in castrated women, *N. Engl. J. Med.*, 268, 1123, 1963.
13. **Robinson, W. R.**, Use of estrogen therapy in the treatment of coronary heart disease, *Geriatrics*, 20, 87, 1965.
14. **Svanborg, A. and Vikrot, O.**, The effect of estradiol and progesterone on plasma lipids in oophorectomized women, *Acta Med. Scand.*, 179, 615, 1966.
15. **Parrish, M. H., Carr, C. A., Hall, D. G., and King, T. M.**, Time interval from castration in premenopausal women to development of excessive coronary atherosclerosis, *Am. J. Obstet. Gynecol.*, 99, 155, 1967.
16. **Steiner, A., Payson, H., and Kendall, F. E.**, Effects of estrogenic hormone on serum lipids in patients with coronary atherosclerosis, *Circulation*, 11, 784, 1955.
17. **Furman, R. H., Palmer, R. H., Novica, L. N., and Keaty, E. C.**, The influence of androgens, estrogens and related steroids on serum lipids and lipoproteins, *Am. J. Med.*, 24, 80, 1958.
18. **Rivin, A. U. and Dimitroff, S.**, The incidence and severity of atherosclerosis in estrogen-treated males, and in females with a hypoestrogenic or a hyperestrogenic state, *Circulation*, 9, 533, 1954.
19. **Stamler, J., Pick, R., and Katz, L. N.**, First interim report on the effects of estrogen therapy in males under 50 years of age with a previous single myocardial infarction. Paper presented at the Proc. 8th Annu. Meet. Am. Soc. Stud. Arteriosclerosis, Chicago, October 21 to November 1, 1954, *Circulation*, 10, 587, 1954, abstr.
20. **Stamler, J., Pick, R., Katz, L. N., Kaplan, B., Kaplan, E., Baker, L. A., O'Connor, W. R., Lewis, L. A., Page, I. H., Berkson, P., Carnow, B. W., Cogen, S., and Frankel, J.**, Interim report on the effects of long-term estrogen therapy in men under 50 years of age with a previous single myocardial infarction. *Proc. Cent. Soc. Clin. Res.*, 28th Ann. Meet., Chicago, November 4 to 5, 1955, *Abstr. J. Lab. Clin. Med.*, 46, 955, 1955.
21. **Robinson, R. W., Higano, N., Cohen, W. D., Sniffen, R. C., and Sherer, J. W.**, Effects of estrogen therapy on hormonal functions and serum lipids in men with coronary atherosclerosis, *Circulation*, 14, 365, 1956.
22. **Robinson, R. W., Cohen, W. D., and Higano, N.**, Serum lipid and estrogenic effects of two new "weak estrogens" in male patients with coronary atherosclerosis, *Circulation*, 14, 489, 1956.
23. **Stamler, J., Pick, R., Katz, L. N., Kaplan, B., and Pick, A.**, Evaluation of estrogen therapy in males with previous myocardial infarction — interim report — four-year follow-up, *Circulation*, 16, 940, 1957.
24. **Cohen, W. D., Higano, N., and Robinson, R. W.**, Serum lipid and estrogenic effects of manvene, a new estrogen analog. Comparison with premarin in men with coronary heart disease, *Circulation*, 17, 1035, 1958.
25. **Marmorstom, J., Lewis, J., Kuzma, O., Magidson, O., and Jacobs, R.**, Hormonal treatment in arteriosclerotic disease, in *Hormones and Atherosclerosis*, Pincus, G., Ed., Academic Press, New York, 1959, 449.

26. **Marmorston, J., Moore, F. J., Magidson, O., Kuzma, O., and Lewis, J. J.,** Effects of long-term estrogen therapy on serum cholesterol and phospholipids in men with myocardial infarction, *Ann. Int. Med.,* 51, 972, 1959.

27. **Stamler, J., Pick, R., Katz, L. N., Pick, A., and Kaplan, B. M.,** Interim report on clinical experiences with long-term estrogen administration to middle-aged men with coronary heart disease, in *Hormones and Atherosclerosis,* Pincus, G., Ed., Academic Press, New York, 1959, 423.

28. **Danemann, H. A., Pick, R., and Katz, L.N.,** Paper electrophoresis as a method of evaluating the effect of long-term estrogen therapy on serum lipoproteins in man, *J. Lab. Clin. Med.,* 55, 682, 1960.

29. **Marmorston, J., Moore, F. J., Lewis, J. J., Magidson, O., and Kuzma, O.,** Estrogen therapy in men with myocardial infarction: occurrence of lipid changes before feminization, *Clin. Pharmacol. Ther.,* 1, 449, 1960.

30. **Stamler, J., Katz, L. N., Pick, R., Lewis, L. A., Page, I. H., Pick, A., Kaplan, B. M., Berkson, D. M., and Century, D. F.,** Effects of long-term estrogen therapy on serum cholesterol-lipid-lipoprotein levels and on mortality in middle-aged men with previous myocardial infarction. Paper read at 14th Annu. Meet. Am. Soc. Stud. Atherosclerosis, St. Louis, October 20, 1960, *Circulation,* 22, 658, 1960, abstr.

31. **Marmorston, J., Moore, F. J., Kuzma, O. T., and Weiner, J. M.,** Effect of estrogen therapy on interlipid relationships in men. Paper presented at the Am. Heart Assoc., 34th Sci. Sessions, October 20 to 22, 1961, *Circulation,* 24, 989, 1961, abstr.

32. **Stamler, J., Katz, L. N., Park, R., Lewis, L. A., Page, I. H., Pick, A., Kaplan, B. M., Berkson, D. M., and Century, D.,** Effects of long-term estrogen therapy on serum cholesterol-lipid-lipoprotein levels and on mortality in middle-aged men with previous myocardial infarction, in *Proc. Symp. Drugs Affecting Lipid Metabolism,* Garattini, S., and Paoletti, R., Eds., Elsevier, Amsterdam, 1961, 432.

33. **Voyles, C. M. and Evans, I.C.,** Modification of plasma lipoproteins by estrogen in elderly males, *J. Fla. Med. Assoc.,* 48, 158, 1961.

34. **Marmorston, J., Lewis, J., Kuzma, O., Magidson, O., and Jacobs, R.,** Hormonal treatment in atherosclerotic disease, in *Hormones and Atherosclerosis,* Pincus, G., Ed., Academic Press, New York, 1959, 449.

35. **Marmorston, J., Moore, E. J., Kuzma, O. T., Magidson, O., and Weiner, J. M.,** Effect of estrogens of interlipid relations in men with myocardial infarction, *Proc. Soc. Exp. Biol. Med.,* 113, 357, 1963.

36. **Musa, B. U., Seal, U. S., and Doe, R. P.,** Elevation of certain plasma proteins in man following estrogen administration: a dose-response relationship, *J. Clin. Endocrinol.,* 25, 1163, 1965.

37. **Stadel, B. V. and Weiss, N.,** Characteristics of menopausal women: a survey of King and Pierce Counties in Washington, 1973—1974, *Am. J. Epidemiol.,* 102 (3), 209, 1975.

38. **Curiel, P.,** Some aspects of cardiocirculatory pathology in postmenopause, *J. Gerontol.,* 14, 331, 1966.

39. **Manchester, J. H., Herman, M. V., and Gorlin, R.,** Premenopausal castration and documented coronary atherosclerosis, *Am. J. Cardiol.,* 28, 33, 1971.

40. **Blanc, J. J.,** Menopause and myocardial infarction, *Nouv. Presse Med.,* 3 (34), 2173, 1974.

41. **Wurterle, A.,** Estrogen secretion before and after ovariectomy in women in the pre- and post-menopause periods, *Arch. Gynaekol.,* 203, 184, 1966.

42. **Hunter, R. E.,** The significance of estrogen levels in the post-menopausal women, *Trans. New Eng. Obstet. Gynecol. Soc.,* 20, 25, 1966.

43. **Preibsch, W.,** The relationship between menopausal symptoms and the excretion of oestrogen and 5-hydroxyindole acetic acid, *Ger. Med. Mon.,* 12, 81, 1967.

44. **Sideri, L.,** Qualitative and quantitative changes of urinary gonadotropins in the post-menopause, *Atti Accad. Med. Lomb.,* 22, 576, 1967.

45. **Burckhardt, D.,** Estrogen-like effect of digitalis. Its effect on gonadotropin excretion in post-climacteric women, *Schweiz. Med. Wochenschr.,* 98, 1250, 1968.

46. **Sedlis, A., Turkell, W. V., and Stone, D. F.,** Estrogen activity in post-menopausal women, *Bull. N. Y. Acad. Med.,* 45, 271, 1969.

47. **Rogers, J.,** Estrogens in the menopause and post-menopause, *N. Engl. J. Med.,* 280, 364, 1969.

48. **Williams, D.,** Oestrogens in oophorectomized women, *Br. Med. J.,* 2, 400, 1971.

49. **Rader, M. D., Flickinger, G. L., DeVilla, G. O., Jr., Mikuta, J. J., and Mikhail, G.,** Plasma estrogens in post-menopausal women, *Am. J. Obstet. Gynecol.,* 116, 1069, 1973.

50. **Schon, J., Jansen, J., and Avidberg, E.,** Decreased concentration of tissue cortisol in the menopause by a new micromethod, *Nature,* 215, 202, 1967.

51. **Kaplan, H. G., and Hreshchyshyn, M. M.,** Gas-liquid chromatographic quantitation of urinary estrogens in nonpregnant women, post-menopausal women and men, *Am. J. Obstet. Gynecol.,* 111, 386, 1971.

52. Jirkalova, V., Gonadotropins in women in menopause, *Cesk. Gynekol.,* 39(8), 575, 1974.
53. Poliak, A., Jones, G. E. S., and Woodruff, J. D., The effect of human chorionic gonadotropin on castrated postmenopausal women, *Am. J. Obstet. Gynecol.,* 109, 555, 1971.
54. Oashi, K., Proceedings: changes in gonadotropin secretion potentials due to aging in post-menopausal and castrated women, *Folia Endocrinol. Jpn.,* 50(2), 577, 1974.
55. Nillius, S. J., Effects of progesterone on the serum levels of FSH and LH in post-menopausal women treated with oestrogen, *Acta Endocrinol.* (Copenhagen), 67, 362, 1971.
56. Pelkonen, R., Metabolic aspects of the climacterium, *Acta Obstet. Gynecol. Scand. Suppl.,* 9, 16, 1971.
57. Kaulhausen, H., Pattern of renin activity in plasma during the menstrual cycle and in post-menopausal women, *Klin. Wochenschr.,* 52, 33, 1974.
58. Gasparri, F., Tryptamine in urine during spontaneous and surgical menopause in relation to the incidence of severe vasomotor disorders. *Boll. Soc. Ital. Biol. Sper.,* 42, 953—954, 1966.
59. Rolli, G. P., The elimination of vanyl-mandelic acid in post-climacteric hypertension, *Minerva Med.,* 59, 2653, 1968.
60. Gnirs, L., Treatment of the climacteric syndrome with conjugated estrones, *Med. Welt.,* 36, 1923, 1966.
61. Cupr, Z., Iatrogenic problems in long-term estrogen therapy of women in the climacterium, *Cesk. Gynekol.,* 33, 154, 1968.
62. Akaike, A., On the combination of conjugated estrogens with librium (menrium) for treatment of climacteric complaints, *Ther. Ggw.,* 107, 1222, 1968.
63. Solaro, F., Observations on therapy with natural conjugated estrogens in pre- and post-menopausal subjects, *J. Gerontol.,* 17, 709, 1969.
64. Gaudefroy, M., Treatment of menopause disorders by the sequential method, *J. Sci. Med. Lille,* 88, 637, 1970.
65. Potocki, J., Effects of estrogen treatment on coronary insufficiency in women after menopause, *Pol. Tyg. Lek.,* 26, 1812, 1971.
66. Radocha, K., Treatment of climacteric symptoms after artificial menopause, *Cas. Lek. Cesk.,* 111, 158, 1972.
67. Imparato, E., Use of an estrogen-progesterone-testosterone combination in control of the menopausal syndrome. Double-blind clinical studies, *Am. Obstet. Gynecol. Med. Perinat.,* 94(5), 361, 1973.
68. Czygan, P. J., Plasma levels of FSH and LH during the menstrual cycle, the menopause and in the course of gestagen treatment, *Acta Endocrinol. (Copenhagen) Suppl.,* 152, 2, 1971.
69. Friedman, S., Clinical uses of serum FSH and LH measurements, *Obstet. Gynecol.,* 39, 811, 1972.
70. Robinson, R. W. and Lebeau, R. J., Effect of conjugated equine estrogens on serum lipids and the clotting mechanism, *J. Atheroscler. Res.,* 5, 120, 1965.
71. Hamman, R. F., Bennet, P. H., and Miller, M., The effect of menopause on serum cholesterol in American (Pima) Indian women, *Am. J. Epidemiol.,* 102, 164, 1975.
72. Cottafavi, M., Serum lipids and early menopause, *Attual. Ostet. Ginecol.,* 11, 619, 1965.
73. Hallberg, L., and Sranborg, A., Cholesterol, phospholipids and triglycerides in plasma in 50 year old women. Influence of menopause, body weight, skinfold thickness, weight gain, and diet in a random population sample, *Acta Med. Scand.,* 181, 185, 1967.
74. Pyorala, K., Glucose tolerance, plasma insulin and lipids in postmenopausal women during sequential oestrogen-progestin treatment, *Acta Obstet. Gynecol. Scand. Suppl.,* 9, 28, 1971.
75. Goretzlehner, G., Effect of menstranol on carbohydrate and lipid metabolism in the post-menopause period, *Z. Alternsforsch.,* 29(1), 89, 1974.
76. Russ, E. M., Eder, H. A., and Barr, D. P., Influence of gonadal hormones on protein-lipid relationship in human plasma, *Am. J. Med.,* 19, 4, 1955.
77. Lewis, L. A., Green, A. A., and Page, I. H., The alpha and beta lipoprotein pattern of normal and pathological human sera, *Fed. Proc. Fed. Am. Soc. Exp. Biol.,* 10, 84, 1951.
78. Gofman, J. W., Jones, H. B., Londgren, F. T., Lyon, T. P., Elliott, H. A., and Strisower, B., Blood lipids and human atherosclerosis, *Circulation,* 2, 161, 1950.
79. DeLalla, O. F., Elliott, H. A., and Gofman, J. W., Ultracentrifugal studies of high density lipoproteins in clinically healthy adults, *Am. J. Physiol.,* 179, 333, 1954.
80. Havel, R. J., Eder, H. A., and Bragdon, J. H., The distribution and chemical composition of ultracentrifugally separated lipoproteins in human serum, *J. Clin. Invest.,* 34, 1345, 1955.
81. Nikkila, E., Studies on the lipid-protein relationships in normal and pathological sera and the effect of herapin on serum lipoproteins, *Scand. J. Clin. Lab. Invest. Supp.,* 5(8), 1, 1953.
82. Barclay, M., Barclay, R. K., Skipski, V. P. Terebus-Kekish, O., Mueller, E. S., and Elkins, W. L., Fluctuations in human serum lipoproteins during the normal menstrual cycle, *Biochem. J.,* 96, 205, 1965.

83. **Furman, R. H., Alupovic, P., and Howard, R. P.,** Effects of androgens and estrogens on serum lipids and the composition and concentration of serum lipoproteins in normolipemic and hyperlipemic states, *Progr. Biochem. Pharmacol.,* 2, 215, 1967.

84. **Notelowitz, M. and Southwood, B.,** Metabolic effect of conjugated estrogens (USF) on lipids and lipoproteins, *S. Afr. Med. J.,* 48, 2552, 1974.

85. **Opplt, J., Skamenova, B., and Misak, J.,** Some remarks on the metabolism of adipose women, *Cas. Lek. Cesk.,* 101, 16/17, 516, 1962.

86. **Novotny, A., Opplt, J., Dvorak, V., and Schreiber, B.,** The changes in the spectrum of serum proteins and lipoproteins in women after gynecological operations, *Acta Univ. Carol. Med. Suppl.,* 19, 129, 1964.

87. **Novotny, A., Dvorak, V., and Opplt, J.,** Dysproteinemia after chirurgical castration of women, *Cas. Lek. Cesk.,* 104, 44, 1221, 1965.

88. **Novotny, A., Dvorak, V., and Opplt, J.,** The influence of surgical castration on the occurrence of vascular disease in women, *Acta Univ. Carol. Med.,* 12, 6/7, 443, 1966.

89. **Opplt, J.,** The evaluation of serum-lipoprotein analyses in atherosclerosis, *Acta Univ. Carol. Med.,* 12, 6/7, 425, 1966.

90. **Opplt, J., Dvorak, V., and Novotny, A.,** Dyslipoproteinemia after surgical castration in women, *Cas. Lek. Cesk.,* 105, 14, 569, 1966.

91. **Opplt, J.,** Lipoproteide im Blutplasma und Die Semiologie Ihrer Fraktionen. Lipid Res. Symp. held in Czechoslovakia, 1965, *Acta Univ. Carol. Med. Suppl.,* 16:194-212, 1966.

92. **Opplt, J., Dvorak, V., and Novotny, A.,** The changes of blood concentration of ATP, ADP, and AMP after different types of operational stress in gynecology, *Cesk. Gynecol.,* 32, 1-2, 40, 1967.

93. **Novotny, A., Dvorak, V., and Opplt, J.,** Dyslipoproteinemia posle operativnoj Kastracii u zenscin, *Ceskoslovackoje medicinskeye obozrenije,* 13, 160, 1967.

93a **Novotny, A., Dvorak, V., and Opplt, J.,** Dyslipoproteinemia After Surgical Castration of Women, *Ceskoslovackoje medicinskeye obrozrenije,* 13, 3, 151, 1967.

94. **Dvorak, V., Novotny, A., and Opplt, J.,** The changes of blood proteins after gynecological operations, *Cas. Lek. Cesk.,* 106, 11, 310, 1967.

95. **Opplt, J., Dvorak, V., and Novotny, A.,** 1-Glutamic-oxalacetic transaminase in serum after gynecological operations, *Cas. Lek. Cesk.,* 32, 10, 734, 1967.

96. **Novotny, A., Dvorak, V., and Opplt, J.,** The changes in total levels and electrophoretic spectrum of serum proteins and lipoproteins occurring after operations of gynecological malignancies, *Cas. Lek. Cesk.,* 32, 10, 734, 1967.

97. **Dvorak, V., Opplt, J., and Novotny, A.,** The changes in cholesterol, triglycerides and beta-lipoproteins after gynecological operations, *Cas. Lek. Cesk.,* 107, 34, 1068, 1968.

98. **Cameo, G., Acquatella, H., Waich, S., and Lalaguna, F.,** Relationship between low density lipoprotein structure and its interaction with arterial wall components: a process probably contributing to human atherosclerosis, *Atherosclerosis IV,* Schettler, G., Gotto, Y., Hata, Y., and Klose, G., Eds., Springer-Verlag, Berlin, 1977, 70.

99. **Hatta, Y. and Ishii,T.,** Identification of five existence of lipids accumulating in atherosclerosis, *Atherosclerosis IV,* Schettler, G., Gotto, Y., Hata, Y., and Klose, G., Eds., Springer-Verlag, Berlin, 1977, 67.

100. **Bondjers, G. et al.,** High density lipoprotein (HDL) dependent elimination of cholesterol from normal arterial tissue in man, *Atherosclerosis IV,* Schettler, G., Gotto, Y., Hata, Y., and Klose, G., Eds., Springer-Verlag, Berlin, 1977, 71.

101. **Smith, E. B., and Smith, R. H.,** Early changes in aortic intima, *Atherosclerosis Rev.,* 1, 119, 1976.

Lipoprotein Changes in Disease

ELECTROPHORESIS OF SERUM LIPOPROTEINS IN PROVEN CORONARY ARTERY DISEASE*, **

Jan J. Opplt, Robert C. Bahler, and Marie A. Opplt

INTRODUCTION

The positive correlation between levels of serum lipids and lipoproteins and the incidence of clinical events secondary to coronary atherosclerosis has been clearly demonstrated.[1,2] However, it is important to note that in the Framingham study[11,12] the majority of males who developed coronary heart disease had lipid levels that were only modestly elevated; cholesterol levels were less than 250 mg% at the initial examination of 252 of the 323 subjects who subsequently developed coronary heart disease. Similarly, in a study of 500 survivors of acute myocardial infarction only 31% had elevated serum levels of either cholesterol or triglycerides.[3]

If there was a practical method of determining levels of LDL_1- and LDL_2-cholesterol rather than only total cholesterol, it would be of considerable assistance in predicting the likelihood of subsequent coronary events. Unfortunately, none of the present routine methods offers such practicality.

Cholesterol levels of LDL classes, roughly separated by preparative ultracentrifugation and cholesterol content in β-electrophoretic fraction, have a very significant correlation, and both these analytic approaches could be employed to provide better information about the possible presence of some metabolic disorders associated with a high incidence of CAD. In contrast, recent data indicate that the content of cholesterol in HDL is inversely related to the incidence of coronary heart disease and its prognostic value is independent of total serum cholesterol level.[9-12] It is fair to consider the high levels of $HDL_{2,3}$ (or the cholesterol content in these lipoprotein classes) as an inversely related factor in the incidence of CAD. Perhaps one can extend these observations to the interpretation of the levels of electrophoretic α_1-lipoprotein fractions.

There are data available regarding the electrophoretic distribution of lipoproteins and the prevalence of CAD. Since the electrophoretic fractions bear a quantitative relationship to the levels of the ultracentrifugal classes of lipoproteins, one would anticipate that a semiquantitative and qualitative study of lipoprotein electrophoreograms could provide clues to abnormalities of lipoprotein metabolism that often are not reflected in serum lipid levels.

Electrophoresis of serum lipoproteins in agarose-gel reveals three to six fractions that vary in their relative proportions of different lipoprotein families[33] and their associations. Whether these fractions and their relative proportions are related to the presence or absence of CAD or another form of atherosclerosis needs to be critically

* Abbreviations:

CAD	=	coronary artery disease
Chm	=	chylomicrons (or CHYL)
VVLDL	=	very very low density lipoproteins (large molecules, possible remnants)
VLDL	=	very low density lipoproteins
LDL_1	=	low density lipoproteins (possible intermediates or remnants)
LDL_2	=	low density lipoproteins
HDL	=	$HDL_{2,3}$ = high density lipoproteins

** Supporting factors in the analysis of dentsitometric patterns of agarose-gel lipoprotein electrophoresis in patterns with proven coronary artery disease.

examined, chiefly, with respect to the correct composition and pathophysiologic significance of single electrophoretic fractions and subfractions.

We have examined the relationship between quantitative abnormalities of both serum lipids and electrophoretically separated lipoproteins and CAD by using a more precise diagnostic tool, coronary angiography, for patient selection. This approach excludes subjects who have had either a myocardial infarction or angina pectoris precipitated through some mechanism other than obstructive CAD; consequently, the prevalence of elevated levels of serum lipids and lipoproteins ranges from 40 to 90%,[4-6] and may even be higher in young subjects.[6-8]

This report describes the authors' experience with the interpretation of densitometric scans of standardized agarose-gel electrophoresis of serum lipoproteins. This technique was utilized in 176 patients in whom CAD was documented or excluded by coronary angiography.

METHODS

Patient Selection

A consecutive group of patients, who were referred to the cardiac catheterization laboratory for diagnostic coronary angiography, were selected. The study population included subjects with chest pain of uncertain etiology, candidates for possible coronary bypass surgery, and patients with valvular heart disease undergoing preoperative coronary angiography. Patients with liver disease, thyroid disease, recent myocardial infarction, alcoholism, in the New York Heart Association (NYHA) class IV, or who were taking drugs to lower serum lipids were excluded.

Coronary Angiography

Coronary arteriograms were performed using the Sones technique and were interpreted prior to any knowledge of the lipid studies. The major coronary arteries were individually graded according to the following criteria:

1. 0 = no visible irregularity
2. 1 = definite irregularity but no narrowing greater than 49%
3. 2 = 50 to 74% reduction of vessel diameter
4. 3 = 75 to 99% reduction of vessel diameter
5. 4 = total occlusion

Subjects were then divided into three groups according to the results of coronary angiography:

Group I   = no coronary artery disease (grade 0 in all vessels)
Group II   = minimal coronary artery disease (grade 1 lesions in one or more vessels)
Group III   = obstructive coronary artery disease (grade 2 or greater lesions in one or more vessels)

Analytic Methods

Micromethod for total serum triglyceride estimation—Triglyceride Reagent Set, DOW-Diagnostics, was employed, using alkaline chemical hydrolysis in methanol at 37°C. Interfering substances were precipitated with $MgCl_2$ and removed by centrifugation. Glycerol in the supernatant reacted enzymatically (with glycerol kinase, dehydrogenase, and diaphorase) to produce colored formazan. The absorbance was recorded at 500 nm and is directly proportional to triglycerides concentration. This micromethod was used for control purposes.

Semi-automated adaptation of the DADE triglyceride reagent system—This method, based on the modification of Royer and Ko, was employed for routine analyses. Triglycerides were selectively extracted from serum by a mixture of solvents (nonane-isopropanol 2 to 3½) that would not extract phospholipids. Glycerol was liberated from the triglycerides by transesterification using sodium ethoxide and was then oxidized with sodium periodate to formaldehyde. Color was developed with acetylacetone. The dimethyldihydrolutidine, thus formed, fluoresced strongly and had a yellow color with a definite absorption peak at 415 nm. The amount of both color and fluorescence was directly proportional to triglyceride concentration.

Micromethod for total serum cholesterol estimation—The DOW-Diagnostics method for determination of cholesterol was selected. It is based on Wybenga's method, which agrees with the reference procedure of Abell. Cholesterol, free and esterified, reacted with a reagent composed of ethyl acetate, sulfuric acid, and ferric perchlorate. The resultant purple color was measured photometrically at 595 nm.

Agarose-gel electrophoresis of plasma lipoproteins— Electrophoresis was performed on the Pfizer Pol-E-Film System®, with no modifications in equipment (Cassette Electrophoresis Cell, Power Supply — 639 E) or electrophoresis buffer. All procedures, recommended by the Pfizer Diagnostic Division, were followed exactly; electrophoresis, staining (with Fat Red-7B stain), fixing of films, clearing-destaining, and drying.

The electrophoretograms on agarose-gel Pol-E-Films were scanned on a Beckman Microzone® Digital Integrator, Model R-111. Relative percents (reported as a percentage of total area of entire scan) and total counts per delimited area (reported as concentration units of single fractions and of total lipoproteins) were obtained.

Mobility of electrophoretic fractions—Mobility of isolated molecular filtration fractions was measured in electric field on agarose-gels (Pfizer Pol-E-Film System®), using basically two methods; according to Rapp and Kahlke and according to Wille. Methodical details are fully described elsewhere in this book.[28,29]

RESULTS

Interpretation of densitograms obtained from agarose-gel electrophoresis patterns of serum lipoproteins

Our studies regarding electrophoretic mobilities,[28] molecular size,[29] and composition[28] of single classes and subclasses of serum lipoproteins in normal and pathologic states has led us to the view that agarose-gel electrophoresis is a semiquantitative, but valuable, assay for serum lipoproteins.

The agarose-gel electrophoretic pattern usually exhibits three fractions in the physiologic (normal) state, but five to six fractions may appear in different pathologic conditions. The seventh electrophoretic fraction, composed of chylomicrons, does not occur under physiologic conditions. The six electrophoretic fractions, observed in patients with proven CAD, are the β-fraction with β_2-globulin or slightly slower mobility; the pre - β_1-fraction with a mobility slightly faster than that of the β_1-globulins; the pre - β_2 - fraction with a mobility in the range of β_1 - α_2 - globulins; and the α_2-fraction with the mobility of α_2 - globulin. The $\alpha_{1,2}$-fraction ("x") with mobility slower than α_1-globulin; and $\alpha_{1,1}$-fraction ("y") with a mobility faster than α_1-globulin (Figure 1). To complete the description of the electrophoretic pattern, the electrophoretic mobility of very large lipoproteins, called chylomicrons (CHY), should be considered. These particles are characterized by a negative sedimentation coefficient of density 1.21:$S^{1.21}$ > 1,000 and often > 10,000 and have a mean particle size 1000 to 4000 Å.

DIFFERENT SUBTYPES OF AGAROSE-GEL-ELECTROPHORESIS OF SERUM LIPOPROTEINS

DIFFERENT CONSTELLATIONS OF PATHOLOGIC PATTERNS

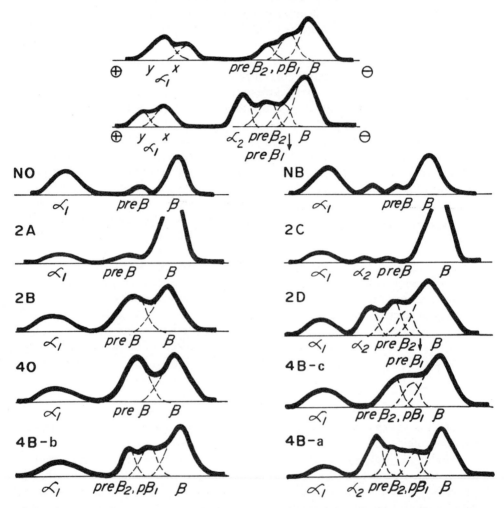

FIGURE 1.   The different constellations of physiologic and pathologic electrophoretic patterns of serum lipoproteins and the analysis of their fractions and subfractions are illustrated. The upper and central patterns demonstrate the most frequent constellations of "double pre-β-" and "triple pre-β-" fractions. These patterns also depict the subdivision of the α_1-electrophoretic fraction into two subfractions: $\alpha_{1,1}$-(fast) or "y" and $\alpha_{1,2}$-(slow) or "x". The first pair of electrophoretograms (line 3) illustrates the usual appearance of normal patterns with one pre-β- = (NO) or two pre-β- fractions (NB). Note the high concentrations of — α_1 and low concentrations of pre-β-fractions. Subtypes 2A (line 4) are closely related to the phenotype IIA and may have one (2A) or two (2C) pre-β-subfractions, although often there is no apparent pre-β-fraction. Note very low concentrations of the pre-β-, as well as -α_1-electrophoretic fractions. The "double pre-β-" fraction is infrequent. Subtypes 2B (line 5) are related to phenotype IV and can have either one (2B), two, or three (2D) elevated pre-β-fractions. Subtypes 4 (last two lines) are related to phenotype IV and can present any of the following patterns: one pre-β-fraction (4O); two pre-β-subfractions, but usually not apparent in the electrophoretic pattern (4B-c); two pre-β-subfractions that are well separated by electrophoresis (4B-b); or the α_2-, pre-β_2-, and pre-β_1-subfractions (4B-a).

These fractions of relatively huge particles, when present in huge amounts, move slowly or remain at the application point (trailing phenomenon). They may form type V patterns, for example, in acute pancreatitis or in diabetic keto acidosis. In addition to the usual trailing characteristic, they also may possess a rapid electrophoretic mobility, resembling (when present in low amounts) the pre-β_2 - or more often, the α_2-electrophoretic fraction.

In order to rapidly and clearly describe the electrophoretic distribution of the lipoproteins, the authors of this paper[28] described new symbols for typing of dyslipoproteinemias. These symbols are composed of a number indicating the relationship to standard phenotypes,[35] an initial capital letter determining the quality of electrophoretic separation of the pre-β-fraction, and an additional letter (a, b, c) indicating the overall configuration of the pre-β-fraction, as depicted in Figures 1 and 2 and defined in Table 1. Since dyslipoproteinemia in certain syndromes (including CAD) may only be evident by abnormalities in the lipoprotein electrophoreotograms, and not reflected in serum lipid levels, it becomes important to designate new symbols characterizing these minor changes.

Abnormalities in the electrophoretogram, expressed by these proposed symbols, have been further characterized by more sensitive physicochemical methods.[28-30,34] As will be described in the discussion, qualitative and quantitative changes in the β-α_2 electrophoretic complex are related to corresponding changes in the molecular distribution of lipoprotein classes.[28,29]

Results and Statistical Significance in Single Groups of Patients

Coronary angiography demonstrated normal coronary arteries in 37 patients (group I), minimal coronary artery disease in 17 (group II), and severe obstructive coronary artery disease in 122 (group III). The mean age of group I was slightly less than those of groups II or III; females predominated in groups I and II.

Within each group, there were no major differences between lipid levels for males and females, except in group I where the triglyceride level in females was significantly less than in males (p < 0.001). Although serum cholesterol levels were somewhat higher in groups II and III as compared to group I, the differences were not striking. The serum triglyceride level was substantially higher in groups II and III (Figure 3). Mean values of total serum cholesterol were 208 mg% (SD = ± 33) in group I, 214 mg% (SD = ± 41) in group II, and 221 mg% (SD = ± 45) in group III. Mean values of total triglycerides were significantly different between the groups: 105 mg% (SD ± 55) in group I, 157 mg% (SD ± 78) in group II, and 174 mg% (SD = ± 101) in group III.

The densitometric curves, representing the relative distribution of electrophoretic fractions of serum lipoproteins, were classified according to the proposed symbols. Figure 4 illustrates the prevalence of abnormal electrophoreograms in each patient group:

Group I patients exhibited the following electrophoretic patterns:

- 60.0% (n = 22) normal pattern (NO or NB)
- 17.1% (n = 6) very mild type 4O-c
- 5.7% (n = 3) very mild type 4B-b
- 14.2% (n = 5) very mild type 2A-c
- 2.8% (a diabetic patient) with type 4B-a

Both $\alpha_{1,y}$ and $\alpha_{1,x}$-fractions of lipoproteins were observed in 13 patients (33.2%).

A greater proportion of group II patients demonstrated significant abnormalities in their electrophoretic patterns:

INTERPRETATION OF THE COMPLEX
$\alpha_2 - \text{pre} - \beta_2 - \beta_1 - \beta$
IN DENSITOMETRIC PATTERNS
OF AGAROSE – GEL – ELECTROPHORESIS
OF SERUM LIPOPROTEINS

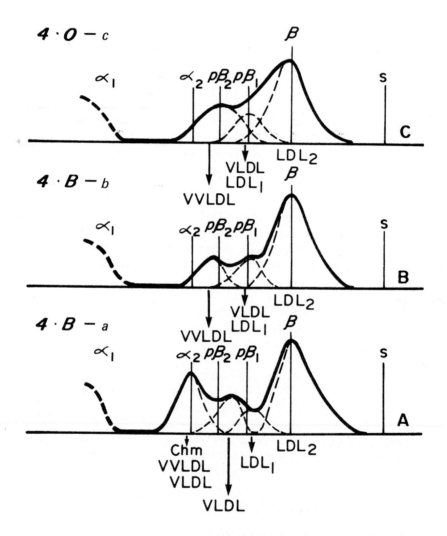

FIGURE 2.   First line — An asymmetry of the anodic descendent of the usual bell-shaped (Gaussian) curve of the β-fraction can imply the existence of the pre-β_1-subfraction. It is usually composed of the VLDL and LDL$_1$ classes of serum lipoproteins. Lp(a) may also be present in very small concentrations. The pre-β_2-subfraction consists of larger (VVLDL) and smaller (VLDL) particles which are characterized by the flotation coefficient — S$_{1-21}$ 1,000 — 70 and by particle size ranging from 1500 Å to 350 Å. Second line — The two pre-β-subractions separate well in an electric field, if the concentrations of VVLDL and VLDL or of lp(a) and LDL$_1$ reach levels favorable for electrophoretic separation. Third line — The α_2-subfraction of the complex separates visibly from the pre-β_2 and pre-β_1 subfractions, if the concentration of chylomicrons (-S$_{1-21}$ 10,000 — 1,000) and VVLDL (-S$_{1-21}$ 1,000 — 400) reaches high levels, but is still <15% of the total serum lipoproteins.

Table 1
SYMBOL KEY TO ELECTROPHORETIC SUBTYPES

Symbol	Meaning
2A	Subtype related to phenotype IIA with none or normal proportion of a single pre-β-fraction.
2B	Subtype related to phenotype IIB with an increased proportion of a single pre-β-fraction.
4O[a]	Subtype related to phenotype IV with an increased proportion of a single pre-β-fraction.
2C	Subtype related to phenotype IIA with low levels of two or three pre-β-fractions.[b]
2D	Subtype related to phenotype IIB with elevated levels of two or three pre-β-fractions.[b]
4B	Subtype related to phenotype IV with elevated levels of two or more pre-β-fractions.[b]
NO[c]	Subtype considered physiological.
NB	Subtype considered physiological with two pre-β-fractions.

Note: Letters a, b, c represent symbols for pre-β-fraction configurations: a for α_2-, β_2-, β_1-subfractions; b for β_2- and β_1-subfractions; c for one visible pre-β_1-fraction.

[a] Subtypes 4O and 2B differ in the delimitation of their pre-β- and β-areas. Subtype 4O is characterized by a significant elevation of only the pre-β-area; whereas, subtype 2B shows an increase in both fractions.

[b] In some cases an α_2-subfraction is also present.

[c] Subtypes 2A differs from subtype NO in having a significant elevation of the delimited β-area.

- 29.4% (n = 5) normal pattern; however, four had nearly abnormal pre-β_1- and pre-β_2-fractions, indicating that only 5.9% of the patterns were entirely physiological
- 23.5% (n = 4) type 4O-c
- 23.5% (n = 4) type 4B-b
- 17.6% (n = 3) type 2D-b
- 5.9% (n = 1) type 2A-c

Both $\alpha_{1,y}$ and $\alpha_{1,x}$ fractions of lipoproteins were observed in 5 patients (29.4%).

In group III, almost all patients had severe configurational changes in the lipoprotein electrophoretic spectrum.

Normal patterns—18%:

- 6.6% (n = 8) type NO (normal with one slowly migrating pre-β-fraction)
- 1.6% (n = 2) type NO-o (normal without any pre-β-fraction)
- 9.0% (n = 11) type NO-a (close to normal, with two pre-β-fractions of subtype b)
- 0.8% (n = 1) type NB-b (close to normal, with two pre-β-fractions; the pre-β_2-subfraction being higher)

Clearly pathologic patterns, related to phenotype IV—38%:

- 7.4% (n = 9) type 4O-c
- 13.9% (n = 17) type 4B-a
- 14.7% (n = 18) type 4B-b (pre-β_2 or pre-β_1 predominates)
- 1.6% (n = 2) type 4B-c (very similar to type 4O-c with two poorly separated pre-β-fractions, usually detectable only by resolving the pre-β-curve)

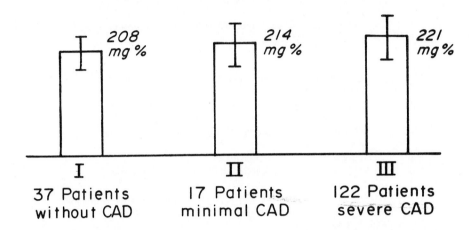

VALUES OF TOTAL SERUM CHOLESTEROL IN PATIENTS WITH AND WITHOUT CAD

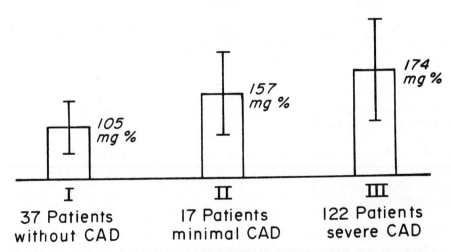

VALUES OF TOTAL SERUM TRIGLYCERIDES IN PATIENTS WITH AND WITHOUT CAD

FIGURE 3.   Mean values of total serum cholesterol and total serum triglycerides (± SD) are presented for the three groups of patients. The differences between groups I and II are statistically significant (p <0.001) for the serum triglycerides.

PATHOLOGIC DENSITOMETRIC PATTERNS OF ELECTROPHOREOGRAMS OF SERUM LIPOPROTEINS IN PATIENTS WITH AND WITHOUT CAD

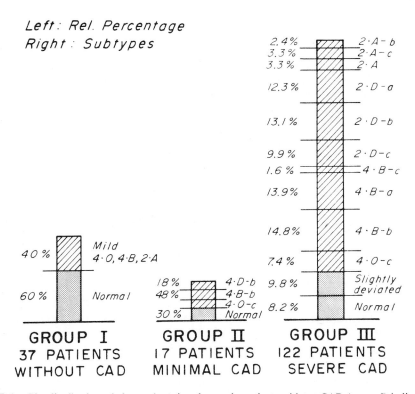

THE HEIGHT OF THE COLUMN CORRESPONDS
TO THE TOTAL NO. OF PATIENTS IN EACH GROUP

Left: Rel. Percentage
Right: Subtypes

FIGURE 4. The distribution of electrophoretic subtypes in patients without CAD (group I) indicated that 60% had normal (physiologic) patterns and 40% had mild pathologic patterns, including all the main subtypes (40, 4B, and 2A). In patients with minimal CAD (group II), the distribution of patterns shifted significantly from normal (30%) to pathologic (70%). In patients with severe CAD (group III), these changes in pathophysiology of the lipoprotein electrophoretic patterns were even more distinguishable: only 18% of these patients had normal patterns and 38% demonstrated subtypes 4 (80% of these were 4B, which means there were two or more pre-β-subfractions). A similar proportion (35%) had subtype 2D (with two or more pre-β-subfractions). Only 9% were subtype 2A. Sixty eight percent had two or more pre-β-subfractions, well separated in the lipoprotein electrophoretograms.

Clearly pathologic patterns, related to phenotype IIB—35%:

- 9.8% (n = 12) type 2D-c
- 13.1% (n = 16) type 2D-b
- 12.3% (n = 15) type 2D-a

Clearly pathologic patterns, related to phenotype IIA—9%:

- 3.3% (n = 4) type 2A-o (high in β-fraction without any pre-β-fraction)
- 3.3% (n = 4) type 2A-c
- 2.5% (n = 3) type 2A-b

Both $\alpha_{1,y}$ and $\alpha_{1,x}$ lipoprotein electrophoretic fractions were found in 61 cases (50%).
Double or triple pre-β-fractions (pre-β_1 and pre-β_2 and/or α_2) were present in 11
subjects (65%) in group II and in 83 subjects (69%) in group III; whereas, only one
subject from group I (a diabetic patient whom the lipoprotein pattern is influenced by
a different metabolic disorder) clearly exhibited this phenomenon (p < 0.001).

In addition to the qualitative changes in the electrophoretic spectrum of serum lipoproteins, one can use the values of the relative distribution of the β-, pre-β-, and α_1-
fractions to describe further qualitative changes in the relationship between the pre-β
and β-fractions of the lipoprotein distribution. The best analytic method would also
include an evaluation of the total area of the lipoprotein spectrum, expressed in units,
i.e., using correction factors that correlate the electrophoretic fraction with the ultracentrifugal class.[26] The simple densitometric analysis of lipoproteinograms offers only
semiquantitative results that should be interpreted with great care. Despite this draw-
back, consideration of the relative distribution of fractions, in conjunction with determination of the electrophoretic types, is recommended, since we have documented
significant differences between patients with and without CAD.

In patients without proven CAD (group 1), the electrophoretic type is normal (physiological), the symbol of type being N. The relative value, in percent, of the pre-β-
fraction does not exceed half of the relative value of the β-fraction, and the level of
the α_1-fraction is not lower than 25 relative percent (for both subfractions x and y
together).

Any pathologic change in the spectrum is reflected in the relationship of fractions.
First, and not specific for any clinical syndrome, is a reduction of the α_1-lipoproteins.
Group II and group III patients showed some decrease in mean values (30.4% ± 6 and
27.2% ± 5, respectively) in comparison to group I patients without CAD (33.9% ±
6). These mean values are not reliable indicators of dyslipoproteinemia. The distribution of α_1-hypolipoproteinemia in single groups offers little more information than the
mean levels. In group II, 3 patients (17.6%) were found with significantly low α_1-lipo-
proteins; in group III, 41 patients (33.6%). This is in contrast to group I, as only 1
patient (2.7%) demonstrated this phenomenon. Practically, only 30% of patients with
proven CAD demonstrated a really significant decrease of α_1-lipoproteins.

Since CAD patients most commonly had dyslipoproteinemia of either types 4O and
4B or 2B and 2D, we also evaluated the relative proportion of the total pre-β-fraction
in the electrophoretogram. The mean value in group I was 18% ± 10, and a prevalence
of elevated values of 27%; in group II, it was 25% ± 4 with a prevalence of 72%; and
in group III, 25% ± 6 with a prevalence of 66%.

An α_1-lipoprotein proportion of less than 25% or a double or triple pre-β-band
(more accurately, an α_2-pre-$\beta_{1,2}$-complex) occurred almost exclusively in groups II and
III (Figure 4). Also, an association of modest abnormalities of serum triglyceride levels
and an abnormality of lipoprotein electrophoresis was almost entirely a feature of

groups II and III (Figure 3). Within group III, severity of CAD did not correlate with serum lipid levels or electrophoretic distribution of lipoproteins. However, very mild CAD (group II) was associated with significantly lower levels of triglycerides and slightly lower proportions of α_2-pre-$\beta_{1,2}$-complex. The proportion of α_1-lipoproteins was higher than that of group III (Figures 3 and 4).

DISCUSSION

Coronary angiography is a precise diagnostic tool that demonstrates early atherosclerotic lesions that have not yet caused any clinical evidence of coronary artery disease. Recognition of this minimal disease group is essential in attempts to characterize metabolic changes in patients with atherosclerosis. The authors' data indicate that this minimal disease group has a prevalence of serum lipid and lipoprotein abnormalities, greater than that of subjects with normal coronary vessels and somewhat less than that of patients with advanced coronary artery obstructions. The importance of these metabolic abnormalities is underscored by the observation that subjects with even minimal coronary irregularity have a higher incidence of subsequent coronary events than those with entirely normal vessels.[13] Furthermore, the authors' data reemphasizes the very low prevalence of serum lipoprotein abnormalities in those subjects in whom coronary angiography outlines entirely smooth vessels.

When coronary artery disease is documented by coronary angiography, the coronary disease groups consistently exhibit higher lipid levels than the controls, but the prevalence, as well as severity, of these quantitative abnormalities varies between studies.[4,6,14] The highest serum lipid levels tend to be associated with more severe coronary disease.[14-17] Indeed, Page et al.[8] have clearly demonstrated an increasing prevalence of coronary obstructions, as either age, serum cholesterol, or serum triglyceride increase. Such data would have been anticipated in view of the known relationship of coronary heart disease to serum lipid levels in a number of epidemiologic studies. However, in up to one half of the victims of coronary disease, there are either no or minimal deviations in the quantities of serum lipids. Therefore, one must search for other abnormalities that are associated with documented CAD.

In our study of lipoprotein electrophoretic patterns in patients with proven CAD, attention has been focused mainly on the configuration of electrophoretic complexes of β- and pre-$\beta_{1,2}$-, and possibly α_2-fraction in conjunction with the levels of total cholesterol and triglycerides in serum. A review of the factors influencing these configurations is appropriate in order to understand the pathological patterns.

The β-α_2-complex may present different configurations in the densitographic evaluations of agarose-gel electrophoretograms, as seen in Table 1. The specific configuration depends[28-30] on changes in concentrations of all the major lipoprotein classes. The physical properties of these classes (vide infra) are relevant to an understanding of the electrophorteograms.

- Chylomicrons are characterized by $-S_{1.21}$ 10,000 to 1,000, a mean particle size of 1000 to 4000 Å, and the electrophoretic mobility of α_2. In high concentrations, this class exhibits a remarkable trailing effect, in which huge particles accumulate at the starting point and in the space between it and the β-fraction.
- Large or very, very low-density lipoproteins (VVLDL)* are characterized by $-S_{1.21}$ 1000 to 400, a mean particle size of 900 to 600 Å, and electrophoretic mobility of pre-β_2.
- Very low-density lipoproteins (VLDL), are characterized by $-S_{1.21}$ 400 to 70, a

---

* Large molecules, possibly remnants.

mean particle size of 600 to 400 Å, and the electrophoretic mobility of pre-β_2 to pre-β_1.

- LDL$_1$,* which is characterized by $-S_{1.21}$ 70 to 40 (42), has a mean particle size of 400 to 250 A, and the electrophoretic mobility of pre-β_1 to β.
- Lp(a), often present in this group of lipoproteins (flotating β-fraction), may have similar physicochemical characteristics, in some cases similar to VLDL and in others closer to LDL$_1$).
- LDL$_2$ is characterized by subclasses $-S_{1.21}$ 40 to 25, 25 to 20, and 20 to 10, and has a mean particle size of 250 to 200 Å, and the electrophoretic mobility of $-\beta$.
- HDL, (HDL$_{2,3}$) is characterized by $-S_{1.21}$ 10 to 1, a mean particle size of 150 to 180 Å, and the electrophoretic mobility of α_1.

Indeed, none of the above lipoprotein classes, although characterized by specific ranges of flotation rates, molecular sizes, and electrophoretic mobilities, represent a single lipoprotein which is determined by its individual apolipoprotein. All classes and all electrophoretic fractions contain associations of lipoprotein families, i.e., lipoproteins with certain apolipoproteins.[33]

Lipoprotein fractions (electrophoretic) are even less homogeneous than classes (ultracentrifugal). They consist of affiliations of lipoprotein classes, subclasses, and their remnants that are all bound together by certain, not yet defined forces, most probably van der Waals forces. These affiliations may be broken to some degree in a gravitational field (over 100,000 × g), but not by molecular filtration or an electric field during electrophoresis. Nevertheless, the electrophoretic fractions are very reproducible, although complex, groups of lipoproteins. Their complexity may vary with the degree of dyslipoproteinemia and to some extent with the nature of each metabolic block.

The electrophoretic method is in our version only semiquantitative, revealing relative distributions of single electrophoretic fractions in the total pattern. It is handicapped further from two basic methodological drawbacks: inaccessible information about the total concentration of serum lipoproteins, and difficulty in standardizing the staining intensity of single fractions because of their broad inhomogeneity, particularly in different pathologic conditions. Despite these drawbacks, some investigators have successfully related the main electrophoretic fractions to the main ultracentrifugal classes: α_1- to HDL, pre-β to VLDL, and β- to LDL. Small numbers of correlations have been performed, and variable success with different fractions has been demonstrated.[18,25,26] The α_1- and β-fractions appear to be the most difficult to correlate with ultracentrifugal classes. Concentration units have been proposed for single electrophoretic fractions on the basis of ultracentrifugal analyses. At the extreme, there has been an attempt to reject electrophoretic analyses of serum lipoproteins as redundant.[27] Limitations and drawbacks may accompany lipoprotein electrophoresis, but the authors propose that one can utilize the relative quantitative changes in the spectrum in conjunction with the careful observation and proper interpretation of the pre-$\beta_{1,2}$ to α_2- complex for a more detailed elucidation of pathologic lipoprotein metabolism in patients with CAD.

The pre-β to α_2- complex is far more complicated[28] than has been reported.[25,26]

The most mobile component, possessing the α_2-mobility in most cases and a mobility greater than pre-β_1- in others, is composed of CHYL, VVLDL, and some VLDL. The apparent separation of α_2-electrophoretic fraction depends, according to our previous research,[28] on the amount of participating chylomicrons- CHYL characterized by a mean size of particles > 750 Å and by a flotation rate of $-S_{1.21}$ 10,000 to 1,000.

The VLDL ($-S_{1.21}$ 400 to 70; size 600 to 400 Å; mobility pre-β_2 and faster) forms a

* Possibly intermediates or remnants.

"plateau" fraction, or creates an independent pre-β_2-fraction, if present in considerable amounts ($> 15\%$).

Lp (a) may also cause a moderate elevation of the pre-β fraction in a relatively large percentage of the population (75 to 80%).[36] It is an "abnormal" lipoprotein, characterized (unlike Lp-β containing apo-B and forming the β_2-fraction) by the apoproteins: apo-B, apo-Lp (a), and possibly apo-D, and albumin.[37] Generally this lipoprotein is considered a minor factor in the genesis of dyslipoproteinemia. In cases of high concentration of Lp (a), the pre-β_2-fraction may be apparent but the triglyceride level should remain low (< 105 mg%).

LDL$_1$ is the major class forming the pre-β_1-fraction and is a constant component of the spectrum: $-S_{1.21}$ 70 to 40; size 400 to 250 Å; electrophoretic mobility pre-β_1. VLDL may also participate in the formation of the pre-β_1-fraction, mainly if present in low concentrations. Small amounts of LDL$_1$ and VLDL may even fail to separate completely from the main β-fraction, causing an asymmetrical descent (oriented towards the anode) instead of the normal Gaussian curve.

Clear separation of subfractions α_2- and pre-$\beta_{1,2}$ (or marked asymmetry in the electrophoretic complex) is indicative of an altered sequence of lipoprotein metabolism that results in the presence of large lipoproteins of $-S_{1.21} > 400$. It is important to stress the rarity of this abnormality in individuals without coronary disease.

Although Papadopoulos et al.[20] described a greater prevalence of separating of double pre-β in normal subjects, data from Scandinavia and the U.S. support the authors' findings that the abnormal pre-β_1 or -$\beta_{1,2}$-complex identifies a metabolic abnormality that is closely associated with CAD.[21-23]

The β-fraction (more precisely corresponding to β_2-globulin by electrophoretic mobility) usually contains LDL$_2$, which is composed of three ultracentrifugal subclasses and is not as homogeneous as was believed in the past 2 decades. The configuration of this fraction should be observed carefully. Mild asymmetry is most often indicative of a poorly separated pre-β_1-fraction. If however, the asymmetry in the " anodic portion" of the curve is very broad and characterized by a great deal of trailing, or if the mobility of the whole (unseparated) β-fraction is greater than that which corresponds to β_2-globulin and basically broad, the presence of so-called β-VLDL, which is typical for phenotype III, is very possible. To confirm such a finding, ultracentrifugation, polyacrylamid-gel electrophoresis, and specific precipitation techniques may be employed. Often, it is very difficult to recognize this anomaly in agarose-gel electrophoretic patterns, but it is extremely important in evaluating patients with CAD. Besides assessing qualitative changes in the β_2-fraction, it is important to evaluate the area of the β-peak (even semiquantitatively according to height and width, measured at a level half of the height). High β-peaks, usually accompanied by low or missing pre-β-fraction(s), indicate subtypes 2A or 2C.

The α_1-lipoprotein fraction was usually reduced in CAD patients and separation of the "x" and "y" electrophoretic bands was often poor or absent (in 50% of the cases), as it occurs in other syndromes.[38-40] The authors were not able to relate the electrophoretic behavior of α_1-fraction to any significant pathologic changes in the α_2- pre-$\beta_{1,2}$-complex in our patients with CAD.

The physicochemical analysis and interpretation of the entire electrophoretic pattern of serum lipoproteins in clinically normal subjects, as well as in 45 patients with CAD, is based on the authors' separate study[28,30] of electrophoretic mobilities of lipoprotein fractions, isolated by molecular filtration.

Each of the isolated fractions was characterized by particle-size measurements using laser light-scattering spectroscopy and flotation rate determinations using micro-ultracentrifugation. We therefore could precisely demonstrate the electrophoretic position of individual lipoprotein classes and/or their complex affiliations. The electrophoretic mobilities that we have identified correspond with the observations of other authors who used entirely different analytic approaches in identifying the components of the electrophoretic spectrum.[19,20,25,38-40] Because the electrophoretic mobilites of pathologic lipoprotein fractions have a profound influence on the configuration of the α_2-pre-$\beta_{1,2}$-complex, the need for new terminology to accurately describe this complex was evident.

CONCLUSION

A total of 176 patients in whom coronary arteriography had been performed were studied. Coronary artery disease was excluded in 37 patients, but documented as minimal in 17, and as severe obstructive disease in 122.

Using routine measurements of total serum cholesterol and triglyceride levels and agarose-gel electrophoresis of serum lipoproteins, the authors attempted to recognize pathologic changes in the densitometric scanning patterns of electrophoretograms by interpreting the integrated areas under each lipoprotein peak and also identifying different configurations of the electrophoretic α_2 + pre-$\beta_{1,2}$ + β-complex (or pre-β_2, pre-β_1 and β-complex). Symbols were suggested for the different subtypes, to express qualitative and quantitative changes more accurately. Single subtypes were defined as indicators of pathology in the electrophoretic lipoprotein pattern.

Interpretation of changes in single lipoprotein subfractions was based on previous studies of patients with proven CAD, involving very complex physicochemical analyses of lipoproteins.[24,28-30] From these measurements, we suggest that in order to reliably detect subtle evidence of dyslipoproteimia, one should adhere to the following guidelines:

- The highest normal value in total serum cholesterol level is 200 mg%.
- The highest normal value in serum triglyceride levels is 105 mg%.
- The agarose-gel electrophoretogram of serum lipoproteins should be examined as to the relative proportions and the configurations of both the α_2-pre-$\beta_{1,2}$ complex and the β-fraction, since evidence of dyslipoproteinemia is demonstrable in this portion of the spectrum in 80% of our patients with proven CAD.
- The α_1-fraction of the electrophoretogram should be evaluated in a semiquantitative approach in order to recognize reduced levels.

REFERENCES

1. **Kannel, W. B., Castelli, W. P., Gordon, T., and McNamara, P. M.,** Serum cholesterol, lipoproteins and the risk of coronary heart disease. The Framingham study, *Ann. Intern. Med.,* 74, 1, 1971.
2. **Roseman, R. H. and Brand, R. J.,** Coronary heart disease in the western collaborative group study. Final follow-up experience of 8½ years, *J.A.M.A.,* 233, 872, 1975.
3. **Goldstein, J. L. and Hazzard, W. L.,** Hyperlipidemia in coronary heart disease. I. Lipid levels in 500 survivors of myocardial infarction, *J. Clin. Invest.,* 52, 1533, 1973.
4. **Falsetti, H. L. and Schnatz, J. D.,** Serum lipids and glucose tolerance in angiographically proved coronary artery disease, *Chest,* 58, 111, 1970.
5. **Allard, C. and Ruscito, O.,** Preoperative serum lipid profile in originally tested patients with coronary arteriosclerosis, *Surg. Gynecol. Obstet.,* 133, 807, 1971.
6. **Heinle, R. A. and Levy, R. I.,** Lipid and carbohydrate abnormalities in patients with angiographically documented coronary heart disease, *Am. J. Cardiol.,* 24, 178, 1969.
7. **Schoonmaker, F. W., King, S. B., III, and Viay, N. K.,** Hyperlipidemia in predicting coronary atherosclerosis by arteriography, *Rocky Mount. Med. J.,* 70, 32, 1973.

8. **Page, I. H. and Berrettoni, J. N.**, Prediction of coronary heart disease based on clinical suspicion, age, total cholesterol, and triglyceride, *Circulation*, 42, 625, 1970.

9. **Rhoads, G. G., Gulbrandsen, L. L., and Kagan, A.**, Serum lipoproteins and coronary heart disease in a population study of Hawaii Japanese men, *N. Eng. J. Med.*, 294, 293, 1976.

10. **Castelli, W. P., Doyle, J. T., Gordon, T., Hanes, C. G., Hjortland, M. L., Hulley, S. B., Kagan, A., and Zukel, W.**, HDL cholesterol and other lipids in coronary heart disease. The cooperative lipoprotein phenotyping study, *Circulation*, 55, 767, 1977.

11. **Gordon, T., Castelli, W. P., Hjortland, M. L., Kannel, W. B., and Dowber, T. R.**, High density lipoprotein as a protective factor against coronary heart disease. The Framingham study, *Am. J. Med.*, 62, 707, 1977.

12. **Gordon, T. and Verter, J.**, The Framingham study. Serum cholesterol, systolic blood pressure, and the Framingham relative weight as discriminators of cardiovascular disease, NIH study, National Institute of Health, Bethesda, Maryland, sect. 23.

13. **Brusche, A. V. G., Proudfit, W. L., and Sones, F. M.**, Clinical course of patients with normal, and slightly or moderately abnormal coronary arteriograms. A follow-up study on 500 patients, *Circulation*, 47, 936, 1973.

14. **Valek, J., Grafnetter, D., Fabian, J., and Belan, A.**, Analysis of lipid disturbances in patients with angiographically confirmed coronary artery disease, *Nutr. Metab.*, 16, 193, 1974.

15. **Barboriak, J. J., Rimm, A. A., Anderson, A. J., Tristani, F. E., Walker, J. A., and Flemma, R. J.**, Coronary artery occlusion and blood lipids, *Am. Heart J.*, 87, 716, 1974.

16. **Farrehi, C., Perley, A. M., Malinow, M., and Judlain, M. P.**, Quantitative relation between coronary atherosclerosis and blood lipids, *Chest*, 63, 409, 1973.

17. **Murray, R. G., Tweddel, A., Third, J. L. H. C., Hutton, I., Hillis, W. S., Lorimer, A. R., and Laurie, T. D. V.**, Relation between extent of coronary artery disease and severity of hyperlipoproteinemia, *Br. Heart J.*, 37, 1205, 1975.

18. **Wong, R. A., Banchero, P. G., Jensen, L. C., Pan, S. S., Adamson, G. L., and Lindgren, F. T.**, Automated microdensitometry and quantification of lipoproteins by agarose gel electrophoresis, *J. Lab. Clin. Med.*, 89, 1341, 1977.

19. **Hatch, F. T., Lindgren, F. T., Adamson, G. L., Jensen, L. C., Wong, A. W., and Levy, R. I.**, Quantitative agarose gel electrophoresis of plasma lipoproteins: a simple technique and two methods for standardization, *J. Lab. Clin. Med.*, 81, 946, 1973.

20. **Papadopoulos, N. M. and Bedynek, J. L.**, Serum lipoprotein patterns in patients with coronary atherosclerosis, *Clin. Chim. Acta*, 44, 153, 1973.

21. **Dahlen, G., Ericson, C., Furberg, C., Lundkvist, L., and Svardoudd, K.**, Angina of effort and an extra pre-beta lipoprotein fraction, *Acta Med. Scand. Suppl.*, 531, 11, 1972.

22. **Frick, M. H., Dahlen, G., Furberg, C., Ericson, C., and Wiljasalo, M.**, Serum pre-β,-1 lipoprotein fraction in coronary atherosclerosis, *Acta Med. Scand.*, 195, 337, 1974.

23. **Insull, W., Napimi, M., and Wloedman, D. A.**, Pre-β-lipoprotein subfractions in diagnosis of coronary artery disease, *Circulation Suppl.*, II, 1055, 1972.

24. **Opplt, J. J.**, Analytical problems in specific estimation of $HDL_{2,3}$ levels in serum, to be submitted.

25. **Noble, R. P., Hatch, F. T., Mazrimas, J. A., Lindgren, F. T., Jensen, L. C., and Adamson, G. L.**, Comparison of lipoprotein analysis by agarose gel and paper electrophoresis with analytical ultracentrifugation, *Lipids*, 4, 55, 1969.

26. **Hulley, S. B., Cook, S. G., Wilson, W. S., Nichaman, M. Z., Hatch, F. T., and Lindgren, F. T.**, Quantitation of serum lipoproteins by electrophoresis on agarose gel, *J. Lipid Res.*, 12, 420, 1971.

27. **Immarino, R. M.**, Lipoprotein electrophoresis should be discontinued as a routine procedure, *Clin. Chem.*, 21, 200, 1975.

28. **Opplt, J. J.**, Electrophoretic characteristics of lipoproteins, separated by molecular distribution, in *Handbook of Electrophoresis*, Lewis, L. A. and Opplt, J. J., Eds., CRC Press, Boca Raton, 1979.

29. **Opplt, J. J.**, Agarose gel electrophoresis of plasma lipoproteins, in *Hanbook of Electrophoresis*, Lewis, L. A. and Opplt, J. J., Eds., CRC Press, Boca Raton, 1979.

30. **Opplt, J. J., Bahler, R. C., and Opplt, M. A.**, Molecular pathology of serum lipoproteins, in *Atherosclerosis IV*, Schettler, G., Goto, Y., Hata, Y., and Klose, G., Eds., Springer-Verlag, Berlin, 1977, 247.

31. **Bahler, R. C. and Opplt, J. J.**, Lipoprotein metabolism on molecular level during atromid-S therapy, in *Atherosclerosis IV*, Schettler, G., Goto, Y., Hata, Y., and Klose, G., Eds., Springer-Verlag, Berlin, 1977, 545.

32. **Carlson, L. A. and Ericson, M.**, Quantitative and qualitative serum lipoprotein analysis. Part I: Studies in healthy man and woman, *Atherosclerosis*, 21, 417, 1975.

33. **Alaupovic, P., Lee, D. M., and McConathy, W. J.**, Studies on the composition and structure of plasma lipoproteins. Distribution of lipoprotein families in major density classes of normal human plasma lipoproteins, *Biochim. Biophys. Acta*, 260, 689, 1972.

34. **Opplt, J. J. and Opplt, M. A.,** Separation of plasma lipoproteins according to molecular size, *Clin. Chem.,* 20, 906, 1974.
35. **Fredrickson, D. S. and Levy, R. I.,** Familial hyperlipoproteinemia, in *The Metabolic Basis of Inherited Disease,* Stanbury, J. B., Wyngaarden, J. B., and Fredrickson, D. S., Eds., McGraw-Hill, New York, 1972, 545.
36. **Albers, J. J. and Hazzard, W. R.,** Immunochemical quantification of human plasma Lp (α) lipoprotein, *Lipids,* 9, 15, 1974.
37. **Jurgens, G. and Kostner, G. M.,** *Immunogenetics,* 1, 560, 1975.
38. **Papadopoulos, N. M. and Kintzios, J. A.,** Varieties of human serum lipoprotein pattern. Evaluation by agarose gel electrophoresis, *Clin. Chem.,* 17, 427, 1971.
39. **Papadopoulos, N. M. and Herbert, P. N.,** The β-lipoprotein doublet in type III hyperlipoproteinemia, *Clin. Chem.,* 23, 978, 1977.
40. **Lindgren, F. T., Adamson, G. L., Jensen, L. C., and Wood, P. D.,** Lipid and lipoprotein measurements in a normal adult American population, *Lipids,* 10, 750, 1975.

CHANGES IN THE PLASMA LIPOPROTEIN SYSTEM DUE TO LIVER DISEASE

H. Wieland and D. Seidel

The liver is the major site of the synthesis of plasma lipoproteins and has central functions in lipoprotein metabolism. Disturbances of liver function are often accompanied by abnormal serum lipid concentrations.[1] Thus, it seems to be reasonable to assume that liver disease may result in alterations of the plasma lipoprotein system. These alterations find their expression mainly in the occurrence of different, abnormal plasma lipoproteins in the different density classes. Improved techniques for isolation and characterization of these abnormal plasma lipoproteins have stimulated many investigators to define more precisely the nature of serum lipid disturbances in liver disease.

Concerning electrophoretic bands and density classes, the "family concept" and "ABC-nomenclature" introduced by Alaupovic et al.[2] will be used to describe the qualitative and quantitative changes of the plasma lipoprotein system in patients suffering from liver disease. The family concept involves the density classes or electrophoretic bands and categorizes lipoproteins in free or associated forms into families. A lipoprotein family may be represented by lipoproteins with only one homogenous apoprotein in its lipid-binding form. A lipoprotein family containing only apo B as a protein moiety is then called Lp-B, which has been successfully isolated as has Lp-A, Lp-C, Lp-D, and Lp-E. In this review Lp-E is referred to as to the lipid-binding form of the arginine-rich peptide.

From a biological point of view, the family concept seems to be more useful to describe lipid transport in plasma than previous ideas about electrophoretic bands or density classes as physicochemical and biological entities. The concept enables us to understand why, for instance, in liver disease many different lipoproteins may occur in only one density class and have the same electrophoretic mobility.

The best-known lipid disturbance in liver disease is hypercholesterolemia of cholestasis.[3] Since it has been investigated most extensively, much emphasis will be placed on disturbances of the plasma lipoprotein system occurring in cholestasis.

Flint (1862) was the first to describe deviation of cholesterol values in the plasma of patients suffering from jaundice. Since then, numerous investigations by several groups have been instituted in order to elucidate the nature of this lipid disturbance.[4-7]

Elevation of total cholesterol is caused by a rise in free cholesterol, which results in a rise in the proportion of free cholesterol to total cholesterol and persists as long as the liver is not further restricted in its function. It is not disputed today that hepatic cholesterol synthesis rises in obstructive jaundice; however, the cause of this rise has not been completely elucidated up to now. The majority of experimental data available indicates enhanced cholesterol synthesis in the liver under conditions of cholestasis[8,9] The same applies to cholesterol synthesis in the intestinal wall in obstructive jaundice.[10]

While plasma triglyceride values do not show any clear deviation from the norm in cholestasis, there is very frequently a rise in the phospholipid concentration of plasma with a fall in the proportion of total cholesterol to phospholipids. Lecithin increases the greatest amount in both relative and absolute terms. Phospholipids (including lecithin) are normally the main constituent of bile and are probably retained because of obstruction of the bile duct.[11] The extent of this retention could be sufficient to explain the rise in plasma phospholipids. Increased synthesis of lecithin in the liver in obstructive jaundice could not be definitely demonstrated up to now. Ahrens and Kunkel[12] and Friedman and co-workers[13] were the first to suggest that elevation of phospholipids in the plasma is specifically responsible for the hyperlipoproteinemia in cholestasis. It was assumed that phospholipids increase the stability of the plasma lipoprotein complexes and thus enhance their binding capacity for cholesterol.

Using zone electrophoresis, Kunkel and Ahrens[14] and Kunkel and Slater[1] found a correlation between the rise in total lipids and the concentration of the β-globulins. Gofman[15] was the first to describe the characteristic rise in concentration of the low-density lipoproteins (LDL), [16-19] S/f 0—10 and 10—20 (d 1.006 to 1.063 g/ml). Several other groups[16-19] described, in addition, a fall in concentration of the high-density lipoproteins (HDL) in plasma of patients with obstructive jaundice. The importance of this fact will be discussed later.

HDL are normally found in Cohn fractions IV to VI and represent the main lipoprotein class of this fraction. Russ and co-workers[20] found, on the other hand, in Cohn fractions IV to VI of patients suffering from cholestasis a characteristic rise in free cholesterol and phospholipids which has been attributed to the occurrence of an abnormal lipoprotein. It belongs to the LDL class but does not react immunologically with an antiserum raised against apo B. Likewise, Switzer[21] described in the LDL fraction of patients with obstructive jaundice a lipoprotein which did not react against anti-β-lipoprotein serum and was characterized by an unusually low protein content (for a low-density lipoprotein), but a high phospholipid and free cholesterol content.

The LDL fraction of patients with cholestasis reacts immunologically against anti-HDL serum. This, together with the well-known fact of the high phospholipid content of normal α-lipoproteins, led to the assumption that the rising concentration of phospholipids in the LDL fraction in cholestasis could be caused by a displacement of the α-lipoproteins from the HDL fraction to the LDL fraction.[22] Burstein and Caroli[23] similarly reported a β-lipoprotein in the plasma of patients with obstructive jaundice, which showed an α_1-globulin mobility after total delipidation in paper electrophoresis. Since apo B is insoluble under normal conditions but not apo-A, the authors also came to the conclusion that the abnormal low-density lipoproteins could be a "special" (abnormal) form of an α-lipoprotein. This idea was disproved by the complete fractionation of the LDL fraction from patients with cholestasis and the isolation of an abnormal lipoprotein using a combination of ultracentrifugation, Cohn fractionation, and polyanion precipitation by Seidel and coworkers.[24] The abnormal lipoprotein has been designated Lp-X. Its synthesis, structure, and metabolism shall be the subject of the following chapter.

THE ABNORMAL LIPOPROTEIN IN CHOLESTASIS (LP-X)

The protein and lipid composition of LP-X (Figure 1) is very constant and obviously independent of the absolute concentration of Lp-X in patients' plasma. It is characterized by a high content of phospholipids and free cholesterol as well as by a protein content, very low for a low-density lipoprotein, thus resulting in an unusually high phospholipid/protein ratio of 11. Lp-X has proven to be a good antigen for immunization and consequent production of antibodies. Electrophoretically, Lp-X migrates slightly slower in most of the common media than the β-lipoproteins. As a result of its size (300 to 700 Å) and its low protein content, it is transported towards the cathode in supporting materials exhibiting a strong electroendosmosis, for instance, in agar gel. No other lipoprotein shows this behaviour, which is typical for Lp-X. This unusual mobility provides the basis for a simple clinical and chemical test for Lp-X.[25] Previously, Lp-X had to be demonstrated in its typical position by immunoprecipitation using a specific antibody. Today, it is visualized in its typical position at the cathode

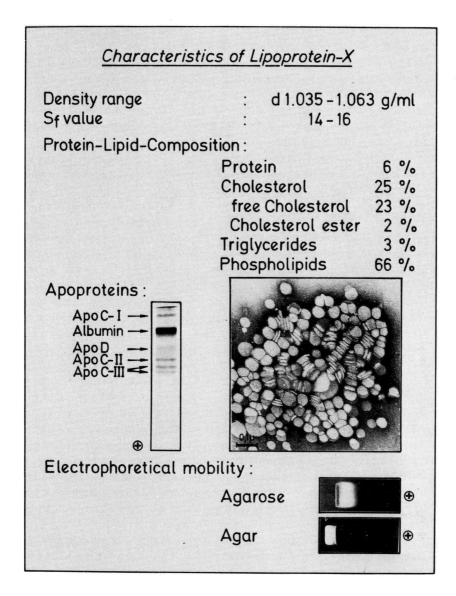

FIGURE 1. Characteristics of lipoprotein-X isolated from the plasma of patients suffering from obstructive jaundice. Apoproteins are demonstrated after separation in 7% polyacrylamide gel electrophoresis, 8 M urea.

in agar gel by polyanion precipitation. The advantage of the latter technique consists in its quickness and its independence of a specific antiserum which has led to the detection of Lp-X in different animal species. Production of Lp-X in the animal experiment made investigations concerning the origin and the metabolic fate of Lp-X possible.[26]

Anti-Lp-X serum always reacts immunologically with pure, isolated Lp-X and consistently with human serum albumin. On the other hand, isolated and pure Lp-X in its intact form does not react with anti-human serum. However, after partial delipidization, Lp-X reacts with anti-albumin serum, which implies that Lp-X contains albumin, either located in the core of the lipoprotein and inaccessible to the antibody or that the antigenic determinants appear to be blocked by lipids in the intact molecule.

The albumin-free protein part of Lp-X is identical with apo C, which is the main lipoprotein of the very low density lipoprotein (VLDL) fraction. All three peptide chains belonging to apo C could be demonstrated in apo X.[24] Most probably, only C-I is located on the surface. Another protein located on the surface is apo-D (see Figure 1). In addition, there are indications for the occurrence of other not yet characterized apoproteins in the protein part of Lp-X.[27] Most probably, Lp-X is transporting an isoenzyme of alkaline phosphatase, EC, 3.1.3.1.[28] It is conceivable that additional enzymes are located on Lp-X. Anti-Lp-X serum is immunologically reactive to hepatitis-b antigen,[29] which does not necessarily imply that Lp-X is transporting hepatitis-B antigen. It is only an indication that there are similar proteins participating in the structure of the two proteins.

It is remarkable that albumin constantly makes up 40% of the protein in Lp-X. The importance of albumin to the structure of Lp-X was recently elucidated and led to important information concerning the origin of Lp-X.[30] Cholestasis can be defined as a condition in which substances normally excreted in bile appear in the blood stream. Typical Lp-X could never be demonstrated in bile; therefore, it can not be regarded as a substance which is normally excreted in bile. The lipid composition of Lp-X, isolated from serum, differs significantly from normal plasma lipoproteins but shows great similarity to lipids found in normal bile.[31] Almost all of the biliary cholesterol exists in unesterified form as in Lp-X. Phospholipid concentration is approximately twice as high as the concentration of cholesterol in both bile and Lp-X. Even the phosphatide distribution of bile is very similar to that present in Lp-X. Moreover, the phospholipid fatty acids of hepatic bile and those of Lp-X show a close similarity.[32] These facts taken together may suggest common origin and relationship of bile lipids and Lp-X.[30] It was tempting to consider Lp-X as a lipoprotein that was, or parts of which were, normally a constituent of bile but refluxing into the plasma in cholestasis. In vitro incubation of bile and serum or bile and albumin resulted in formation of Lp-X, which could be isolated by common methods and was indistinguishable from normal Lp-X isolated from the plasma of patients suffering from cholestasis[30] (Figure 2).

Lipoprotein electrophoresis of bile shows a lipoprotein band in the albumin position: this band is reacting immunologically, and with anti-albumin serum and can be the stained with lipid dyes (Figure 3). It is possible to isolate by ultracentrifugation or polyanion precipitation a fraction from bile containing all lipids occurring in the original bile. This fraction has a density corresponding to LDL but after isolation no longer migrates in agar electrophoresis. No immunological reactions are detectable with antisera against apo A, B or C. Incubation of this fraction with serum does not lead to formation of Lp-X. Addition of bile salts will drastically change the properties of the bile lipids. Like the fraction from the original bile, they now electrophoretically migrate in the albumin position and show immunological reaction with anti-albumin serum. This allows the conclusion that the lipids in bile are present as lipoproteins.[30,33,34]

Partial delipidization of isolated bile lipoprotein using *n*-heptane allows investigation of the protein moiety in the lipoprotein. After this delipidization, albumin seems to be the major apoprotein that is immunologically detectable. The presence of apo B could not be demonstrated.

Incubation of bile lipoproteins with bile salts and serum leads to complete conversion of the former into Lp-X. The bile lipoprotein and bile salts can therefore be regarded as the biliary components necessary for the formation of LP-X.

Mixing of bile with VLDL, LDL, HDL, or isolated γ-globulins does not yield Lp-X. Mixing of bile and albumin leads to formation of a lipoprotein, having the same electrophoretic mobility as Lp-X but not reacting with antisera against the apoproteins

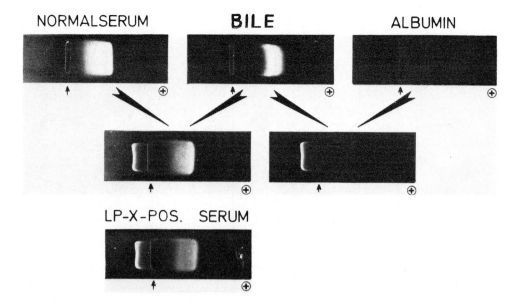

FIGURE 2. In vitro formation of Lp-X: Incubation of bile with normal serum or albumin leads to formation of Lp-X. It is, in both cases, demonstrable in its typical position towards the cathode by polyanion precipitation after electrophoresis in 1% agar. Incubation of normal serum and bile yields an electrophoretic pattern very similar to Lp-X positive serum.

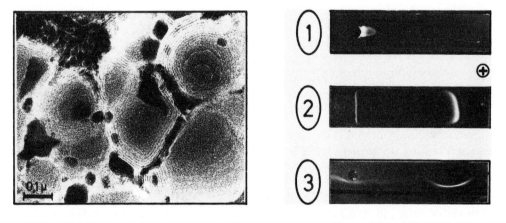

FIGURE 3. Electron microscopic and electrophoretic behaviour of native bile. On the electron micrographs, some lamellar structures which may assume a myelinic shape are apparent. Lipoprotein electrophoresis in 1% agar (1) and 0.8% agarose gel (2) clearly reveals the presence of a lipoprotein in native bile, which can be visualized by polyanion precipitation. The electrophoretic mobility of the bile lipoproteins is in the albumin position as judged by immunoelectrophoresis against antihuman serum (3). The immunoprecipitation in the albumin position can be stained for lipids.

unlike Lp-X, which will immunologically react with anti-C and anti-D. Like Lp-X, it does not react with anti-albumin serum; the albumin component, however, appears after partial delipidization. Radioactively labeled albumin could be demonstrated as part of the Lp-X-like particle formed after addition of bile. The ratio of labeled to unlabeled albumin in that particle was like that in the bile, now devoid of lipid. This indicates true incorporation of albumin into the newly formed Lp-X.

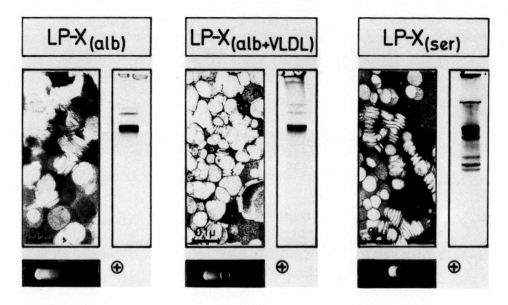

FIGURE 4.   Electrophoretic (1% agar gel and 7% PAGE) and electron microscopic comparison of various forms of Lp-X. Lp-X$_{(alb)}$ has only albumin as an apoprotein as demonstrated in 7% polyacrylamide gel (PAG) and migrates toward the cathode on agar gel electrophoresis further than it does after incubation with VLDL. The latter causes a transfer of apo C from VLDL to Lp-X$_{(alb)}$ as demonstrated by electrophoresis in 7% PAG of Lp-X$_{(alb + VLDL)}$ and a diminution of particle size. Incubation of bile and serum leads to Lp-X$_{(ser)}$ which shows a much higher content of apo C and apo D (PAG), smaller particle size, and reduced migration toward the cathode.

Lp-X^s having only albumin as a protein moiety is larger in size than the Lp-X occurring in the plasma of patients with cholestasis and has a higher protein/lipid ratio. It is stable for only a short time in the very pure form. Addition of serum, VLDL, or HDL leads to diminution of particle size and decrease of protein content. The common mechanism of stabilization in these cases is transfer of all three C peptides, maybe in the form of Lp-C, to the Lp-X particle (Figure 4).

From a theoretical point of view, it is very remarkable that none of the major apoproteins like apo A or apo B are transferred. Apo C is known to associate readily with apo B, forming lipoprotein complexes which are responsible for the major part of triglyceride transport. Lp-X, however, contains remarkably little triglyceride. In which form C-peptides are exchanged, as Lp-C or apo C, is not yet known. Only recently has the physicochemical possibility of the existence of Lp-C been demonstrated.[35] The HDL fraction is not enriched in phospholipids during the lipolytic processes in the plasma. Since apo C is accepted during this lipolytic phase, it seems to be unlikely that it is being transferred as a phospholipid-protein complex. Therefore, some authors postulate exchange in the form of apoproteins.[81] The component of serum necessary to change bile lipoprotein into an Lp-X-like particle seems to be albumin. The amount of albumin necessary is dependent on the concentration of bile salts in the bile. In every case, however, it is possible to transform the total amount of bile lipoprotein present in bile into this kind of Lp-X.

From all these in vitro experiments, it seems justified to expect that Lp-X is formed when bile lipoprotein, together with bile salts, which are usually excreted by the liver into bile, refluxes into the blood because of cholestatic conditions. Whenver these components come into contact with albumin, Lp-X is formed with albumin as apoprotein. This complex then accepts apo C and D, stabilizing the particle by decreasing its size.

LP-X   LP-X+BS

LP-X + BS + ALB

FIGURE 5. Effect of bile salts on the electrophoretic behaviour of Lp-X. Addition of bile salts (BS) to Lp-X abolishes cathodal migration. Lp-X now migrates toward the anode. The original mobility of Lp-X can be restored by addition of excess albumin.

Recently, formation of Lp-X from a cholesterol-lecithin emulsion after sonication was successfully tried, i.e., without bile salts. This method of synthesis is certainly not representative of conditions in vivo. In the latter case, bile salts seem to play an important role as emulsifying factors. The bile salt-free synthesis, however, has some importance for investigation of protein-lipid interactions.

In vitro-formed Lp-X was the first lipoprotein which was assembled completely from its components. It has all the properties of naturally occurring lipoprotein. Addition of increasing amounts of bile salts leads to different physicochemical properties in Lp-X. It will no longer migrate toward the cathode in agar electrophoresis but to the anode, i.e., it is not detectable in the Lp-X test. In lipoprotein electrophoresis in agar gel, it migrates like the bile lipoprotein in original bile or after isolation with consecutive addition of bile salts to the albumin position. Immunoreactivity against anti-albumin serum appears after addition of bile salts. All these changes are most probably caused by binding of bile salts to Lp-X which changes the electric charge of the particle and somehow delipidizes albumin revealing antigenic determinants hitherto blocked by lipids. Excessive addition of albumin, which has a higher affinity for bile salts than Lp-X, causes complete abolition of the bile salt effect, i.e., Lp-X now migrates toward the cathode and albumin is no longer detectable on its surface. It, therefore, seems advisable to add albumin to sera containing high amounts of bile salts, e.g., after prolonged cholestasis, to detect also any physicochemically changed Lp-X in the Lp-X test (Figure 5).

The hypothesis concerning formation of Lp-X is supported by animal experiments. After insertion of the common bile duct into the vena cava of dogs, Lp-X was demonstrated as early as 3 hr after the operation in the four animals treated in this way. A rapid increase in Lp-X concentration during the first 8 hr was paralleled by increases in both cholesterol and phospholipids. Plasma triglycerides remained almost unchanged during the course of the experiment. The initial rapid increase of Lp-X and cholesterol was followed by a period of decreasing concentration. Although there is at present no explanation for this observation, it is interesting to note that the increase in phospholipids and cholesterol was considerably greater than one would expect or calculate from the bile lipids entering the blood. The increase in Lp-X in cholestasis corresponds to that of refluxed bile lipids. The unexpected high increase in phospholipid and cholesterol, in relation to the refluxed bile lipids under experimental cholestasis, has been previously demonstrated.[36]

Structure of Lp-X

According to electron microscopic investigations, Lp-X appears after negative staining as a round particle with an average diameter of 300 to 700 A and a strong tendency to aggregate and form discs.[37] (See Figure 1.) This so-called rouleau formation is not typical for Lp-X. Similar lamellar structures were described in artificial phospholipid-protein and phospholipid-apolipoprotein complexes.[38] They have also been demonstrated in the plasma of cholesterol-fed guinea pigs[39] and in the LDL and HDL fractions of patients suffering from LCAT deficiency.[40] Since these formations have not been found inside the liver cells of mice after ligation of the common bile duct but in the space of Disse, they probably originate in the plasma.[41]

The structure of the wall of the electron microscopically demonstrated lipoprotein seems to be a bilayer. This and the fact that water-soluble phosphotungstic acid and succinic anhydride are able to penetrate into the particle have led to the conclusion that Lp-X represents a vesicle.[42] Small-angle X-ray diffraction studies have led to the same conclusion. The polar-head groups of phospholipids are located outside at a distance of 45 Å from each other. Fatty acids and cholesterol are on the inside of the membrane directed toward the contents of the vesicle. SH groups are not detectable on the surface using ferritin. Addition of ammonium molybdate, which destroys the particles, and the immunological proof for apoproteins indicate the presence of proteins on the surface. Since phospholipids are also located on the surface of the particle, it is easily destroyed by phospholipases.[43]

The exact location of ablumin is not yet known. Among the proteins so far identified on the surface of Lp-X are C-I and apo D. The content of apo A as reported by Picard and Veissiere[44] is probably due to contamination of the anti-α-lipoprotein serum used by these investigators with anti-D. In polyacrylamide gel electrophoresis, C-I, C-II, and C-III in its polymorphic forms are easily detected (Figure 1).

Lp-X as a Substrate for the Lecithinolesterol Acyltransferase

A lipoprotein particle, very similar to Lp-X, is usually found in the LDL fraction of patients suffering from a very rare, familial metabolic disease in which the LCAT-enzyme is not present.[45,46] This enzyme normally catalyzes esterification of plasma cholesterol. The presence of cholestatic conditions or disturbances of liver function have not yet been unequivocally proven in these patients. On the other hand, LCAT is produced by the liver;[47] therefore, severe liver disease usually causes a decrease in the activity of this enzyme in plasma.[48]

In cholestasis there are frequently normal or even increased activities of this enzyme. Therefore a deficiency of LCAT can be excluded as the primary cause for the occurrence of Lp-X in cholestasis. Theoretically, Lp-X should be a good substrate for this enzyme since it has a cholesterol/phospholipid ratio very similar to HDL and its cholesterol is almost exclusively present in unesterified form. Incubation with normal serum leads, according to Patsch et al.[49] and Ritland and Gjone,[50] to a diminution of the added Lp-X. This has been demonstrated using quantification of Lp-X in its typical position either after electrophoresis, presuming an unchanged mobility for Lp-X or in the elution pattern after zonal centrifugation. From these results, the conclusion was drawn that Lp-X is degraded by LCAT.

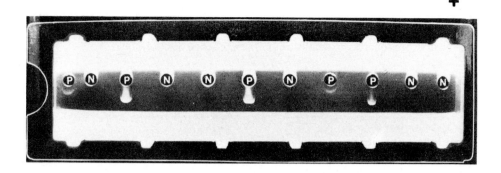

FIGURE 6. The Lp-X test performed in the Rapidophor system. P = Lp-X positive serum, N = Lp-X negative serum. Lp-X is visualized in its typical position after electrophoretic separation by polyanion precipitation (0.1 M MgCl$_2$, 0.15% sodiumheparin, 1.3% NaCl).

Other investigators have come to opposite conclusions. Not only does Lp-X not serve as substrate for LCAT,[48] but addition of Lp-X to normal serum inhibits initial esterification rate as indicated by measurements of the formation of cholesterylesters using the double enzymatic method for cholesterol measurement. According to Morriset,[51] only rather fluid molecules like HDL can serve as substrates for LCAT. Current ideas about the structure of lipoproteins imply localization of protein, phospholipids, and cholesterol on the outside of the molecule, cholesterylesters and triglycerides on the inside. To enable the enzyme to interact with it, the substrate should exhibit a certain mobility. This is the case in HDL already at 25°C. Lp-X, however, is a very rigid particle and does not show any increase in the mobility of its lipids during heating, up to 60°C.

Clinical Importance of the LP-X Test

The fact that Lp-X never occurs in the sera of patients not suffering from cholestasis (except LCAT deficiency) made it reasonable to assume that a very specific test for detection of cholestasis could be devised. These expectations have been completely fulfilled. For this purpose, series of well-devised studies have been instituted in recent years in various countries, independent of each other[52-58] in order to establish the clinical and chemical accuracy of the Lp-X test. A test system based on polyanion precipitation is now available for general use (Rapidophor All-in, Immuno, Vienna, Austria)[25] (Figure 6). Results of these studies unanimously and clearly show the high accuracy of this test for demonstration or exclusion of cholestasis when a clear anatomical finding is employed as the criterion. However, it is not possible to distinguish between intra-and extrahepatic cholestasis on the basis of Lp-X demonstration.

The only study not agreeing with the other studies,[59] used an immunological procedure for detection of Lp-X. The antiserum, however, was open to criticism, since it had been absorbed with normal serum. Up to now, all investigators agree that the immunological reaction of Lp-X with antiserum is mainly due to the presence of C peptides. Since these peptides are also present in normal serum, the major antibody to the Lp-X in the antiserum had been removed by absorption. The immunological reaction of Lp-X with that antiserum must have been based on the existence of a specific antigenic determinant of Lp-X, not present in whole serum. Recently, there have been indications that this specific determinant exists. It seems to be secreted in the bile of certain individuals and is by no means present on every Lp-X molecule. In the case of the existence of this specific determinant, double immunodiffusion would be sufficient for the detection of Lp-X.[60] In addition, a combination of two unspecific parameters like bilirubin and alkaline phosphatase which are used as criteria for cholestasis do not yield more specific information on the existence of cholestasis.

Animal Models for the Investigation of the Synthesis of Lp-X

Up to now, successful production of Lp-X has been reported in mice, rats, and dogs.[41,61,62] It occurs most often in the LDL fraction, and its protein composition is similar to that of human Lp-X.

According to the literature, it is not possible to produce Lp-X in swine;[63] however, the authors have demonstrated the presence of Lp-X in a mixture of human bile and pig serum and one of pig bile and human serum. It is, thus, clear that the pig possesses the necessary humoral conditions for Lp-X formation. This has been proven by successful incubation of bile and serum from the pig. Production of Lp-X in the mini-pig by ligation of the common bile duct was also successfully attempted. Lp-X was detectable in the serum for 7 days after ligation. Following implantation of the common bile duct into the vena renalis dextra, Lp-X appears after only 3 days. The reason for its later appearance after ligation may be a special anatomical condition or the high bile-salt content of pig bile. It is well known that pigs almost never have gallstones, despite the fact that they are on purpose fat and fertile.

Methods for Quantification of Lp-X

During the last 3 years, five methods for quantification of Lp-X have been developed. The first one relies on electrophoretic isolation of Lp-X and subsequent phosphorus determination.[64] A piece of gel on the cathodal site of electrophoresis, where Lp-X is presumably located, is punched out, and the total phosphorus content of that piece of gel is determined which then allows one to draw conclusions about the Lp-X concentration of the serum under investigation. The chief disadvantages of this method are the complicated phosphorus determination and the danger of not punching out all the Lp-X since one only knows its presumed location in electrophoresis. This method results in reproducibility and sensitivity that is too low.

In the second method,[65] one determines the fraction in Lp-X of radioactively labeled cholesterol, which has been previously added to the serum. Equilibration time is 4 hr. Lp-X is then electrophoretically separated and visualized by polyanion precipitation. In this method, it is not necessary to apply exact amounts of serum since only relative amounts are determined. The disadvantage of this method lies in the necessity to work with radioactive substances and its high consumption of time and labor.

The third method[66] consists of quantitative immunoelectrophoresis using anti-C-serum. The apo B-apo C complexes, like VLDL, LDL$_1$, and part of LDL$_2$, are removed by immunoprecipitation with anti-B either in a test tube or in an antiserum-containing agarose strip. If one uses the agarose strip, it should be placed so that the serum under investigation has to cross over it before it can reach the anti-C-containing agarose. This method exhibits a very high sensitivity, and only minute amounts of serum are necessary. The main disadvantage is the necessity of anti-C-serum and a stable standard, both of which are not commercially available. Since the unassociated Lp-C, present in plasma is also measured, the amount of Lp-X is easily overestimated.

In the elution patterns of the zonal centrifuge, the LDL fraction of patients suffering from cholestasis displays not only one peak, like normal LDL, but two. One of these represents Lp-X. The fourth method relies on quantitative isolation of this peak.[49] This method requires an amount of serum which is not feasible for clinical chemical investigations and very expensive tools which are restricted to a few research laboratories. In addition, the isolated lipoprotein fraction has to be analyzed and characterized each time to exclude the presence of lipoproteins other than Lp-X.

The fifth method uses densitometric scanning of Lp-X precipitated after electrophoretic separation[67] (Figure 6). The advantages of the method lie in its high sensitivity and specificity. The necessary standardization of the system is easily achieved by using filters. This method is the fastest method of all. It is necessary, however, to apply exact amounts and to use an expensive densitometer.

Catabolism of Lp-X

Using the method of quantification by scanning the Lp-X band, it was possible to follow the concentration of Lp-X in the serum of animals and to determine its biological half life.[62] Injection of Lp-X in homologous control animals, upon which no operation was performed, resulted in biological half lives of 10 hr for the rat and 37 hr for the dog[62] — data which have also been described for normal plasma lipoproteins. Lp-X appears within 20 hr after ligation of the common bile duct in the plasma of most animals, except the pig. Previous cholecystectomy accelerates its appearance.[65] A period of marked increase, which is followed by a decrease of lipoprotein, is detectable on the cathodal site. The decrease does not imply that catabolism of Lp-X exceeds production. As seen previously, bile salts in increased concentrations may cause a change in the electrophoretic mobility of Lp-X. This phenomenon may also be produced by free fatty acids.[68] Incubation of Lp-X with postheparin plasma leads to a gradual decrease of lipoprotein migrating toward the cathode, followed by its disappearance. One can reconstitute completely the Lp-X concentration at its usual location on the cathode by adding albumin. Incubation of postheparin plasma without addition of Lp-X at 37°C for 3 hr causes an immediate change in the mobility of the added Lp-X. In this case also, addition of albumin leads to reappearance of Lp-X at its usual location. Studies concerning degradation of Lp-X in postheparin plasma should be interpreted very carefully.

It is not clear which organs are responsible for the true catabolism of Lp-X. Lp-X-like particles were detected within Kupffer cells after experimental cholestasis.[41] Therefore, it is reasonable to assume participation of the RES.

REGULATION OF CHOLESTEROL SYNTHESIS IN CHOLESTASIS: POSSIBLE IMPORTANCE OF LP-X

Increased hepatic cholesterol synthesis in cholestasis can be regarded as a well-established fact. Similarly it is beyond any doubt that the increased free cholesterol is transported in Lp-X, the concentration of which is, therefore, responsible for the degree of hypercholesterolemia in cholestasis.

The mechanisms responsible for the increased hepatic cholesterol synthesis are not yet completely elucidated. Since intestinal lipoproteins are able to exhibit negative feedback regulation on hepatic cholesterol synthesis,[69] it was tempting to assume that lack of intestinal lipoproteins due to impaired fat absorption in the case of cholestasis might be responsible for the defective feedback regulation.[70] This situation occurs in the rat with bile fistula where the defect can easily be overcome by infusion of physiological amounts of intestinal lipoproteins. This, however, is not possible in the rat after experimental biliary obstruction. In this case, more than twice the physiological amount of cholesterol transported in intestinal lipoproteins has to be infused during a 24-hr period in order to achieve feedback regulation.[71] Since Lp-X is responsible for a high cholesterol level in patients suffering from cholestasis, it would be interesting to know whether it is capable of feedback inhibition. For this purpose, rats have been infused with unphysiologically high amounts of human Lp-X and LDL. Whereas human LDL did exhibit feedback inhibition, Lp-X did not.[72] This fact explains the persistent high cholesterol synthesis of the liver despite the high cholesterol levels in plasma. If newly synthesized cholesterol appears in LDL, feedback inhibition would probably work.

In order to study the effect of cholestasis on cholesterol synthesis in normal liver, only the median lobe was obstructed.[71] In this way the unobstructed lobe was also exposed to cholestatic conditions in the plasma. Cholesterol synthesis was increased but feedback regulation was intact. Therefore, there had to be an agent in cholestatic plasma which enhanced hepatic cholesterol synthesis. This agent could well have been Lp-X since it has been shown that HDL causes an increase of cholesterol synthesis in isolated rat hepatocytes. The cholesterol/phospholipid ratio in the Lp-X of all plasma lipoproteins is most similar to that of HDL.

Since increased cholesterol synthesis in the obstructed lobe cannot be explained by a defective feedback mechanism and stimulation by a certain cholesterol/phospholipid ratio in plasma alone, one has to look for other mechanisms. In this context it would be interesting to know in which form newly synthesized cholesterol is excreted in the plasma. Under normal conditions all cholesterol secreted into bile is derived from lipoprotein catabolism;[73] newly synthesized cholesterol leaves the liver in the form of VLDL most often and less often as LDL and HDL. To discover whether normal secretion mechanisms are maintained, one has to determine VLDL and LDL turnover in cholestatic patients. If normal cholesterol secretion is not maintained, newly synthesized cholesterol would have to be used for synthesis of bile lipoprotein, which would then reach the plasma and be converted to Lp-X. Further studies regarding these aspects seem to be warranted.

HYPERTRIGLYCERIDEMIA IN LIVER DISEASE

Hypertriglyceridemia as a consequence of alcoholic liver damage is a well-known phenomenon.[74] The existence of hypertriglyceridemia due to nonalcoholic liver dysfunction has been recognized only in the last few years. Clinical studies have indicated that hypertriglyceridemia is frequently accompanied by severe cholestasis.[75-78] The fasting plasma of these patients does not contain chylomicrons. Concentrations of VLDL are inconsistent, increasing above normal or decreasing. These fluctuations, alone, are unlikely to be responsible for this form of hypertriglyceridemia since the protein-lipid composition of the VLDL fraction is normal.[79] The main increase in triglyceride concentration is found in the LDL_2 class. With this information, it is possible to isolate three different lipoprotein peaks using zonal ultracentrifugation:[77] 1. Lp-X, 2. normal β-lipoproteins, and 3. triglyceride-rich lipoproteins.

Lipoprotein electrophoresis in agar gel of the LDL_2 fraction reveals only a single band in the β-position,[79] despite the fact that this fraction contains three different lipoprotein classes (Figure 7). It is, however, possible to separate these classes by agar gel electrophoresis[79] (Figure 8). Lp-X migrates toward the anode, and the triglyceride-rich low-density lipoprotein, an abnormal lipoprotein, does not migrate and reacts immunologically with anti-C and -B. This fact, in conjunction with their typical electrophoretic mobility, makes it possible to detect this abnormal lipoprotein, already in the whole serum of patients suffering from cholestasis with hypertriglyceridemia, by immunoelectrophoresis against anti-B and -C. After Cohn fractionation it is found in fractions I to III and can be isolated by immunoabsorption with anti-C due to its content of apo C[79] (Figure 8). Recently it has been isolated by chromatography on hydroxyapatite and Biogel® A 50 M.[80] Like Lp-X, C-I seems to be the only C peptide present on the surface; C-II and C-III are apparently not accessible to the antibody.

FIGURE 7. Lipo-
protein electrophore-
sis of the LDL frac-
tion of a patient
suffering from ob-
structive jaundice and
hypertriglyceridemia.
The LDL fraction
looks rather homoge-
neous despite the fact
that it contains three
different lipoprotein
fractions: 1. Lp-X, 2.
normal β-lipoproteins,
and 3. β_2-lipoproteins.

It also seems to contain apo D. Apo E was not demonstrable either in intact or delipidized lipoprotein. Using the family concept, it would be regarded as an association of
Lp-E, Lp-C, and Lp-D, present in the LDL$_2$ fraction.

Because of the electrophoretic mobility of the abnormal lipoprotein in agarose gel,
it has been designated β_2-lipoprotein (β_2-Lp).[79] The protein-lipid composition of β_2-Lp
is markedly different from that of normal β-lipoproteins despite the same hydrated
density. Having a similar protein content which is qualitatively very different, it has
more than a threefold amount of triglycerides and only a third of the cholesterol transported by normal LDL. The ratio of esterified to unesterified cholesterol is 1; whereas,
normal LDL shows a ratio of 5. The phospholipid content of β_2-Lp is only 2/3 of that
of normal LDL.

Normally there are only traces of an Lp-B- Lp-C complex in the LDL$_2$ fraction.[35]
In the case of the β_2-Lp, there is probably an abnormal increase of a lipoprotein,
normally occurring only in trace amounts but now accumulating because of overproduction, decreased catabolism, or a combination of both.

As mentioned earlier, there is no doubt about increased cholesterol synthesis in cholestasis. Since increased VLDL concentrations in fasting plasma are always accompanied by increased hepatic cholesterol synthesis,[73] it is conceivable that increased

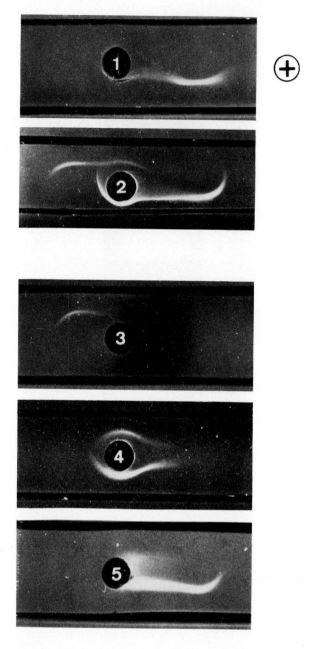

FIGURE 8.   Immunoelectrophoresis patterns in 1% agar gel
against anti-apo C (upper trough) and anti-apo B (lower trough):
1. control LDL, 2. LDL of a patient with acute cholestatic hepatitis, 3. isolated Lp-X from sample, 4. isolated β_2-lipoprotein
from sample, 5. isolated normal β-lipoproteins from sample.

VLDL synthesis may partially contribute to accumulation of β_2-Lp in the serum of
patients suffering from cholestasis. Dietary studies in patients with hypertriglyceridemia due to β_2-Lp indicate the important role of disturbed lipoprotein catabolism to
the appearance of hypertriglyceridemia.

A carbohydrate-rich diet, low in fat, usually increases VLDL synthesis and consequently plasma triglyceride concentration. This diet, however, caused in the authors'

patients a fall in the concentration of plasma triglycerides. If such a diet is given for 5 days, the triglyceride level usually drops about 30% and the ratio of VLDL triglycerides to d>1.006 triglycerides — a good marker for the concentration of β_2-Lp, normally 2 — is 0.53 in these patients. After taking the diet mentioned above, the ratio increases to 1.0. With a regular diet, it drops again and the plasma triglyceride concentration increases. Apparently, concentration of β_2-Lp occurring in the plasma of cholestatic patients is influenced by dietary fat intake, provided that production of chylomicrons is not severely impaired because of a lack of bile salts arriving in the intestine. Therefore, the main reason for excessive β_2-Lp seems to lie in the defective chylomicron catabolism. Before discussing this aspect, VLDL catabolism will be outlined briefly according to current knowledge.

The impact of lipolytic enzymes on VLDL causes formation of intermediate lipoproteins in a very short time. Further catabolism of these intermediate lipoproteins is a more time-consuming process. Mechanisms leading to formation of the LDL originating from the intermediate lipoproteins are not yet known. Most probably, a triglyceride lipase, stemming from the liver, and active in the circulation of the liver, plays an important role. This triglyceride lipase does not need cofactors like C-I or -II for its activity. It is present in postheparin plasma and cannot be inhibited by high salt concentrations or protamine sulfate. Therefore it has been designated protamine-insensitive lipase.

The intermediate lipoproteins stemming from VLDL catabolism belong to the LDL_1 fraction (d 1.006 to 1.019 g/ml). Their protein-lipid composition resembles that of β_2-lipoprotein. Although they have the same phospholipid and triglyceride content qualitatively, their main difference consists in a protein content in β_2-Lp twice as high as that of the LDL_1 fraction, mainly on account of the cholesterylesters which increase its hydrated density. It is likely that there are similar intermediate lipoproteins stemming from chylomicron catabolism which may be related to, or identical with, the β_2-Lp. The ideas about chylomicron catabolism are mainly derived from studies using the rat as model.[81]

A two-step catabolism has been observed in rats. In the first step, chylomicrons are deprived of the major part of their triglycerides by the action of lipolytic enzymes which are activated by C peptides. This enzymatic action occurs on the surface of the capillary endothelium and maybe in the circulation as well. In the second step, the major part of the resulting cholesterol- and protein-rich remnants is taken up by the liver shortly after hydrolysis of the cholesterylesters on the surface of the liver cell. Binding and incorporation are probably mediated by receptors.;[81] Chylomicron catabolism in man is unlikely to happen this way. Remnants circulate in plasma for a much longer time, are probably degraded by lipases to LDL like the intermediate lipoprotein stemming from VLDL catabolism and do not reach the liver in appreciable amounts.

The above-mentioned dietary study indicates that β_2-Lp can be regarded as a product of the catabolism of chylomicrons. According to the authors' experiments, postheparin lipolytic activity (PHLA) of patients with hypertriglyceridemia due to β_2-Lp is almost completely inhibited by protamine. This is an indication that the lipoprotein lipase from the liver is markedly decreased in these patients.

The biological function of this enzyme has not yet been elucidated. Unlike C-activated lipoprotein lipases, it is not active in postprandial plasma.[82] This may indicate that it plays a physiological role mainly on the cell surface. According to Assmann et al.,[83] it is localized in the outer membrane of the hepatocyte. The fact that it does not need C peptides for its activation makes it especially suited for degradation of triglyceride-rich lipoproteins, relatively poor in apo C. It is conceivable that in the case of hypertriglyceridemia due to β_2-Lp, intermediate lipoproteins are recognized by specific receptors and bound to the liver cell, where most of the cholesterylesters are hydrolyzed. Due to a lack of this lipase, triglycerides cannot be removed and further catabolism to normal LDL is blocked.

It seems clear that C peptides can be easily exchanged between chylomicrons or VLDL and HDL. The authors' investigations, discussed above, show clearly that during formation of Lp-X, C peptides are bound and accumulate in the newly formed particle. Sine β_2-Lp occurs in appreciable concentrations almost exclusively in the plasma of patients suffering from cholestasis, it is imaginable that it represents a catabolic product of VLDL or chylomicrons almost devoid of apo C which are therefore catabolized in an abnormal way. This product could not be further converted to LDL by the hepatic triglyceride lipase, which is not present in sufficient amount. This theory is supported by the fact that hypertriglyceridemia due to β_2-Lp is already found in patients not having severe liver dysfunction. On the other hand, the effect of bile salts, which increase in the plasma of cholestatic patients, may cause impaired degradation of the triglyceride-rich lipoproteins by hepatic as well as extrahepatic lipoprotein lipases.

In conclusion, hypertriglyceridemia in cholestatic patients is due to accumulation of triglyceride-rich low-density lipoprotein, poor in cholesterol. This lipoprotein consists of an association of Lp-B and Lp-C, normally present only in trace amounts in the LDL$_2$ fraction, and accumulates as a consequence of the deficiency in the protamine-sensitive lipoprotein lipase. The LDL fraction of these patients is a good example for the heterogeneity of lipoprotein fractions since they have in common only physical properties like density and electric charge.

CHANGES OF THE VLDL AND HDL FRACTION OCCURRING IN LIVER DISEASE

Besides the already mentioned abnormalities of the LDL fraction, changes may also be found in the other density classes.[84] Using the analytical ultracentrifuge, one can demonstrate diminution of the HDL fraction;[15] lipoprotein electrophoresis frequently shows only one broad band in the β-position with the pre-β- and α-bands lacking[84] (Figure 9). Although this electrophoretic phenomenon is predominantly found in liver disease, it has no diagnostic significance. Absence of these lipoprotein bands in electrophoresis does not necessarily mean that the corresponding lipoproteins are not present in the plasma.

Usually a VLDL fraction can be isolated by preparative ultracentrifugation at d 1.006 g/ml. The VLDL fraction may even increase. Isolated VLDL exhibit not pre-β-, but β-mobility, in lipoprotein electrophoresis (Figure 10). There is probably no universal reason for altered electrophoretic mobility of lipoprotein fractions in liver disease. Incubation of normal VLDL with artificially produced Lp-X from albumin and bile leads to an uptake of C peptides by these Lp-X-like particles which then become true Lp-X. The VLDL lose their apo C and develop β-mobility. In the case of cholestasis this fact may contribute to the altered mobility of VLDL[85] (Figure 10).

The missing α-lipoprotein band is usually accompanied by a decrease in HDL concentration, about 75% on the average. The combined concentration of both A-I and A-II in plasma is not found to be decreased that much; therefore, one has to assume that they are present in the d 1.21 infranatant. The immunoelectrophoretic pattern against anti-α-lipoprotein serum shows in the corresponding positions strong precipitin lines, which indicate the presence of at least two different protein molecules[84] (Figure 11A and 11B). Studies based on double immunodiffusion have shown that the following lipoprotein families are present in the HDL fraction:[85,86] Lp-A consisting of A-I and A-II with all lipids; Lp-A-I, which has only A-I as its protein moiety; Lp-D in

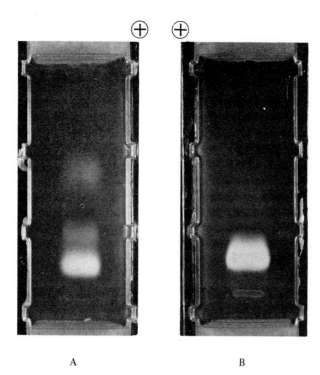

A   B

FIGURE 9.   Lipoprotein electrophoresis in 1% agarose gel
(Lipidophor-System, Immuno, Vienna) of: A. Normal serum B.
Serum of a patient suffering from cholestasis due to acute hepatitis. Note the absence of the pre-β- and α-bands. Bands are visualized by polyanion precipitation.

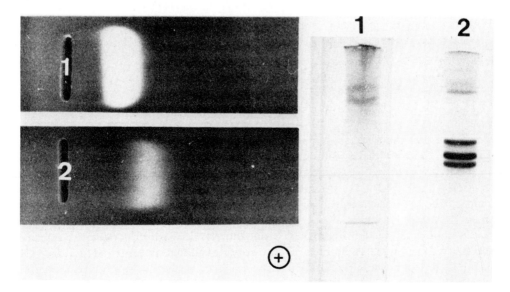

FIGURE 10.   Comparison of a VLDL fraction from a patient suffering from liver disease (1) with normal
VLDL (2). The patient's VLDL do not exhibit pre-β-mobility but rather β-mobility as demonstrated by
polyanion precipitation after electrophoretic separation in 1% agarose. The C-II and C-III peptides are
completely lacking as shown in 7% PAG at pH 8.6; the protein bands present have not yet been identified.
Apo B does not enter the gel. Normal VLDL shows a strong concentration of C-II and C-III peptides. C-
I cannot be identified by this technique.

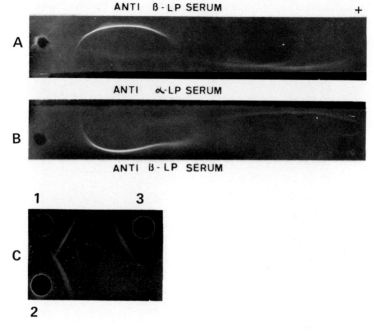

FIGURE 11.   Immunoelectrophoretic comparison of: A. serum of a patient suffering from liver disease and B. normal serum. C. A demonstration
of Lp-A-II in the HDL fraction of the patient. Despite the fact that there
is no α-lipoprotein band visible in lipoprotein electrophoresis (Figure 9),
strong precipitin lines against anti-α-lipo protein serum can be demonstrated. This line, however, is split unlike normal serum (B). Because to a
lack of pre-β-lipoproteins, no further elongation of the precipitin line
against anti-β-lipoprotein serum may be observed in the patient's serum
(A). The isolated HDL fraction of the paient (C), present in the center well,
reacts immunologically with anti-A-I (1), anti-A-II (2), and anti-D (3). Reactions against anti-A-I and anti-A-II show a pattern of complete noni-
dentity, indicating the presence of Lp-A-I and Lp-A-II in the HDL fraction
of the patient. Normal HDL may also contain Lp-A-I, but never Lp A-II.

free form and associated with Lp-A and -C; and Lp-C in free form and associated
with Lp-A and -D.

In liver disease the HDL fraction hardly contains any cholesterol or triglycerides but
does contain phospholipids. Immunological studies have revealed the presence of Lp-
A-II which seems to be the major protein-lipid association of that HDL fraction (Figure 11C). There is no Lp-C detectable either in association or free form. The other
lipoprotein complexes described for HDL are present in very small concentrations and
contain only trace amounts of cholesterol and triglycerides. In addition, there has been
isolated another abnormal lipoprotein from the HDL fraction of patients suffering
from severe, prolonged cholestasis. It has been designated as Lp-E, having only the
arginine-rich peptide as its apoprotein.[91] Interestingly enough, it exhibits rouleau formations like Lp-X.

Normal distribution of the apoproteins of the HDL fraction, in the case of liver
disease, is apparently disturbed in the following manner: there is a dissociation of A-I
and A-II and the single apoproteins are apparently not able to bind neutral lipids. The
amounts of apoprotein A-I and A-II in plasma may be diminished but certainly not
as strongly as one would assume by estimating only the HDL fraction or the α-lipoprotein band.

Since A-I and A-II cannot bind neutral lipids they are hardly detectable in lipoprotein electrophoresis. Lack of neutral lipids may also cause the deficiency of HDL in Lp-C. Another condition which also shows the existence of Lp-A-II and a deficiency of Lp-C in the HDL fraction is the so-called Tangier disease, a familiar Lp-A deficiency. In this case, cholesterylesters are stored in many tissues of the body. The current hypothesis is that lack of high-density lipoproteins that serve as a reservoir for the Lp-C produced during chylomicron and VLDL catabolism leads to abnormal lipoproteins, which are then stored. In this disease, a true deficiency in A-I and a strong diminution of A-II are observed, along with a series of resulting changes of the lipoprotein system.[87]

To clarify the nature of the structural abnormality of HDL in liver disease, much work has to be done. The concentrations of A-I and A-II in the plasma of these patients and in the HDL fraction as well should be determined to find out whether these changes should be regarded as a kind of Tangier disease, secondary to liver damage. Lp-A-II should be isolated and characterized, especially regarding its protein-lipid composition. Since Lp-A-II is found in the HDL fraction and not in the d 1.21 infranatant and most of the A-I is found in the d 1.21 infranatant, it seems likely that the primary defect in binding capacity for neutral lipids is found in the A-I peptide. For this purpose it is especially important to isolate the A-I peptide from the d 1.21 infranatant in order to achieve complete characterization.

In the case of cholestasis, the HDL fraction is almost devoid of apo C, which is presumably bound by Lp-X. If a lack of apo C could cause the described changes in the HDL fraction, Lp-C would have to play a central role regarding the stability of the lipoprotein associations present in HDL. Since up to 60% of normal HDL consists of Lp-A, i.e., a lipoprotein having only A-I and A-II as its protein moiety, any structural importance of Lp-C is unlikely to exist.

Changes of the electrophoretic mobility of VLDL in liver disease without cholestasis are difficult to understand. As a possible explanation, it was proposed earlier: either the presence of structurally impaired apo A in VLDL or a complete lack of apo A in the VLDL fraction, which usually contains apo A in trace amounts. Incubation with normal HDL is followed by complete restoral of the pre-β-mobility of this VLDL fraction. This may be caused by transfer of apo A and possibly Lp-C.

The exact location where apo C is produced is not yet known. Nascent chylomicrons obtain apo C from the circulation; nascent VLDL do not contain apo C.[88] It is likely that it is made by the liver; therefore, severe liver disease may cause a lack of apo C and subsequently production of VLDL with changed mobility but normal size, which are measured by polyacrylamide gel electrophoresis and electron microscopy.[84]

Like in Tangier disease, absence of a reservoir for Lp-C may lead to disturbed catabolism of chylomicrons and VLDL. This fact, together with a deficiency of the hepatic triglyceride lipases, may cause the appearance of β_2-Lp.

Besides changes in the HDL fraction, Lp-X, β_2-Lp, and VLDL with β-mobility have been found in patients suffering from LCAT-deficiency.[45,46,89] These patients show diminished concentrations of apo B and A-I in their plasma.[89] The HDL fraction was found diminished; Lp-A-I and Lp-C were present. In addition, Lp-E can always be found and seems to be responsible for the electron microscopically observed rouleau formations of the HDL fraction of these patients. Lp-E has been isolated by zonal centrifugation[90] and column chromatography on Sepharose® 4B. In combination with other mechanisms, a deficiency of LCAT, secondary to liver disease, could cause the described lipoprotein abnormalities. It is certainly on very rare occasions the only cause.

The sole change in the plasma lipoprotein system, specific to a certain disease, is the occurrence of Lp-X. The other abnormal lipoproteins are probably due to a variety of different disease states.

It is possible that more refined methods, like quantitative immunoelectrophoresis, quantitative polyacrylamide gel electrophoresis, and radioimmunoassays, will allow detection and characterization of changes in the lipoprotein system with much higher sensitivity and accuracy leading to the possibility of drawing diagnostic conclusions. Apparently, the lipoprotein system represents a very sensitive marker for disturbances of the function of the liver.

REFERENCES

1. **Kunkel, H. G. and Slater, R. J.**, Lipoprotein patterns of serum obtained by zone electrophoresis, *J. Clin. Invest.*, 31, 677, 1952.
2. **Alaupović, P., Lee, D. M., and McConathy, W. J.**, Studies on the composition and structure of plasma lipoproteins; distribution of lipoprotein families in major density-classes of normal human plasma lipoproteins, *Biochim. Biophys. Acta*, 260, 689, 1972.
3. **Flint, A., Jr.**, Experimental researches into a new excretory function of the liver, consisting in the removal of cholesterine from the blood, and its discharge from the body in the form of stercorine, *Am. J. Med. Sci.*, 44, 305, 1862.
4. **Feigl, J.**, Neue Untersuchungen über akute gelbe Leberatrophie. III. Fette und Lipide des Blutes. Chemische Beitrage zur Kenntnis der Entwicklung und Charakteristik spezifischer Lipämien, *Biochem. Z.*, 86, 1, 1918.
5. **Rothschild, M. A. and Felsen, J.**, The cholesterol content of the blood in various hepatic conditions, *Arch. Intern. Med.*, 24, 520, 1919.
6. **Thannhauser, S. J. and Schaber, H.**, Über die Beziehung des Gleichgewichtes Cholesterin und Cholesterinester in Blut und Serum zur Leberfunktion, *Klin. Wochenschr.*, 5, 252, 1926.
7. **Epstein, E. Z.**, Cholesterol of the blood plasma in hepatic and biliary diseases, *Arch. Intern. Med.*, 50, 203, 1932.
8. **Fredrickson, D. S., Loud, A. V., Hinkelman, B. T., Schneider, H. S., and Frantz, J. D.**, The effect of ligation of the common bile duct on cholesterol synthesis in the rat, *J. Exp. Med.*, 99, 43, 1954.
9. **Kattermann, R. and Reimold, W. V.**, Leberschaden und Lipidstoffwechsel, *Acta Hepato Splenol.*, 17, 75, 1970.
10. **Dietschy, J. M. and Siperstein, M. D.**, Cholesterol synthesis by the gastrointestinal tract: localization and mechanism of control, *J. Clin. Invest.*, 44, 1311, 1965.
11. **Quarfordt, S. H., Oelschlager, H., Krigbaum, W. R., Jakobi, L., and Davis, R.**, Effect of biliary obstruction on canine plasma and biliary lipids, *Lipids*, 8, 522, 1973.
12. **Ahrens, E. H. and Kunkel, H. G.**, The relationship between serum lipids and skin xanthomata in 18 patients with primary biliary cirrhosis, *J. Clin. Invest.*, 28, 1565, 1949.
13. **Friedman, M. S., Byers, O., and Roseman, R. H.**, Lipogenic hypercholesterolemia. A guide for reorientation in the consideration of lipid-cholesterol relationships, *Arch. Intern. Med.*, 116, 807, 1965.
14. **Kunkel, H. G. and Ahrens, E. H., Jr.**, The relationship between serum lipids and the electrophoretic pattern, with particular reference to patients with primary biliary cirrhosis. *J. Clin. Invest.*, 28, 1575, 1949.
15. **Gofman, J.**, The serum lipoprotein transport system in health, metabolic disorders, atherosclerosis and coronary artery diseases, *Plasma*, 2, 484, 1954.
16. **Furman, R. H., Conrad, L. L., and Howard, R. P.**, A serum lipoprotein pattern characteristic of biliary obstruction, with some comments on "jaundice due to methyltestosterone," *Circulation*, 10, 586, 1954.
17. **Havel, R. J., Eder, H. A., and Bragdon, J. H.**, The distribution and chemical composition of ultra-centrifugally separated lipoproteins in human serum, *J. Clin. Invest.*, 34, 1345, 1955.
18. **Furman, R. H. and Conrad, L. L.**, Ultracentrifugal characterization of the lipoprotein spectrum in obstructive jaundice: studies of serum lipid relationships in intra- and extrahepatic biliary obstruction, *J. Clin. Invest.*, 36, 713, 1957.
19. **Lindgren, F. T. and Nichols, A. V.**, Structure and function of human serum lipoproteins, in *The Plasma Proteins*, Vol. 2, Putnam, F., Ed., Academic Press, New York, 1960, 1.

20. Russ, E. M., Raymunt, J., and Barr, D. P., Lipoproteins in primary biliary cirrhosis, *J. Clin. Invest.*, 35, 133, 1956.

21. Switzer, S., Plasma lipoproteins in liver disease. I. Immunologically distinct low-density lipoproteins in patients with biliary obstruction. *J. Clin. Invest.*, 46, 1855, 1967.

22. Fredrickson, D. S., Levy, R. I., and Lees, R. S., Fat transport in lipoproteins — an integrated approach to mechanisms and disorders, *N. Engl. J. Med.*, 276, 273, 1967.

23. Burstein, A. and Caroli, J., Isolement et étude des lipoprotéines sériques anormales au cours des ictères par rétention après floculation par le polyvinyl-pyrrolidone. *Rev. Fr. Etud. Clin. Biol.*, 13, 387, 1968.

24. Seidel, D., Alaupović, P., Furman, R. H., and McConathy, W. J., A lipoprotein characterizing obstructive jaundice. II. Isolation and partial characterization of the protein moieties of low density lipoproteins, *J. Clin. Invest.*, 49, 2396, 1970.

25. Wieland, H. and Seidel, D., Eine neue und vereinfachte Methode zum Nachweis des Lp-X, eines cholestasespezifischen Lipoproteins, *Dtsch. Med. Wochenschr.*, 98, 1474, 1973.

26. Seidel, D., Büff, H. K., Fauser, U., and Bleyl, K., On the metabolism of lipoprotein- X (LP-X). *Clin. Chim. Acta*, 66, 195, 1976.

27. Seidel, D. and Wieland, H.,unpublished results, Heidelberg, W. Germany, 1977.

28. Brocklehurst, D., Lahte, G. H., and Aparicio, S. R., Serum alkaline phosphatase, nucleotide pyrophosphatase, 5′-nucleotidase and lipoprotein-X in cholestasis, *Clin. Chim. Acta*, 67, 269, 1976.

29. Neurath, R. A., Prince A. M., and Lippin, A., Hepatitis B antigen: antigenic sites related to human serum proteins revealed by affinity chromatography, *Proc. Natl. Acad. Sci.,U.S.A.*, 71, 2663, 1974.

30. Manzato, E., Fellin, R., Baggio, G., Neubeck, W., and Seidel, D., Formation of lipoprotein-X: its relationship to bile compounds, *J. Clin. Invest.*, 57, 1248, 1976.

31. Quarfordt, S. H., Oelschlager, H. U., and Krigbaum, W. R., Liquid crystalline lipid in the plasma of humans with biliary obstruction, *J. Clin. Invest.*, 51, 1979, 1972.

32. Picard, J., Veissière, F., and Bereziat, G., Composition en acides gras des phospholipides dans les lipoproteins sériques anormales de la cholestase, *Clin. Chim. Acta*, 36, 247, 1971.

33. Nalbone, G., Lafont, H., Domingo, N., Lairon, D., Pautrat, G., and Hauton, J., Ultramicroscopic study of the bile lipoprotein complex, *Biochimie*, 55, 1503, 1973.

34. Seidel, D. and Baggio, G., Origin and metabolism of lipoprotein-X, paper presented to the Eur. Soc. Study Liver, Barcelona, Spain, September 11 to 13, 1975.

35. McConathy, W. J., Wieland, H., and Alaupović, P., Fractionation of Lipoprotein Density Classes on Immunosorbers. Evidence for the Lipoprotein Heterogeneity, paper presented at the 4th Int. Symp. Atherosclerosis, Tokyo, Japan, 1976, Abstr.

36. Byers, S. O. and Friedman, M., The relation of biliary retention of cholesterol, distension of the biliary tract, the shunting of bile of the vena cava, and removal of the gastro-intestinal tract to the hypercholesterolemia consequent of biliary obstruction, *J. Exp. Med.*, 95, 19, 1952.

37. Seidel, D., Agostini, A., and Müller, P., Structure of an abnormal plasmalipoprotein (LP-X) characterizing obstructive jaundice, *Biochim. Biophys. Acta*, 260, 146, 1972.

38. Hoff, H. F., Morrisett, J. D., and Gotto, A. M., Jr., Interaction of phosphatidylcholine and apoprotein-alanine: electronmicroscopic studies, *Biochim. Biophys. Acta*, 265, 471, 1972.

39. Sardet, C., Hansma, H., and Ostwald, R., Characterization of guinea pig plasma lipoproteins: the appearance of new lipoproteins in response to dietary cholesterol, *J. Lipid Res.*, 13, 624, 1972.

40. Uterman, G., Menzel, H. J., and Langer, K. H., On the polypeptide composition of an abnormal high-density lipoprotein (Lp-E) occurring in LCAT-deficient plasma, *FEBS Lett.*, 45, 29, 1974.

41. Stein, O., Alkan, M. D., and Stein, Y., Obstructive jaundice lipoprotein particles studied in ultrathin sections of livers of bile duct ligated mice, *Lab. Invest.*, 29, 176, 1974.

42. Jonas, A. and Seidel, D., Properties of the abnormal human plasma lipoprotein (LP-X) characteristic of cholestasis after chemical modification with succinic anhydride, *Arch. Biochem. Biophys.*, 163, 200, 1974.

43. Seidel, D., Structure of lipoprotein-X, *Expo. Annu. Biochim. Med.*, 31, 2, 1972,

44. Picard, J. and Veissière, D., Separation des lipoproteines sériques anormales dans la cholestase, *C. R. Acad. Sci.* (Paris), 270, 1845, 1970.

45. Seidel, D., Gjone, E., Blomhoff, J. P., and Geisen, H. P.,Plasma Lipoproteins in Patients with Familial Plasma Lecithin: cholesterol Acyltransferase (LCAT) Deficiency — Studies on the Apolipoprotein Composition of Isolated Fractions with Identification of LP-X, paper presented at the Int. Symp. Lipid Metabolism, Obesity, and Diabetes Mellitus: Impact upon Atherosclerosis, Wiesbaden, W. Germany, April, 1972, Suppl. Series No. 4.

46. Torsvik, H., Berg, K., Magnani, H. N., McConathy, W. J., Alaupović, P., and Gjone, E., Identification of the abnormal cholestatic lipoprotein (LP-X) in familial lecithinolesterol acyltransferase deficiency. *FEBS Lett.*, 24, 165, 1973.

47. **Simon, J. B. and Boyer, J. L.,** Production of lecithin: cholesterol acyltransferase by the isolated perfused rat liver, *Biochim. Biophys. Acta,* 215, 549, 1971.
48. **Wengeler, H., Greten, H., and Seidel, D.,** Serum cholesterol esterification in liver disease. Combined determination of lecithinolesterol acyltransferase and lipoprotein-X, *Eur. J. Clin. Invest.,* 1, 372, 1971.
49. **Patsch, W., Patsch, J., and Sailer, S.,** In Vitro Degradation of LP-X by Normal Plasma, paper presented at the 10th Annu. Meet. Eur. Soc. Clin. Invest., 1976, abstr. 242.
50. **Ritland, S. and Gjone, E.,** Quantitative determination of LP-X in familial LCAT deficiency and during cholesterol esterification, *Clin. Chim. Acta,* 59, 109, 1975.
51. **Morrisett, J. D.,** personal communication.
52. **Fischer, M., Falkensommer, C., Barouch, G., Wuketich, S., Kronberger, O., and Schnach, H.,** Zur Diagnostik der Cholestase: Lipoprotein-X (LP-X). *Wien. Klin. Wochenschr.,* 87, 524, 1975.
53. **Mayr, K.,** Der Wert des Nachweises von Lipoprotein-X zur Feststellung einer Cholestase: ein Vergleich mit anderen klinisch-chemischen Untersuchungen, *Dtsch. Med. Wochenschr.,* 100, 2193, 1975.
54. **Mordasini, R. C., Berthold, S., Schlumpf, E., Keller, H., and Riva, G.,** Lipoprotein-X bei Hepatobiliaren Erkrankungen, *Schweiz. Med. Wochenschr.,* 105, 863, 1975.
55. **Poley, J. R., Alaupović, P., Seidel, D. and McConathy, W. J.,** Differential diagnosis between neonatal hepatitis and extrahepatic biliary arteria in infants: a new test, *Gastroenterology,* 58, 983, 1970.
56. **Prexl, H. J. and Petek, W.,** Die Bedeutung des Lipoprotein-X and der Serumcholinesterase in der präoperativen Diagnostik des Verschlussikterus, *Chirurgie,* 44, 310, 1973.
57. **Rittland, S., Blomhoff, J. P., Elgjo, K., and Gjone, E.,** Lipoprotein-X (LP-X) in liver disease, *Scand. J. Gastroenterol.,* 8, 155, 1973.
58. **Seidel, D., Schmitt, E. A., and Alaupović, P.,** An abnormal low density lipoprotein in obstructive jaundice. II. Significance in differential diagnosis of jaundice, *Dtsch. Med. Wochenschr.,* 15, 671, 1970.
59. **Vergani, C., Pietrogrande, M., Grondana, M. C., and Pizzolato, A.,** Studio do una lipoproteina anomala (LP-X) caratteristica della colestasi, *Clin. Med.,* 64, 1461, 1973.
60. **Petek, W., Kostner, G., and Holasek, A.,** Untersuchungen zur Methodik der immunologischen Lipoprotein-X (LP-X) Bestimmung, *Z. Klin. Chem.,* 11, 415, 1973.
61. **Müller, P., Fauser, U., Fellin, R., Wieland, H., and Seidel, D.,** Isolation and characterization of lipoprotein-X (LP-X) from canine, *FEBS Lett.,* 38, 53, 1973.
62. **Seidel, D., Büff, H. K., Fauser, U., and Bleyl, K.,** On the metabolism of lipoprotein-X (LP-X)., *Clin. Chim. Acta,* 66, 195, 1976.
63. **Danielson, B., Ekman, R., Johansson, B. G., and Petersson, B. G.,** Zonal ultracentrifugation of plasma lipoproteins from normal and cholestatic pigs, *Clin. Chim. Acta,* 65, 187, 1975.
64. **Magnani, H. N. and Alaupović, P.,** A method for the quantitative determination of the abnormal lipoprotein (LP-X) of obstructive jaundice, *Clin. Chim. Acta,* 38, 405, 1972.
65. **Rittland, S.,** Quantitative determination of the abnormal lipoprotein of cholestasis, LP-X, in Liver Disease, *Scand. J. Gastroenterol.,* 10, 5, 1975.
66. **Kostner, G. M., Petek, W.,,and Holasek, A.,** Immunochemical measurement of lipoprotein-X, *Clin. Chem.,* 20, 676, 1974.
67. **Neubeck, W. and Seidel, D.,** Direct method for quantitative determination of lipoprotein-X (LP-X)., *Clin. Chem.,* 21, 853, 1975.
68. **Seidel, D. and Baggio, G.,** unpublished results, Heidelberg, W. Germany, 1977.
69. **Bhattachiry, E. and Siperstein, M.,** Feedback control of cholesterol synthesis in man, *J. Clin. Invest.,* 42, 1613, 1963.
70. **Weis, H. J. and Dietschy, J. M.,** Presence of an intact cholesterol feedback mechanism in the liver in biliary stasis, *Gastroenterology,* 61, 77, 1971.
71. **Cooper, A. D. and Ockner, K.,** Studies of hepatic cholesterol synthesis in experimental acute biliary obstruction, *Gastroenterology,* 66, 586, 1974.
72. **Liersch, M., Baggio, G., Heuck, C. C., and Seidel, D.,** Effect of Lp-X on cholesterol biosynthesis in rat liver, 10th Annu. Meet. Eur. Soc. Clin. Invest. 1976, abstr. 121.
73. **Sodhi, H. S. and Kudchodkar, B. J.,** Correlating metabolism of plasma and tissue cholesterol with that of plasma-lipoproteins, *Lancet,* 1, 513, 1973.
74. **Baraona, E. and Lieber, L. S.,** Effects of chronic ethanol feeding on serum lipoprotein metabolism in the rat, *J. Clin. Invest.,* 49, 769, 1970.
75. **Alcindor, L. G., Infante, R., and Caroli, J.,** Plasma VLDL catabolism in cholestasis, paper presented at the 5th meet. Int. Assoc. Study Liver, Versailles, France, July 1972, abstr.
76. **Fellin, R. and Seidel, D.,** Behaviour of Serum Lipoproteins in Cholestasis, paper presented at the 1st Int. Symp. Cholestasis, Florence, Italy, June 1973.
77. **Klör, U., Ditschuneit, H. H., Rakow, D., and Ditschuneit, H.,** Further characterization of dyslipoproteinemia in hepatic disease, *Eur. J. Clin. Invest.,* 2, 291, 1972, abstr.

78. **Pearson, A. J. G.**, Triglycerides in Obstructive Liver Disease, paper presented at the 5th meet. Int. Assoc. Study Liver, Versailles, France, July 1972, abstr.

79. **Müller, P., Fellin, R., Lambrecht, J., Agostini, B., Wieland, H., Rost, W., and Seidel, D.**, Hypertriglyceridemia secondary to liver disease, *Europ. J. Clin. Invest.*, 4, 419, 1974.

80. **Kostner, G. M., Laggner, P., Prexl, J. H., and Holasek, A.**, Investigation of the abnormal low-density lipoproteins occurring in patients with obstructive jaundice, *Biochem. J.*, 157, 401, 1976.

81. **Eisenberg, S. and Levy, R. J.**, Lipoprotein metabolism, in *Advances in Lipid Research*, Vol. 13, Academic Press, New York, 1975, 1, 89.

82. **Wieland, H., Ganesan, D., McConathy, W. J., and Alaupovic, P.**, Lipoproteins and lipolytic activities in the postprandial state, 1977, in preparation.

83. **Assmann, G., Krauss, R. M., Fredrickson, D. S., and Levy, R. J.**, Characterization, subcellular localization and partial purification of a heparin released triglyceride lipase from rat liver, *J. Biol. Chem.*, 248, 1922, 1973.

84. **Seidel, D., Greten, H., Geisen, H. P., Wengeler, H., and Wieland, H.**, Further aspects on the characterization of high and very low density lipoproteins in patients with liver disease, *Eur. J. Clin. Invest.*, 2, 359, 1972.

85. **Wieland, H., Baggio, G., and Seidel, D.**, unpublished results, Heidelberg, W. Germany, 1976.

86. **McConathy, W. J. and Alaupovic, P.**, personal communication.

87. **Assmann, G., Fredrickson, D. S., Herbert, P., Forte, T., and Heinen, R.**, An A-II lipoprotein particle in Tangier disease, *Circulation*, 49, Suppl. 3, 259, 1974.

88. **Windmueller, H. G., Herbert, P. H., and Levy, R. J.**, Biosynthesis at lymph and plasma lipoprotein apoproteins by isolated perfused rat liver and intestine, *J. Lipid Res.*, 14, 215, 1973.

89. **McConathy, W. J., Alaupovic, P., Curry, M. D., Magnani, H. N., Torsvik, H., Berg, K., and Gjone, E.**, Identification of lipoprotein families in familial lecithin:cholesterol acyltransferase deficiency, *Biochim. Biophys. Acta*, 326, 406, 1973.

90. **Danielson, B., Ekman, R., and Petersson, B. G.**, An abnormal high-density-lipoprotein in cholestatic plasma isolated by zonal ultracentrifugation, *FEBS Lett.*, 50, 180, 1975.

91. **Uterman, G., Menzel, H. J., and Langer, K. H.**, On the polypeptide composition of an abnormal high-density lipoprotein (Lp-E) occurring in LCAT-deficient plasma, *FEBS Lett.*, 45, 29, 1974.

LIPOPROTEIN CHANGES IN RENAL DISEASES

Lena A. Lewis

INTRODUCTION

Renal diseases and their associated serum lipoprotein and lipid changes received the attention of the clinical chemist and clinician soon after techniques became available for evaluation of lipoproteins. Development of more microtechniques and greatly broadened knowledge concerning lipoprotein structure, regulation, and function has resulted in recent further study. The advent of the use of dialysis procedures for prolongation of life after renal failure and the application of renal transplantation in treatment of these patients has provided an entirely new perspective of the study of lipoproteins in relation to the kidney.

EARLY STUDIES AND THE NEPHROTIC SYNDROME

In 1952, Kunkle and Slater [1] reported the use of starch powder pressed into blocks as a support media during electrophoretic analysis of serum proteins. They cut the starch-block at fixed points after completion of the electrophoretic run and eluted the proteins and analyzed them for lipid and protein. The nephrotic pattern showed a highly concentrated lipoprotein fraction with mobility of α_2-β_1-n globulin which appeared to have two or three poorly resolved components. With spontaneous remission of disease and increase in serum albumin, the lipoprotein pattern became nearly normal.

The same year Swahn[2] described the separation of the serum lipoproteins on filter paper by electrophoresis and showed densitometric scans of the well-resolved α_1-, α_2- and −Sd β-lipoproteins after staining them with Sudan black. Patterns obtained on hyperlipemic sera showed greatly augmented levels of α_2- and β-lipoproteins. [3]Nikkila [4] extended the studies of lipoproteins using electrophoresis and filter paper as support media and reported chemical analyses of the resolved fractions. This study included analyses of the sera of two nephrotic patients, whose lipoprotein patterns showed greatly increased amounts of cholesterol in the α - and β-globulin lipoprotein fractions and a decrease in that of α-globulin mobility.

Early electrophoretic studies showed that speed of migration of the lipoproteins in some hyperlipemic sera was greater than that of normal lipoproteins. The important studies of Gordon [5] showed that altered mobilities could largely be explained by binding of free fatty acid (FFA) to the lipoprotein which would result in an accelerated migration rate. β-Lipoproteins were first saturated with FFA, after which the α-lipoprotein migration rate was altered. These observations provide an important aid in interpretation of lipoprotein electrophoretic patterns in the presence of greatly varying amounts of FFA, in many pathologic and extreme physiologic conditions. The importance of FFA in determining electrophoretic mobility of lipoproteins was confirmed and further studied by Laurell.[6]

Nys[7] in 1954 described two types of lipoprotein pattern in patients with active nephrosis. The first type showed a broad, intensely lipid-stained area, apparently consisting of two or three poorly resolved components migrating in the β-α_2-globulin area, This seems to correspond to the pattern obtained on starch-block by Kunkel and Slater. [1] The second type, in contrast, had well-resolved, highly concentrated fractions of α_2- and β-lipoprotein mobility; there was appreciable trailing from the point of application to the β-lipoprotein area, probably due to high concentration of neutral lipid. The degree of lipoprotein abnormality decreased with remission of the disease and increase of serum albumin levels.

Studies of nephrotic children's serum lipoproteins were made by Gittlin et al.[8] using ultracentrifugal techniques, and the disappearance of specific radioiodinated lipoproteins from their plasma was studied after the children were given tracer doses of the labeled lipoproteins. The exact electrophoretic properties and analyses of the lipoproteins which were labeled were determined. The nephrotic children showed greatly increased concentration of S_f 10 — 200, low-density β- lipoprotein. By use of combined ultracentrifugal and electrophoretic techniques, it was demonstrated that the fractional catabolic rate of catabolism of α- and of β-lipoprotein was somewhat elevated in nephrotic children with ascites, but only significant amounts of labeled proteins were excreted in the urine. Both normal subjects and nephrotic children showed significant conversion of S_f 10 — 200 iodinated proteins to S_f 3 — 10. Conversion was decreased in the nephrotic children.

Similarly, Raynaud et al.[9,9a] emphasized the correlation in nephrotic patients' serum between the decrease in albumin concentration and the increase in migration rate of the β-lipoprotein fraction. They noted that a decrease in concentration of serum α-lipoprotein was characteristic of all nephrotics. The change in electrophoretic mobility of the β-lipoprotein was comparable to that observed when free fatty acid levels were increased following heparin injection by post- heparin lipoprotein lipase activity or increased after eating due to increased "clearing factor" lipase concentration.

The type of plasma lipoprotein abnormality which occurs in nephrotic patients [10] was investigated in a group of 96 adult patients who had careful histologic and clinical evaluations of their renal status. All types of lipoprotein abnormality (using the nomenclature suggested by W.H.O.* except type I were observed, though types IIA, IIB and V were most frequent and observed in approximately similar numbers. No particular histologic diagnosis was associated with only one specific lipoprotein type. The exact frequency of abnormal lipoprotein patterns among nephrotic patients could not be stated, since the lipoprotein studies were only made when cholesterol or triglyceride or both were abnormal.

Until recently, vigorous treatment of the hyperlipidemia of the nephrotic syndrome was not pursued. It has, however, become more and more obvious, since improved treatment now available has effectively prolonged the lives of the nephrotics by 10 to 20 years, that an accelerated atherosclerosis occurs in an alarming proportion of these patients. Since one of the recognized factors associated with early atherosclerosis, hyperlipoproteinemia, is a characteristic finding in this syndrome, recognized methods for treating hyperlipidemia have been investigated in these patients.[13] Cholestyramine or tryptophan together with CPIB (chlorophenoxyisobutyrate) was found to be effective in controlling the hyperlipidemia in a number of patients. An appropriate lipid-lowering diet is also recommended.[12] Reversal of pre-β-hyperlipoproteinemia was partially achieved by heparin infusion.

LIPOPROTEIN PATTERNS IN RENAL DISEASE IN CHILDHOOD

While earlier studies of nephrotics included investigations of children, a general discussion of the problem in childhood has only relatively recently appeared.[14] As reported in earlier studies, the reviews of recent investigations of the nephrotic syndrome in childhood found that nephrotic children's lipoprotein patterns were characterized by increase in concentration of β- and/or preβ-lipoprotein and sometimes chylomicron

---

* In this chapter, the classification of hyperlipoproteinemic patterns as recommended by the W.H.O. publication of 1970[12] will be used.

also, so that the patterns could appear to have characteristics of type II, IV, or V (W.H.O. nomenclature) hyperlipoproteinemia.[12] Serum lipids and lipoproteins returned to normal when the proteinuria of the patient had been completely corrected.

Serum cholinesterase and low-density, i.e. β-lipoproteins were studied in nephrotic children in the acute and convalescent stages of the disease.[19] The greatly elevated levels of both components found in the acute phase decreased markedly with clinical improvement. It was suggested that a strong, previously reported, affinity of cholinesterase for β-lipoprotein may stabilize the molecules and retard their metabolism in the child.

Recently, investigators have made use of newer techniques in an attempt to understand changes that occur at cellular and subcellular levels in juvenile renal disease. Faulk et al.[20] conducted electronmicrosocopic and immunofluorescent studies on renal biopsies from 84 children representing eight diagnostic categories of kidney disease. The purity of β-lipoprotein antigen prepared by dextran sulfate precipitation and chromatography was checked by immunoelectrophoresis. Patients with the nephrotic syndrome were known to have abnormal plasma β-lipoprotein values. The glomeruli of five patients — two of whom had glomerulonephritis; one, membranous glomerulonephritis, one, hemolytic uremic syndrome; and one, lipoidal nephrosis — contained β-lipoprotein. The level of serum β-lipoprotein was not necessarily related to presence or absence of β-lipoprotein in the kidney, nor was the deposition of β-lipoprotein in renal glomeruli clearly associated with the clinical condition or biochemical findings on the patent.

LIPOPROTEIN CHANGES IN FAMILIAL LECITHIN: CHOLESTEROL ACYLTRANSFERASE DEFICIENCY WITH RENAL FAILURE

In the unusual, familial, genetically determined syndrome known as lecithin-olesterol acyltransferase deficiency (LCAT), among other characteristics demonstrated by the subjects, is proteinuria. In some of the subjects, renal involvement becomes a major problem.[22] Two members of the original family studied developed renal failure which necessitated renal transplantation. While some characteristics during the course of the disease were suggestive of the nephrotic syndrome, the pathologic changes and some clinical findings were unique to the LCAT-deficient subject. The sera of LCAT-deficient subjects had greater than normal concentrations of Lp-X, a lipoprotein characterized by high content of free cholesterol and of phospholipid. When the two subjects at the time of renal transplantation received transfusions from healthy donors of blood containing LCAT enzyme, concentrations of Lp-X in their plasma decreased. The studies showed that Lp-X could be a source of free cholesterol for the esterification reaction. The same conclusions was reached whether Lp-X was from the genetically LCAT-deficient subjects in good health or from the two subjects following renal transplantation.

In addition to the abnormal Lp-X which was present in the sera of the LCAT-deficient patients, whether they appeared to be healthy or were the two who developed renal failure, their serum lipoprotein patterns showed no pre-β- or α-lipoprotein. A high concentration of β-lipoprotein was present and there was trailing into the chylomicron position[23] when the serum lipoproteins were studied by paper electrophoresis.

P3 lipoprotein changes in patients during terminal renal failure, dialysis, and following renal transplantation

In terminal renal failure when urinary protein loss is minimal, serum lipoprotein and protein patterns are much less abnormal than in the protein- wasting stage.[24]

Serum α-lipoprotein levels were low in both undialyzed and dialyzed patients. Following renal transplantation when kidney function was good, serum α-lipoprotein levels were within the normal range.[24]

Bagdade et al.[25,26] have observed increased plasma pre-β-lipoprotein and triglyceride levels in 13 undialyzed and 26 dialyzed patients with nonnephrotic uremia. With very intensive efforts to regulate the uremia, lipoprotein abnormalities were minimized. Increased immunoreactive insulin in the serum of dialyzed patients and decreased postheparin lipoprotein lipase activity suggested that the high pre-β-lipoprotein and triglyceride levels were due to hepatic overproduction and decreased efficiency of catabolism.

Following renal transplantation serum pre-β-lipoprotein levels continued to be elevated in some patients. These elevated levels may partially result from the immunosuppressive drug therapy and also from the relatively high dietary carbohydrate intake of renal transplant patients. Study of serum lipids and lipoproteins of patients with non-renal-related disease who received prednisolone showed elevated triglyceride levels similar to those of renal transplant patients.[29]

Hypertriglyceridemia and hyper-pre-β-lipoproteinemia have been observed in patients with chronic renal failure who were not on a dialysis program,[30] as well as in those being dialyzed regularly.[32] The general nutritional condition of the patients and hormonal imbalances[26,27] associated with renal disease may contribute to the frequent occurrence of abnormal lipoprotein patterns in these patients.

Khanna et al.[37] found that of 25 patients with chronic dialysis and 125 postrenal transplant patients, approximately one third had normal serum lipoprotein and lipid levels. Of the two thirds showing abnormalities, as has also been observed by others,[27,29] hypertriglyceridemia and increased pre-β-lipoproteinemia occurred most frequently. The potential importance of the lipid-lipoprotein abnormality in these patients and the need for its regulation were emphasized, since atherosclerosis occurred much more frequently in these patients than in people of similar age without renal disease.

Methylprednisolone administered intravenously was found to result in production of aberrant serum lipoprotein fractions when used in treatment of human beings during threatened renal transplant rejection.[38] Paper electrophoretograms showed the marked changes during prolonged treatment with methylprednisolone by appearance of a lipid-stainable band between the normal pre-β-lipoprotein and α-lipoprotein position. These changes were further evaluated by analytical ultracentrifugation at d 1.21. A peak of flotation rate, $-S$ 10—20, not present in demonstrable concentration in normal serum, was found in the sera which had the unusual electrophoretic pattern. The $-S$ 10—20 component was of a density between that of high-density lipoprotein (HDL) and low-density lipoprotein (LDL) of normal serum. After withdrawal of methylprednisolone, the unusual lipoprotein fraction disappeared from the patients' sera, and the lipoprotein pattern returned to one typical of the patient.

SUMMARY AND GENERAL CONCLUSIONS

Hyperlipoproteinemia and hyperlipidemia are of frequent occurrence in the sera of patients with renal disease. Since the development of improved treatment procedures for the primary disease and prolongation of life by hemodialysis and renal transplantation, it is evident that atherosclerotic processes are becoming of primary importance as disabling and lethal factors to these patients. More adequate control of serum lipoprotein and lipid levels may greatly reduce one of the risk factors. As a guide to the efficacy of lipoprotein regulation, electrophoretic procedures can be of great help.

Table 1
ELECTROPHORETIC TECHNIQUES

Author and ref. no.	Year	Medium, special conditions	Buffer	Cell and special equipment	Electrical conditions	Time of electrophoresis (hr)	Temp	Special handling	Staining	Evaluation	Densitometry	Methodological error
Kunkle, H. G. and Slater, R. J.[1]	1952	Filter paper, starch-block	Barbital, pH 8.6, μ 0.1; improved o resolution with lower ionic strength buffer	Blocks of starch placed between glass plates; connected to electrode vessels with plastic sponges	25 — 60 mA 100 — 500 V	24	Cold room	—	Lipid and protein analysis of eluates of segments	Mobility followed by visual inspection for bilirubin of stained albumin	—	—
Swahn, B.[2,3]	1952 1953	Apparatus had advantage of two strips being able to be electrophoresed simultaneously; Whatman* #1 filter paper strip wetted and blotted to point when shiny surface turned dull.	Barbital, 0.05 μ.	Paper strips between two glass plates; strips extending into buffer vessels	3 — 4 mA 300 V	3 — 6	Strips dried at 60°C	Test made which demonstrated the very specific staining of lipid by Sudan black stain	Lipid stain Sudan black in 50% alcohol; protein stain, bromphenol blue	Dye eluted from cut areas of filter paper by means of acidified alcohol were evaluated also by visual inspection or scanning	Intensity of color in eluted fractions measured by photoelectric colorimeter	This was considered a preliminary report and dye binding-power of different lipid classes was not determined
Nikkila, E.[4]	1953	Munktel Swedish filter paper 20/150	Barbital, pH 8.6, υ 0.05	Durrum type cell	300V	4	—	Electrical endosmosis flow corrected by raising surface of cathode buffer slightly	Sudan black	Chemical lipid extraction ethanol:) ether 3:1	—	—
Gordon, R. S., Jr.[5]	1955	—	phosphate, pH 7.8, υ 0.16, sera diluted 2.7 times	Anumc2® portable electrophoresis, paper electrophoresis; Procedure of Swahn[2,3]	—	—	1 — 2°C	—	—	All results were compatible with idea that a β-globulin was temporarily acquiring higher mobility so that it was migrating as o-globulin or albumin; an additional component not originally separated from albumin increased its mobility to be faster than albumin; addition of eleate in vivo to serum caused accelerated mobility of β-globulin similar to that observed in serum of hyper-lipemic patients after injection of heparin and resulting release of FFA; association of FFA with β-lipo was reversible	—	—

Table 1 (continued)
ELECTROPHORETIC TECHNIQUES

Author and ref. no.	Year	Medium, special conditions	Buffer	Cell and special equipment	Electrical conditions	Time of electrophoresis (hr)	Temp	Special handling	Staining	Evaluation	Densitometry	Methodological error
Laurell, S. et al.[6]	1955	Whatman No. 1 filter paper 78 ×350mm	Michaelis acetate-veronal, pH 8.6	LKB paper electrophoresis equipment	120 V	12	—	1.7 mg disodium ethylene diamine-tetracetic acid (EDTA) was added per milliliter of serum	Sudan black[3]	Pattern showed a fast-migrating fraction of mobility between that of alb. and α-glob; a β-lipo fraction with mobility, between β-glob and α-glob; in this procedure, γ-glob migrated anodally ca. 20 — 25 min	—	—
Nys, A.[7]	1954	Whatman 1 filter paper 10×34 cm	Sodium veronal, pH 8.6, 0.05M	Sample applied near middle of strip (so that γglob always migrated to cathode)	6 — 8 mA	•	—	Two strips simultaneously; one for lipid, one for protein stain	Sudan black	Swahn'[?] direct photometric	—	—
Gitlin, D. et al.[8]	1958	Starch-block 4×30 cm zonal	Barbiturate, pH 8.6, γ/2 = 0.1	—	15V/cm	18	4°C	starch cut in segments	Protein eluted from starch	Read at 280 ml	—	—
Raynaud, R. et al.[9,?]	1954 1959	Paper zonal							Histochemical	—	—	—
Newmark, S. R. et al.[10]	1975	Paper from Lees and Hatch[?] as modified by Ellefson et al.,[?] polyacrylamide 7.5% and 3.75%; 2.0% spacer	Barbital, pH 8.6, EDTA 0.371 g/ℓ	Albumin added only to buffer wetting strips	—	—	—	—	Oil red 0	—	—	—
Ellefson, R. D. et al.[11]	1971	Paper	Barbital, pH 8.6, wetting solution contained 1.0% alb	Durrum type cell	100V	12	Room temp	During staining, strips agitated	Oil red 0, 2 hr, room temp; aqueous acetone	Visual; chemical analyses of eluted materials	—	Correlated with ultra centrifugal lipo determination results
Edwards, R. D. G.[12]	1972	Review of previously reported lipoprotein studies by electrophoresis and ultracentrifugation										
Lloyd, J. K.[13]	1972	Review article, no techniques										
Salt, H. B. and Wolff, O. H.[14]	1957	Paper electrophoresis	Reference given in Lloyd[13] with no details									

Author	Year	Support/Method	Buffer	Apparatus	Standardization	Temp	Quantitation	Stain	Interpretation	Comments
Lees, R. S. and Hatch, F. T.[?]	1963	Paper or agarose								Reference given in Lloyd[?] with no details
Immarino, R. M. et al.[?]	1969	—								Only reference given in #14
Fletcher, M. H. and Styliou, M. H.[?]	1970	—								Only reference given in #14
Fauk W. P. et al.[?]	1974									Immunoelectrophoresis used to check purity of β-lipoprotein antigen used; no details given
Way, C. et al.[?]	1975	Used method of Bickering and Ellefson;[?] cellulose acetate							Lp-X migrates to cathode	No other details of methodology in article
Ritland, S. and Gjone, E.[?]	1975	Agar gel								No other details of methodology in article
Gjone, E. and Norum, K. R.[?]	1968	Paper[?]								No other details of methodology in article
Lewis, L. A. et al.[?]	1966	Paper; 30 λ serum applied to paper	Barbital, pH 8.6	Durrum type cell	Method standardized by comparing with quantitative ultracentrifugal values using[?] 1.21	18	percent of total lipid per peak estimated by scanning stained strip	Oil 0	Visual inspection and scanning	Scanning to obtain relative percent of different fractions
Ghosh, P. et al.[?]	1973	Cellulose acetate; 12-hr fasting venous blood sample	Gelman high resolution, 0.00238ml, pH 8.9	—	—	—	FFA conc also determined	Schiff's reagent	—	—
Brons, M. et al.[?]	1972	Paper	Barbital	—	—	—	—	Oil red 0	According to original terminology of Fredrickson et al.[?]	—
A/ora, K. K. et al.[?]	1973	Agarose gel according to Fredrickson et al.[?]	—	—	—	—	—	—	Visual qualitative	—

Table 1 (continued)
ELECTROPHORETIC TECHNIQUES

Author and ref. no.	Year	Medium, special conditions	Buffer	Cell and special equipment	Electrical conditions	Time of electrophoresis (hr)	Temp	Special handling	Staining	Evaluation	Densitometry	Methodological error
Wada, M. et al.[13]	1975	Paper Polyacrylamide[14] gel disk; polyacrylamide[15] block, Immunoelectrophoresis[16] including quantitation by Laurell rocket technique	Barbital	—	—	—	—	Patterns in some sera were correlated with ultracentrifugally and chemically determined data	—	Intensity of lipoprotein stain scored visually	—	—
Khanna, R. et al.[17]	1975	Paper[18]	Barbital, pH 8.6, 1% albumin solution wetting strips	Durrum type cell	5mA/cell 110 — 115 V	18	Room temp	Oven-dry at 100° C for 30 min	Oil red 0	Visual; standardized by comparison with quantitative values determined by ultracentrifugation at d 1.21[42]	—	—
		Starch gel[43]	TRIS-borate double buffer system[44]	Durrum type cell	—	18	Room temp	—	Oil red 0 Sudan black B	—	—	—
Lewis, L. A.[45]	1973	Paper	Barbital (albuminated), pH 8.6	Durrum type cell	—	—	—	—	Oil red 0	Visual evaluation; special estimation of mobilities of fractions	—	—

* These abbreviations will be used in Table 1: lipo = lipoprotein; glob = globulin; FFA = free fatty acid, alb = albumin; μ = ionic strength.

Table 2
APPLICATIONS OF ELECTROPHORESIS[1]

Author and ref. no.	Year	Pattern	Type or character	Normal values	SD	Reproducibility	Pathology	±SD	Character of subject, age, sex, race	Summary of results
Kunkel, H. G. and Slater, R. J.[1]	1952	One large irregular peak of one to three components in region between α-β-globulin, α is markedly reduced	Normal β-peak nearly entirely absent; large peak of fast β-mobility	Separation of normal lipoproteins into two main peaks of α- and of β-mobility in agreement with chemical fractionation methods; larger free: ester chol in α-lipo	—	—	Markedly elevated peak of intermediate mobility between α-β present in nephrotic with hypoalb; after apparently spontaneous remission of disease, loss of edema, and retention of albumin, lipoprotein pattern became essentially normal	—	Four nephrotic children	Markedly elevated lipoprotein of fast β-mobility associated with extreme hypoalbuminia in nephrotic
Swahn, B.[2]	1952	Faster migrating fraction about 30% of lipid-stainable material; slower peak with mobility α-globulin about 3% of total; peak with mobility β-globulin (most intensely stained) about 50% of total slow or nonmigrating band, "probably chylomicron"	Ten healthy blood donors pooled	—	—	Good reproducible results both by elution and by scanning techniques and results by methods are interchangeable; the relative mobility of lipoprotein compared with that of proteins was very constant and reproducible	—	—	Normal blood donors	Separation of serum lipoprotein by electrophoresis and specific lipid-staining resulting fractions provide an excellent means of demonstrating types of lipoproteins in normal and pathologic sera including the hyperlipoproteinemia of nephrotic syndrome and of diabetic nephropathy
Swahn, B.[3]	1953	Evaluation of lipid-stainable material: very greatly increased β-lipo and chylomicron; decreased α-lipo	—	—	—	—	Diabetes; serum total cholesterol, 510 mg/dl	—	Patient with diabetic nephropathy	—
		Greatly increased β-lipoprotein and chylomicron; decreased α-lipo	—	—	—	—	Nephrotic; serum total cholesterol, 350 mg/dl	—	Nephrotic syndrome	—
Nikkila, E.[4]	1953	Well-resolved α- and β-lipoprotein fractions in normal serum	—	Normal beta-lipo contained 2/3 chol and 1/2 PL, α-lipo contained 1/4 chol, 1/3 PL, percent of chol in β increased with age;	—	—	In nephrotic highly abnormal lipoprotein pattern with increased chol in α- and β-glob region; α-chol decreased	—	Two nephrotic patients	Nephrotic serum lipoprotein pattern showed greatly increased α-1 and β-lipo and decreased α-lipo

Table 2 (continued)
APPLICATIONS OF ELECTROPHORSIS[1]

Author and ref. no.	Year	Pattern	Type or character	Normal values	SD	Reproducibility	Pathology	±SD	Character of subject, age, sex, race	Summary of results
				alpha$_2$- globulin contained 1/10 chol greater than 1/10 PL; most variable fraction for β-lipo: C/PL = 1.24 and for alpha- lipo: C/PL = 0.63						
Gordon, R. S., Jr.[5]	1955	Mobility of lipoprotein and β-lipo greatly accelerated by high FFA conc	—	—	—	—	—	—	—	
Laurell, S.[6]	1955	Normal lipoprotein pattern similar to that described by Swahn[2]	—	—	—	—	Electrophoretic mobility of lipoproteins greatly dependent on FFA content, this migration being greatly accelerated when FFA levels are high	—	Normal healthy subjects	When plasma treated with postheparin lipoprotein lipase with resulting increase in FFA, mobility of lipoproteins accelerated so identification and evaluation difficult
Nys, A.[7]	1954	In nephrotic, two types of lipoprotein pattern	First type — broad concentrated lipoprotein band extending from β to α_2- position having triple components Second type — two well- resolved concentrated lipoproteins of β-lipo and α_2-lipo mobility	Normal values in agreement with those of alcohol fractionation method	—	—	Emphasizes need for simultaneous protein and lipoprotein study for correct evaluation; in active phase of nephrosis changes in lipoprotein pattern of nephrotic are beyond those of range of normal values	—	Nephrotic and normal subjects	Pattern of active disease showed extreme hyperlipemia with two types of pattern; greatly improved with remission of disease

Author	Year		Method / Comments			Findings	Subjects		Results
Gittlin, D. et al.*	1958	—	Lipoprotein fraction studied before[14] I labeling; fractions prepared by ultracentrifugation, mobility properties specifically considered	S_f 3 — 9 migrated with mobility between α_2- and β-glob	At no time either in vivo or in vitro was evidence found of shift of label from S_f 3 — 9, S_f 10—200, or faster-migrating α_1-lipo	Nephrotic serum, lipid, and lipoprotein pattern has high chol and TG and greatly increased very low- density (preβ) lipo	Nephrotic children, 49 60 months; normal subjects, 8 males and females, 21 — 29 years	—	The fractional rate of catabolism of S_f 3 — 9 lipoprotein was increased in the nephrotic and was not due to loss in urine, but about 1/4 of α- lipoprotein was excreted; a significant part of S_f 10 to 200 lipoprotein was converted to S_f 3 — 9
Raynaud, R. et al.***	1954, 1959	—	With low alb in nephrotic serum, β-lipo mobility increased due to loading with FFA* Noted similar changes in normal subjects β- lipoprotein mobility after high-fat meal when clearing factor lipase increased **	Degree of lactescence directly related to alb conc; most lactescent had fastest migrating β- lipo; decreased α in nephrotic	—	Change in mobility of β- lipoprotein greater in "pure" nephrotic than in any amyloid nephrotic or nephrotic phase of nephritis	Nephrotic patients — no details given	—	Mobility of β-lipo in nephrotic accelerated and most altered in extreme hypoalbuminemia.
Newmark, S. R. et al.[10]	1975	—	All lipoprotein types (W.H.O. classification, modified from Fredrickson) except type I observed in nephrotic; types IIA, IIB, and IV about equal in number, type III rare	—	—	Percent of total nephrotics having abnormal values not available since lipo determination only done if chol or TG abnormal; no specific histologic diagnosis was associated with any one lipoprotein type	Average age, 40 years for 96 adult nephrotics; one group of ten had lupus nephrosis	—	Fairly frequent occurrence of chylomicronemia in nephrotics suggests depressed clearing factor activity; no specific type of lipoprotein change typical of nephrotic
Edwards, K. D. G.[13]	1972	—	Review article; some discussions of use of diet and diet plus drugs in control of hyperlipoproteinemia of nephrotic	—	—	Nephrotic hypercholesteremia may be part of mixed hyperlipemia in acute nephrotic syndrome, or be the only abnormality in mild or chronic nephrotic syndrome with raised β-lipo; hyperlipidemia,	—	—	It is proposed that hypoalb uminemia and high FFA/alb ratio may be of primary importance in increased lipogenesis and build- up of β — and preβ- lipo in nephrotic

Table 2 (continued)
APPLICATIONS OF ELECTROPHORSIS[1]

Author and ref. no.	Year	Pattern	Type or character	Normal values	SD	Reproducibility	Pathology	±SD	Character of subject, age, sex, race	Summary of results
							hyper-pre-β-lipoproteinemia, and chylomicronemia in nephrotics may be due to loss of α-phospholipoprotein which is known to activate lipoprotein lipase			
Lloyd, J. K.[14]	1972	Review article cited references, used paper, agarose, cellulose acetate[15,16,17,18]	Hyperlipoproteinemia of childhood in nephrotic may have type II, IV, or V	—		—	—	—	Pediatric subjects	Emphasized proper preparation of subjects; infants, 8 hr without milk; children, fasting 10 hr; pubertal girls, know exact date of last menstrual cycle; exact mechanism of the hyperlipoproteinemia in nephrotic syndrome is still not fully understood
Lewis, L. A., et al.[24]	1966	Renal transplantation 64 patients	Low α-lipoprotein levels were observed in M patients with renal failure and during dialysis; after renal transplantation, canc of serum α-lipo increased when renal function was good	—		Patterns checked by comparison with ultracentrifugally determined (d 1.21) lipoprotein patterns	β-Lipoproteins are relatively normal in persons with renal failure and during dialysis treatment; α-lipoproteins are low; with good renal function, lipoprotein pattern becomes normal and remains so as long as renal function is good	—	Renal transplant subjects of both sexes; dialysis patients studied just before dialysis	Good renal function necessary for maintenance of normal lipoprotein pattern, especially α-lipo.
Bagdade, J. D.,[25]	1975	In uremia, hyper-pre β-lipoproteinemia, hypertriglyceridemia, often a type IV pattern	Agarose electrophoresis of supernatant of serum d 1.006 after preparative ultracentrifugation to test for type III lipoprotein abnormality[28]	—		Methods of Lipid Research Clinics Program used[28]	Plasma lipid and lipoprotein alterations in chronic uremia due to disturbance in both production and catabolism	—	Review-type article	Hypertriglyceridemia and hyperlipoproteinemia (type IV) associated with uremia was improved by an intensive dialysis program

Investigators	Year	Lipoprotein results		Pathology / findings		Analytical methods		Subjects		Patient details		Conclusions
Bagdade, J. et al.[27]	1976	Lipoprotein pattern of each group included many patients with type IV characteristics	—	Group (see column "Pathology" for classification): 1. Diet low in sodium, 2. Dialysis treatment 20 hr/wk, 3. Transplant patients receiving prednisolone and azathioprine patients' diets' estimated calories: 40% fat 15% protein 45% carbohydrate	—	Analytical methods for lipoprotein-lipids; Lipid Research Clinics Program[34] agarose-gel electrophoresis	—	Chronic renal disease: 1. Undialyzed — 6 males, 7 females 2. Dialyzed — 11 males, 3 females 3. Patients following successful renal transplantation — 23 patients, 25 + 17 months after transplantation	—	—	—	Hypertriglyceridemia and hyper-pre-β-lipoproteinemia which were observed in chronic uremia persisted after transplantation, possibly partially due to immunosuppressive drug therapy; the lipoprotein pattern of transplant patients was very varied
Ghosh, P. et al.[29]	1973	—	—	Serum cholesterol and triglyceride of renal allograft group and of asthmatics receiving corticosteroids were elevated; in renal failure only triglyceride was elevated	—	—	—	Original diagnosis before transplantation: 14 glomerulonephritis 14 pyelonephritis 2 polycystic kidney 1 sarcoidosis All transplant patients receiving prednisolone and azathioprene Renal failure: 12 nonnephrotic, renal failure prior to dialysis Asthmatics 17 receiving prednisolone	—	Renal transplant patients, ages 17 — 51: 19 male 13 female	—	Increased triglyceride and pre-β-lipoproteins in allograft patients may be in part due to effect of prednisolone, as group with asthma but normal renal function when receiving drug showed similar lipid changes
Brons, M. et al.[30]	1972	Abnormal patterns were had by 17 of 25 patients: Type II — 1 patient, Type III — 1 patient, Type IV — 13 patients, Type V — 2 patients None showed decreased α-lipoprotein	—	Renal disease appeared to be primary cause of altered lipoprotein pattern; three patients with chronic glomerulonephritis all had normal lipoproteins	—	—	—	There were 25 patients with chronic renal failure, nonnephrotic; studies made after 12-hr fast	—	Patients excreting more than 3.5 g protein/24 hr were excluded	—	Patients with chronic renal failure frequently had abnormal lipoprotein patterns the most frequent pattern being that of type IV
Arora, K. K. et al.[32]	1973	Lipoprotein pattern of six to eight dialyzed patients showed hyper-pre-β-lipoproteinemia and hypertriglyceridemia	—	Glucose levels of patients were higher than controls, but no significant difference in their insulin levels, either fasting or following glucose ingestion	—	—	—	Diet 60 g protein/day; carbohydrate 40 — 45 % of total calories	—	Eight patients age 22 — 49, with end-stage renal failure on intermittent dialysis 2 times week, 18 hr/week Six females two males Five healthy males	—	Patients with chronic renal failure and on dialysis program show hyper-pre-β-lipoproteinemia which seems not to be due to modified levels of insulin, as these patients' insulin levels did not vary from that of controls

Table 2 (continued)
APPLICATIONS OF ELECTROPHORSIS[1]

Author and ref. no.	Year	Pattern	Type or character	Normal values	SD	Reproducibility	Pathology	±SD	Character of subject, age, sex, race	Summary of results
Wada, M.[33]	1975	Broadβ-lipoprotein band with some characteristics of type III was found in about half of the patients after initiation of dialysis — may be due to heparin; α-lipoprotein tended to be low when determined immediately before dialysis was started; in some patients α-lipo increased after dialysis; conc of characteristic rich β-lipoprotein increased during dialysis	—	—	—	—	Diagnosis by renal biopsy or clinical tests; dialysate contained 4.08 g glucose/1000 mℓ; heparin, 910 IU/hr, was infused during dialysis; patients on 2200-cal diet: protein, 70 g, carbohydrate, 295 g. fat, 83 g	—	19 patients on hemodialysis for more than 7 months 6 — 6.5 hr/wk, two or three times/week; 15 patients with renal disease	Shifts in lipoprotein pattern observed in patients on chronicdialysis program may be partially the result of chronic heparinization; broad lipoprotein-band similar to that of type III
Khanna, R., et al.[37]	1975								Total of 210 patients with renal disease — 25 undialyzed with renal failure, i. e., CRF; 65 dialyzed, i. e., CID; 125 renal transplant, i. e., RT	Of 31 patients who had cinearteriography, 19 had greater than 50% narrowing of arteries; 10 of 19 had hypertension; mean serum triglyceride level of group was elevated; cholesterol level below mean normal; hyperpreβ-lipoproteinemia was of common occurrence; 6 of the 19 patients were under 35 years of age
Lewis, L. A.[39,39a]	1973	Lipid-stained band of mobility between that of pre-β- and α-lipoprotein in sera of patients during methylprednisolone therapy	—	—	—	—	Patients following renal transplantation threatening rejection	—	One male, one female patient who had renal transplant	Intravenous methylprednisolone treatment of patients in threatened renal transplant rejection is accompanied by appearance of an abnormal lipoprotein in their serum; the

unusual fraction migrate between pre-α2 and α-lipoprotein; after withdrawal of the drug the abnormal fraction disappeared

| Bertone, E.[44] | 1966 | Lipoprotein patterns in patients with nephrotic syndrome, acute nephritis, or uremia showed large deviation from normal; the changes in none of the groups were specific enough to be of direct differential diagnostic value | — | — | Electrophoresis using Grasman technique, veronal buffer, pH 8.6; Whatman M 1B filter paper; Sudan black B stain for lipoprotein, PAS for glycoprotein | Representative cases with various types and stages of renal disease evaluated; glocyproteins and lipoproteins studied | — | Not given | While large deviation from normal was observed in both lipoprotein and glycoprotein patterns of serum in patients with various types of renal disease, the changes were not specific enough to be a differential diagnostic aid |

LIPOPROTEIN PATTERN

Treatment*	No. of subjects	Normal %	IIA	IIB	IV	V	With chylo. IIB	Period of follow-up (months-average)
CRF	25	36	4	16	40	0	4	6
CID	65	36.9		1.5 3 .1	36.9	13.9	7.7	14.2
RT	125	39.2	12	12.8	28.8	4.0	3.2	20

CRF = chronic renal failure, no dialysis; CID = chronic intermittent dialysis; RT = renal transplant

* These abbreviations will be used in Table 2: lipo = lipoprotein; chol = cholesterol; alb = albumin; hypoalb = hypoalbumin; PL = phospholipid; glob = globulin; FFA = free fatty acid; TG = triglyceride; C/PL = cholesterol over phospholipid ratio; IU = internation unit; chylo = chytomicron; v = ionic strength.

REFERENCES

1. **Kunkel, H. G. and Slater, R. J.,** Lipoprotein patterns of serum obtained by zone electrophoresis, *J. Clin. Invest.,* 31, 677, 1952.
2. **Swahn, B.,** A method for localization and determination of serum lipids after electrophoretical separation on filter paper. *Scand. J. Clin. Lab. Invest.,* 4, 98, 1952.
3. **Swan, B.,** A method for localization and determination of serum lipids after electrophoretical separation on filter paper, *Scand. J. Clin. Lab. Invest.,* (Suppl. 95), 1953.
4. **Nikkila, E.,** Studies on the lipid-protein relationships in normal and pathological sera and the effects of heparin on serum lipoproteins, *Scand. J. Chem. Lab, Invest.,* (Suppl. 8), 5, 1, 1953.
5. **Gordon, R. S., Jr.,** Interaction between oleate and the lipoproteins of human serum, *J. Clin. Invest.,* 34, 477, 1955.
6. **Laurell, S.,** The effect of free fatty acids on the migration rates of lipoproteins in paper electrophoresis, *Scand. J. Clin. Lab. Invest.,* 7, 28, 1955.
7. **Nys. A.,** L'étude des Protéines et des lipoprotéines séreques par électrophorese sur paper, en particulier dans les syndromes hepatiques, néphrotiques et dans l'atherosclérose, *Rev. Belge Pathol. Exp.,* 23, 329, 1954.
8. **Gittlin, D., Cornwell, D. G., Nakasato, D., Oncley, J. L., Hughes, W. J., Jr., and Janeway, C. A.,** Studies on the metabolism of plasma proteins in the nephrotic syndrome. II. The lipoproteins, *J. Clin. Invest.,* 37, 172, 1958.
9. **Raynaud, R., D'Eshouguea, R., and Pasquet, P.,** L'électrophorèse sur papier dans les syndromes néphrotique, *Sem. Hop. Paris,* 30, 4065, 1954.
9a. **Raynaud, R., Miniconi, P., and Pasquet, P.,** Données actuelles sur le métabolisme lipido-protidique. Application aux hyperlipidémies rénales, *Presse Med.,* 67, 7, 1959.
10. **Newmark, S. R., Anderson, C. F., Donadid, J. V., Jr., and Ellefson, R. D.,** Lipoprotein profiles in adult nephrotics, *Mayo Clin. Proc.,* 50, 359, 1975.
11. **Ellefson, R. D., Jimenez, B. J., and Smith, R. C.,** Preβ(or α_2) lipoprotein of high density in human blood, *Mayo Clin. Proc.,* 46, 328, 1971.
12. **Beaumont, J. L., Carlson, L. A., Cooper, G. R., Fejfar, Z., Fredrickson, D. S., and Strasser, T.,** Classification of hyperlipidemias and hyperlipoproteinemias, *Bull. W. H. O.,* 43, 891, 1970.
13. **Edwards, K. D. G.,** Antilipaemic drugs and nephrotic hyperlipidaemia. Drugs affecting kidney function and metabolism, *Prog. Biochem. Pharmacol.,* 7, 370, 1972.
14. **Lloyd, J. K.,** Hyperlipoproteinemia in childhood, *Aust. Paediatr. J.,* 8, 266, 1972.
15. **Salt, H. B. and Wolff, O. H.,** The application of serum lipoprotein electrophoresis in pediatric practice, *Arch. Dis. Child,* 32, 404, 1957.
16. **Lees, R. S. and Hatch, F. T.,** Sharper separation of lipoprotein species by paper electrophoresis in albumin-containing buffer, *J. Lab. Clin. Med.,* 61, 518, 1963.
17. **Immarino, R. M., Humphrey, M., and Antolik, P.,** Angar gel lipoprotein electrophoresis: A correlated study with ultracentrifugation, *Clin. Chem.,* 15, 1218, 1969.
18. **Fletcher, M. H. and Styliou, M. H.,** A simple method for separating serum lipoproteins by electrophoresis on cellulous acetate, *Clin. Chem.,* 16, 362, 1970.
19. **Way, C., Hutton, C. J., and Kutty, K. M.,** Relationship between serum cholinesterase and low density lipoproteins in children with nephrotic syndrome, *Clin. Biochem.,* 8, 103, 1975.
20. **Faulk, W. P., McCormick, J. N., Goodman, J. R. , and Piel, C. F.,** Glomerular beta-lipoprotein in childhood renal disease, *Scand. J. Immunol.,* 3, 655, 1974.
21. **Bickering, R. E., Jr. and Ellefson, R. D.,** A rapid method for lipoprotein electrophoresis using cellulose acetate as support medium, *Am. J. Clin. Pathol.,* 53, 84, 1970.
22. **Ritland, S. and Gjone, E.,** Quantitative studies of lipoprotein-X in familial lecethin: Cholesterol acyltransferase deficiency and during cholesterol esterification, *Clin. Chim. Acta,* 59, 109, 1975.
23. **Gjone, E. and Norum, K. R.,** Familial serum cholesterol ester deficiency, *Acta Med. Scand.,* 183, 107, 1968.
24. **Lewis, L. A., Zuehlke, V., Nakamota, S., Kolff, W. J., and Page, I. H.,** Renal regulation of serum α-lipoproteins. Decrease of α-lipoproteins in the absence of renal function, *N. Engl. J. Med.,* 275, 1097, 1966.
25. **Bagdade, J. D.,** Disorders of carbohydrate and lipid metabolism in uremia, *Nephron,* 14, 153, 1975.
26. **Bagdade, J. D., Porte, D., and Bierman, C. L.,** Hypertriglyceridemia, a metabolic consequence of chronic renal failure, *N. Engl. J. Med.,* 279, 181, 1968.
27. **Bagdade, J., Caseretto, A., and Albers, J.,** Effects of chronic uremia, hemodialysis and renal transplantation on plasma lipids and lipoproteins in man, *J. Lab. Clin. Med.,* 87, 37, 1976.
28. *Lipid Research Clinics Program. Manual of Laboratory Operations,* Vol. 1, DHEW No. (NSH)75-628, USDHEW, 1975.

29. Ghosh, P., Evans, D. B., Tomlinson, S. A., and Calne, R. Y., Plasma lipids following renal transplantation, *Transplantation,* 15, 521, 1973.
30. Brons, M., Christenson, N. C., and Horder, M., Hyperlipoproteinemia in patients with chronic renal failure, *Acta Med. Scand.,* 192, 119, 1972.
31. Fredrickson, D. S., Levy, R. J., and Lees, R. S., Fat transport in lipoproteins. Integrated approach to mechanisms and disorders, *N. Engl. J. Med.,* 276, 34, 94, 148, 215, 1967.
32. Arora, K. K., Atkinson, M. K., Trafford, J. A. P., and Sheldon, J., Changes in glucose tolerance, insulin and lipoproteins in patients with renal failure on intermittent haemodialysis, *Postgrad. Med. J.,* 49, 293, 1973.
33. Wada, M., Minamisono, T., Fujii, H., Morita, T., Akamatsu, A., Mise, J., Nakamoto, S., and Naito, H., Studies on the effects of hemodialysis on plasma lipoproteins, *Trans. Am. Soc. Artif. Intern. Organs,* 21, 464, 1975.
34. Naito, H. K., Wada, M., Ehrhart, L. A., and Lewis, L. A., Polyacrylamide gel disc electrophoresis as a screening procedure for serum lipoprotein abnormalities, *Clin. Chem.,* 19, 228, 1973.
35. Wada, M., Naito, H. K., Ehrhart, L. A., and Lewis, L. A., Polyacrylamide gel block electrophoresis of plasma lipoproteins, *Clin. Chem.,* 19, 235, 1973.
36. Laurell, C. B., Quantitative estimation of proteins by electrophoresis in agarose gel containing antibodies, *Anal. Biochem.,* 15, 45, 1966.
37. Khanna, R., Braun, W. E., and Lewis, L. A., Lipid and lipoprotein abnormalities in renal failure, hemodialysis and transplantation, in *Proc. 6th Int. Congr. Nephrology,* Florence, Italy, 1975, in press.
38. Khanna, R., Braun, W. E., and Lewis, L. A., unpublished data, Cleveland Clinic Foundation, Cleveland , 1966 — 1975.
39. Lewis, L. A., Unusual serum lipoprotein patterns in human beings in threatened renal transplant rejection and in healthy canines produced by methylprednisolone, in *Proc. Conf. Serum Lipoproteins,* University of Graz, Graz, Austria, 1973.
39a. Naito, H. K. and Lewis, L. A., Recognizing unusual serum lipoprotein electrophorenic patterns, *Clin. Chem.,* 19, 644, 1973.
40. Lewis, L. A., Thin-layer starch-gel electrophoresis, a simple accurate method for characterization and quantitation of protein components, *Clin. Chem.,* 12, 596, 1966.
41. Poulik, M. D., Starch-gel electrophoresis in a discontinuous system of buffers, *Nature,* 180, 1477, 1957.
42. Lewis, L. A., Green, A. A. and Page, I. H., Ultracentrifuge lipoprotein pattern of serum of normal, hypertensive and hypothyroid animals, *Am. J. Physiol.,* 171, 391, 1952.
43. Lewis, L. A., Brown, H. B., and Page, I. H., 10 Years of dietary treatment of primary hyperlipidemia, *Geriatrics,* 25, 64, 1970.
44. Bertone, E., La determinazione delle frazioni proteiche, delle glico e delle lipoproteine nel siero di sangue. Contributo sperimentale per la diagnostica delle affezioni renali, *Arch. Sci. Med.* (Torino), 12, 248, 1966.

PLASMA LIPOPROTEINS IN DIABETES

Jan J. Opplt

INTRODUCTION

It is an established belief that diabetes mellitus is characterized by increased levels of plasma lipids and lipoproteins. Metabolic disorders in lipids and lipoproteins constitute an integral part of pathologic changes in this syndrome, but significant findings may not be observed in every diabetic patient, nor in exactly all phases of the development of the disease, or in its different degrees of severity.

The pathogenesis of diabetic hyperlipemia was first studied by Fisher (1903) who postulated a disorder in the lipolysis. Later, Klemperer (1907) emphasized cytolytic processes as a source, and Reicher (1911) indicated a lack of hepatic glycogen, coupled with an increase in lipids moving from adipose tissue to the liver. Bang (1918) then stressed the possible impairment in the uptake of circulating fat by the liver. Certain unifying aspects subsequently were proposed by Allen et al., (1924) who founded a modern research in pathogenesis of diabetes and its complications.

Methodology played an important role. From 1916 (Bloor) until 1930 (Hartmann), newly invented methods for the analyses of blood cholesterol also were employed in the study of diabetic hyperlipidemia. Groups studied included diabetics treated with insulin, juvenile diabetics and diabetics in comas, (Adler and Imric) and compensated diabetes (Albrink and Man). More complex methods for analyzing serum lipids were used by Joel (1924), who called the postprandial hyperlipemia a "retained lipemia". Others included Geelmuydem (1928), who hypothesized neurohormonal disorders as the cause of diabetes and Burger (1928), who pursued a theory of chronic intoxication of the diabetic due to acetone. Of other early theories, the thesis of Katsch and Krainick (1939), concerning differentiation between "a primer hyperlipemia" and "a secondary hyperlipemia" — created by acute decompensation of a diabetic patient — should be mentioned.

Although lipoproteins were discovered by Macheboeuf in 1932, it took almost 25 years before two novel methods — ultracentrifugation (Gofman and coworkers, 1950) and electrophoresis (Swahn, 1952)— were applied to the analysis of plasma lipoproteins in diabetes. Only then was progress made in the research of the pathogenesis of diabetes and of the clinical significance of dyslipoproteinemia for diagnostic purposes.

PLASMA LIPIDS AND LIPOPROTEINS IN DIABETES

Specific reasons, applicable to diabetes and its complications, forced researchers to rely mostly on a triad of standard analytic procedures: (1) chemical analyses of plasma lipids (mostly cholesterol-free and esterified, triglycerides; free fatty acids; phospholipids; and total lipidemia), (2) electrophoretic analyses of serum-lipoproteins (sometimes replaced by precipitation techniques using methods discovered by Burstein), and (3) ultracentrifugal flotation analyses of plasma lipoproteins, which accompanied all major studies.

In 1966[246] (and later in a monography in 1970), Syllaba and Opplt[827] emphasized their long experience in the research of lipoprotein in diabetes by pointing out practical analytical guidelines for a safe clinical orientation concerning hyperlipemia: free fatty acids, plasma cholesterol, triglycerides, and the β-fraction of plasma lipoproteins, estimated electrophoretically or by precipitation technique, using heparine in the presence of calcium salts according to Burstein (however including a sum of lipoprotein classes VLDL and LDL, characterized by the flotation rate at $d_{20} < 1.063 = S°_f$ 0-400).[822]

The so-called "total lipemia" or "total lipidemia" often may not be determined because of the difficult and time consuming analytical procedures that are always involved. However, this figure offers basic quantitative information concerning the extent of the metabolic disorder involving lipids and lipoproteins in diabetes. Excellent correlations also have been discovered between the total level of plasma lipids (total lipemia) and total concentration of plasma lipoproteins.[829]

Triglycerides are diagnostically very sensitive and therefore valuable indicators of metabolic disorder in lipids and lipoproteins, reflected so often in diabetic patients. Their higher concentrations in the blood usually show up before any changes are detectable in the concentrations of cholesterol and phospholipids. Triglyceride levels correlate with the concentrations of the chylomicrons, very very low-density lipoproteins (VVLDL) and very low-density lipoproteins (VLDL),[829] as well as with the level of the pre-β-complex (= subfractions) in agarose-gel electrophoretograms of plasma lipoproteins.[830] The concentrations of triglycerides also are reliably related to the severity and brittleness of the diabetes. High plasma concentrations are reported for so-called "hepatic diabetics".[843] These levels also increase during the course of the development of the diabetic's vascular complications, particularly diabetic retinopathy and nephropathy. Indeed, the duration of the clinically diagnosed diabetes plays an important role. In the case of diabetic nephropathy one also should consider the relative increase of plasma density,[846] and the pathologic changes in the biosynthesis of lipoproteins.[826] Undiagnosed atherosclerosis also may be characterized and caused by a type of dyslipoproteinemia, which brings about similar changes in plasma triglycerides.[850]

From the practical and clinical point of view, the use of plasma triglycerides to estimate the degree of diabetic lipolysis in adipose tissue is not recommended. This avoids analyzing the concentration of free fatty acids (FFA) in the plasma of a diabetic subject twice.[827] In a compensated diabetic patient, even if he is suffering from highly developed nephropathy and atherosclerosis, the level of free fatty acids remains normal. However in a decompensated patient, or with a brittle diabetes and/or with ketoacidosis, the blood levels of FFA are high, reflecting an acute lipolysis in adipose tissue, while the metabolism of glucose is deeply disordered. The increasing levels of FFA indicate an imminent ketoacidosis and are very high during a diabetic coma.[848] Another major problem in diabetic ketosis is the impairment in the "combustibility" of FFA. This is caused by an inhibition of the enzyme catalyzing the condensation of citric acid, in excess of FFA, activated by CoA. Its activity, therefore, is much lower than the activity of an oxydizing enzyme should be. In this way, the entry of CoA into the Krebs-cycle is further impaired; the CoA accumulates, condensates, and constitutes the synthesis of the acetoacetic acid. In the liver cell, the activity of the citric acid cycle is regulated by the degree of the oxidation of FFA. All the above-mentioned metabolic disorders are responsible for a further increase in the levels of FFA during the diabetic ketoacidosis and coma, primarily initiated by lipolysis.[827]

Increased levels of plasma cholesterol also contribute to the general pattern of diabetic hyperlipidemia. Nevertheless, a well controlled, metabolically compensated diabetic subject (with mild diabetes) may have normal cholesterol levels, although seldom up to 200 mg% only. The criterion of such a critical borderline — 200 mg% — was emphasized by Opplt et al.[830] in a study of dyslipoproteinemia in patients with proven coronary artery disease. Even in compensated, mild forms of diabetes, those levels may be expected to be higher (mean value 225 mg% ± 20 in 55 subjects with mild, uncomplicated diabetes).[827] The level of the cholesterolemia depends on the severity of the diabetes, on its brittleness, as well as on its duration. In diabetic subjects with diabetic retinopathy, the level of plasma cholesterol was found to be significantly higher than in diabetic patients without clinically apparent microangiopathy.[844] Highest concentrations of plasma cholesterol were observed in some diabetic patients with diabetic nephropathy (in some cases over 400 mg%).[826] The study of cholesterol levels in the clinical course of diabetes is of particular value because, in this syndrome, they correspond well (r = 0.71) to the levels of LDL classes of serum lipoproteins and to the concentrations of β-electrophoretic fractions.[845] The diagnostic situation here is normally simpler than in other syndromes because the HDL (and, therefore, the HDL-cholesterol) are usually decreased and do not change significantly later during the course of the disease.[827] Thus, the increase of the total cholesterol reflects in diabetes fairly well the increase of the LDL_2 class of lipoproteins, which constitute the β-electrophoretic fraction (using agarose-gel electrophoresis).[829]

DYSLIPOPROTEINEMIA IN DIABETES

The first systematic study of plasma lipoproteins in diabetes using the analytical ultracentrifuge was initiated in 1952 by Keiding, Root, Lowry, and Marble[375] and later in 1955 Jahnke and Scholtan[346,347] used the electrophoresis on paper process as a support media.

Following the development of information concerning the plasma lipoproteins in diabetes, the reader must carefully consider the methods used by the investigators, allowing different separation of lipoprotein classes and lipoprotein electrophoretic fractions. It may be further confusing to the reader as regard the nomenclature of lipoprotein classes as compared to the lipoprotein electrophoretic fractions, which depends on the particular method used in the reported research.* To ease the reader to avoid the confusions concerning the proper understanding of the nomenclature, the following abbreviations, for the methods employed in particular citations will be used: LPE stands for lipoprotein electrophoresis on paper; LPEA, lipoprotein electrophoresis on paper, with albumin in the buffer system; LAGE, lipoprotein electrophoresis on agarose-gel; and LPAAE, lipoprotein electrophoresis on polyacrylamide-gel; IE, immunoelectrophoresis. The most commonly employed ultracentrifugal analytic methods will be specified as follows: according to Gofman and DeLalla (UG), using flotation coefficients at d = 1.063: $S°_f$[164-166] and at d = 1.21: $-S_{1-21}$.[435-438] according to Lewis (UL).

In metabolically well compensated diabetic subjects with mild diabetes, the electrophoretic patterns of plasma lipoproteins are not significantly changed in comparison to the healthy population of comparable age.[4,138,191,345] However, Castiglioni et al.[138] observed that the β-electrophoretic fraction (LPE) increased with the duration of the diabetes, even in those patients with mild, compensated disease. The total concentration of plasma lipoproteins may not be significantly elevated in subjects with a mild or even a severe diabetes as long as their disease is metabolically well compensated. Individual exceptions have been noticed. Despite this, the β-electrophoretic fractions in the spectrum of their plasma lipoproteins are increased significantly and the α_1-fractions significantly decreased (LPE). This brings about an internal shift in the spectrum for LDL and VLDL classes of plasma lipoproteins.[844] Subjects with decompensated diabetes or patients who are seldomly clinically controlled and poorly treated, always demonstrate significant elevations of the total lipoprotein concentrations as well as β-electrophoretic fractions.[845] The findings of an increase in the β-electropho-

---

* Electrophoretic, as well as ultracentrifugal methods and nomenclatures are described in detail in other chapters in this handbook. Please refer to them first.

retic fraction in diabetics may depend on the patient selection techniques used by different authors. This may explain the different views of different researchers: a statistically insignificant increase in the β-electrophoretic − complex-fraction (LPE) in diabetics was reported by Adlersberg,[4] Szüds,[723] and Lippmann;[451] while a significantly increased β-fraction was found by Castiglioni et al.,[138] Jahnke et al.,[345] and Syllaba and Opplt.[847,849]

Pomeranze and Kunkel[601] reported that within a group of 273 diabetic patients individual findings varied greatly: approximately 50% of the patients were found to have pathologically high levels of plasma cholesterol, triglycerides, and phospholipids that were unrelated to the severity, duration, or course of the disease (even without a relation to the frequency of the clinical control and to the type of treatment). Wolff and Salt[770] claimed that the concentration of fasting glucose-200 mg% may represent a "critical borderline," indicating the degree of a metabolic disorder reflected in the metabolism of plasma lipids and lipoproteins, regardless of all other clinical criteria of diabetes.

In diabetic ketoacidosis and in coma, all the above changes in plasma lipids (including the free fatty acids) and lipoproteins are enormously enhanced. Some report that γ-globulin fraction is significantly increased in the electrophoretograms of plasma proteins (PE), simultaneously with the β-fraction of plasma lipoproteins (LPE). Others[37,431,432] confirmed those changes, including a significant decrease of the α_1-electrophoretic fraction of plasma lipoproteins (LPE). Besides all these changes, Syllaba and Opplt[844] described an additional phenomenon: an enormous trailing effect of serum lipoproteins in diabetic ketoacidosis (LPE), indicating a load of the diabetic plasma with chylomicrons and large particles of VVLDL. This effect, depicted in the lipoprotein electrophoretograms on paper (LPE) as supporting media, often as a single pathologic sign, was later classified as Type V[821] as soon as the techniques of the lipoprotein electrophoresis improved (LPEA, LAGE).

The electrophoretic lipoprotein pattern was found to be very significantly changed in patients with diabetic angiopathies, mainly during a rapid progression of the disease or at times when the disease was poorly controlled. The total level of plasma lipoproteins, as well as the pre-β and β-electrophoretic fractions (LPE) were found to be significantly increased[844] and the α_1-lipoprotein fraction further decreased to 10 to 18 relative%.[345] The β-fraction was always found to be pathologically elevated if the electrophoretogram (LPE) offered only the separation of two fractions.[191,502] The most impressive elevation of the pre-β-electrophoretic fraction (LPE) was reported by Vernet and Smith.[687] In contrast, Castiglioni et al.[138] were unable to find any significant differences between a group of 24 diabetic patients without any clinically apparent angiopathies and a group of 56 diabetic subjects with developed and clinically detectable diabetic vascular complications.

In diabetic patients with diabetic nephropathy (syndrome Kimmelstiel-Wilson), the pathologic changes in lipoprotein electrophoretic patterns are the most significant. Decreased levels of the α_1-electrophoretic fractions generally were observed, but the degree of such changes (from 100 to 20%)[191,345,723,844,845] were reported differently in relation to the mean values, which were obtained from equivalent healthy population. The total lipoproteinemia is elevated; 100 to 150% higher values in comparison to the normal values, were found perhaps because the β-electrophoretic fraction (LPE) was most significantly elevated, as described in 22 diabetic patients with Kimmelstiel-Wilson syndrome in 1958.[844] A significantly high pre-β and β-electrophoretic fraction (LPE) is stressed for this syndrome by Szücs and Csapo[723] and the pathologically high β-fraction (LPE) only, by Djurič et al.[174]

The electrophoretic results (LPE) offered a fairly good description of general pathologic patterns of changes in plasma lipoproteins in diabetic patients according to the severity and duration of the diabetes and its vascular complications. However, information concerning a finer mode of impact of diabetes on single lipoprotein classes and subclasses, was still lacking. Therefore researchers turned with hope to the results of investigations performed with the analytical ultracentrifuge. Keiding et al.[375] presented basic information about diabetic dyslipoproteinemia. According to them, the duration of the diabetes is most influential on all low- and very low-density lipoproteins that are characterized by a broad flotation coefficient $S^\circ_f > 12$ (UG). According to Barach and Lowy,[458] all diabetics who are poorly clinically controlled (by insulin therapy and diet), or uncontrolled and untreated, present with significantly higher concentrations of plasma lipoprotein classes of S°_f 0-12, 12-20, and 20-100 (UG), so that the increased levels of total cholesterol may fairly represent these metabolic changes in lipoproteins. Tuller et al.[501] described that the lipoprotein class of S°_f 12-20 (UG) reacts quickly to the treatment with insulin in previously untreated diabetic subjects, but that the other lipoprotein classes may change in late sequence.

In diabetic coma and ketoacidosis, there is a massive increase of classes, identified by S°_f 100-400 and 20-100, representing VLDL and VVLDL. This phenomenon was confirmed otherwise by electrophoresis of plasma lipoproteins.

Different authors have proven that the lipoprotein class of S°_f 12-20 is most significantly increased in diabetics with diabetic retinopathy.[4,375,482-484,826]

Very significant changes in plasma lipoproteins, investigated by the analytical ultracentrifuge, were obviously found in diabetic patients with diabetic nephropathy (Kimmelstiel-Wilson's syndrome). Lowy and Barrach[458] and, later, Opplt and Syllaba[820] described significantly elevated lipoprotein classes of S°_f 12-20 and S°_f 20-100 (UG). These findings were confirmed in separate studies by Adlesberg[4] and Engelberg.[193] However, Gofman et al.[273] reported changes in larger sizes of lipoprotein classes: S°_f 12-20, 20-35, and 35-100. Even with this, the more detailed ultracentrifugal separation of lipoprotein classes and subclasses does not alter the basic findings concerning classes in the range of S°_f 12-20 (UG). Contrary to these supposedly specific characteristics in dyslipoproteinemia in diabetics with diabetic nephropathy, Hanig and Laufer[309-311] reported that such changes in elevation of the class of S°_f 12-20 may be found in all diabetics, with or without diabetic vascular complications, when they are compared with clinically healthy people of equal age. This scientific doubt concerning specificity of lipoprotein pathology in diabetic vascular complications in general, and in diabetic nephrophathy in particular, does not seem to be shared by Lewis and Page's[437] study in 1957, which confirmed that in the Kimmelstiel-Wilson syndrome, both classes: characterized by $-S_{1-21}$ 400-70 (UL), (corresponding to VLDL) and by $-S_{1-21}$ 70-40, (corresponding to LDL), are significantly increased. The VLDL class was electrophoretically defined as faster β-lipoprotein.

In this stage of the research on plasma lipoproteins in diabetes and diabetic vascular complications it was still not very clear how the severity of the disease, its duration, and clinical management really influences the metabolic disorders in lipids and lipoproteins and if this pathologic process is really significantly enhanced in diabetic vascular complications.

THE AUTHOR'S OWN RESEARCH IN LIPIDS AND LIPOPROTEINS IN DIABETES AND IN DIABETIC VASCULAR COMPLICATIONS

The author's prime research aim was to estimate, then verify, the possibly significant or insignificant differences in lipid and lipoprotein metabolism in diabetic patients in relation to (1) the severity of their disease, (2) the type and extent of diabetic vascular complications, and (3) to the underlying atherosclerosis. A secondary aim was to establish the most practical and available methods for the proper clinical control of diabetic hyperlipidemia and dyslipoproteinemia.*

Clinical Material for the Electrophoretic Study

In cooperation with Professor J. Syllaba, M.D., ScD., (The Chairman of the IInd Department of Medicine, University Hospital, School of Medicine and Hygiene, Prague, Czechoslovakia), a selection of patients was made as follows:

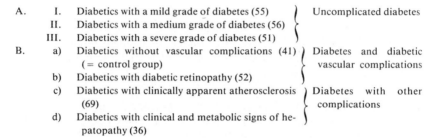

The groups of diabetic patients were arranged according to the severity of the diabetes (I to III), and according to the complications (a, b, c, d). A group of 110 diabetics with diabetic nephropathy were enrolled into a separate group C.

- Group I — (mild grade of diabetes) the patients did not need insulin or antidiabetics.
- Group II — (medium grade of diabetes) the patients needed up to 50 IU of insulin per day or peroral antidiabetics.
- Group III — (severe grade of diabetes) the patients were treated with more than 50 IU of insulin per day.

All patients were hospitalized in the IInd Department of Medicine, clinically completely examined (including special eye examinations), and their metabolic profile was established.

The 110 patients in Group C were divided in subgroups according to the clinical stages established by Syllaba:[849]

Clinical evaluation of diabetics	Kimmelstiel-Wilson Syndrome-stage of development	Classification of the stage	Examined patients in groups (KW)
KW I to II	Preclinical stage	KW I	
KW II	Clinical stage	KW II	KW I, II, III[a,b]
KW II to III	Prenephrotic stage	KW III	
KW III	Nephrotic stage	KW IV	
KW III to IV	Preterminal stage	KW V	KW IV, V, VI[a,b]
KW IV	Terminal stage	KW VI	

[a]   With clinically apparent atherosclerosis.
[b]   Without clinically apparent atherosclerosis.

*   This research was accomplished at the Charles University, Prague, in the Department of Clinical Chemistry, University Hospital, School of Medicine and Hygiene, Prague, Czechoslovakia (Chairman: Docent Jan J. Opplt, M.D., Ph.D., Sc. Acad. D.)

Partial complications (retinopathy, neuropathy, arteriosclerosis, arteriolosclerosis, cholecystopathy, and obesity), also were taken into account.

Diabetics with KW-syndrome and with additional liver disease were placed into a special group, as were those with a complicated renal disease (pyelitis, cystopyelitis, or pyelonephritis).

Patients with or without KW-syndrome, who were suffering from other infectious disease or malignant tumors were excluded from the research group.

Clinical Material for the Ultracentrifugal Study*

A. A control group: (20) clinically healthy subjects, approximately equal in age and sex to the group of diabetics.

B. I. Diabetic patients with a mild grade of diabetes (12).
    II. Diabetic patients with a medium grade of diabetes (20).
    III. Diabetic patients with a severe grade of diabetes (5).
    a) Diabetic patients without diabetic vascular complications (6).
    b) Diabetic patients with proven diabetic retinopathy (21).
    c) Diabetic patients with clinically apparent atherosclerosis (10).

C. Diabetic patients with diabetic glomerulosclerosis, proven either by renal biopsy or by autopsy (12).

Laboratory Methods

Electrophoretic separation of plasma lipoproteins was performed according to the modification of Lees and Hatch[425] using Whatman No. 1 analytical paper as a supporting medium and with albumin enriched (= 1%) veronal-oxalate-citrate buffer (μ = 0.66 and pH = 8.6) in a vertical version of a "Zeiss-Jena" electrophoretic cell at 21°C. The electrophoresis was accomplished on strips (4 × 30 cm) of paper, in a stabilized, homogenous electric field, characterized by 70 V and 0.12 mA/cm of the medium, for 16 hr, in an air-conditioned laboratory. Plasma (20 λ) was applied by a calibrated micropipet. After 16 hr there was a reproducible distance of 4.5 cm between the α_1-electrophoretic band and the point of application of the sample.

The strips were dried in air at laboratory temperature, stained (16 hr) in a staining bath (1 mg% solution of Sudan-Black-B in 96% ethylalcohol, diluted with distilled water 1:5), washed (3 × 30 min) in 50% ethanol, and dried at laboratory temperature in a dark space.

The electrophoretograms (strips 4 × 15 cm) were evaluated densitometrically on a "Zeiss-ERI-10" extinction densitometer with an integrator. The densitograms were first integrated by calculations of relative distribution of electrophoretic fractions (in relative %) in the patterns, second by determination of single areas corresponding to each fraction in the spectrum. These values, related to concentrations of standards, analyzed simultaneously, were used for calculations of concentrations of fractions in the sample (and further in the plasma) and expressed in mg%. Mean values and ± SD, as well as standard errors were estimated for each of multiple (5) analyses. The sum of values of single fractions provided the value of total level of plasma lipoproteins. The influence of any possible inhomogenicity in the paper was controlled by use of a so-called "permanent standard".

The above described electrophoretic method was carefully standardized and all available modifications studied prior to its being used in the research.

----

* Combined with the electrophoretic study and the clinical pathological research of relationships between lipid and lipoprotein analyses.

The particularly analyzed different dyes — Sudan-Black-B (Gurr), Oil-Red-Oil (Gronow), Sudan-III, and Sudan IV — also were checked for homogenicity. Sudan-Black-B is composed from ten colored fractions,[67] Oil-Red-Oil from four fractions, (Jeneks, 1955) and Sudan-III from ten fractions.[372] All fractions of Sudan-Black-B have an affinity to the lipoproteins, but this varies in each of them. The staining potential of the mixture is dependent on its final concentration in the staining bath. The optimal absorbtion is at 580 to 600 nm. The esters of cholesterol are most intensively stained (I = 2), as are fatty acids (unsaturated acids are stained most intensively). Triglycerides are stained less intensively (I = 1). The phospholipids are not stained (I = 0). Different classes of plasma lipoproteins are stained differently. Therefore, specific indexes for proper corrections of lipoprotein concentrations were used: HDL = 1.06, LDL in the β-fraction = 0.84, and VLDL in the pre-β-fraction = 0.98.

The use of triolein as a "permanent standard" was first suggested by Swahn.[720] Moinat et al.[501] proposed a preanalyzed serum, and Erb[197] constructed a "standard artificial solution" using immersion oil, dissolved in a mixture of ethanol and acetic acid. Wunderly and Wiem[779-782] attempted to use triolein in a modified manner. Opplt and Urbaskova-Bergmanova[818] prepared a modified system, consisting of a standardized mixture of albumin and triolein in veronal-oxalate-citrate-buffer, using the emulgator Tween-80 for the preparation of a proper, homogeneous emulsion. This emulsion used to be applied on the medium, employed for the electrophoretic analyses of plasma lipoproteins (0.003 mg → 0.1 mg of triolein value); a constant, standardized current was applied for a short time (15 min); the spots of the standard were stained in an alcoholic bath containing Sudan-Black-B and dried; a densitometric evaluation followed; the planimetric values, corresponding to the concentrations of triolein, were linear in a practical range and were utilized for the calculation of relations of concentrations of lipoprotein fractions to the concentrations of triolein. In this way, the constituted "triolein number" demonstrated a triglyceride-equivalent (I = 1), which helped to express the concentrations of electrophoretic fractions of plasma lipoproteins in a standardized and comparable manner. (It may be noted that Tween-80, a sorbitate of palmic acid, is not stained by Sudan-Black-B.)

A new permanent standard was prepared in the second stage of our research.[821] This was an ultracentrifugally isolated, purified, and analyzed (by electrophoresis and flotation analysis) lipoprotein class defined by S_f^o 0-400. It offered a "direct standard" for the β- as well as pre-β-electrophoretic fractions of plasma lipoproteins (with correlated values of LDL and VLDL) so that required correlations were made for the standard itself and, accordingly, for all analyses of patients. Analytical values of this standard were based on accurate ultracentrifugal analyses of the low- and very-low-density lipoproteins (and later of the high-density lipoproteins). Because of the limited durability of such natural "primary" standard, various experiments were undertaken to preserve permanent values of the LDL vs. β and VLDL vs. pre-β relationships:

1. A new permanent standard was isolated and analyzed by help of preparative and analytical ultracentrifuge; a restandardization of the electrophoresis was performed weekly.
2. Once electrophoretically and ultracentrifugally analyzed, the standard was preserved as an electrophoretogram, embedded into special resin and — being unchanged for years — reused with every routine or research analysis (the durability of such standard was controllable by an artificial — second — marker, but possible corrections for staining variations were of course not available).
3. Ten years ago, no computerization of the densitometry had been done; therefore, a simple procedure (which could be done today with ease) was not at hand:

the insertion of the corresponding standard data into the memory of a computer capable of performing calculations and integrations of densitometric data for routine analyses was not at hand.

The ultracentrifugal "permanent standard" described above was also analyzed for the total concentration of extractable lipids (Bloor), esterified fatty acids (Saphiro), free fatty acids (Doole), total cholesterol (Bloor), and its fractions (Bloor). The precipitation technique for lipoproteins (according to Burnstein) and the radial diffusion technique for quantitation of the Apo-β-apolipoprotein (Opplt and Vakocova) were standardized in a similar way. Using the second "permanent standard" and an analysis of variance, we calculated standard deviations for every electrophoretic fraction:

α_1-Fraction	7—8%
α_2-Fraction	20—23%
or pre-β-fraction	
β-Fraction	6—7%
Chylomicrons	12—14%
Sum of fractions	4%

As part of the standardization (to modify the prescription for the preparation of the staining bath and control its extinction coefficient), 0.2 g of Sudan-Black-B was dissolved in 300 mℓ of a mixture of 45 mℓ glycerol + 75 mℓ distilled water + 180 mℓ methanol. After cooking and filtrating, the mixture was diluted by an additional portion, up to 1000 mℓ. Albumin was added to obtain the final 1% solution. The standard concentration of the staining bath was expressed in the transmission value of 38 at 595 nm (Specol, Zeiss-Jena), measured against a blank (mixture without Sudan-Black-B dye).

The radial diffusion technique was performed in a modification[818,821] of the original method by Mancini et al.,[476] using calculations suggested by Darcy.[163]

The precipitation of plasma lipoproteins, using heparin and Ca^{++}, was done by the unmodified, original method, described by Burstein and Samaille,[119-122] and compared with its modification, published by Ledvina and Kellen.[422] The standardization was maintained and developed by Opplt[821] and Opplt et al.[817]

Immunoelectrophoresis and immunodiffusion were used in the original description, according to Ouchterlony,[571,572] Grabar and Uriel.[740,741]

Total concentrations of extractable lipids (= total lipidemia) were estimated according to Sperry's[696] modification and gravimetrically according to Kien's.[793a]

Total concentrations of esterified fatty acids were determined according to Stern and Saphiro.[793b]

Total concentrations of free fatty acids were analyzed using the method of Dole.[175]

Total cholesterol and its fractions (free and esterified) were analyzed according to Pearson et al.[793c] and controlled with the method of Schoenheimer and Sperry[793d] according to Bloor.[793e]

Total triglycerides analyses were accomplished according to Carlson and Waldstrom.[793f]

Apolipoproteins were isolated from ultracentrifugally separated lipoprotein classes by the "cold extraction technique" according to Scanu et al.[658]

Ultracentrifugal separation of plasma lipoproteins was performed according to the original method of DeLalla and Gofman.[165]

Preparative Flotation at d < 1.07 g/mℓ
Five mℓ of blood serum were mixed with 4 mℓ of sodium salt solution of density d = 1.1315 g/mℓ ± 0.0005 g/mℓ at 20°C. The final density of the serum + salt mixture was 1.063 g/mℓ ± 0.0005 g/mℓ at 20°C. The preparative rotor P-2 was used and the ultracentrifuge MOM·G·120 employed for subsequent preparation at G = 79,640 g ∼ 37,540 RPM, for 16 hr at 14 to 21°C. One to two mℓ of a flotant were separated from the preparative cuvette and this volume was exactly measured.

Analytical Flotation at d = 1.063 g/mℓ
The analytical, double sector cuvette (cell) was employed. As a reference sodium salt solution of d = 1.063 g/mℓ ± 0.0005 g/mℓ at 26°C was used. The analytical ultracentrifugation was accomplished in a vacuum: 10^{-4} mmHg, at 26.0°C ± 0.5°C. During the analytical run of the ultracentrifuge, the following parameters were measured, while the ultracentrifugal film was recorded:

- t_f = up-to-speed time (during period of acceleration)
- t_o = 0 min
- t_1 = 6 min (4 sec exposure at 30° angle)
- t_2 = 12 min
- t_3 = 22 min
- t_4 = 30 min
- t_5 = 38 min

The evaluation of ultracentrifugal patterns was completed on copies, 10× enlarged, so that the distance between reference baselines was 16 cm. The actual analytical ultracentrifugal film record (see the original photographs) demonstrated four flotation classes of plasma lipoproteins:

S_f^o	400-100	t = 0 min
S_f^o	100-20	t = 6 min
S_f^o	20-12	t = 30 min
S_f^o	12-0	t = 30 min

The determination of the rate of flotation of a lipoprotein was based upon the application of the Svedberg identity:

$$S = \frac{d_x/d_t}{\omega^2 x}$$

Where S = migration rate per unit centrifugal field, x = distance of lipoprotein from center of rotation, t = time of centrifugation at full speed with the equivalent of t_f time (1/3 U.T.S.); ω = angular velocity. The analysis is limited to the determination of all lipoproteins between arbitrary flotation rate limits. These limits, determining individual lipoprotein classes, were determined according to this equation:

$$\lg R_{S_f} = s_f \cdot \omega^2 \cdot \Delta^t - \lg R_{max}$$

R_{sf}	= a distance between chosen flotation rate limits and the axis of rotation (cm)
s_f	= flotation coefficient (S_f^o)
ω^2	= angular velocity (rad sec^{-1})
	= 2 π n/60
n	= RPM
t	= (t + 1/3 t_f × 60)............ in sec
	t = time in min from t_o → t_x

t_f = up-to-speed time

R_{max} = a distance of the cell-bottom from the axis of rotation (in cm; in MOM·G·120 = 7.13 cm)

A practical help was offered by precalculated tables compiled by Kyncl (Chem. Listy, 1968).

The calculations of areas, corresponding to single lipoprotein classes, were accomplished by help of planimetry. The calculations of the concentrations of individual lipoprotein classes in the spectrum were performed according to the equation:

$$C = \frac{A \cdot tg\alpha \cdot 1000}{L \cdot T \cdot M \cdot m \cdot N \cdot E^2 \cdot \Delta n} \cdot k \ (mg\%)$$

A = area, corresponding to the lipoprotein band (mm²)

tgα = tg of the angle of the diagonal wire element with respect to slit image

C = concentration of lipoprotein in the original serum in mg%

L = optical level arm (= 757 mm)

M = magnification of cell height (= 1.95)

m = magnification of cylindrical lens system (= 2)

N = factor by which the lipoproteins have been concentrated in the preparative procedure

E = linear magnification of the enlarger used in preparing the tracing (10:1.95)

Δ n = specific refractive increment for the lipoproteins encountered (= 0.00154)

T = height of the fluid-column in the cell (12 mm)

k = dilution of the flotant in the analytical cell

The corrections of C for the radial-concentration effects of the sector-shaped cell and in the inhomogeneous centrifugal field (C_{cor}) is represented by the equation:

$$C_{cor} = C \cdot \left(\frac{X_t}{X_o}\right)^2$$

where X_t = distance from center of rotation to the "peak" position and X_o = distance from the center of rotation to cell base.

The Johnston-Ogston consideration:

$$C = \frac{R \cdot C_f^m \cdot C_s^m}{N + R \cdot C_f^m}$$

$$R = k \cdot S_s^o$$

where k = constant of the flotation rate vs. concentration (= 0.310 in g^{-1} = 310 in mg^{-1}). S_s^o = the flotation rate at infinite dilution of the slower migrating species, C_f^m = measured concentration of the faster migrating species, and C_s^m = measured concentration of the slower migrating species.

$$C_f^m = C_f \text{ and } C_s^m - C_s$$

where C_f = calculated concentration of the faster migrating species (S_f 12-20), corrected for concentration effect; C_s = calculated concentration of the slower migrating species (S_f 0-12), corrected for concentration effect; N = $S_f^\alpha - S_f^\beta$; S_f^α = the measured flotation rate of the faster migrating species; and S_f^β = the measured flotation rate of the slower migrating species.

$$S_s^o = 5.9 \, S_f^o$$

The final equation may be in this form:

$$\Delta C = \frac{k \cdot S_s^o \cdot C_f \cdot C_s}{S_f^o \, (1 - k \cdot C_f) - 5.9 \, (1 - k \cdot C_s) + k \cdot S_s^o \cdot C_f}$$

The value ΔC is added to the C_{cor} of the faster migrating species ($S°_f$ 12-20) and substrated from C_{cor} of the slower migrating species ($S°_f$ 0-12):

$$C_f^{cor} = C_f + \Delta C$$

$$C_s^{cor} = C_s - \Delta C$$

RESULTS OBTAINED BY ELECTROPHORETIC RESEARCH ON PLASMA LIPOPROTEINS

The level of total plasma lipoproteins was found significantly elevated (statistically, at 5% level of significance, $p < 0.05$) in diabetic patients with a severe grade of diabetes (526 ± 46*), in comparison with patients with a mild grade of diabetes (389 ± 44) or medium grade of diabetes (447 ± 44). The total lipoproteins are estimated according to the total concentrations of lipids, stained by an alcoholic solution of Sudan Black B, in each electrophoretic fraction.

It was interesting to observe, that the α_1-electrophoretic fraction was significantly higher in patients with a severe grade of diabetes (128 ± 15) than with all others with medium (104 ± 15) or mild (78 ± 15) grades of the disease.

The fraction, (corresponding mostly to VLDL class of plasma lipoproteins, later established by the symbol pre-β, but providing significant trailing effect on paper impregnated by albumin, although usually well separable on 0.6% agarose-gels with or without albumin) was only insignificantly higher in severe diabetes (150 ± 21), but no significant difference was found in the mild or medium grade of the disease.

However the β-electrophoretic fraction provided a significant increase (248 ± 22 vs. 190 ± 21) in severe grade of the disease, if compared with the mild grade.

When the patients with different grades of severity of diabetes, but without complications, were investigated (n = 41, symbols I, II, IIIa) and compared with patients with different grades of disease — but wih diabetic retinopathy (n = 52, symbols I, II, IIIb), or atherosclerosis (n = 69, symbols I, II, IIIc) — no significant differences were found in any of the electrophoretic fractions or in the total lipoproteinemia.

Not only did the quantitative evaluation of dyslipoproteinemia in the severe grades of diabetes offer significant differences vs. other milder grades of the disease, but there also was a more frequent finding of this metabolic disorder in severe diabetes: the hyperlipoproteinemia (in 80% vs. 30%) and the hyper-β-lipoproteinemia (60% vs. 20%), but the hyper-pre-β-lipoproteinemia 0% vs. 50% (= the dyslipoproteinemia is changing character in severe diabetes, becoming closer related to the phenotype IIA).

Diabetic patients (n = 36) with an additional complication of a liver disease (chronic hepatitis or cirrhosis of the liver) were mostly classified as diabetics with a severe grade of diabetes ($\sim$ 75%) and less often as diabetics with a medium grade of the disease ($\sim$ 25%), based on statistical findings in their plasma lipoproteins.

---

* The concentrations are given in mg% according to the correction factors estimated by the analytical ultracentrifuge; ± t.$S_{\bar{x}}$, are calculated 95% borderlines of reliability (= confidence intervals for the mean).

Plasma lipoproteins in diabetics with the Kimmelstiel-Wilson syndrome are characterized by significantly higher levels of both the lipoproteins and the β-electrophoretic fraction, and also with significantly low α_1-fraction. The patients with this syndrome have higher levels of total lipoproteins (506 ± 76 vs. 451 ± 35 mg%) and higher levels of β-lipoproteins (271 ± 35 vs. 245 ± 39 mg%) in the advanced stages of the disease (the nephrotic and terminal stages) than in the initial stages (preclinical and clinical). These differences are statistically significant ($p = < 0.05$).

The concentration of the electrophoretic β-fraction in the circulation of the diabetics with K. W. syndrome was found higher, but not significantly different, in subgroups of patients with and without the syndrome and clinically evident atherosclerosis. However, these differences become clearly significant by comparison of groups of diabetic patients with fully developed K. W. syndrome and atherosclerosis with groups of diabetic subjects with mild or even severe degrees of diabetes. The β-lipoproteins, therefore, are significantly higher in diabetic patients with diabetic nephropathy and atherosclerosis than in diabetic patients with the atherosclerosis only. It also may be of interest that patients with diabetes, K. W. syndrome, and an additional complication — chronic infection of kidneys (chronic pyelonephritis, cystopyelitis, etc.) — presented the highest levels of plasma β-lipoproteins.

It has been mentioned that the levels of the α_1-lipoproteins are higher in severe diabetes than in mild diabetes. This trend of an increase of the electrophoretic α_1-fraction, which is correlative to the grade of the severity of diabetes, also was observed in diabetics with the K. W. syndrome (patients with the more developed syndrome, mainly reaching the nephrotic stage of the glomerulosclerosis, exhibit an increase in the electrophoretic lipoprotein fraction). Further research of this phenomenon perhaps will demonstrate if such an increase occurs because of the HDL, or less probably because of other proteins (i.e., serum albumin binding free fatty acids, or the dye taking up protein, or the albumin unspecifically binding the Sudan Black B).

For the clinical evaluations of the lipoprotein electrophoretograms of the above described type (LPEA), one may consider further conclusions on the basis of our research.

An increase of the total lipoproteinemia should not be expected in the course of development of diabetic vascular complications (= diabetic retinopathy and nephropathy), nor in the development or presence of diabetic neuropathy.

The pre-β-electrophoretic fraction, although increased in severe diabetes, as well as in the K.W. syndrome (more in the prenephrotic stage and later), was observed to be the most elevated component in the electrophoretic pattern in K. W. syndrome, complicated by clinically apparent atherosclerosis and/or additional kidney disease. Otherwise, this electrophoretic fraction undergoes few remarkable changes if the diabetes is constantly severe for many years.

It also should be noted that the electrophoretic fraction (= pre-β-fraction), corresponding mainly to the VLDL classes of lipoproteins, is elevated in some diabetics (in approximately 10% of all studied diabetic patients), but much more so in a severe grade of the disease and in K. W. syndrome (observed increased in 59% of the cases), but does not change significantly in relation to the severity of the diabetes, nor in connection with its vascular complications. If atherosclerosis develops, an increase of this fraction is often observed (in 82% of our cases).

The α_1-fraction is significantly decreased in diabetics with severe dyslipoproteinemia, but increases somehow illogically, if the severity of diabetes deepens, as described above.

It seems apparent that the dyslipoproteinemia (characterizing the severe diabetes), once established, remains relatively stable during the course of the disease and does not grow worse until the patient reaches the prenephrotic and nephrotic stage due to developing diabetic glomerulosclerosis.

RESULTS OBTAINED BY ULTRACENTRIFUGAL RESEARCH ON PLASMA LIPOPROTEINS

This part of the original study should elucidate findings already obtained by electrophoretic analyses of plasma lipoproteins.

1. Basically, we were first interested in an experiment that would confirm or exclude significant differences between the diabetic patients (without diabetic vascular complications, but with different grades of the severity of diabetes) and clinically healthy subjects of comparable age and sex: D I, II, III, a,c (16) vs. N (20). These analyses revealed that the differences in lipoproteinemia between normal and diabetic patients in general are very significant (p = 0.01). The total values of lipoprotein classes, flotating at d < 1.063 g/mℓ and characterized by flotation coefficients $S°_f$ 0-400, are significantly elevated (p = 0.01), reaching in our group of diabetics a mean value 300% greater than normal.

The significance of such a prominent difference is lessened by a large standard deviation, disclosing huge individual variations in the analytical findings of single patients. Similar results are seen in the class $S°_f$ 0-12, corresponding to the LDL_2 lipoproteins and to a small amount of HDL_1 and in the class $S°_f$ 12-20 that includes mainly the Lp (a) and LDL_1 lipoproteins. The next, less significantly elevated lipoprotein class of $S°_f$ 20-100 (p = 0.05), embodies — without any doubt — the VLDL. It would correspond well to our previous electrophoretic investigations, concerning the pre-β-fraction of plasma lipoproteins in diabetic subjects, which proved only an insignificant increase of VLDL. This has now been confirmed by a firm insignificance in the class of $S°_f$ 100-400.

A comparison of the mean values (x) of concentrations (in mg%) of standard lipoprotein classes for diabetic patients with different grades of severity of diabetes with mean values for normal subjects (y). The comparison was performed with the aid of the t-test; ts_x = 0.95 confidence interval = CI; s = standard deviation of measurements. N.S. = not significant.

D I, II, III a,c   vs.   N
(n = 16)   (n = 20)

Total concentrations of standard classes
$S°_f$ 0—400

x =	958,963	y =	297,070	
ts_x =	±352,920	ts_y =	±26,439	t = 3,983
s =	662,7	s =	56,570	(p<0.01)

Standard classes $S°_f$ 0—12

x =	475,353	y =	211,615	
ts_x =	±180,73	ts_y =	±26,942	t = 3,073
s =	339,41	s =	57,65	(p<0.01)

Standard classes $S°_f$ 12—20

x =	77,284	y =	22,960	
ts_x =	±38,346	ts_y =	±10,088	t = 3,002
s =	72,02	s =	21,585	(p<0.01)

Standard classes $S°_f$ 20—100

x =	250,743	y =	44,425	
ts_x =	±170,208	ts_y =	±12,461	t = 2,575
s =	319,64	s =	26,66	(p<0.05)

Standard classes S_f° 100—400

$x =$	155,568	$y =$	18,085	
$ts_x =$	±160,070	$ts_y =$	±10,776	$t = 1,825$
$s =$	300,62	$s =$	23,06	(N. S.)

The precise ultracentrifugal analytical results firmly demonstrate where the pathologic metabolic blocks are located in diabetic patients with the progressive diabetic retinopathy (proven and differentiated from atherosclerotic retinopathy). We have documented a significant involvement of VLDL, flotating at d = 1,063 g/mℓ as two distinguishable classes: S_f° 20-100 and S_f° 100-400. The first strikingly elevated lipoprotein class contains 10 to 15% of apolipoproteins (LP-B, LP-C, and LP-E: in both classes S_f° 12: LP-C = 5%, LP-B: LP-C = 30% and LP-B: LP-C: LP-E = 45%) and 85-90% of lipids, while the second lipoprotein class, slightly less elevated (p = 0.05) seems to be composed from less protein (∼ 5%) and 95% of lipids. In case of dyslipoproteinemia in diabetic retinopathy it is represented mainly by Apo-B, C-III, and Apo-E constitutive polypeptides. Although the "atherogenic" activity of both classes of VLDL (pathologically elevated in diabetic nephropathy) was not confirmed, their possible direct or indirect connections with the pathogenesis of diabetic retinopathy cannot be fully excluded at our present level of knowledge.

It was further necessary to confirm our preceding electrophoretic studies, indicating a significant deteriorating of dyslipoproteinemia, characterized by a further elevation of LDL and VLDL, in diabetic patients with a developing (or progressing diabetic retinopathy:

D II, II, III b	vs.	N
(21)		(20)

Total concentrations of standard classes
S_f° 0—400

$x =$	676,121	$y =$	297,070	
$ts_x =$	161,369	$ts_y =$	26,439	$t = 4,844$
$s =$	353,86	$s =$	56,570	(p<0.01)

Standard classes S_f° 0—12

$x =$	406,807	$y =$	211,615	
$ts_x =$	106,381	$ts_y =$	26,945	$t = 3,718$
$s =$	233,23	$s =$	57,65	(p<0.01)

Standard classes S_f° 12—20

$x =$	75,881	$y =$	22,960	
$ts_x =$	27,912	$ts_y =$	10,088	$t = 3,727$
$s =$	61,20	$s =$	21,585	(p<0.01)

Standard classes S_f° 20—100

$x =$	141,557	$y =$	44,425	
$ts_x =$	45,928	$ts_y =$	12,461	$t = 4,266$
$s =$	100,70	$s =$	26,66	(p<0.01)

Standard classes S_f° 100—400

$x =$	51,881	$y =$	18,085	
$ts_x =$	27,634	$ts_y =$	10,776	$t = 2,382$
$s =$	60,58	$s =$	23,06	(p<0.05)

2. The final research aim consisted of a comparative ultracentrifugal study that would demonstrate possible differences in dyslipoproteinemia in diabetics: without diabetic vascular complications (D I, II, III a,c), with diabetic retinopathy (D I, II, III, b), and with proven diabetic glomerulosclerosis (KW): D I, II, III, a,c — 16 patients; D I, II, III, b — 21 patients; KW — 12 patients. The analysis of variance and the Duncan's test was used for the statistical evaluation of the ultracentrifugal results.

Total concentrations of all standard classes S^o_f 0-400 (in mg%)
Analysis of variance

Variation	Sum of squares	Degree of freedom	Variance	F (Variance ratio)
Between sets	782.414	2	391.207	1,635
Within sets	11,007.085	46	239.284	(N.S.)
Total	11,789.499	48		

Group	D I, II, III, b	KW	D I, II, III, a,c
x	676,121	877,000	958,963
ts_x	161,369	265,012	352,920
s	353,86	417,20	662,70
N	21	12	16

(Duncan's test does not apply since the analysis of variance shows no difference between sets).

Standard classes S^o_f 0-12
Analysis of variance

Variation	Sum of squares	Degree of freedom	Variance	F
Between sets	256.777	2	128.388,5	1,569
Within sets	3,763.684	46	81.819,2	(N.S.)
Total	4,020.461	48		

Standard classes S^o_f 12-20
Analysis of variance

Variation	Sum of squares	Degree of freedom	Variance	F
Between sets	236.042	2	118.021	3,061
Within sets	1,773.805	46	38.561	(N.S.)
Total	2,009.847	48		

Standard classes S^o_f 100-400
Analysis of variance

Variation	Sum of squares	Degree of freedom	Variance	F
Between sets	109.738	2	54.869	1,727
Within sets	1,461.670	46	31.775	(N.S.)
Total	1,571,408	48		

The results of analytical flotation analyses and their statistical evaluations solved long-standing dilemmas of many from the past 2 decades:

1. Which class or classes of plasma lipoproteins are different, and how significantly, in patients with diabetic vasculopathy, compared to diabetics free of this frequent complication.
2. If some of the lipoprotein classes were to have a pathogenic role in the development of diabetic vascular disease.

Our answer became surprisingly simple. None of the plasma lipoprotein classes was significantly changed, nor did the total lipoprotein levels show any significant differences (within the groups). Following a detailed inspection one can conclude that the diabetic patients with KW-syndrome have the highest concentrations of lipoproteins of $S°_f$ 0-12, 20-100, and 100-400, hence of LDL and VLDL. But the statistical significance of these findings is weakened by large values of the standard deviation(s). It may be possible that some researchers, studying by coincidence more favorably assembled groups of diabetic subjects, would be able to prove statistical significances for one, or even both, lipoprotein classes. We failed to provide such proof.

A similar statistical analysis of the relative distributions of all standard lipoprotein classes in the spectrum (expressed in %) deceived in the same attempt.

Despite these circumstances, metabolic disorders in individuals with diabetes and diabetic angiopathies, may be reflected most probably in significant elevations of LDL and/or VLDL. Even diabetics, in whom one would expect more uniform metabolic abnormality, may constitute general conditions apt to develop dyslipoproteinemias closely related to the main phenotypes described by Fredrickson, Levy and Leese.[241,243]

Therefore, we tried to estimate phenotypes from the electrophoretical and ultracentrifugal analyses of diabetics both with and without Kimmelstiel-Wilson syndrome. In 37 diabetic patients (with "controlled" diabetes, classified in groups I, II, III, a,b,c) we determined:

- 20 patients with dyslipoproteinemia Type IIA
- 9 patients with dyslipoproteinemia Type IIB
- 8 patients with dyslipoproteinemia Type IV

The 12 KW-diabetic patients with proven diabetic glomerulosclerosis all demonstrated the dyslipoproteinemia Type II.

It also was important to learn if any of the known phenotypes were connected with any type of diabetic vascular complications, and if the electrophoretic patterns (LPEA) truly reflect the distribution of ultracentrifugal classes in lipoprotein spectrum (Figures 1 to 4).

Therefore, we used again the statistical method of analysis of variance and Duncan's test for such verification. We evaluated the possible significance of differences in concentrations of all lipoprotein classes as well as a significance in relative distribution (expressed in relative %) of those classes in the lipoprotein patterns.

Significance of differences in concentrations of lipoprotein classes $S°_f$
0-400 in different phenotypes

Analysis of variance

Variation	Sum of squares	Degree of freedom	Variance	F
Between sets	1,606.965,72	2	803.482,86	3,326
Within sets	8,212.528,60	34	241.544,96	(p<0.05)
Total	9,819.494,32	36		

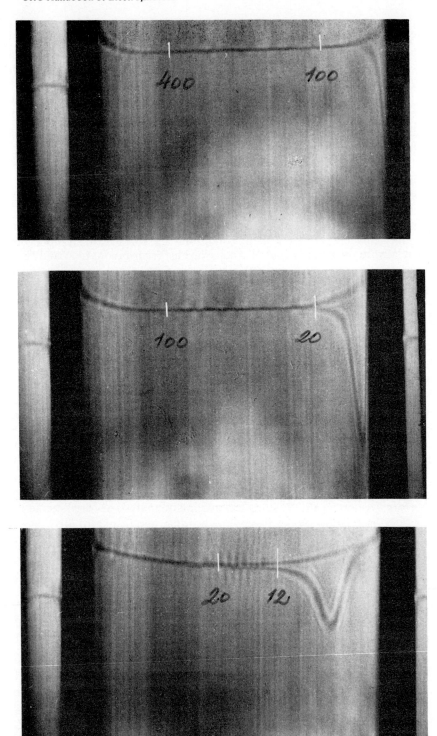

FIGURE 1.   Analytical ultracentrifugal film record demonstrating standard lipiprotein classes commonly present in normolipemic human plasma. Arbitrary flotation rate limits are determining the sum of concentrations of lipoprotein classes in the range of S_f 0—12 to S_f 100—400.

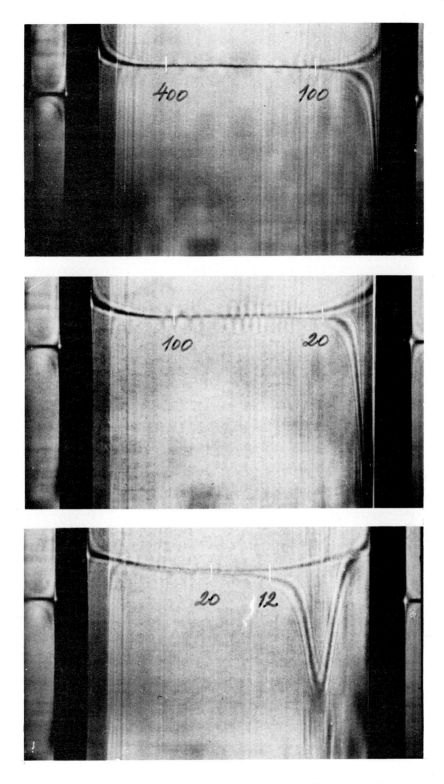

FIGURE 2. Analytical ultracentrifugal film record demonstrating standard lipoprotein classes present in Type IIA dyslipoproteinemia in a diabetic patient with a severe grade of diabetes. Arbitrary flotation rate limits are determining the sum of concentrations of lipoprotein classes in the range of S_f 0—12 to S_f 100—400.

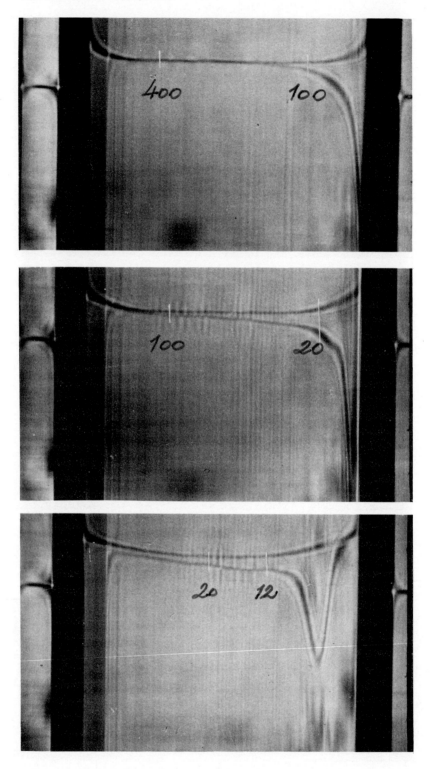

FIGURE 3.   Analytical ultracentrifugal film record demonstrating standard lipoprotein classes present in Type II—IV (IIB?) dyslipoproteinemia in a diabetic patient with diabetic retinopathy. Arbitrary flotation rate limits are determining the sum of concentrations of lipoprotein classes in the range of S_f 0—12 to S_f 100—400.

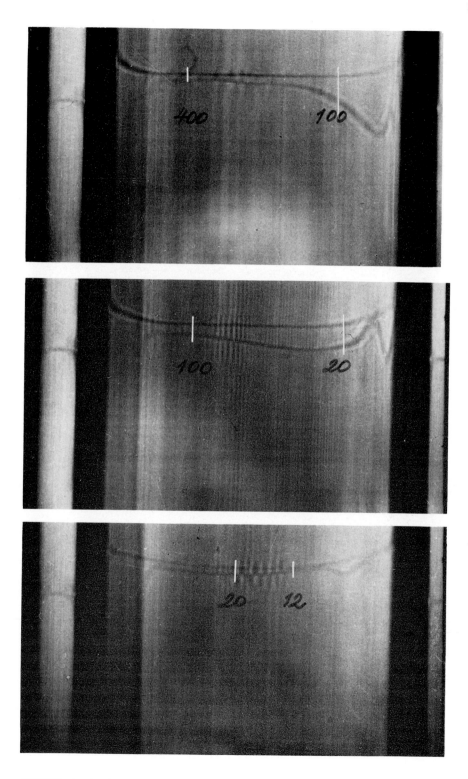

FIGURE 4. Analytical ultracentrifugal film record demonstrating standard lipoprotein classes present in Type IV dyslipoproteinemia in a diabetic patient with a brittle diabetes. Arbitrary flotation rate limits are determining the sum of concentrations of lipoprotein classes in the range of S_f 0—12 to S_f 100—400.

Duncan's test

Group	Type IIA	Type IIB (II—IV)	Type IV
$\bar{x}$	664,43	748,37	1,189,76
$ts_{\bar{x}}$(CI)	±195,92	±284,75	±613,06
s	419,23	369,81	734,80
N	20	9	8

Note: Significant differences in total concentrations of lipoprotein classes S°_f 0-400 were found between Type IIA
and IV (p = 0.05).

See Figure 5.

Significance of differences in concentrations of lipoprotein classes S°_f
0-12 in different phenotypes

Analysis of variance

Variation	Sum of squares	Degree of freedom	Variance	F
Between sets	153.562,68	2	76.781,34	0,965
Within sets	2,704.985,20	34	79.558,39	(N.S.)
Total	2,858.547,88	36		

Duncan's test: not applicable.

See Figure 6.

Significance of differences in concentrations of lipoprotein classes S°_f
12-20 in different phenotypes

Analysis of variance

Variation	Sum of squares	Degree of freedom	Variance	F
Between sets	3.949,36	2	1.974,68	0,451
Within sets	148.765,08	34	4.375,44	(N.S.)
Total	152.714,44	36		

Duncan's test: not applicable.

See Figure 7.

Significance of differences in concentrations of lipoprotein classes S°_f
20-100 in different phenotypes

Analysis of variance

Variation	Sum of squares	Degree of freedom	Variance	F
Between sets	892.925,56	2	446.462,78	18,288
Within sets	950.690,34	34	27.961,48	(p<<0.01)
Total	1,843.615,90	36		

Duncan's test

Group	Type II	Type IIB (II—IV)	Type IV
$\overline{x}$	84,54	163,00	478,11
$ts_{\overline{x}}$ (CI)	±37,45	±61,64	±278,08
s	80,14	80,05	333,25
N	20	9	8

See Figure 8.

Significance of differences in concentrations of lipoprotein classes S°_f 100-400 in different phenotypes

Analysis of variance

Variation	Sum of squares	Degree of freedom	Variance	F
Between sets	521.992,73	2	260.996,37	8,833
Within sets	1,004.671,03	34	29.549,15	(p<0.005)
Total	1,526.663,76	36		

Duncan's test

Group	Type II	Type IIB (II—IV)	Type IV
$\overline{x}$	19,58	70,22	319,39
$ts_{\overline{x}}$ (CI)	10,77	40,07	311,07
s	23,04	52,03	372,81
N	20	9	8

Differences are significant.

See Figure 9.

Significance of differences in relative distribution (Expressed in % of total) of classes S°_f 0-12 in the patterns of different phenotypes

Analysis of variance

Variation	Sum of squares	Degree of freedom	Variance	F
Between sets	10.952,89	2	5.476,45	68,064
Within sets	2.735,57	34	80,46	(p<<0.01)
Total	13.688,46	36		

Duncan's test

Group	Type IV	Type IIB (II—IV)	Type IIA
$\overline{x}$	30,39	56,69	73,86
$ts_{\overline{x}}$ (CI)	8,70	3,42	4,57
s	10,43	4,44	9,78
N	8	20	9

Differences are significant.

See Figure 10.

Significance of differences in relative distribution (Expressed in % of total) of classes S°_f 12-20 in the patterns of different phenotypes

Analysis of variance

Variation	Sum of squares	Degree of freedom	Variance	F
Between sets	127,96	2	63,98	1,899
Within sets	1.145,57	34	33,69	(N.S.)
Total	1.273,53	36		

See Figure 11.

Significance of differences in relative distribution (expressed in % of total) of classes $S°_f$ 20-100 in the patterns of different phenotypes
Analysis of variance

Variation	Sum of squares	Degree of freedom	Variance	F
Between sets	4.501,97	2	2.250,99	43,115
Within sets	1.775,01	34	52,21	(p<<0.01)
Total	6.276,98	36		

Duncan's test

Group	Type IIA	Type IIB (II—IV)	Type IV
$\overline{x}$	12,01	22,28	39,96
$ts_{\overline{x}}$ (CI)	3,68	2,84	6,98
s	7,87	3,68	8,36
N	20	9	8

Differences are significant.

See Figure 12.

Significance of differences in relative distribution (Expressed in % of total) of classes $S°_f$ 100-400 in the patterns of different phenotypes
Analysis of variance

Variation	Sum of squares	Degree of freedom	Variance	F
Between sets	2.300,55	2	1.150,28	20,971
Within sets	1.864,81	34	54,85	(p<<0.01)
Total	4.165,36	36		

Duncan's test:

Group	Type II	Type IIB (II—IV)	Type IV
$\overline{x}$	2,74	9,58	22,76
$ts_{\overline{x}}$ (CI)	1,13	4,85	11,96
s	2,41	6,30	14,33
N	20	9	8

Differences are significant.

See Figure 13.

From the statistical evaluations and graphical illustrations it follows that the electrophoretograms (LPEA) in diabetes indicate correctly the real concentrations of VLDL classes of $S°_f$ 20-100 and $S°_f$ 100-400, which differ significantly between the phenotypes IIA, IIB, and IV (p = 0.001). The highest being, of course, in the Type IV. However, the classes of $S°_f$ 0-12, characteristically elevated in dyslipoproteinemia of phenotypes

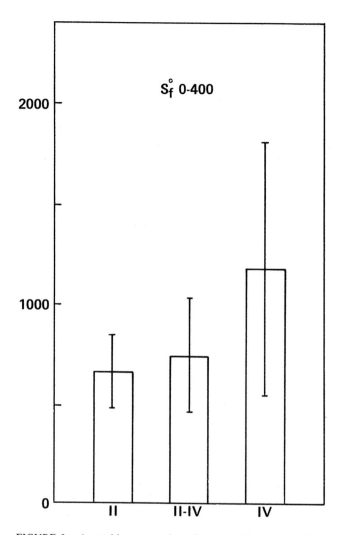

FIGURE 5. A graphic presentation of concentrations of total lipoprotein classes in phenotypes IIA, IIB, and IV in patients with diabetes. Axis x shows the symbols for phenotypes IIA (II), IIB (II—IV),
and IV. Axis y shows the concentrations of lipoprotein classes S_f°
0—400 (in mg %). The vertical bars represent the mean (x) ± 0.95%
C.I. (= confidence interval, $t_{0.975}$ S_x).

IIA, do not exhibit significant differences between all three types apparently because
of a general tendency of the metabolic disorder in diabetes that leads to the enhancement of circulating LDL. Although these lipoproteins in Type IIA are significantly
higher (p = 0.01), when observed in the patterns as relative components (estimated in
relative %). The same is true as regards the distribution of VLDL classes (S_f° 20-400)
in the electrophoretic or ultracentrifugal patterns of diabetic patients, classified as
phenotype IV (p = 0.01).

A logical supposition was that there had to be some significant differences between
the phenotypes. Nevertheless, our detailed study using combined (electrophoretic and
ultracentrifugal) methods proved that the only safe approach to the reliable classification of phenotypes and their subtypes has to be based on a consideration of concentrations of standard classes or fractions of plasma lipoproteins.

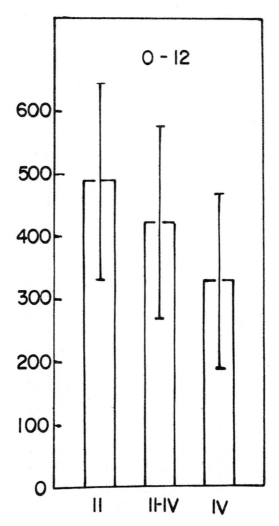

FIGURE 6.   A graphic presentation of concentrations of classes $S_f°$ 0—12 in phenoypes IIA, IIB, and IV in patients with diabetes. Axis x shows the symbols for phenotypes IIA (H), IIB (II—IV), and IV. Axis y shows the concentrations of lipoprotein classes $S_f°$ 0—12 (in mg %). The vertical bars represent the mean (x) ± 0.95% C.I. (= confidence interval, $t_{0.975}$ s_x).

RESULTS OBTAINED FROM A STUDY OF RELATIONSHIPS BETWEEN PLASMA LIPIDS AND PLASMA LIPOPROTEINS ANALYZED BY ELECTROPHORESIS AND ULTRACENTRIFUGATION

Lipoprotein analyses usually are evaluated by standardized lipid analyses, and usually for the purpose of typing and for clinical diagnostic use. Although some relationships already have been described,[829] they have not been evaluated for diabetes and its specific complications.

Statistical evaluations of analytical data were performed with the estimation of correlation coefficients. Graphical illustrations of the correlations were accomplished by insertion of analytical data and regression lines (expressing interdependence of y on x

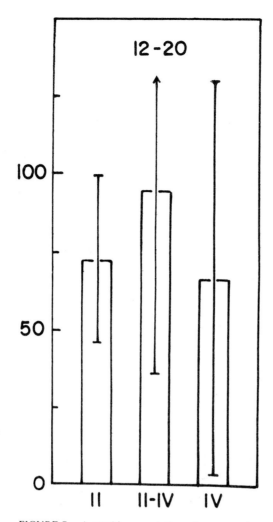

FIGURE 7. A graphic presentation of concentrations of classes $S_f°$ 12—20 in phenotypes IIA, IIB, and IV in patients with diabetes. Axis x shows the symbols of phenotypes IIA (II), IIB (II-IV), and IV. Axis y shows the concentrations of lipoprotein classes $S_f°$ 12—20 (in mg%). The vertical bars represent the mean (x) ± 0.95% CI(= confidence interval, $t_{0.975}$ S_x).

(full lines) and x on y (broken lines) in the graphs.) According to the statistical results, we evaluated the correlations as moderate (r < 0.5), good (r = 0.5-0.7), close (r = 0.7-0.9) or very close (r > 0.9).

The correlations between the levels of total plasma lipids and a sum of lipoprotein classes of $S_f°$ 0-400 were found in diabetics only good (r = 0.78, N = 55) and even worse, if classes of $S_f°$ 12-400 were considered (r = 0.70, N = 55).

The correlation between the values, obtained by precipitation of plasma lipoproteins according to Burnstein,[110-122] and total plasma lipids was evaluated as close (r = 0.82, N = 55). Perhaps most of LDL and VLDL were involved in the precipitation.

The correlation between the concentrations of total plasma cholesterol and a sum of classes of plasma lipoproteins, characterized by $S_f°$ 0-400, was surprisingly close (r = 0.86, N = 57) for diabetic patients without and with diabetic vascular complications (Figures 14 and 15). This unusual result may be related to our previous experience of

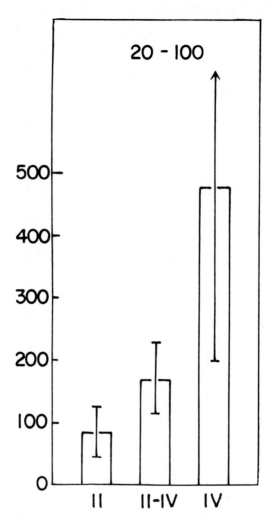

FIGURE 8. A graphic presentation of concentrations of classes $S_f°$ 20-100 in phenotypes IIA, IIB, and IV in patients with diabetes. Axis x shows the symbols for phenotypes IIA (II), IIB (II—IV), and IV. Axis y shows the concentrations of lipoprotein classes S_f 20-100 (in mg%). The vertical bars represent the mean (x) ± 0.95% IC(= confidence interval, $t_{0.975}$ S_x).

finding a preponderance of the phenotypes IIA and IIB in the lipoprotein electrophoretic as well as in ultracentrifugal patterns. To demonstrate this statement, we selected lipoprotein patterns classified as Type II (IIA and IIB) and correlated their concentrations of $S_f°$ 0-12 classes (in mg%) with the corresponding concentrations of plasma cholesterol. As expected, the correlation was very close (r = 0.95, N = 20). The failure of the same classes of lipoproteins ($S_f°$ 0-12) to correlate with corresponding total cholesterol levels was proven in selected cases with dyslipoproteinemia Type IV (r = 0.08, N = 17). These more or less constrained correlations, were calculated mostly for the demonstration of limited diagnostic value of the total plasma cholesterol levels as a screen of dyslipoproteinemia in diabetes (as well as in any other clinical syndrome).

The values of concentration of esterified fatty acids in plasma of diabetics correlated well with the corresponding concentrations of total lipoprotein classes, flotating at d = 1.063 g/mℓ ($S_f°$ 0-400). The correlation coefficient was close (r = 0.83, N = 56),

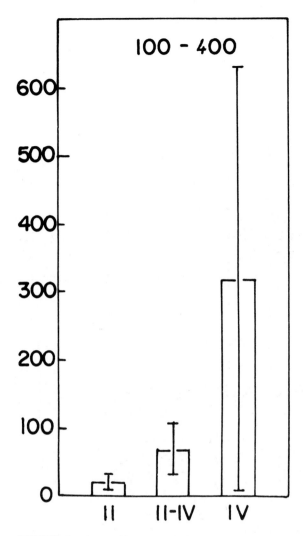

FIGURE 9. A graphic presentation of concentrations of classes of S_f° 100-400 in phenotypes IIA, IIB, and IV in patients with diabetes. Axis x shows the symbols for phenotypes IIA (II), IIB (II—IV) and IV. Axis y shows the concentrations of lipoprotein classes S_f° 100-400 (in mg%). The vertical bars represent the mean (x) ± 0.95% C.I. (= confidence interval, $t_{0.975}$ s_x).

probably because of a relative compensation in distribution of esters of fatty acids throughout the lipoprotein spectrum: classes characterized by low S_f° values are rich on cholesterol-esters; classes of high S_f° values are loaded with glycerol-esters (= triglycerides), so that the fraction of esterified fatty acids may be fairly balanced in all of them. Using values for classes S_f° 12-400 for the above discussed correlation, the results are equal (r = 0.86, N = 56).

The concentrations of plasma triglycerides are very closely correlating with the flotating fraction of plasma lipoproteins at d < 1.063 (r = 0.95, N = 19) as well as with its VLDL-fraction of S_f° 20-400 (r = 0.93, N = 19) (Figure 16).

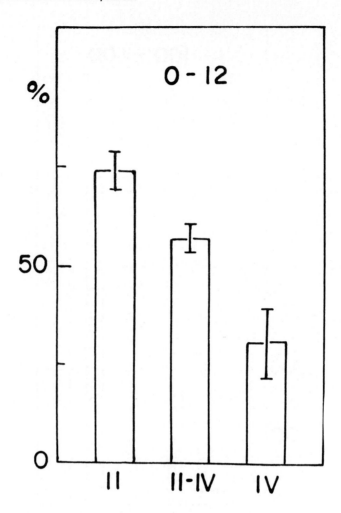

FIGURE 10. A graphic presentation of relative distribution of classes S_f° 0-12 in the patterns of different phenotypes. Axis x shows the symbols for phenotypes IIA (II), IIB (II—IV), and IV. Axis y shows the relative distributions of classes S_f° 0-12 in the patterns (in relative %). The vertical bars represent the mean (x) ± 0.95% C.I. (= confidence interval, $t_{0.975}$ s_x).

RESULTS OBTAINED FROM COMPARISON OF DIFFERENT METHODS FOR THE ESTIMATION OF TOTAL PLASMA LIPOPROTEINS

The correlation of other methods used for the estimation of plasma lipoproteins or their fractions also deserves attention.

The values gained by Burstein's precipitation method (using heparin and Ca^{++}) correlated well with the levels of total plasma lipids (r = 0.82, N = 55), but not with the total fraction of plasma lipoproteins flotating at d = 1.063 g/mℓ and characterized by S_f° 0-400 (r = 0.34, N = 55). See Figure 17. Such divergence, which had not been expected, required further analyses. It appeared that the discrepancy specifically accompanies the flotation classes of S_f° 0-20 (r = 0.27, N = 56) equal to LDL classes of plasma lipoproteins, where the classes of S_f° 20-400, containing mostly the VLDL, correlate well (r = 0.82, N = 55). These measurements are important for the proper

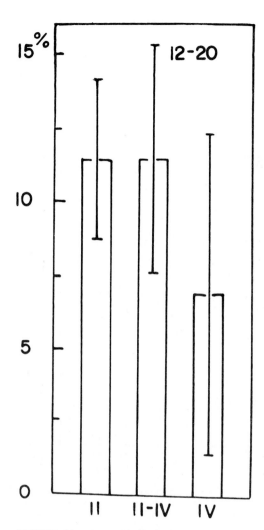

FIGURE 11. A graphic presentation of relative distribution of classes S_f 12-20 in the patterns of different phenotypes. Axis x shows the symbols for phenotypes IIA (II), IIB (II—IV), and IV. Axis y shows the relative distributions of classes $S_f°$ 12-20 in the patterns (in relative %). The vertical bars represent the mean (x) ± 0.95% C.I. (= confidence interval, $t_{0.975}$ S_x).

evaluation of the clinical values of Burstein's modification, which was used in this study.

Of no less importance was the next correlative study, employing the measurements of concentrations of the Apo-B fractions of apolipoproteins, which constitute the main portion of LDL_2-apoproteins and are another steady component of VLDL-apoproteins. The method of radial diffusion in our own modification of the original proposition of Mancini et al.[476] and calculations according to Darcy[163] and Augener[26] was very suitable. By using specific antibodies (antisera "Anti-B" manufactured by Behringwerke-Marburg Lahn, DBR*) we were able to estimate the statistical analysis of variance, ± SD, and mean values along with their 95% confidence intervals (ts$_x$) (for concentrations 500 mg% ± 20 to 30 mg%, for concentrations 1000 mg% ± 50 to 80 mg%). The relationship between the values of radial diffusion (d in 0.1 mm/log c in

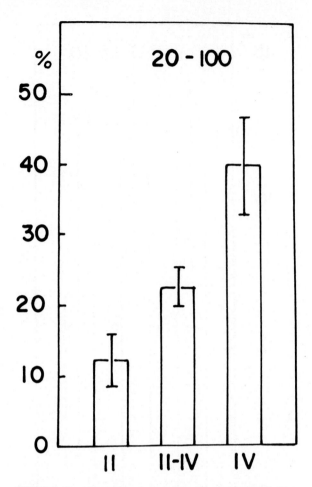

FIGURE 12.   A graphic presentation of relative distribution
of classes $S_f°$ 20—100 in the patterns of different phenotypes.
Axis x shows the symbols for phenotypes IIA (II), IIB (II—IV)
and IV. Axis y shows the relative distrubitons of classes $S_f°$
20—100 in the patterns (in relative %). The vertical bars represent the mean (x) ± 0.95% C.I. (= confidence interval, $t_{0.975}$
s_x).

mg%) and the concentrations of the sum of lipoprotein classes of $S_f°$ 0-400, containing
Apo-B, is very close (r = 0.97, N = 51). See Figure 18. Beside the ultracentrifugation,
the determination of the concentrations of Apo-β in the floating plasma lipoproteins,
appeared to be the best laboratory method for clinical evaluation of their total levels.
(Radial diffusion is approximately 100 times faster and 100 times less expensive than
ultracentrifugation and one technologist can handle 200 specimens per day).

The total concentrations of flotating lipoproteins — if precisely determined — indicate very sensitively pathologic changes in the plasma lipoproteins, wherever they occur: in the $S_f°$ 0-20 classes or in LDL_2 (and in lesser extent in LDL_1 and HDL_1), and/
or in the $S_f°$ 20-400 classes, corresponding closely to the VLDL. The primary clinical
screening or follow-up does not necessarily require detailed analyses of plasma lipoproteins, nor the exact typing of the dyslipoproteinemia.

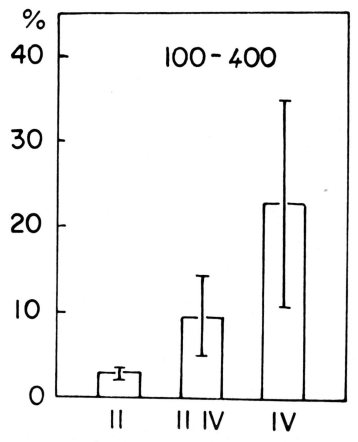

FIGURE 13. A graphic presentation of relative distribution of classes S_f° 100-400 in the patterns of different phenotypes. Axis x shows the symbols for phenotypes IIA (II), IIB (II—IV), and IV. Axis y shows the relative distributions of classes S_f° 100—400 in the patterns (in relative %). The vertical bars represent the mean (x) $\pm$ 0.95% C.I. (= confidence interval, $t_{0.975} s_x$).

RESULTS OBTAINED FROM A STUDY OF THE RELATION BETWEEN ELECTROPHORETIC FRACTIONS AND ULTRACENTRIFUGAL CLASSES OF PLASMA LIPOPROTEINS

By staining isolated lipoprotein fractions with the help of electrophoresis, we detect mostly the esters of fatty acids with glycerol (= triglycerides) and/or with cholesterol distributed relatively equally throughout the α_2-β-complex. The sum of all stained electrophoretic fractions of flotating lipoproteins at d = 1063 g/ml, or α_2-β-complex, expressed in units of areas delimited by the corresponding peaks, was compared with the values (in mg%) of lipoprotein classes of S_f° 0-400. This relation was close (r = 0.83, N = 55). It was interesting to observe that similar regression coefficients were achieved by comparing the same sum of lipoprotein fractions (α_2-β-complex) with the total levels of esterified fatty acids (r = 0.83) and even better with the total levels of plasma triglycerides (r = 0.95). See Figures 19 and 20.

Encumbrance was demonstrated in relating the flotation fraction (S_f° 0-12) with the electrophoretic fraction-β (r = 0.28, N = 56). This relationship was found in diabetic patients with and without diabetic vascular complications (LPEA), contrary to the findings of other authors[803,832,854,856] for patients with atherosclerosis or for subjects without any clinically demonstrable disease (LAGE).[829]

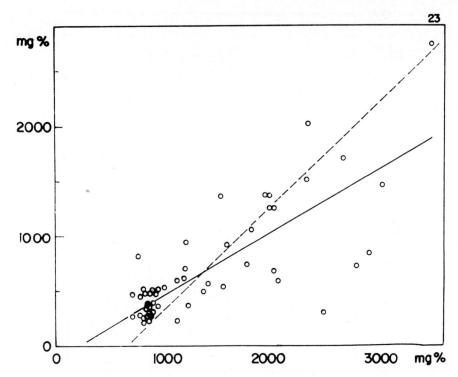

FIGURE 14.   Graphical presentation of the relation between concentrations of flotating lipoproteins at $d = 1.063$ g/mℓ (S$_f^\circ$ 0—400) and the total lipidemia in plasma of diabetics.

Description of the graph:
    Unbroken line: regression line, expressing the relation of the variable y on the variable x (y
        $= 0.58 x - 129.66$).
    Broken line: regression line, expressing the relation of the variable x on the variable y
        (x $= 1.062 y + 661.58$).
    r $= 0.789$ (the difference from zero is statistically significant at p<0.01)
    N $= 55$
    On the axis x: concentrations of the total lipidemia
    On the axis y: concentrations of the flotation classes S$_f^\circ$ 0−400
    Result: the correlation was found "close"

However, a substantially closer relationship was indicated by a comparison of the
electrophoretic fraction α_2- and/or pre-β (LPEA) with the ultracentrifugal class S$_f^\circ$ 20-
400 (r $= 0.78$, N $= 56$). This finding (LPEA) agrees better with results of
authors[829,856,858] studying patients with atherosclerosis or subjects who are healthy
(LAGE).

These measurements have an important practical impact on the daily work of clinical
pathologists and diabetologists. They indicate a degree of confidence that has to be
known before any evaluation or correlation of lipid and lipoprotein analyses can be
safely applied in a diabetologic practice.

CONCLUSIONS FROM THE AUTHOR'S STUDY OF LIPIDS AND LIPOPROTEINS IN DIABETES

This author's basic research of different electrophoretic methods included the electrophoresis of the lipoproteins, using paper as a support media both with and without
the buffer system in which either albumin (to a concentration of 1%) is added to the

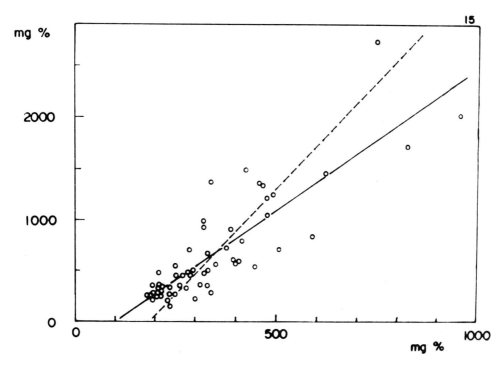

FIGURE 15. Graphical presentation of the relation between concentrations of flotating lipoproteins at d = 1.063 g/mℓ (S$_f$° 0—400) and the total plasma cholesterol in diabetics.

Description of the graph:

    Unbroken line: regression line, expressing the relation of the variable y on the variable x (y = 2.754 x − 284.19).

    Broken line: regression line, expressing the relation of the variable x on the variably y (x = 0.268 y + 166.90).

    r = 0.859 (the difference from zero is statistically significant, at p <0.01)

    N = 57

    On the axis x: concentrations of total cholesterol

    On the axis y: concentrations of lipoprotein classes S$_f$° 0—400

    Result: the correlation was found "close"

buffer solution or a "TEB"-buffer is used. The agarose-gel electrophoresis and the electrophoresis of lipoproteins on acetylated cellulose also was investigated. The electrophoretic mobilities of lipoprotein fractions were determined on agarose gels and in the cuvette (by use of so-called free electrophoresis). The isoelectric points of lipoprotein electrophoretic fractions were determined by electrophoretic convexion. Immunoelectrophoresis was used to identify single lipoproteins, and the radial diffusion technique was used to determine the concentrations of lipoprotein classes containing the Apo-β-lipoprotein. The electrophoretic method using paper as a support media, the veronal buffer-system with added albumin (LPEA), and the ultracentrifugal analysis of lipoproteins according to Gofman and DeLalla (UG) were employed throughout the entire research of dyslipoproteinemia in diabetics with and without diabetic vascular complications. All results were statistically evaluated and graphically illustrated. The results were compared with findings obtained by plasma lipid analyses (of total lipidemia, esterified fatty acids, triglyceridemia, cholesterolemia, and phospholipidemia), by precipitation techniques according to Burstein, and by radial diffusion techniques. Finally, the results from electrophoretic and ultracentrifugal research were correlated. All methodical approaches were statistically and graphically evaluated for

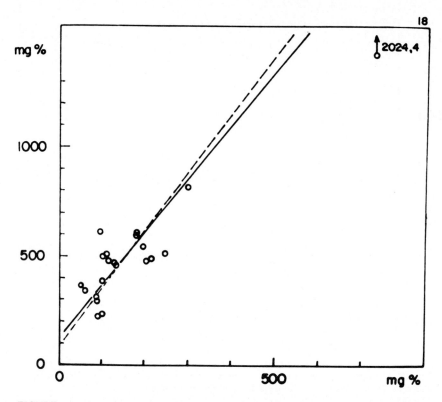

FIGURE 16.   Graphical presentation of the relation between concentrations of flotating lipoproteins at d = 1.063 g/ml (S$_f^o$ 0—400) and the total plasma triglycerides in diabetes.

Description of the graph:

   Unbroken line: regression line, expressing the relation of the variable y on the variable x (y = 2.419 x + 133.68).

   Broken line: regression line, expressing the relation of the variable x on the variable y (x = 0.371 y − 32,33).

   r = 0.948 (the difference from zero is statistically significant at p<0.01)

   N = 19

   On the axis x: concentrations of total plasma triglycerides

   On the axis y: concentrations of lipoprotein classe S$_f^o$ 0—400

   Result: the correlation was found "very close"

groups totaling 162 diabetics with different grade of severity of diabetes, groups totaling 52 diabetics with proven diabetic retinopathy, and groups totaling 110 diabetics with proven diabetic glomerulosclerosis, examined in different stages of development of the nephropathy.

Patients with a severe grade of diabetes have significantly higher levels of total plasma lipoproteins, caused mainly by a significant increase of the β-electrophoretic fraction, despite a significant decrease of the α$_1$-lipoproteins. However, the pre-β fraction increases only insignificantly.

No significant differences in the electrophoretic patterns were found between diabetics with various degrees of severity of the disease and diabetics with diabetic retinopathy and/or atherosclerosis. However, diabetics with diabetic nephropathy can clearly be distinguished from the others by their significantly higher levels of electrophoretic fractions (pre-β and β-, or the β-fraction alone), while the α$_1$-fractions remain very significantly depressed. These changes are more striking in the advanced stages of this syndrome, particularly in the nephrotic stage. Patients with an advanced dia-

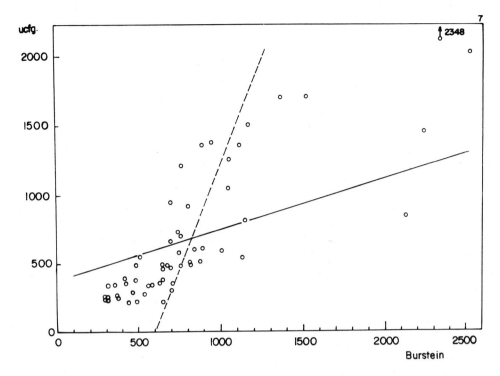

FIGURE 17. Graphical representation of the relation between concentrations of flotating lipoproteins at d = 1.063 g/ml (S, 0—400) and the fractions of plasma lipoproteins, which are in precipitated by heparin, the presence of CaCl₂, according to Burnstein.[117] Diabetic patients with different grades of the severity of diabetes and different diabetic vascular complications (55).

betic nephropathy have significantly more severe dyslipoproteinemia than their counterparts with clinically apparent atherosclerosis only.

The diabetics with an additional hepatic disorder were more often (75% of the time) classified as having a severe grade of diabetes, accompanied by a severe dyslipoproteinemia, and diabetics with complicating kidney infections evidenced the highest levels of electrophoretic β-fraction in their lipoprotein patterns.

Although the levels of α₁-lipoproteins are significantly low in diabetics, they have always been higher in patients with severe diabetes either with or without glomerulosclerosis. The author believes this phenomenon is caused by an unspecific increase of HDL in the α₁-lipoprotein fraction.

The pre-β-electrophoretic fraction used to be mostly elevated in the Kimmelstiel-Wilson syndrome (in 59% of cases) and even more so in diabetics with clinical complications of atherosclerosis (in 82% of cases), but it shows few changes during the development of these complications.

The ultracentrifugal research of plasma lipoproteins in diabetes confirmed that the difference between total lipoprotein levels in diabetics and healthy people of similar age and sex is highly significant (approximately 3 × higher), but the variations in individual values are high.

The increases in lipoprotein classes S°, 0-12 and 12-20 contribute most significantly to this phenomenon (at p = 0.01 level), and where the classes of S°, 20-100 are generally less elevated (significantly, but only at p = 0.05 level) the class of S°, 100-400 remains insignificantly changed.

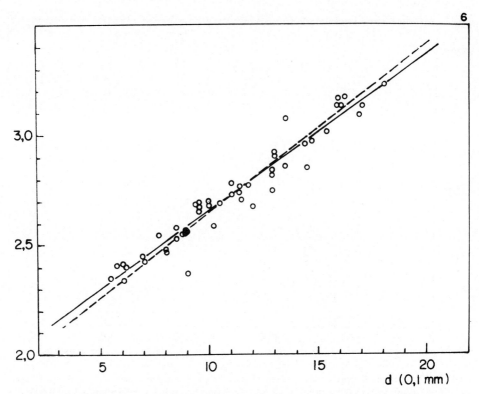

FIGURE 18.   Graphical representation of the relation between the concentrations of Apo-B in plasma, determined by the radial diffusion techniques in mg% and the sum of lipoprotein classes (S$_f$° 0—400) containing Apo-B. Diabetic patients (51) with different grade of severity of diabetes and different diabetic vascular complications.

Description of the graph:

   Unbroken line: regression line, expressing the relation of the variable x (y = 0.0709 x + 1.95).
   Broken line: regression line, expressing the relation of the variable y (x = 13.170 y − 24,98).
   r = 0.967 (the difference from zero is significant at p<0.01)
   On the axis x: values of the radial diffusion (d in mm)
   On the axis y: log c (= concentration) of plasma lipoproteins (S$_f$° 0—400)
   Result: "very close" correlation

However, the patterns change in diabetics with proven diabetic retinopathy where the ultracentrifugal research indicated the strikingly significant elevation in the class of S°$_f$ 20-100 (at p = 0.01), with an additional increase in the class of S°$_f$ 100-400 (at p = 0.05 level). Such a metabolic disorder was not previously apparent from the electrophoretic research. It should be noted that the total concentrations of plasma lipoproteins (for S°$_f$ 0-400), as well as of the classes defined by S°$_f$ 0-12 and 12-20, remain significantly elevated (p = 0.01) in comparison to the "normal" values (found in the clinically healthy population).

Subsequent ultracentrifugal research revealed that diabetic patients with proven diabetic glomerulosclerosis do not have significantly different dyslipoproteinemia than diabetics with proven diabetic retinopathy, and that both groups of patients do not significantly differ from the group of diabetics without diabetic vascular complications but with various grades of severity of the disease.

This means that diabetics with a specific vascular disease demonstrate more severe dyslipoproteinemia than diabetics without such a disease, but the differences are not statistically significant.

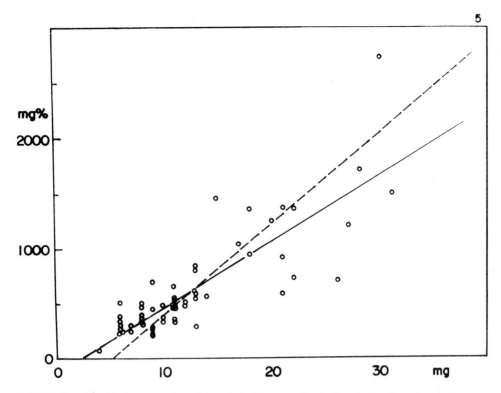

FIGURE 19.   Graphical presentation of the relation between electrophoretic fractions, forming a complex of α_2-, pre-β and β-fractions, and the sum of ultracentrifugal classes defined by $S_f°$ 0—400 in diabetics with and without diabetic vascular complications.

Description of the graph:
    Unbroken line: regression line, expressing the relation of the variable y on the variable x (y = 59.425
        x − 117.70).
    Broken line: regression line, expressing the relation of the variable x on the variable y (x = 0.012 y
        + 5.33).
    r = 0.833 (the difference from zero is statistically significant at p<0.01).
    N = 55
    On the axis x: weight of areas, corresponding to the electrophoretic fractions α_2−, pre-β and β- (in
        mg)
    On the axis y: concentrations of lipoprotein classes of $S_f°$ 0—400 (in mg%)
    Result: the correlation was found "close"

All diabetic patients, with and without diabetic vascular complications, most often have Type-IIA (= 54%), or Type-IIB (= 24%) dyslipoproteinemia and relatively infrequently Type IV (= 22%). Type II dyslipoproteinemia was found in all patients with proven diabetic glomerulosclerosis.

It was assumed that phenotypes may be more difficult to determine in diabetes than in other syndromes. Therefore, extensive research was undertaken in this direction. It appears that for the correct determination of phenotypes, one has to know the concentrations of LDL and VLDL, because the relative distribution of ultracentrifugal classes and electrophoretic fractions in the lipoprotein patterns may not be (and were found not to be) significantly distinct. Otherwise, it has been proven that the phenotypes in diabetes correspond to all criteria found in the general field of lipoprotein disorders.

Furthermore, various correlations between the plasma lipids and lipoprotein classes and fractions were studied. The correlations between the total plasma lipids and the sum of lipoprotein classes ($S_f°$ 0-400) were found to be good only for r = 0.78 —

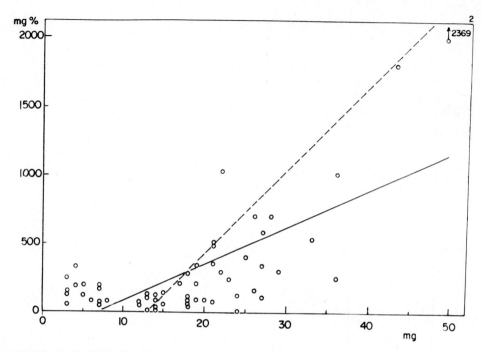

FIGURE 20.   Graphical presentation of the relation between electrophoretic-α_2 and pre-β-fractions and the sum of ultracentrifugal classes defined by $S_f°$ 20—400 in diabetics with and without diabetic vascular complications.

Description of the graph:
>   Unbroken line: regression line, expressing the relation of the variable y on the variable x (y = 27.768 x − 194.11).
>   Broken line: regression line, expressing the relation of the variable x on the variable y (x = 0.017 y + 12.79).
>   r = 0.784 (the difference from zero is statistically significant at p<0.01)
>   N = 56
>   On the axis x: weight of areas, corresponding to the electrophoretic fractions α_2 and pre-β (in mg)
>   On the axis y: concentrations of lipoprotein classes of S_f 20—400 (in mg%)
>   Result: the correlation was found "close"

worse, if the classes $S_f°$ 12-400 were considered (r = 0.70). If correlated with precipitated fractions of plasma lipoproteins (by heparine and Ca^{++} according to Burstein), the relation was closer (r = 0.82).

The total plasma cholesterol levels correlated well (r = 0.86) with ultracentrifugal classes ($S_f°$ 0-400) of plasma lipoproteins perhaps because the classes $S_f°$ 0-20 are the major components of the flotation fraction at d = 1.063 g/ml (correlations with them were very close: r = 0.95, but not with the classes of $S_f°$ 20-400: r = 0.08).

The total concentrations of esterified fatty acids correlated with the sum of plasma lipoprotein classes of $S_f°$ 0-400 (r = 0.83) and even better with its component of $S_f°$ 12-400 (r = 0.86). A close correlation was observed also with the total concentrations of the sum of electrophoretic fractions α_2-, pre-β, and -β (r = 0.83).

The concentrations of total plasma triglycerides revealed similar close relationships to the whole fraction of flotating lipoproteins at d < 1.063 g/ml (r = 0.95), as well as to its portion of $S_f°$ 20-400 (r = 0.93) and to the sum of electrophoretic fractions α_2-, pre-β, and -β (r = 0.95). The sum of electrophoretic fractions α_2-, pre-β, and -β correlated closely with the sum of flotating lipoproteins at d < 1.063 ($S_f°$ 0-400): r = 0.83.

However, there was a very inferior correlation between the concentrations of electrophoretic β-fraction and the concentrations of the lipoprotein class of S°_f 0-12 (r = 0.28).

In contrast, the relationship between the concentrations of the electrophoretic subfractions α_2- and pre-β, and the concentrations of the ultracentrifugal class of S°_f 20-400 proved to be closer (r = 0.78).

For direct screening purposes of dyslipoproteinemia in diabetes, basically two methods were evaluated:

1. The precipitation procedure according to Burstein, which disclosed very good correlation with the levels of total plasma lipids (r = 0.82), but an inferior one with the values of total floating lipoproteins of S°_f 0-400 (r = 0.34) because of the specific relationships to their classes (S°_f 20-400), consisting of VLDL (r = 0.82). No significant relationships were found to LDL, which constitute the main components of classes of S°_f 0-20 (r = 0.27).
2. In contrast, the radial diffusion technique, employed for the determination of concentrations of the Apo-β-lipoprotein (forming almost an entirety of the LDL-apolipoprotein and a fraction of the VLDL-apolipoproteins), proved to be an excellent analytical tool for indirect measurements of total levels of plasma lipoproteins, floating at d = 1.063 g/ml (S°_f 0-400). The significance of the relationship of both analytical values is very high (r = 0.97). This method is very recommendable for practical clinical use.

RECENT ADVANCES IN THE RESEARCH OF PLASMA LIPOPROTEINS IN DIABETES MELLITUS

Further research in lipids and lipoproteins in diabetes basically confirmed the findings and relationships discussed in the previous sections. Gligore et al.,[803] using agarose-gel electrophoresis reached similar conclusions, based on a research of 50 diabetics and 20 healthy individuals. All diabetics with angiopathy had hyperlipoproteinemia. Serum pseudocholineesterase was most active in Type IV hyperlipoproteinemia. The authors evaluated the diabetic dyslipoproteinemia as a valuable criterion for preventive measures in diabetes mellitus, especially with respect to vascular complications and atherosclerosis (LAGE).

Chance and Albutt[800] are in favor of the classification of diabetic dyslipoproteinemias according to phenotypes, but they oppose equal criteria applied by Roe and Walsh[438] for "untreated" as well as "poorly controlled" diabetics. The differentiation in the typing of dyslipoproteinemia as II and III and "mixed" is considered less helpful in clinical practice, although Type III may occur in poorly controlled diabetes. Type IV dyslipoproteinemia in diabetes may be insulin sensitive, but may respond to caloric deprivation in mild diabetes with or without obesity (LAGE).[801] Type IV dyslipoproteinemia is induced by transformation of carbohydrates into fat (VLDL) and by excess lipolysis in diabetics with ketosis or severe diabetes, according to Bazex et al.[795] Insulin and clofibrate, or nicotinic acid are needed for treatment. This type is generally found in labile, insulin-dependent diabetics under age 40, sometimes overdosed by insulin.[841] In contrast, Type II dyslipoproteinemia was detected in the highest incidence in diabetics with obliterating angiopathy. Type III was mostly associated with diabetics suffering from coronary artery disease (the broad-β-variant, or the Type II to IV), according to Strat.[841]

The α_1-electrophoretic fraction is generally severely depressed in diabetics,[797,842] more in those treated with diet only, or with peroral antidiabetics. Significantly low levels[805,814] were observed in diabetics with atherosclerosis, obesity, and with hepatic cirrhosis (LAGE).

A similar relationship between the complications of diabetes and the significant increase of the electrophoretic pre-β-fraction (LAGE)[797,806,814] was found between diabetics treated with diet or peroral antidiabetics and diabetics treated with insulin, but not in diabetics with hepatic cirrhosis (showed low levels).[842] The β-electrophoretic fraction was evaluated in diabetes as always significantly increased (LPE,[794,808,813,833,841] LPEA,[822,827] and LAGE[797,806,814]) but was increased also only insignificantly.[842] After therapy with insulin, it was observed that this fraction either decreased,[833,853,856] or remained high.[842]

Hyperlipoproteinemia was often found in diabetics (in 60% of 400 diabetics according to Reimer, 1973[826]) and Types II and IV in equal proportions (24% each), which is impressive. However, some authors are reporting that Type IIA occurs even more often (equal to 54%) and Types II-IV which may be equivalent to Type IIB in 24% of all diabetic dyslipoproteinemias[826] and are supporting their electrophoretic findings by ultracentrifugal analyses. Indeed, the distributions of types may vary in different groups of randomly selected diabetics. There is always a problem of how to differentiate the primary from the secondary hyperlipemia (first distinguished by Katch and Krainick in 1939). Schwandt[836] understands that whenever the patient is stabilized by insulin, the persisting hyperlipoproteinemia should be the primary one. That may be the case only in some diabetics according to findings of other investigators.[802,842,851] The primary lipid disorder should be treated with a diet and biguanidines rather than with insulin.[836] The typing in-diabetics should be performed with care. Some authors[815] do not classify fractions in lipoproteinograms correctly, some[856] employ with great care the measurements of electrophoretic mobilities of β- and pre-β fractions. The later mode of evaluation is recommended, together with a consideration of the absolute concentrations of electrophoretic fractions.

The Type IV, apparently common in brittle diabetes, worsens in precoma. The electrophoretic pre-β-fraction (LAGE) usually broadens, and later, in ketoacidosis (which may develop further a coma) Type V dyslipoproteinemia appears: the pre-β-fraction remains high and broad and the trailing chylomicrons occupy the starting spot on the electrophoretogram; the β-fraction may even be overlapped by the pre-β-fraction.[831] The therapy with insulin usually reverses this pathologic metabolic disorder in the exact opposite direction. Therefore, the electrophoretic examination of plasma lipoproteins belongs (in present methodical development) among the most valuable diagnostic tools for the clinical evaluation of diabetic ketoacidosis (LAGE).

Attention also is devoted to the α-electrophoretic fraction in diabetes, which concentration also varies during the therapy. Some authors[801,857] expressed the opinion that dye impurities rather than the FFA bound to albumin may cause falsely elevated values of α-lipoprotein. Wille and Aarseth[857] reported values of 320 to 390 mg% in three diabetic patients with a maturity onset diabetes well controlled (normal values: 200 ± 45 mg%), but they did not offer any explanation. This phenomenon deserves special investigation.

Research employing the analytical ultracentrifuge (UL) was during the last decade rarely performed. Haller et al.[804] published a study about patients (n = 21) with hypertriglyceridemia and premanifest diabetes (n = 12) and diabetes with hypertriglyceridemia (n = 9), compared with clinically healthy subjects (n = 12). The two stages of metabolic disorder displayed no significant difference of lipid and lipoprotein pattern, but both showed extremely high elevations of VLDL (class of $-S_{1.21} > 50$). This lipoprotein class was defined in the original paper by mistake as "low density lipoprotein fraction", although it correlated closely with triglyceride levels and demonstrated all physical characteristics of VLDL.

Patients (n = 81) with coronary artery disease, not proven by coronary arteriography but who suffered from previous myocardial infarction, were studied[858] together with 37 nonobese, healthy subjects (14 men and 23 women). Of these, 83% had one of the following abnormalities: diabetes or glucose intolerance (15%), diabetes or glucose intolerance with hyperlipoproteinemia (33%) and simple hyperlipoproteinemia (33%). Type IV dyslipoproteinemia was determined in 39%, Type II in 25%, while Type III and Type V were uncommon. It remains uncertain, if those figures can be considered as representative for diabetes (LAGE).

Wille and Aarseth[856] studied a group of 43 diabetics (from which 11 suffered from coronary heart disease) using agarose-gel electrophoresis and lipid analyses. Both diabetics and diabetics with CAD had significantly higher levels of β- and pre-β-fractions; diabetics with poor control presented mostly Type IV; diabetics with maturity onset diabetes often had Type IV dyslipoproteinemia, despite good clinical control. Patients in poor control presented mostly with Type II and IV. "Broad-beta-disease" was found in one case, Type IIB in two cases. Three patients had hyper-α-lipoproteinemia. Type-III-like patterns appeared in ketoacidosis; after an insulin injection, there was a decrease in lipid levels within 1 hr and a normalization after 24 hr (LAGE).

Richetti et al.[832] found significantly higher levels of triglycerides in diabetics (n = 97) in comparison to nondiabetic patients (n = 98). This metabolic abnormality was related to significantly higher incidence of coronary insufficience, recent myocardial infarction, and angina. Diabetics with fully developed symptomatology (n = 48) had significantly higher cholesterol (22), even if triglycerides were normal (21). Type IIB dyslipoproteinemia was prevalent (63). Type IIA (25) and Type IV (31) were less common in the whole group (n = 195) of studied subjects (LPE). No specific information is given that would be related especially to diabetes mellitus. Skopichenko[839] found in 40 diabetics with diabetic nephropathy a significantly higher β-electrophoretic fraction (LPE) and total cholesterol levels in comparison to 39 patients with diabetes associated with atherosclerosis and hypertensive disease.

Mazovetsky et al.[811] confirmed findings of Skopichenko in 41 diabetics with and without diabetic nephropathy, compared with 35 patients with obliterating atherosclerosis of the lower extremities. The control group included 51 persons. In patients with diabetes in comparison with healthy subjects a decrease in the electrophoretic α-fraction and rise in β-electrophoretic fraction were established. In the presence of angiopathy the fraction of β-lipoproteins was increased by 22% as compared to patients with diabetes but without angiopathy (LPE). As compared to patients with obliterating atherosclerosis, in patients with diabetes and angiopathy, an increase of α-lipoprotein fraction, a decrease in pre-β-fraction (β-fast-lipoprotein), and an increase of β-fraction (β-slow-lipoprotein) were observed.

Kremer[809] found in 247 diabetics the following distribution of phenotypes: Type IIA in 15%, Type IIB in 15%, Type IV in 29%, and Type V in 0.8%. The significant dyslipoproteinemia was estimated in 60% of all diabetics (LAGE).

Wahl et al.[854] reported a very unusual frequency of the phenotype IV in 181 diabetics (81%), while Type IIA was relatively rare (9%) as well as the other types: IIB (6.6%), III (0.6%), and V (2.8%). From 491 diabetics, 49 male (27.5%) and 132 female patients (42.2%) had hyper-lipoproteinemia. Diabetics with insulin therapy had lowest triglyceride levels. Otherwise, the body weight was closely related to the concentrations of total plasma cholesterol and triglycerides.

Kumar and Gupta[810] investigated two groups of subjects: 83 untreated diabetics and 52 healthy persons. The diabetics demonstrated significantly elevated β-electrophoretic fractions of plasma lipoproteins (LPE). Diabetic retinopathy and nephropathy, or coronary artery disease were associated with significantly high concentrations of plasma cholesterol and β-lipoproteins. The authors expressed the opinion that diabetic neuro-pathy may be unrelated to changes in plasma lipids.

Schonfeld et al.[835] investigated a relation between concentrations of apolipoprotein β and lipoprotein levels in 97 diabetics and 72 nondiabetics. Among the 97 diabetics, 71 were normolipemic, 12 were Type IIA, and 14 were Type IV. Most interesting is the observation that LDL and HDL were enriched in triglycerides (p < 0.01) in each patient of the diabetic group. Thus the presence of diabetes itself appears to result in altered low and high density lipoprotein compositions. The triglyceride-enriched LDL could mean that a less dense moiety of low density lipoproteins may be accumulating in diabetic plasma which could result from altered rates of turnover of Apo-B-contain-ing lipoproteins.

Infants of diabetic mothers have higher α_1-lipoprotein, according to Davidsen.[802] Serum lipids and lipoproteins have been investigated in 135 diabetic children prior to treatment. Serum total lipids were elevated in 64% of the patients, and serum total cholesterol in 43%. Abnormal lipoprotein patterns were found in 77%, the commonest abnormality being an increase in pre-β-lipoprotein (LPE).[801] Roe and Walsh[833] pointed out how confusing the interpretation of pre-β-electrophoretic fraction in diabetics may become and how problematic the subsequent typing of dyslipoproteinemias may be, mainly in the area resembling Type IV, V, and III. Chiefly the "broad-beta-band" presence in diabetes is a tough problem and should not be estimated without help of other methodical verifications. Brown's "mixed type", properly defined as II-IV type, caused another diagnostic problem. As Opplt[826] claimed, most of the above described difficulties may be avoided, if only the concentrations of classes and electrophoretic fractions of plasma lipoproteins would be considered for typing purposes.

Kobierska-Szczepanska[808] concluded from a study of 50 children with adequately treated diabetes that the β-lipoproteins were significantly elevated and the α-lipopro-teins decreased (LPE), mainly after 3 years of the disease, in children with poorly controlled diabetes and with suspected angiopathy.

In contrast, Murthy et al.[813] did not find differences between 68 diabetic and 22 normal subjects' lipoproteidograms (LPE), except in actual ketoacidosis. However, he admitted that sera of diabetic patients have been observed to have pre-β-lipid band increased on electrophoresis (using "Beta-L-screening test" simultaneously).

A decreased serum beta/alpha lipoprotein ratio has been reported in diabetics (n = 30) after treatment with chlorpropamide and phenetyl-l-biguanide,[812] but no significant changes after chlorpropamide and butamide.[838] Unchanged β-lipoproteins also were reported[834] after biguanide therapy (estimated by precipitation with dextransulfate).

Tiengo et al.[851] described the highest levels of triglycerides in patients with reduced insulin secretion, which may in turn provoke a reduction of lipoprotein lipase activity. This phenomenon could be explained by continuous hyperstimulation of β-cells, partly reinforced by the associated obesity and partly by carbohydrate rich diet, and may lead very often to insulin-dependent diabetes.

However, the lipoprotein lipase activity was not found altered in a group of 80 dia-betics, according to Bhan et al.[796] Significantly increased triglycerides and pre-β-lipo-proteins together with β/α-ratio may result from a primary lipid disturbance and may not be carbohydrate dependent nor associated with the lipoprotein lipase activity.

CONCLUSIONS

The recent research of dyslipoproteinemia in diabetes revealed general aims, highly important in the clinical management of the disease. The majority of authors selected one of the electrophoretic methods and used it for research purposes in diabetic patients, adults, and children, randomly assembled in groups according to the type of diabetes, clinical management, treatment, and, in a few cases, according to the degree of severity of the disease and according to different vascular complications. In this last decade of research, there has been a common trend to distinguish dyslipoproteinemias in diabetes by typing them according to Fredrickson and coworkers. The majority of researchers employed the electrophoresis of plasma lipoproteins in semi-quantitative versions and mostly electrophoresis on paper. The agarose-gel electrophoresis was utilized less often and the ultracentrifugation only exceptionally. Therefore, only very few researchers displayed quantitative values of lipoprotein classes or electrophoretic fractions.

The results, derived from the research of plasma lipoproteins in the last decade, demonstrate a great variability in the prevalence of different phenotypes in treated and untreated diabetics, with different degrees of severity of the disease. Such typing does not seem to reveal greater homogenity in juvenile diabetes. On the other hand, observations are fairly uniform in diabetic ketoacidosis and in diabetic nephropathy.

The reader will notice the above mentioned variance in findings, presented by different authors. He should also carefully study the form and criteria for selection of diabetic patients, the diagnostic tools used for proof of vascular complications, and finally, statistical methods employed for evaluation of results. Beside the arrangement of the clinical material, a close and critical attention has to be focussed on the laboratory methods used in any particular study. Attempts, based on the typing of dyslipoproteinemias in diabetes using visual or even densitometric evaluations of relative distribution of electrophoretic fractions in lipoprotein patterns may be very misleading. The only reliable evaluation of phenotypes has to be based on consideration of concentrations of single lipoprotein fractions. Measurements of their mobilities in electric field are also helpful.

Studies based on the above defined clinical and laboratory criteria contribute to our present knowledge of dyslipoproteinemia in diabetes and its vascular complications. These findings also guarantee correct evaluation of lipoprotein patterns under clinical conditions associated with different types and degrees of metabolic disorder associated with diabetes.

ACKNOWLEDGMENT

The author is exceptionally grateful to Professor Jiri Syllaba for the original clinical classification of diabetics and for the release of clinical material for the part of university dissertation. He and professors A. F. Richter, V. Hlavacek, J. Sula, and so many others, sustained constructive criticism, comments, constant encouragements, meritorious discussions, and mainly friendship, which meant so much to him.

The author is deeply indebted to his wife Mimi for her great self sacrifice, patience, continuous lovely support, and close cooperation throughout the entire 12-year period of the research.

The author gratefully acknowledges exceptionally friendly, sincere, and highly professional laboratory cooperation and devotion to this research, performed by Heda Vakocova, Pavel Minarik, Ing. F. Kyncl, and Ing. K. Marcan.

Dr. Jan Misak and all the author's coworkers in the clinical laboratories are highly appreciated for their generous, professional work in clinical chemical evaluations of the patients, with an excellent and serene assistance of Ing. B. Fiserova, Dr. E. Bergmanova, and Dr. E. Marova. Statistical help was provided by Ing. Z. Roth.

The author is also obliged to Patricia Mysyk and Vicky Belair for their administrative help.

To CRC Press and its editors Paul Gottehrer and Tom Jolley go recognition for their accomplishment in the editing and printing of this publication.

REFERENCES

1. Ackermann, P. G., Toro, G., and Kountz, W. B., *J. Lab. Clin. Med.*, 44, 517, 1954.
2. Ackermann, P. G., Toro, G., and Kountz, W. B., *Circulation*, 14, 494, 1956.
3. Adair, G. S. and Adair, M. E., *J. Physiol.*, 102, 17, 1943.
4. Adlersberg, D., Diabetes, 5, 116, 1956.
5. Ahrens, E. H. and Kunkel, H. G., *J. Clin. Invest.*, 28, 1565, 1949.
6. Akedo, H. and Christensen, H. N., *J. Biol. Chem.*, 237, 118, 1962.
7. Aladjem, F., *Nature (London)*, 209, 1003, 1966.
8. Aladjem, F. and Campbell, D. H., *Nature (London)*, 179, 204, 1957.
9. Aladjem, F., Lieberman, M., and Gofman, J. W., *J. Exp. Med.*, 105, 49, 1957.
10. Aladjem, F. and Rubin, L., *Am. J. Physiol.*, 178, 267, 1954.
11. Albrink, M. J., *J. Clin. Invest.*, 40, 536, 1961.
12. Allison, A. C. and Blumberg, B. S., *Lancet*, I, 634, 1961.
13. Allison, A. C. and Blumberg, B. S., in *Progress in Medical Genetics*, Vol. 4, Steinberg, A. G. and Bearn, A. C., Eds., Grune, New York, 1965, 176.
14. Amenta, J. S. and Waters, L. L., *J. Biol. Med.*, 33, 112, 1960.
15. Anfinsen, Ch. B., *Fed. Proc. Fed. Am. Soc. Biol.*, 15, 894, 1956.
16. Anfinsen, Ch. B., in *Symposium on atherosclerosis*, Natl. A. S. N. R. C. Publ., No. 338, 217, 1955.
17. Anfinsen, Ch. B., in *Fat Metabolism: A Symposium on the Clinical and Biochemical Aspects of Fat Utilisation in Health and Disease*, Najjar, V. A., Ed., John Hopkins Press, Baltimore, 1954.
18. Anfinsen, Ch. B., Boyle, E., and Brown, R. K., *Science*, 115, 583, 1952.
19. Antoniades, H. N., Tullis, J. L., Sargeant, H., Pennel, R. B., and Oncley, J. L., *J. Lab. Clin. Med.*, 51, 630, 1958.
20. Antonini, F. M. and Sodi, A., *Sperimentale*, 106, 423, 1956.
21. Asamer, H., Gabl, F., and Reiner, G., *Wien. Klin. Wochenschr.*, 46, 853, 1967.
22. Ashworth, L. A. E. and Green, C., *Biochem. J.*, 89, 561, 1963.
23. Ashworth, L. A. E. and Green, C., *Biochim. Biophys. Acta*, 70, 68, 1963.
24. Ashworth, L. A. and Green, C., *Biochim. Biophys. Acta*, 84, 182, 1964.
25. Auerswald, W., Doleschel, W., and Müller-Hartburg, W., *Klin. Wochenschr.*, 41, 580, 1963.
26. Augener, W., *Proc. 12th Colloq., Bruges, 1964*, Elsevier, Amsterdam, 1965, 363.
27. Augustin, R. and Hayward, B. J., *Nature (London)*, 187, 129, 1960.
28. Augustyniak, J. et al., *Biochim. Biophys. Acta*, 84, 721, 1964.
29. Avigan, J., *J. Biol. Chem.*, 226, 957, 1957.
30. Avigan, J. and Anfinsen, C. B., *Biochim. Biophys. Acta*, 31, 249, 1959.
31. Ausser, G., Jahnke, K., Heinzler, F., and Breitbarch, A., Diabetes mellitus, *3rd Congr. Int. Diabetes Fed., Düsseldorf, 1958*, Georg Thieme Verlag, Stuttgart, 1959, 107.
32. Avigan, J., Redfield, R., and Steinberg, D., *Biochim. Biophys. Acta*, 20, 557, 1956.
33. Awai, M. and Brown, E. B., *J. Lab. Clin. Med.*, 61, 363, 1963.
34. Ayrault-Jarrier, M., Levy, G., and Wald, R., *Bull. Soc. Chim. Biol.*, 45, 349, 1963.
35. Ayrault-Jarrier, M., Cheftel, R., and Polonovski, J., *Bull. Soc. Chim. Biol.*, 43, 811, 1961.
36. Ayrault-Jarrier, M., Wald, R., and Polonovski, J., *Bull. Soc. Chim. Biol.*, 41, 753, 1959.
37. Azérad, E., Ghata, J., and Lewin, J., *Presse Med.*, 66(44), 995, 1958.
38. Badin, J., *Presse Med.*, 73, 2659, 1965.
39. Bahl, A. N. et al., *Indian J. Med. Sci.*, 20, 485, 1966.
40. Baker, K. J., *Proc. Soc. Exp. Biol. Med.*, 122, 957, 1966.
41. Bálint, S., Gööz, K., and Dán, S., *Orv. Hetil.*, 106, 743, 1965.
42. Baltzer, V., *Bull. Schweiz. Akad. Med. Wiss.*, 15(2—3), 193, 1959.
43. Banasyak, L. J. and MacDonald, H. J., *Biochemistry*, 1, 344, 1962.
44. Banerjee, D. et al., *J. Assoc. Physicians India*, 13, 935, 1965.
45. Bansi, H. W., Neth, R., and Schwarting, G., *Verh. Dtsch. Ges. Kreislaufforsch*, 21, 139, 1855.

46. Barclay, M. et al., *Nature (London)*, 200, 362, 1963.

47. Barclay, M. et al., *Clin. Chim. Acta*, 11, 389, 1965.

48. Barclay, M., Barclay, R. K., Terebus-Kekish, O., Shah, E. B., and Skipski, V. P., *Clin. Chim. Acta*, 8, 721, 1963.

49. Barnabei, O. and Ferrari, R., *Boll. Soc. Ital. Biol. Sper.*, 36, 880, 1960.

50. Barr, D. P., Russ, E. M., and Eder, H. A., *Am. J. Med.*, 11, 480, 1951.

51. Barre, R. and Labat, J., *Ann. Biol. Clin. (Paris)*, 20, 679, 1962.

52. Barre, R. and Labat, J., *Ann. Biol. Clin. (Paris)*, 18, 422, 1960.

53. Bastenie, P. A., Pirart, J., and Franckson, J. R. M., in *Diabetes mellitus*, 3rd Congr. Int. Diabetes Fed., Düsseldorf, 1958; Georg Thieme Verlag, Stuttgart, 1959, 81.

54. Becker, W. and Rapp, W., *Z. Klin. Chem. Klin. Biochem.*, 3, 113, 1968.

55. Beilstein, A., 4 ed., Springer, Berlin, 1920.

56. Bennhold, H., *Verh. Dtsch. Ges. Inn. Med.*, 62, 657, 1956.

57. Bennhold, H., *Ergeb. Inn. Med. Kinderheilkd.*, 42, 273, 1932.

58. Bennhold, H., *Verh. Dtsch. Ges. Inn. Med.*, 43, 211, 1931.

59. Berg, K., *Acta Pathol. Microbiol. Scand.*, 59, 369, 1963.

59a. Berg, K., *Acta Pathol. Microbiol. Scand.*, 62, 276, 287, 600, 613, 1964.

59b. Berg, K., *Acta Pathol. Microbiol. Scand.*, 63, 127, 142, 1965.

60. Berg, K., *Bibl. Haematol.*, 23, 365, 1965.

61. Berg, K., *Sangre (Barcelona)*, 9, 24, 1964.

62. Berg, K., *Vox Sang.*, 10, 513, 1965.

62a. Berg, K., *Vox Sang.*, 11, 419, 1966.

63. Berg, K. et al., *Humangenetik*, 1, 24, 1964.

64. Berg, G. and Roller, E., *Blut*, 6, 32, 1960.

65. Berg, G., Scheiffarth, F., and Marwan, G., *Klin. Wochenschr.*, 35/8, 415, 1957.

66. Bermes, E. W. and McDonald, H. J., *Fed. Proc.*, 15, 220, 1956.

67. Bermes, E. W. and McDonald, H. J., *Arch. Biochem. Biophys.*, 70, 49, 1957.

68. Bermes, E. W. and McDonald, H. J., *J. Chromatogr.*, 4, 34, 1960.

69. Bernfeld, P. in *The Lipoproteins*, Homburger, F., Eds., S. Karger, Basel, 1958, 24.

70. Bernfeld, P. and Berkowitz, M. E., *J. Clin. Invest.*, 36, 1363, 1957.

71. Bernfeld, P., *Fed. Proc.*, 14, 182, 1955.

72. Bernfeld, P. and Nisselbaum, J. S., *Fed. Proc.* 15, 220, 1956.

73. Bernfeld, P., Donahue, V. M., and Berkowitz, M. E., *J. Biol. Chem.*, 226, 51, 1957.

74. Bernfeld, P. and Kelley, Th. F., *J. Biol. Chem.*, 239, 3341, 1964.

75. Bernfeld, P. and Nisselbaum, J. S., *J. Biol. Chem.*, 220, 851, 1956.

76. Bernsohn, J., *J. Lab. Clin. Med.*, 49, 478, 1957.

77. Bierman, E. L., Dole, V. P., and Roberts, Th. N., *Diabetes*, 7, 189, 1958.

78. Bierman, E. I., Gordis, E., and Hamlin, J. T., *J. Chem. Invest.*, 42, 2254, 1962.

79. Bierman, E. L., Schwartz, I., and Dole, V. P., *Am. J. Physiol.*, 191, 359, 1957.

80. Bihari-Varga, M. et. al., *Orv. Hetil.*, 104, 2130, 1963.

81. Bihari-Varga, M., Gergely, J., and Gero, S., *Orv. Hetil.*, 104, 1401, 1963.

82. Bihari-Varga, M. et al., *J. Atheroscler. Res.*, 4, 106, 1964.

83. Björklund, R. and Katz, S., *J. Am. Chem. Soc.*, 78, 2122, 1956.

84. Blaney, J. D., Hardwicke, J., and Whitfield, A. G. W., *Lancet*, II, 1208, 1954.

85. Blix, F. G., Lund, Lindstedt, 1925.

86. Blix, F. G., *Acta Med. Scand.*, 64, 142, 234, 1926.

87. Blix, F. G., Tiselius, A., and Svensson, H., *J. Biol. Chem.*, 137, 485, 1941.

88. Bloom, B. and Pierce, F. T., *Metabolism*, 1, 155, 1952.

89. Bloor, W. R., *J. Biol. Chem.*, 24, 447, 1916.

89a. Bloor, W. R., *J. Biol. Chem.*, 49, 201, 1921.

90. Blöck, J. and Graf, E., *Wien. Klin. Wochenschr.*, 65, 20, 1953.

91. Blöck, J., Durrum, L., and Zweig, G., *Paper Chromatography and Paper Electrophoresis*, Academic Press, New York, 1958, 554.

93. Blumberg, B. S., Bernanke, D., and Allison, A. C., *J. Clin. Invest.*, 41, 1936, 1962.

94. Blumenthal, H. T., Goldenberg, S., and Berns, A. W., On the nature and treatment of diabetes, 5th Congr. Int. Diabetes Fed., Toronto, 1964, *Ed. Exc. Med. Found.*, 1965, 397.

95. Bolt, W., Bolte, A., and Lichins, N., *Aerztl. Forsch.*, 14(1), 107, 1960.

96. Bottcher, C. J., Woodford, F. P., Klynstra, F. B., *J. Atheroscler. Res.*, 3, 24, 1963.

97. Boyd, G. S., *Biochem. J.*, 58, 680, 1954.

98. Boyle, E. et al., *J. Lab. Clin. Med.*, 53, 272, 1959.

99. Böhle, E., Böttcher, K., Piekarski, H. G., and Biegler, R., *Dtsch. Arch. Klin. Med.*, 203, 29, 1956.

100. Bragdon, J. H., in *The Lipoproteins, Methods and Clinical Significance,* Homburger, F. and Bernfeld , P., Eds., S. Karger, Basel, 1958, 37.
101. Bragdon, J. H. and Gordon, R. S., *J. Clin. Invest.,* 37, 574, 1958.
102. Bragdon, J. H. and Havel, R. J., *Am. J. Physiol.,* 177(1), 128, 1954.
103. Bragdon, J. H., Havel, R. J., and Boyle, E., *J. Lab. Clin. Med.,* 48, 36, 1956.
104. Bragdon, J. H., *Ann. N.Y. Acad. Sci.,* 72, 845, 1959.
105. Braun, H. J., *Therapiewoche,* 17, 1926, 1967.
106. Braunsteiner, H. et al., *Klin. Wochenschr.,* 44, 116, 1966.
107. Briner, W. W., Riddle, J. W., and Cornwell, D. C., *J. Exp. Med.,* 110, 113, 1959.
108. Burijan, J., Kapetanović, B., Babić, D., and Mičić, J., in *Diabetes mellitus,* 3rd Congr. Int. Diabetes Fed., Düsseldorf, 1958, Georg Thieme Verlag, Stuttgart, 1959, 113.
109. Burstein, M., *Bull. Schweiz. Acad. Med. Wiss.,* 17, 92, 1961.
110. Burstein, M., *C. R.,* 244, 3189, 1957.
111. Burstein, M., *C. R. Acad. Sci.,* 255, 605, 1962.
111a. Burstein, M., *C. R. Acad. Sci.,* 244, 3189, 1957.
112. Burstein, M., *J. Physiol.,* 53, 519, 1961.
112a. Burstein, M., *J. Physiol.,* 54, 647, 1962.
113. Burstein, M., *Life Sci.,* 12, 739, 1962.
114. Burstein, M., *Nouv. Rev. Fr. Hematol.,* 3, 139, 1965.
115. Burstein, M., *Pathol. Biol.,* 8, 1247, 1960.
116. Burstein, M., Proc. 7th Congr. Int. Soc. Blood Transfusion, Rome, 1958.
117. Burstein, M. and Berlinski, M., *Rev. Fr. Etud. Clin. Biol.,* 5, 193, 1960.
117a. Burstein, M. and Berlinski, M., *Rev. Fr. Etud. Clin. Biol.,* 6, 479, 1961.
117b. Burstein, M. and Berlinski, M., *Rev. Fr. Etud. Clin. Biol.,* 9, 105, 1964.
118. Burstein, M. and Praverman, A., *Rev. Med. Chir. Mal. Foie.,* 33(1), 21, 1958.
119. Burstein, M. and Samaille, J., *Clin. Chim. Acta,* 3, 320, 1958.
119a. Burstein, M. and Samaille, J., *Clin. Chim. Acta,* 5, 609, 1960.
120. Burstein, M. and Samaille, J., *Ann. Anat. Pathol.,* 4/2/, 1959.
121. Burstein, M. and Samaille, J., *Presse Med.,* 66/43/, 974, 1958.
122. Burstein, M. and Samaille, J., *Ann. Biol. Clin.* , 17 (1-2), 23, 1959.
122a. Burstein, M. and Samaille, J., *Ann. Biol. Clin.,* 19, 335, 1961.
123. Bücker, T. and Klingenberg, M., *Angew. Chem.,* 70, 552, 1958.
124. Bürger, M., in *Diabetes mellitus,* 3rd Congr. Int. Diabetes Fed. Düsseldorf, 1958, Georg Thieme Verlag, Stuttgart, 1959, 124.
125. Cain, A. J., *Biol. Rev. Cambridge Philos. Soc.,* 25, 73, 1950.
126. Cann, J. R., Kirkwood, J. G., and Brown, R. A., *Arch. Biochem. Biophys.,* 72, 37, 1957.
127. Cann, J. R., Kirkwood, J. G., Brown, R. A., and Plescia, O. J., *J. Am. Chem. Soc.,* 71, 1603, 1949.
128. Carlin, H. and Hechter, O., *Proc. Soc. Exp. Biol. Med.,* 115, 127, 1964.
129. Carlson, L. A., 3rd Int. Conf. Biochem. Probl. Lipids, Brussels, 1956.
130. Carlson, L. A., *Acta Chem. Scand.,* 8, 510, 1954.
131. Carlson, L. A., *Acta Med. Scand.,* 172, 641, 1962.
131a. Carlson, L. A., *Acta Med. Scand.,* 173, 719, 1963.
131b. Carlson, L. A., *Acta Med. Scand.,* 174, 215, 1963.
132. Carlson, L. A., *Acta Physiol. Scand.,* 62, 51, 1964.
133. Carlson, L. A., *Clin. Chim. Acta,* 5, 528, 1960.
134. Carlson, L. A., *Scand. J. Clin. Lab. Invest. Suppl.,* 20, 7, 1955.
135. Carruthers, B. M. and Winegrad, A. I., *Am. J. Physiol.,* 202, 605, 1962.
136. Castaigne, A. et al., *Ann. Biol. Clin.,* 21, 587, 1963.
137. Castaigne, A. and Amselem, A., *Rev. Med. Franc.,* 42, 311, 1961.
138. Castiglioni, C. A., Rocca, F., and Garbino, C., in *Diabetes Mellitus,* Monog., 3rd Kongr. Int. Diabetes Fed., Düsseldorf, 1958; Georg Thieme Verlag, Stuttgart, 1959, 98.
139. Cicvárek, Z., *Bratisl. Lek. Listy,* 40, 129, 1960.
140. Chimenes, H. and Boulin, R., *Diabetes Mellitus,* 3rd Congr. Int. Diabetes Fed., Düsseldorf, 1958, Georg Thieme Verlag, Stuttgart, 1959, 782.
141. Cohn, E. J., Strong, L. E., Hughes, W. L., Mulford, D. J., Ashwort, J. N., Melin, M., and Taylor, H. I., *J. Am. Chem. Soc.,* 68, 459, 1946.
142. Cohn, E. J., Gurd, F. R. N., Surgenor, D. M., Barnes, B. A., Brown, R. K., Deronaux, G., Gillespie, J. M., Kahnt, F. W., Lever, W. F., Lin, C. H., Mittleman, D., Mouton, R. F., Schmid, K., and Uroma, E., *J. Am. Chem. Soc.,* 72, 465, 1950.
143. Colfs, B. et al., *Clin. Chim. Acta,* 18, 325, 1967.

144. Comfort, A., *Biochem. J.,* 59, X, 1955.
145. Cook, W. H. and Martin, W. G., *Biochem. Biophys. Acta,* 56, 362, 1962.
146. Cook, W. H., *Can. J. Biochem.,* 43, 661, 1965.
147. Cook, W. H., *Nature (London),* 195, 1308, 1962.
148. Cornwell, D. G., Kruger, F. A., Hamwi, G. J., and Brown, J. B., *Am. J. Clin. Nutr.,* 9, 24, 41, 1961.
149. Cornwell, D. G. and Kruger, F. A., *J. Lipid Res.,* 2, 110, 1961.
150. Cornwell, D. G. and Kruger, F. A., *Proc. Soc. Exp. Biol. Med.,* 107, 296, 1961.
151. Courtice, F. C., *Arch. Int. Pharmacodyn. Ther.,* 139, 371, 1962.
152. Courtice, F. C., Woolley, G., and Garlick, D. G., *Aust. J. Exp. Biol. Med. Sci.,* 40, 111, 1962.
153. Cramér, K., *Acta Med. Scand.,* 171, 413, 1962.
153a. Cramér, K., *Acta Med. Scand.,* 170, 1, 1961.
154. Cramér, K., Aurell, M., and Pehrson, S., *Clin. Chim. Acta,* 10, 470, 1964.
155. Crowle, A. J., *Immunodiffusion,* Academic Press, New York, 1961.
156. Csapo, G., *Experientia,* 20, 335, 1964.
157. Czuppon, A., *Kiserl. Orvostud.,* 11, 550, 1959.
158. Dangerfield, W. G. and Smith, E. B., *Biochem. J.,* 58, XIII, 1954.
159. Dangerfield, W. G., et al., *Clin. Chim. Acta,* 10, 123, 1964.
160. Dangerfield, W. G. and Smith, E. B., *J. Clin. Pathol.,* 8, 132, 1955.
161. Dangerfield, W. G. and Smith, E. B., *Proc. Biochem. Soc. Biochem. J.,* 58, 13, 1954.
162. Dangerfield, W. G., *Proc. R. Soc. Med.,* 49, 1065, 1956.
163. Darcy, D. A., *Nature (London),* 206, 826, 1965.
164. De Lalla, O. F., Elliott, H. A., and Gofman, J. W., *Am. J. Physiol.,* 179, 333, 1954.
165. De Lalla, O. F. and Gofman, J. W., in *Methods of Biochemical Analysis,* D. Glick, Ed., Vol. I, Interscience, New York, 1954, 459.
166. De Lalla, O. F., *U. C. R. L.,* 8550, 1, 1958.
167. Del Gatto, L., Lingren, F. T., and Nichols, A. V., *Anal. Chem.,* 31, 1397, 1959.
168. Del Gatto, L., Nichols, A. V., and Lindgren, T. T., *Proc. Soc. Exp. Biol.,* 101(1), 1959.
169. Deliamure, L. L., *Vopr. Med. Khim.,* 9, 200, 1963.
170. Deuel, H. J., *Chemistry,* Vol. 1, Interscience Publications, New York, 1951.
170a. Deuel, H. J., *Biochemistry: Digestion, Absorption, Transport and Storage,* Vol. 2, Interscience Publications, New York, 1955.
170b. Deuel, H. J., *Biochemistry: Biosynthesis, Oxidation, Metabolism and Metrional Value,* Vol. 3, Interscience Publication, New York, 1957.
171. Deuel, H. J. and Reiser, R., *Vitam. Horm.,* New York, 13, 29, 1955.
172. Dietrich, F., *Z. Physiol. Chem.,* 302, 227, 1955.
173. Di Leo, F. P., Betti, R., and Solinas, P., *Minerva Med.,* 48, 3714, 1957.
174. Djurič, D. S., Babič, D., Mičič, J. V., and Jančič, M., *C. R.,* 4th Congr. Int. Diabetes Fed., Geneve, 1961, 70.
175. Dole, V. P., James, A. T., Webb, J. P. W., Rizack, M. A., and Sturman, M. F., *J. Clin. Invest.,* 38, 1544, 1959.
176. Donald, S., Fredrickson, D. S., Levy, R. J., and Lees, R. S., *Protides of the Biological Fluids,* Elsevier, New York, 1966, 269.
177. Drevon, B., *C. R. Soc. Biol.,* 153, 1370, 1959.
178. Drevon, B. and Donikian, R., *Bull. Soc. Chim. Biol.,* 37, 605, 1955.
179. Duncan, Ch. H. and Best, M. M., *Circulation,* 14, 491, 1956.
179a. Duncan, Ch. H. and Best, M. M., *Circulation,* 18, 490, 1958.
180. Duncan, L. E. et al., *Science,* 142, 972, 1963.
181. Durrum, E. L., *Am. Chem. Soc.,* 115th Meeting, 22, 1949, *Am. Chem. Soc.,* 72, 4329, 1950.
182. Durrum, E. L., Paul, M. H., and Smith, E. R. B., *Science,* 116, 428, 1952.
182a. Durrum, E. L., Paul, M. H., and Smith, E. R. B., *Science,* 113, 66, 1951.
183. Eboué-Boris, D., Chambant, A. M., Volfin, P., and Clauser, H., *Nature (London),* 199, 1183, 1963.
184. Eder, H. A., et al., *Trans. Assoc. Am. Physicians,* 77, 259, 1964.
185. Eder, H. A., *Am. J. Med.,* 23, 269, 1957.
186. Eder, H. A., Russ, E. M., Pritchett, R. A. R., Wilber, M. M., and Barr, D. P., *J. Clin. Invest.,* 34, 1147, 1955.
187. Edsall, J. T., Gilbert, G. A., and Scheraga, H. A.: *J. Am. Chem. Soc.,* 77, 157, 1955.
188. Edsall, J. T., *Discuss. Faraday Soc.,* 6, 93, 1949.
189. Edsall, J. T., Ed., *Advances in Protein Chemistry,* Vol. 3, Academic Press, New York, 1947, 384.
190. Eiber, H. B., Sang, J. B., and Danishefsky, J., *Proc. Soc. Exp. Biol.,* 92, 700, 1956.
191. Ejarque, P., Marble, A., and Tuller, E. F., *Am. J. Med. ,* 27/2, 221, 1959.

192. **Eley, D. D., Hedge, D. G.,** *J. Colloid Sci.,* 12, 419, 1957.
193. **Engelberg, H., Gofman, J. W., and Jones, H. B.,** *Diabetes,* 1, 425, 1952.
194. **Engström, A. and Fineau, J. B.,** *Biological Ultrastructure,* Academic Press, New York, 1958.
195. **Engström, W. W. and Liebman, A.,** *Am. J. Med.,* 15, 180, 1953.
196. **Epstein, F. H. and Block, W. D.,** *Proc. Soc. Exp. Biol. Med.,* 101, 740, 1959.
197. **Erb, W.,** *Acta Hepato Splenol.,* 8, 205, 1961.
198. **Euer, N. et al.,** *Suvrem. Med.,* 16, 159, 1965.
199. **Evans, R. J.,** *Biochem. Biophys. Res. Commun.,* 19, 171, 1965.
200. **Ewing, A. M., Freeman, N. K., and Lindgren, F. T.,** *Adv. Lipid Res.,* 3, 25, 1965.
201. **Fabian, E. et al.,** *Rev. Czech. Med.,* 12, 92, 1966.
202. **Fabian, E.,** *Cas. Lek. Cesk.,* 98, 513, 1959.
203. **Fagerberg, S. E.,** in *On the Nature and Treatment of Diabetes,* Monogr., 5th Congr. Int. Diabetes Fed., Toronto, 1964, 522.
204. **Farkas, E.,** *Z. Kinderheilkd.,* 82, 335, 1959.
205. **Farstad, M.,** *Clin. Chim. Acta,* 14, 341, 1966.
206. **Fasoli, A.,** *Acta Med. Scand.,* 145, 233, 1953.
207. **Fasoli, A. and Bonelli, M.,** *Arch. Sci. Biol.,* 34, 161, 1950.
208. **Fasoli, A.,** *Lancet,* 262, 106, 1952.
209. **Fasoli, A., Salteri, F., and Cesana, A.,** *Bull. Schweiz. Acad. Med. Wiss.,* 13/1—4/, 200, 1957.
210. **Fasoli, A.,** *Lancet,* 262, 106, 1952.
211. **Fassina, G.,** *G. Biochim.,* 6, 91, 1957.
212. **Feinberg, H., Rubin, L., Hill, R., Entenman, C., and Chaikoff, I. L.,** *Science,* 120, 317, 1954.
213. **Felts, J. M. and Mayes, P. A.,** *Nature (London),* 206, 195, 1965.
214. **Fidanza, F. and Cioffi, L. A.,** *Boll. Soc. Ital. Biol. Sper.,* 35, 1901, 1959.
215. **Fine, J. M., et al.,** *Experientia,* 14, 411, 1958.
216. **Fireman, P., Vaunier, W. E., and Goodman, H. C.,** *Proc. Soc. Exp. Biol. Med.,* 115, 845, 1964.
217. **Fischer, B.,** *Z. Kreislaufforsch.,* 48, 517, 1959.
218. **Fischer, B. and Kroetz, Ch.,** *Bull. Schweiz. Acad. Med. Wiss.,* 13, 268, 1957.
219. **Fisher, W. R. and Gurin, S.,** *Science,* 143, 362, 1964.
220. **Fischer, F. W.,** *Klin. Wochenschr.,* 34, 849, 1956.
221. **Fischer, R. and Monnier, J.,** *Schweiz. Med. Wochenschr.,* 975, 1956.
222. **Fiske, C. H. and Sulbarow, Y.,** *J. Biol. Chem.,* 66, 375, 1925.
223. **Flodin, P. and Killander, J.,** *Biochim. Biophys. Acta,* 63, 403, 1962.
224. **Floyd, J. C., Fajans, S. S., Conn, J. W., Pek, S., Rull, J., and Knopf, R.,** *6th Congr. Int. Diabetes Fed., Stockholm, Sweden, 1967,* Exc. Med. Found., Amsterdam, 1969, 515.
225. **Florsheim, W. H., Dimick, B., and Morton, M. E.,** *Experientia,* 12, 343, 1956.
226. **Florsheim, W. H. and Gonzales, C.,** *Proc. Soc. Exp. Biol. Med.,* 104, 618, 1960.
227. **Fonnesu, V.,** *Progr. Med. (Naples),* 17, 706, 1961.
228. **Forbes, J. C. and Taylor, P. C.,** *Proc. Soc. Exp. Biol.,* 90, 411, 1955.
229. **Forsyth, C. C. et al.,** *Arch. Dis. Child.,* 40, 47, 1965.
230. **Franzini, C.,** *Clin. Chim. Acta,* 14, 576, 1966.
231. **Frazer, A. C.,** *4th Int. Kongr. Biochem.,* Pergamon Press, London, 1958.
232. **Frazer, A. C.,** *Bull. Soc. Chim. Biol.,* 33, 961, 1951.
233. **Frazer, A. C.,** *Brit. Med. Bull.,* 14, 212, 1958.
234. **Frazer, A. C.,** *Discuss. Faraday Soc.,* 6, 81, 1949.
235. **Frazer, A. C.,** *Lect. Sci. Basis Med.,* 4, 311, 1954.
236. **Frank, T. and Lindgren, F. T.,** *Am. J. Clin. Nutr.,* 13, 1, 1961.
237. **Fredrickson, D. S.,** *JAMA,* 164, 1895, 1957.
238. **Fredrickson, D. S.,** *J. Clin. Invest.,* 43, 228, 1964.
239. **Fredrickson, D. S. et al.,** *Circulation,* 31, 321, 1965.
240. **Fredrickson, D. S. and Gordon, R. S.,** *Physiol. Rev.,* 38, 585, 1958.
241. **Fredrickson, D. S. et al.,** *N. Engl. J. Med.,* 276, 215, 1967.
242. **Fredrickson, D. S. and Levy, R. J.,** *Hyperlipoproteinemia, types I, II, III, IV, V, Diagnosis and Treatment,* U.S. National Institutes of Health, Bethesda, Maryland, 1970.
243. **Fredrickson, D. S.,** in *Metabolic Basis of Inherited Disease,* 2nd ed., Stanbury, J. B., Wyngaarden, J. B., and Fredrickson, D. S., Eds., McGraw-Hill, New York, 1966, 486.
244. **Freedlender, A. E., Nelson, C. A., and Rosenberg, S.,** 6th Congr. Int. Diabetes Fed., Stockholm, Sweden, 1967.
245. **Freeman, N. K., Lindgren, F. T., and Nichols, A. V.,** *J. Biol. Chem.,* 203, 293, 1953.
246. **Freeman, N. K., Lindgren, F. T., Nichols, A. V.,** in *Progress in the Chemistry of Fats and other Lipids,* Vol. 6, Holman, R. T., Lundberg, W. O., and Malkin, T., Eds., 1963, 215.

247. Freislederer, W. and Kastner, E., *Klin. Wochenschr.*, 36(3), 177, 1958.
248. Fried, R. and Hoeflmayer, J., *Klin. Wochenschr.*, 41, 246, 1963.
249. Fronescu, E. et al., *Med. Int.*, 16, 569, 1964.
250. Fruton, J. S., in *The Proteins*, Vol. 1, Neurath, H., Ed., Academic Press, New York, 1963, 190.
251. Fry, I. K. and Butterfield, W. J. H., *Lancet*, 2, 66, 1962.
252. Furman, R. H., *Am. J. Clin. Nutr.*, 9, 73, 1961.
253. Furman, R. H., *Circulation*, 16, 507, 1957.
254. Furman, R. H., Norcia, L. N., Fryer, A. W., and Wamack, B. S., *J. Lab. Clin. Med.*, 47, 730, 1956.
254a. Furman, R. H., Norcia, L. N., Fryer, A. W. and Wamack, B. S., *J. Lab. Clin. Med.*, 63, 193, 1964.
255. Gabl, F. Wachter, H., *Protides of the Biological Fluids*, Elsevier, New York, 1966, 359.
256. Gage, S. H. and Fish, P. A., *Am. J. Anat.* 34, 1, 1924.
257. Gagli, V. et al., *Boll. Soc. Ital. Biol. Sper.*, 32, 7, 1956.
258. Geinitz, W., 3rd Int. Conf. Biochemical Problems of Lipids, Ciba Edition, 1957.
259. Gergely, J., *Proc. 8th Colloq., Bruges*, 1960; Elsevier, Amsterdam, 1961, 138.
260. Gilles, G. A., Lindgren, F. T., and Cason, J., *J. Am. Chem. Soc.*, 78, 4103, 1956.
261. Giongo, F., Jato, E., Mononi, G., *Osp. Magg.*, 49, 39, 1961.
262. Gitlin, D. and Cornwell, D., *J. Clin. Invest.*, 35, 706, 1956.
262a. Gitlin, D. and Cornwell, D., *J. Clin. Invest.*, 37, 172, 1958.
263. Gitlin, D., *Pediatrics*, 19, 657, 1957.
264. Gitlin, D., *Science*, 117, 591, 1953.
265. Gitlin, D. and Jeneway, Ch., *Symp. Chemistry of Lipides as Related to Atherosclerosis*, Charles C Thomas, Springfield, Ill., 1958.
266. Gitlin, D., in *The Nephrotic Syndrome, Proc. 7th Annu. Conf.*, National Nephrosis Foundation, New York, 1956, 94.
267. Glazier, F. W., Tamplin, A. R., Strisower, B., De Lalla, O. F., Gofman, J. W., Dawber, T. R., and Philips, E., *J. Gerontol.*, 9, 395, 1954.
268. Gligore, V. et al., *Journ. Annu. Diabetol. Hotel*, Dieu 7, 315, 1967.
269. Glomset, J. A., *Biochim. Biophys. Acta*, 70, 389, 1963.
270. Gloster, J. and Fletcher, R. F., *Clin. Chim. Acta*, 13, 235, 1966.
271. Gofman, J. W., Rubin, L., McGinley, J. P., and Jones, H. B., *Am. J. Med.*, 17, 514, 1954.
272. Gofman, J. W., Jones, H. B., Lindgren, F. T., Lyon, T. P., Elliot, H. A., and Strisower, B., *Circulation*, 2, 161, 1950.
273. Gofman, J. W., Jones, H. B., Lyon, T. P., Lindgren, F. T., Strisower, B., Colman, D., and Herring, V., *Circulation*, 5, 119, 1952.
273a. Gofman, J. W., Jones, H. B., Lyon, T. P., Lindgren, F. T., Strisower, B., Colman, D., and Herring, V., *Circulation*, 14, 691, 1956.
274. Gofman, J. W., Lindgren, F. T., and Elliott, H. A., *J. Biol. Chem.*, 179, 973, 1949.
274a. Gofman, J. W., Lindgren, F. T., and Elliott, H. A., *J. Biol. Chem.*, 182, 1, 1950.
275. Gofman, J. W., Strisower, B., De Lalla, O. F., Tamplin, A. R., Jones, H. B., and Lindgren, F. T., *Mod. Med. (Minneapolis)*, 11, 119, 1953.
276. Gofman, J. W., De Lalla, O. F., Glazier, F., Freeman, N. K., Lindgren, F. T., Nichols, A. V., Strisower, B., and Tamplin, A. R., *Plasma*, 2, 413, 1954.
277. Gofman, J. W., Glazier, F. W., Tamplin, A. R., Strisower, B., and De Lalla, O. F., *Physiol. Rev.*, 34, 589, 1954.
277a. Gofman, J. W., Glazier, F. W., Tamplin, A. R., Strisower, B., and De Lalla, O. F., *Physiol. Rev.*, 45, 747, 1965.
278. Gofman, J. W., Lindgren, F. T., Elliot, H. A., Mantz, W., Hewitt, J., Strisower, B., Herring, V., and Lyon, T. P., *Science*, 111, 166, 1950.
279. Gofman, J. W., *Bull. Schweiz. Akad. Med. Wiss.*, 15, 105, 1959.
280. Gofman, J. W., in *The Lipoproteins*, Homburger, F. and Bernfeld, P., Eds., S. Karger, Basel, 1958.
281. Goodman, D. S., *J. Clin. Invest.*, 43, 2026, 1964.
282. Goodman, D. S. et al., *J. Lipid Res.*, 5, 307, 1964.
283. Goldwater, W. H., Randolph, M. L., Snavely, J. R., and Turner, R. H., *Fed. Proc. Fed. Am. Soc. Exp. Biol.*, 9, 178, 1950.
284. Gottfried, S. P., Pope, R. H., Friedman, N. H., and Akerson, J. B., *Clin. Chem.*, 1, 253, 1955.
285. Gottfried, S. P., Pope, R. H., Friedman, N. H., and di Mauro, S., *J. Lab. Clin. Med.*, 44, 651, 1954.
286. Guillot, M., *Ann. Biol. Clin.*, 14, /1-2/, 1956.
287. Graham, D. M., Lyon, T. P., Gofman, J. W., Jones, H. B., Simonton, J., and White, S., *Circulation*, 4, 666, 1951.

288. Granda, J. L. and Scanu, A., *Fed. Proc. Fed. Am. Soc. Exp. Biol.*, 24, 224, 1965.
289. Grant, G. H. and Everall, P. H., *Protides of the Biological Fluids*, Elsevier, New York, 1966, 321.
290. Green, C., Oncley, J. L., and Karnovsky, M. L., *J. Biol. Chem.*, 235, 2884, 1960.
291. Green, A. A., Lewis, L. A., and Page, J. H., *Fed. Proc. Fed. Am. Soc. Exp. Biol.*, 10, 191, 1951.
292. Greppi, M., Antonini, F. M., and Salvini, L., *Boll. Soc. Ital. Biol. Sper.*, 34(18), 1339, 1958.
293. Gros, H., *Acta Hepatol. (Hamburg)*, 3, 1/173, 1955.
294. Gross, Ph. and Weicker, H., *Klin. Wochenschr.*, 509, 1954.
295. Groulade, J. and Ollivier, C., *Ann. Biol. Clin.*, 18, 577, 1960.
295a. Groulade, J. and Ollivier, C., *Ann. Biol. Clin.*, 17, 428, 1959.
296. Grundy, S. M., Dobson, H. L., Kitzmiller, G. E., and Griffin, A. C., *Am. J. Physiol.*, 200, 1307, 1961.
297. Grundy, S., Dobson, H. L., and Griffin, A. C., *Proc. Soc. Exp. Biol. Med.*, 100, 704, 1959.
298. Gurd, F. R. N., *Lipid Chemistry*, Hanahan, D. J., Ed., John Wiley, New York, 1960, 208.
299. Gustafson, A., *Acta Med. Scand. Suppl.*, 446, 1, 1966.
300. Gustafson, A. et al., *Biochemistry*, 4, 596, 1965.
300a. Gustafson, A. et al., *Biochemistry*, 5, 632, 1966.
301. Gustafson, A. et al., *Biochim. Biophys. Acta*, 84, 767, 1964.
302. Gustafson, A., *J. Lipid Res.*, 6, 512, 1965.
303. Hack, M. H., *Proc. Soc. Exp. Biol. (N.Y.)*, 1, 92, 1956.
304. Haft, D. E., Roheim, P. S., White, A., and Eder, H. A., *J. Clin. Invest.*, 41, 842, 1962.
305. Hagashi, S., Lindgren, F., and Nichols, A., *J. Am. Chem. Soc.*, 81, 3793, 1959.
306. Hagenfeldt, L., *Clin. Chim. Acta*, 13(2), 266, 1966.
307. Hales, C. N., *Colloquia on Endocrinology*, Vol. 15, Ciba Foundation, London, 1964.
308. Hales, C. N., *Proc. Nutr. Soc.*, 25, 61, 1966.
309. Hanig, M. and Shainoff, J. R., *J. Biol. Chem.*, 219, 479, 1956.
310. Hanig, M., Shainoff, J. R., Lowy, A. D., *Science*, 124, 176, 1956.
311. Hanig, M., in *The Lipoprotein*, Homburger, F. and Bernfeld, P., Eds., S. Karger, Basel, 1958.
312. Hardy, W. B., *J. Physiol.*, 33, 251, 1905.
313. Hatch, F. T., Abell, L. L., and Kendall, F. E., *Am. J. Med.*, 19, 48, 1955.
314. Havel, R. J., Eder, H. A., and Bragdon, J. H., *J. Clin. Invest.*, 34, 1345, 1955.
315. Havel, R. J., Felts, J. M., and Van Duyne, M., *J. Lipid Res.*, 3, 297, 1962.
316. Havel, R. J., *Metabolism*, 10, 1031, 1961.
317. Hayashi, S., Lindgren, F. T., and Nichols, A., *J. Am. Chem. Soc.*, 81, 3793, 1959.
318. Hayes, Th. L. and Hewitt, J. E., *J. Appl. Physiol.*, 11, 425, 1957.
319. Halvorson, H., *The Methods in Enzymology*, Vol. 8, Academic Press, New York, 1965.
320. Hayes, Th. L., Marchio, J. C., Lindgren, F. T., and Nichols, A. V., University of California, Radiation Laboratory, Report UCRL, 1959, 8597.
321. Hazelwood, R. N., *J. Am. Chem. Soc.*, 80, 2152, 1958.
322. Heide, K., Schmidtberger, R., and Schwick, G., *Behringwerk Mitt.*, 33, 96, 1957.
323. Herbst, F. S. M., Lever, W. F., Lyons, M. E., and Hurley, N. A., *J. Clin. Invest.*, 34, 581, 1955.
324. Herbst, F. S. M., Lever, W. F., and Hurley, N. A., *J. Invest. Dermatol.*, 24, 507, 1955.
325. Hevelke, G., in *Diabetes mellitus*, 3rd Congr. Int. Diabetes Fed., Düsseldorf, 1958, Georg Thieme Verlag, Stuttgart, 1959, 93.
326. Hewitt, J. E. and Hayes, Th. L., *Am. J. Physiol.*, 185, 257, 1956.
327. Hillyard, L. A., Entenman, C., Feinberg, H., and Chaikoff, I. L., *J. Biol. Chem.*, 214, 79, 1955.
328. Himsworth, H. P., *Lancet*, 2, 1, 65, 118, 17L, 1939.
329. Hirata, Y. et al., *Jpn. J. Clin. Pathol.*, 13, 492, 1965.
329a. Hirata, Y. et al., *Jpn. J. Clin. Pathol.*, 25, 298, 1967.
330. Hirschfeld, J., *Bibl. Haematol.*, 23, 365, 1965.
331. Hjerten, S. and Mosbach, R., *Ann. Biochem.*, 3, 109, 1962.
332. Hjerten, S., *Biochim. Biophys. Acta*, 31, 216, 1959.
333. Hohorst, H. J., Stratmann, O., and Bartels, H., *Klin. Wochenschr.*, 42, 245, 1964.
334. Hood, B., Bedding, P., and Carlander, B., *J. Atheroscler. Res.*, 2, 438, 1962.
335. Horvath, E. et al., *Dtsch. Z. Verdau. Stopfwechselkr.*, 24, 273, 1965.
336. Hoxter, G., *Rev. Hosp. Clin. Fac. Med. S. Paulo*, 16, 410, 1961.
337. Horejsi, J. and Slavik, K., Zaklady Chemickeho vysetrovani v lek, *Stat. Zdrav. Nakl.*, Praha, 1953.
338. Houtsmuller, A. J. et al., *Clin. Chim. Acta*, 9, 497, 1964.
339. Hrabane, J. and Reinis, Z., *Vnitr. Lek.*, 4, 399, 1958.
340. Charman, R. C. et al., *Anal. Biochem.*, 19, 177, 1967.
341. Chen, J. S., *J. Formosa Med. Assoc*, 63, 53, 1964.
342. Chick, H., *Biochem. J.*, 8, 404, 1914.

343. Ireland, J. T., Patnaik, B. K., and Duncan, L. J. P.,6th Congr. Int. Diabetes Fed., Stockholm, Sweden, 1967.
344. Iriarte Ezcurdia J. A., Masden Olleta, S., Balaguer-Vintro, J., and Nolla-Panades, J., *Rev. Clin. Esp.*,73, 401, 1959.
345. Jahnke, K., Heinzler, F., Assener, G., and Breitbach, A., in *Diabetes mellitus, Monogr., 3rd Kongr. Int. Diabetes Fed., Dusseldorf, 1958,* Georg Thieme Verlag, Stuttgart, 1959, 107.
346. Jahnke, K. and Scholtan, W., *Verh. Dtsch. Ges. Inn. Med.*,61, 312, 1955.
347. Jahnke, K. and Scholtan, W., *Z. Ges. Exp. Med.*,122, 39, 1953.
347a. Jahnke, K. and Scholtan, W., *Z. Ges. Exp. Med.*,125, 59, 1955.
348. James, A. T. and Wheatley, V. R., *Biochem. J.*,63, 269, 1956.
349. Janado, M. et al., *J. Lipid Res.*,6, 331, 1965.
350. Jarnefelt, J., *Biochem. Pharmacol.*,5, 381, 1961.
351. Yasuda, J., *Jpn. Circ. J.*,27, 519, 1963.
352. Jeanrenaud, B.,6th Congr. Int. Diabetes Fed., Stockholm, Sweden, 1967.
353. Jencks, W. P. and Durrum, E. L., *J. Clin. Invest.*,34, 1437, 1955.
354. Jencks, W. P., Hyatt, M. R., Jetton, M. R., Mattingly, T. W., and Durrum, E. L., *J. Clin. Invest.*, 35, 980, 1956.
355. Jencks, W. P. and Durrum, E. L., 3rd Congr. Internbiochim., Resumes Communs., Brussels, 1955, 145.
356. Jenkins, D. J., *Lancet,* 1, 91, 13, 1968.
357. Johansson, B. and Rymo, L., *Acta Chem. Scand.*,18, 217, 1964.
358. Johnston, J. P. and Ogston, A. G., *Trans. Faraday Soc.*,42, 789, 1946.
359. Jones, H. B., Gofman, J. W., Lindgren, F. T., Lyon, T. P., Graham, D. M., Strisower, B., and Nichols, A. K., *Am. J. Med.*,11, 358, 1951.
360. Jones, D. P. et al., *Diabetes,* 15, 565, 1966.
361. Jones, H. B., Biggs, M., Graham, D. M., Rosenthal, D., Gofman, J. W., and Kritchevsky, D., *Circulation,* 4, 475, 1951.
362. Jordan, W. J. et al., *Anal. Biochem.*,14, 91, 1966.
363. Kabat, E. A. and Meyer, N. M., in *Experimental Immunochemistry,* 2nd ed., Charles C Thomas, Springfield, Ill., 1948, 22.
364. Kallee, E. and Roth, E., *Z. Naturforsch. Teil B,* 8, 614, 1953.
364a. Kallee, E. and Roth, E., *Z. Naturforsch Teil B,* 7, 661, 1952.
365. Kanabrocki, E. L. et al., *Clin. Chem.*, 4, 382, 1958.
366. Karkkainen, V. J. and Hartel, G., *Naturwissenschaften,* 43, 373, 1956.
367. Katz, L. N., *Circulation,* 5, 101, 1952.
368. Katz, L. N., *Ann. Intern. Med.*,43, 930, 1955.
369. Katz, L. N., *Minn. Med.*,38, 755, 1955.
370. Kauzmann, W., *Adv. Protein Chem.*,14, 1, 1959.
371. Kawamata, A., *Tokushima J. Exp. Med.*,8, 178, 1961.
372. Kay, W. W. and Whitehead, R., *J. Pathol. Bacteriol.*,53, 279, 1941.
373. Kazal, L. A., *Trans. N. Y. Acad. Sci.*,27, 613, 1965.
374. Keen, H. and Smith, R., in *Diabetes Mellitus, Monogr., 3rd Congr. Int. Diabetes Fed., Dusseldorf, 1958,* Georg Thieme Verlag, Stuttgart, 1959, 225.
375. Keiding, N. R., Mann, G. V., Root, H. F., Lowry, E. Y., and Marble, A., *Diabetes,* 1, 434, 1952.
376. Kellen, J., *Acta Hepato Splenol. (Stuttgart),* 11, 144, 1964.
377. Kellen, J. A., *Clin. Chim. Acta,* 9, 138, 1964.
378. Kellen, J. and Pisarcikova, E., *Klin. Wochenschr.*,39, 1028, 1961.
378a. Kellen, J. and Pisarcikova, E., *Klin. Wochenschr.*,41, 200, 1963.
379. Kellen, J., *Z. Ges. Inn. Med.*,19, 43, 1964.
380. Killander, J. and Flodin, P., *Vox Sang.*,7, 113, 1962.
381. Killander, J., Monogr., in *Protides of the Biological Fluids,* Vol. 2, Peeterse, H., Ed., Elsevier, Amsterdam, 1963, 446.
382. Kimmelstiel, P., in *Diabetes mellitus,* 3rd Congr. Int. Diabetes Fed., Dusseldorf, 1958, Georg Thieme Verlag, Stuttgart, 1959, 178.
383. Kipnis, D. M. and Kull, F. J., *Proc. Natl. Acad. Sci. U.S.A.*,50, 493, 1965.
384. Kipnis, D. M. and Noall, M. W., *Biochim. Biophys. Acta,* 28, 226, 1959.
385. Kipnis, D. M., in *On the nature and treatment of diabetes, Monogr., 5th Congr. Int. Diabetes Fed., Toronto, 1964,* Ed. Exc. Med. Found., 1965, 258.
386. Kirkeley, K., *Scand. J. Clin. Lab. Invest.*,18, 437, 1966.
387. Kirkwood, J. G., Cann, J. R., and Brown, R. A., *Biochim. Biophys. Acta,* 5, 301, 1950.
388. Kirkwood, J. G., *J. Chem. Phys.*,2, 767, 1934.
388a. Kirkwood, J. G., *J. Chem. Phys.*,9, 878, 1941.

388b. Kirkwood, J. G., *J. Chem. Phys.*, 18, 54, 1950.
389. Klein, E. and Franken, F. H., *Dtsch. Med. Wochenschr.*, 1808, 1956.
390. Klimov, A. N. et al., *Lab. Delo*, 5, 276, 1966.
391. Kohn, H. I., *Am. J. Physiol.*, 163, 410, 1950.
392. Kohn, J., *Nature (London)*, 189, 312, 1961.
393. Korngold, L. and Lipari, R., *Science*, 121, 170, 1955.
394. Kosek, M., *Cas. Lek., Cesk.*, 98, 1130, 1959.
395. Koshy, P., *Indian Heart J.*, 17, 105, 1965.
396. Krahl, M. E., *Am. J. Physiol.*, 206, 618, 1964.
397. Krahl, M. E., *J. Biol. Chem.*, 200, 99, 1953.
398. Krahl, M. E. and Berstein, J., *Nature (London)*, 173, 949, 1954.
399. Krahl, M. E., Penhos, J. C., and Kraemer, A., in *On the nature and treatment of diabetes, 5th Congr. Int. Diabetes Fed., Toronto, 1964,* Ed. Exc. Med. Found., 1965, 92.
400. Krahl, M. E., *Perspect. Biol. Med.*, 1, 69, 1957.
401. Krahl, M. E., *Recent Progr. Horm. Res.*, 12, 199, 1956.
402. Krahl, M. E., *Science*, 116, 524, 1952.
403. Krahl, M. E., *The Action of Insulin on Cells,* Academic Press, New York, 1961, chap. 4, 183.
404. Kramer, B., Stern, K., and Hellman, L., The Nephrotic Syndrome, Monogr., Proc. 8th Annual Conf., National Nephrosis Foundation, New York, 1957, 30.
405. Kranz, Th. and Heide, K., Protides of the biological fluids, Lipoproteins, Section B, Monogr., Elsevier, 1966, 281.
406. Kulenda, Z. and Vondráček, L., *Vnitrni Lek.*, 4, 629, 1958.
407. Kumar, D. et al., *J. Assoc. Physicians India*, 15, 357, 1967.
408. Kunkel, H. G. and Trautman, R., *Circulation*, 14, 691, 1956.
409. Kunkel, H. G. and Ahrens, E. H., *J. Clin. Invest.*, 28, 1575, 1949.
409a. Kunkel, H. G. and Ahrens, E. H., *J. Clin. Invest.*, 31, 677, 1952.
409b. Kunkel, H. G. and Ahrens, E. H., *J. Clin. Invest.*, 35, 641, 1956.
410. Kunkel, H. G. and Slater, R. J., *Proc. Soc. Exp. Biol. Med.*, 80, 42, 1952.
411. Kunkel, H. G. and Bearn, A. G., *Proc. Soc. Exp. Biol.*, 86, 887, 1954.
412. Kutt, H., McDowell, F., and Pert, J. H., *Proc. Soc. Exp. Biol. Med.*, 102, 38, 1959.
413. Lamy, M., Frezal, J., Polonovski, J., Bard, D., and Rey, J., *C. R. Acad. Sci. (Paris)*, 253, 2135, 1961.
414. Langan, T. A., Durrum, E. L., and Jenks, W. P., *J. Clin. Invest.*, 34, 1427, 1955.
415. Larkey, B. J. and Belko, J. S., *Clin. Chem.*, 5, 566, 1959.
416. Lathe, G. H. and Ruthven, R. J., *Biochem. J.*, 62, 655, 1956.
417. Laurell, S., *Acta Physiol. Scand.*, 41, 158, 1957.
418. Laurell, C. B., *Scand. J. Clin. Lab. Invest.*, 6, 22, 1954.
418a. Laurell, C. B., *Scand. J. Clin. Lab. Invest.*, 7, 257, 1955.
418b. Laurell, C. B., *Scand. J. Clin. Lab. Invest.*, 8, 81, 1956.
419. Lawrence, S. H. and Melnick, P. J., *Proc. Soc. Exper. Biol. Med.*, 107, 998, 1961.
420. Lawrence, S. H. and Shean, F. C., *Science*, 137, 227, 1962.
421. Lawry, E. Y. et al., *Am. J. Med.*, 22(4), 605, 1957.
422. Ledvina, M. and Kellen , J., *Klin. Wochenschr.*, 40, 916, 1962.
423. Ledvina, M., Coufalová, S., and Souček, V., *Clin. Chim. Acta*, 5, 818, 1960.
424. Lees, R. S. and Fredrickson, D. S., *J. Clin. Invest.*, 44, 1968, 1965.
425. Lees, R. S. and Hatch, F. T., *J. Lab. Clin. Med.*, 61, 518, 1963.
426. Lemaire, A., *Ann. Med. (Paris)*, 57, /1-2/, 1956.
427. Leuthardt, F. and Stublfauth, K., in *Medizinische Grundlagenforschung,* Bauer, K. F., Ed., Georg Thieme Verlag, Stuttgart, 1955.
428. Lever, W. F., Gurd, F. R., Uroma, E., Brown, B. K., Barnes, B. A., Schmid, K., and Schultz, E. L., *J. Clin. Invest.*, 30, 99, 1951.
429. Lever, W. F., Herbst, F. S. M., and Lyons, M. E., *Arch. Dermatol. Syphilol,* 71, 158, 1955.
430. Lever, W. F., Smith, P. A. J., Hurley, N. A., *J. Invest. Dermatol.*, 22, 33, 1954.
430a. Lever, W. F., Smith, P. A. J., Hurley, N. A., *J. Invest. Dermatol.*, 27, 325, 1956.
431. Levine, L., Kaufmann, D. L., and Brown, R. K., *J. Exp. Med.*, 102, 105, 1955.
432. Levine, R., in *On the Nature and Treatment of Diabetes,* Monogr., 5th Congr. Int. Diabetes Fed., Toronto, 1964; Ed. Exc. Med. Found., 1965, 250.
433. Levy, R. I. and Fredrickson, D. S., *J. Clin. Invest.*, 43, 1286, 1964.
433a. Levy, R. I. and Fredrickson, D. S., *J. Clin. Invest.*, 44, 426, 1965.
433b. Levy, R. I. and Fredrickson, D. S., *J. Clin. Invest.*, 45, 63, 531, 1966.
434. Levy, M. and Marcus, A., *J. Urol. Med. (Paris)*, 64/3/, 172, 1958.

435. Lewis, L. A. et al., *Am. J. Med.,* 38, 286, 1965.

436. Lewis, L. A., Green, A. A., and Page, I. H., *Am. J. Physiol.,* 171, 391, 1952.

437. Lewis, L. A. and Page, J. H., *Circulation,* 7, 707, 1953.

437a. Lewis, L. A. and Page, J. H., *Circulation,* 16, 224, 1957.

438. Lewis, L. A., *Minn. Med.,* 38, 775, 1955.

439. Lillie, R. D. et al., *J. Histochem.,* 1, 8, 1953.

440. Lindgren, F. T., Nichols, A. V., and Wills, R. D., *Am. J. Clin. Nutr.,* 9, 13, 1961.

441. Lindgren, F. T., Nichols, A. V., Hayes, Th. L., Freeman, N. K., and Gofman, J. W., *Ann. N.Y. Acad. Sci.,* 72, 826, 1959.

442. Lindgren, F. T., Elliott, H. A., Gofman, J. W., and Strisower, B., *J. Biol. Chem.,* 182, 1, 1949.

443. Lindgren, F. T., Elliott, H. A., and Gofman, J. W., *J. Phys. Colloid Chem.,* 55, 80, 1951.

444. Lindgren, F. T., Elliott, T., and Gofman, J. W., *J. Phys. Chem.,* 55, 80, 1951.

445. Lindgren, F. T., Nichols, A. V., and Freeman, N. K., *J. Phys. Chem.,* 59, 930, 1955.

446. Lindgren, F. T., *Clin. Chim. Acta,* 9, 402, 1964.

447. Lindgren, F. T. and Nichols, A. V., in *Plasma Proteins,* Vol. 2, Putnam, F. W., Ed., Academic Press, New York, 1960.

448. Lindgren, F. T. et al., U.S. A.E.C. University of California Radiat. Lab. (Berkeley), 1963, 91.

449. Lins, H., Jahnke, K., and Scholtan, W., in *Diabetes mellitus, Monogr., 3rd Congr. Int. Diabetes Fed., Düsseldorf, 1958;* Georg Thieme Verlag, Stuttgart, 1959, 203.

450. Lipman, F., Harvey Lectures, Series 44, New York, 1950, 99.

451. Lippmann, H., Int. Symp. Diab. Angiopathie, Karlsburg, 1962.

452. Longsworth, L. G., *Anal. Chem.,* 23, 346, 1951.

453. Longsworth, L. G., *J. Am. Chem. Soc.,* 61, 529, 1939.

453a. Longsboeuf, L. G., *J. Am. Chem. Soc.,* 65, 1755, 1943.

453b. Longsworth, L. G., *J. Am. Chem. Soc.,* 69, 1288, 1947.

454. Longsworth, L. G., *J. Phys. Chem.,* 58, 770, 1954.

455. Longsworth, L. G., *Proc. Natl. Acad. Sci. U.S.A.,* 36, 502, 1950.

456. Lošticky̌, C., *Clin. Chim. Acta,* 8, 859, 1963.

457. Lowry, O. H., Rosebrough, N. J., Farr, A. L., and Randall, R. J., *J. Biol. Chem.,* 193, 265, 1951.

458. Lowy, A. D. and Barach, H. J., *Diabetes,* 1, 441, 1952.

459. Lowy, A. D. and Barach, H. J., *Diabetes,* 6, 342, 1957.

460. Luft, R., in *On the Nature and Treatment of Diabetes,* Monogr. 5th Congr. Int. Diabetes Fed., Toronto, 1964, Ed. Exc. Med. Found. 1965, 496.

461. Lukens, F. D. W., in *On the Nature and Treatment of Diabetes,* Monogr. 5th Congr. Int. Diabetes Fed., Toronto, 1964, Ed. Exc. Med. Found., 1965, 324.

462. Lundbaek, K., in *Diabetes Mellitus,* Monogr., 3rd Congr. Int. Diabetes Fed., Düsseldorf, 1958, Georg Thieme Verlag, Stuttgart, 1959, 141.

463. Lundbaek, K., in *On the Nature and Treatment of Diabetes,* Monogr. 5th Congr. Int. Diabetes Fed., Toronto, 1964, Ed. Exc. Med. Found. 1965, 436.

464. Löffler, G., Matschinsky, F., and Wieland, O., *Biochem. Z.,* 342, 76, 1965.

465. McDonald, H. J. and Kissane, J. Q., *Anal. Biochem.,* 1, 178, 1960.

466. McDonald, H. J. and Bermes, E. W., *Biochim. Biophys. Acta,* 17, 290, 1955.

467. McDonald, H. J. and Bermes, E. W., *Clin. Chem.,* 2, 257, 1956.

468. McDonald, H. J. and Banaszak, L. J., *Clin. Chim. Acta,* 6, 25, 1961.

469. McFarlane, A. S., *Biochem. J.,* 29, 660, 1935.

470. McFarlane, A. S., *Ir. J. Med. Sci.,* 441, 423, 1962.

471. McFarlane, A. S., *Proc. Faraday Soc.,* 6, 74, 1949.

472. McFarlane, A. S., *Nature (London),* 149, 439, 1942.

472a. McFarlane, A. S., *Nature (London),* 182, 53, 1958.

473. Macheboeuf, M. A., *Bull. Soc. Chim. Biol. (Paris),* 11, 268, 1929.

473a. Macheboeuf, M. A., *Bull. Soc. Chim. Biol. (Paris),* 14, 1168, 1932.

474. Macheboeuf, M., in *Blood Cells and Plasma Proteins. Their State in Nature,* Tullis, J. L., Ed., Academic Press, New York, 1953, 358.

475. Malmros, H. and Swahn, B., *Nord. Med.,* 48, 1028, 1952.

476. Mancini, G., Vaerman, J. P., Carbonara, A. O., and Heremans, J. F., *Protides of the Biological Fluids,* Elsevier, New York, 1966, 370.

477. Manchester, K. L. and Young, F. G., *Biochem. J.,* 70, 353, 1958.

478. Manchester, K. L. and Krahl, M. E., *J. Biol. Chem.,* 234, 2938, 1959.

479. Manchester, K. L. and Young, F. G., *J. Endocrinol.,* 18, 381, 1959.

480. Manchester, K. L., in *On the Nature and Treatment of Diabetes,* 5th Congr. Int. Diabetes Fed., Toronto, 1964, Ed. Exc. Med. Found., 1965, 101.

481. Manchester, K. L. and Young, F. G., *Vitam. Horm. (New York),* 19, 95, 1961.
482. Mann, G. V., *Bull. N. Engl. Med. Center,* 16(4), 152, 1954.
483. Mann, G. V., *Diabetes,* 4, 273, 1955.
484. Mann, E. B. and Peters, J. P., *J. Clin. Invest.,* 14, 579, 1935.
485. Marble, A., in *Diabetes Mellitus,* Monogr., 3rd Congr. Int. Diabetes Fed., Düsseldorf, 1958, Georg Thieme Verlag, Stuttgart, 1959, 373.
486. Marder, L., Becker, G. H., Maizel, B., Nicheles, H., *Gastroenterology,* 20, 43, 1952.
487. Margolis, S. et al., *J. Biol. Chem.,* 241, 477, 1966.
488. Marner, I. L., *Scand. J. Clin. Lab. Invest. Suppl.,* 21, 7, 1955.
489. Marshall, W. E. and Kummerow, F. A., *Arch. Biochem.,* 98, 271, 1962.
490. May, P., *Ann. Nutr. (Paris),* 13, 567, 1959.
491. Mellander, O., *Biochem. Z.,* 277, 305, 1935.
492. Memeo, S. A., *Minerva Med.,* 56, 2533, 1965.
493. Middleton, E., *Am. J. Physiol.,* 185, 309, 1955.
494. Mihal, C. et al., *Prog. Med. (Paris),* 93, 813, 1965.
495. Michalec, Č., Kořínek, J., Musil, J., and Ružička, J., Elektroforesa na papire a v jinych nosicich, *Monografie,* CSAV, Praha, 1959.
496. Michalec, Č., Šťastný, M., Novakova, E., *Naturwissenschoften,* 45, 241, 1958.
497. Mikol C., et al., *Progr. Med. (Paris),* 93, 813, 1965.
498. Milár, A., Sokol, A., and Milárová, R., *Vet. Cas.,* 4, 353, 1955.
499. Mills, G. L. et al., *Clin. Chim. Acta,* 14, 273, 1966.
499a. Mills, G. L. et al., *Clin. Chim. Acta,* 7, 685, 1962.
499b. Mills, G. L. et al., *Clin. Chim. Acta,* 8, 701, 1963.
500. Mitchell, F. L. et al., *Clin. Chim. Acta,* 14, 1, 1966.
501. Moinat, P., Appel, W., and Tuller, E. F., *Clin. Chem.,* 4, 304, 1958.
502. Mohnike, G., in *Diabetes Mellitus,* Monogr., 3rd Congr. Int. Diabetes Fed., Düsseldorf, 1958; Georg Thieme Verlag, Stuttgart, 1959, 70.
503. Moreton, J. R., *Science,* 106, 190, 1947.
503a. Moreton, J. R., *Science,* 107, 371, 1948.
504. Moore, D. H., *Electrophoresis,* Bier, M., Ed., Academic Press, New York, 1959, 369.
505. Moore, S. and Stein, W. H., *Methods Enzymol.,* 6, 819, 1963.
506. Mori, K., Kanatsuna, T., and Kuzuya, K., 6th Congr. Int. Diabetes Fed., Stockholm, Sweden, 1967.
507. Mouray, H., Moretti, J., and Fine, J. M., *Bull. Soc. Chim. Biol.,* 43, 993, 1963.
508. Murakami, M., *Jpn. Heart J.,* 5, 397, 1964.
509. Myszkowski, L. et al., *J. Chromatogr.,* 17, 615, 1965.
510. Miyamoto, H., *J. Jpn. Soc. Intern. Med.,* 54, 608, 1965.
511. Nabarro, J. D. N., in *On the Nature and Treatment of Diabetes,* Monogr., 5th Int. Diabetes Fed., Toronto, 1964, Ed. Exc. Med. Found., 1965, 545.
512. Narayan, K. A. et al., *Clin. Chim. Acta,* 13, 532, 1966.
513. Narayan, K. A., *J. Lipid Res.,* 7, 150, 1966.
514. Narayan, K. A., *Nature (London),* 205, 246, 1965.
515. Nelson, G. J. and Freeman, N. K., *J. Biol. Chem.,* 235, 578, 1960.
516. Nerking, J., *Arch. Ges. Physiol.,* 85, 330, 1901.
517. Nielsen, L. E. and Kirkwood, J. G., *J. Am. Chem. Soc.,* 68, 181, 1946.
518. Nichols, A. V., *Adv. Biol. Med. Phys.,* 11, 109, 1967.
519. Nichols, A. V. et al., *Biochem. Biophys. Res. Commun.,* 17, 512, 1964.
520. Nichols, A. V., Lindgren, F. T., and Gofman, J. W., *Geriatrics,* 12(2), 130, 1957.
521. Nichols, A. V., et al., U.S. A.E.C. University of California Radiat. Lab. (Berkeley), 113, 1963.
522. Nikkilä, E., *Ann. Med. Exp. Biol. Fenn.,* 30, 331, 1952.
523. Nikkilä, E., Haathi, E., and Pesala, R., *Acta Chem. Scand.,* 7, 1222, 1953.
524. Nikkilä, E., *Scand. J. Clin. Lab. Invest.,* 5, 1, 1958.
525. Nys, A., *Chem. Weekb.,* 51, 643, 1955.
526. Oliveira, J. M. and Grynberg, N., *Hospital (Rio),* 63, 1327, 1963.
527. Oliver, M. F. and Boyd, G. S., *Am. Heart J.,* 44, 348, 1953.
528. Oliver, M. F. and Boyd, G. S., *Lancet,* 1, 124, 1957.
528a. Oliver, M. F. and Boyd, G. S., *Lancet,* II, 1273, 1956.
529. Oliver, M. F. and Boyd, G. S., *Minn. Med.,* 38, 794, 1950.
530. Oncley, J. L. et al., *J. Am. Chem. Soc.,* 71, 541, 1949.
530a. Oncley, J. L. et al., *J. Am. Chem. Soc.,* 72, 458, 1950.
530b. Oncley, J. L. et al., *J. Am. Chem. Soc.,* 79, 4666, 1957.
531. Oncley, J. L. et al., *Am. J. Med.,* 11, 358, 1951.

532. Oncley, J. L., Scatchard, G., and Brown, A., *J. Physiol. Chem.*, 51, 184, 1947.

533. Oncley, J. L., *Chem. Engin. News*, 31, 668, 1953.

534. Oncley, J. L., *Brain Lipids and Lipoproteins*, Folchi, J. and Bauer, H., Eds., Elsevier, Amsterdam, 1963, 1.

535. Oncley, J. L., Gurd, F. R. N., in *Blood Cells and Plasma Proteins. Their State in Nature*, Tullis, J. L., Ed., Academic Press, New York, 1953, 337.

536. Oncley, J. L., Mannick, V. G., Monogr., 5th Int. Congr. Blood Transfusion, Paris, 1954.

537. Oncley, J. L., in *The Lipoproteins*, Homburger, F. and Bernfeld, P., Eds., S. Karger, Basel, 1958.

538. Oncley, J. L., in *Chemistry of Lipides as Related to Atherosclerosis*, Page, I. H., Ed., Charles C Thomas, Springfield, Ill., 1958, 114.

539. Oncley, J. L., in *The Nephrotic Syndrome*, Proc. 7th Ann. Conf. Nephrotic Syndrome, The National Nephrosis Foundation, New York, 1956, 67.

540. Oncley, J. L., Harvey Lectures. Series 50, New York, 1956, 71.

541. Opplt, J., Kutáček, M., Loštický, C., and Čižimský, B., *Cas. Lek. Cesk.*, 92, 23, 624, 1953.

542. Opplt, J. and Marčan, K., *Cas. Lek. Cesk.*, 93, 34, 934, 1954.

543. Opplt, J. and Musil, J., *Cas. Lek. Ces.*, 94, 46, 1254, 1955.

544. Opplt, J. and Musil, J., Monogr.: "Praktické problémy lékařské chemie". Praha, SZdN, 1956, 201.

545. Opplt, J. and Syllaba, J., *Cas. Lek. Cesk.*, 95, 44/45, 1247, 1956.

546. Opplt, J., Musil, J., and Fišerová, B., *Cas. Lek. Cesk.*, 95, (44/45), 1257, 1956.

547. Opplt, J. and Musil, J., *Univ. Carol. Suppl.*, 5, 110, 1956.

548. Opplt, J. and Syllaba, J., *Univ. Carol. Suppl.*, 5, 114, 1956.

549. Opplt, J. and Hlaváček, Vl., *Cas. Lek. Cesk.*, 96, (7), 198, 1957.

550. Opplt, J. and Blekta, M., *Acta Univ. Carol. Medica*, 5, 137, 1958.

551. Opplt, J., *Acta Univ. Carol. Medica*, 9-10, 1175, 1958.

552. Opplt, J. and Syllaba, J., in *Diabetes Mellitus*, Monogr., 3rd Congr. Int. Diabetes Fed., Düsseldorf, 1958; Georg Thieme Verlag, Stuttgart, 1959, 200.

553. Opplt, J. and Syllaba, J., *Cas. Lek. Cesk.*, 99, 16, 500, 1960.

554. Opplt, J. and Vamberová, M., *Vnit. Lek.*, VII-8, 875, 1961.

555. Opplt, J. and Syllaba, J., *C. R. 4th Congr. Féd. Int. Diabéte, Genéve, 1961; Monogr. Ed. Med. Hyg.*, Geneve, 1962, 513-515.

556. Opplt, J., Novotný, A., Dvořák, V., and Schreiber, B., *Acta Univ. Carol. Med. Suppl.* 19, 129, 1964.

557. Opplt, J., *Acta Univ. Carol. Med.*, 12 (6/7), 425, 1966.

558a. Opplt, J., *Cas. Lek. Cesk.*, 105, 14, 569, 1966.

558b. Opplt, J., *Cas. Lek. Cesk.*, 105, 15, 799, 1966.

559. Opplt, J., Lipid Res., Symp. held in Plzen, Czechoslovakia, 1965, Suppl. 16, 1966, 193.

560. Opplt, J., *Přehled Klinické Biochemie*, SPN, Praha, 1955, 370.

561. Syllaba, J., Opplt, J., Zrůstová, M., and Ireland, J. T., Diabetic nephropathy, *Syndrom Kimmelstiel-Wilson*, Universitas Carolina Pragensis, Praha, 1968.

562. Opplt, J., *Diabetes Mellitus*, Foit, R. and Syllaba, J., Eds., Avicenum, Praha, 1970, 51—88.

563. Opplt, J., Abstr. in Communication 5th Int. Congr. Biochemistry, Moscow, Sect. 28, 1961, 541.

564. Opplt, J. and Syllaba, J., *On the Nature and Treatment of Diabetes*, Excerpta Medica Foundation, Amsterdam, 1965, 447—448.

565. Opplt, J., 2nd Meeting of the Fed. of European Bioch. Soc., Vienna, 1965.

566. Orengo, A. and Mancini, M., *Boll. Soc. Ital. Biol. Sper.*, 32, 652, 1956.

567. Orvis, H. H. and Burger, D., *Med. Ann. D. C.*, 32, 44, 1963.

568. Osborn, D. A., *Clin. Chim. Acta*, 5, 777, 1960.

569. Ott, H. and Roth, E., *Klin. Wochenschr.*, 32 (45-46), 1099, 1954.

570. Oudin, J., *J. Immunol.*, 84, 143, 1960.

571. Ouchterlony, O., *Acta Pathol. Microbiol. Scand.*, 26, 507, 1949.

572. Ouchterlony, O., *Prog. Allergy*, 5, 1, 1958.

573. Owen, J. A. and Smith, H., *Clin. Chim. Acta*, 6, 441, 1961.

574. Page, J. H. and Lewis, L. A., *Circulation*, 20, 1011, 1959.

575. Paget, M., *Ann. Biol. Clin. (Paris)*, 19, 329, 1961.

576. Paletta, B., *Clin. Chim. Acta*, 5, 490, 1960.

577. Pappenheimer, J. R., *Physiol. Rev.*, 33, 387, 1953.

578. Park, C. R., Morgan, H. E., Henderson, M. J., Regen, D. M., Cadenas, E., and Post, R. L., *Recent Prog. Horm. Res.*, 17, 493, 1961.

579. Paronetto, F., Wang, Ch. I., and Adlersberg, D., *Circulation*, 14, 502, 1956.

580. Paronetto, F., Wang, Ch. I., and Adlersberg, D., *Science*, 124, 1148, 1956.

581. Paterson, J. C., Cornish, B. R., and Armstrong, E. C., *Can. Med. Assoc. J.*, 74, 538, 1956.

582. Paterson, J. C., Cornish, B. R., and Armstrong, E. C., *Circulation,* 13, 224, 1956.
583. Pazzanese, D. et al., *Arq. Bras. Cardiol.,* 17(73), 88, 1964
584. Pedersen, K. O., *J. Phys. Colloid Chem.,* 51, 156, 1947.
585. Pedersen, K. O., *The Ultracentrifuge,* Almquist and Wiksell, Stockholm, 1945.
586. Peeters, H. and Laga, E., in *Protides of the Biological Fluids,* Peeters, H., Ed., Elsevier, Amsterdam, 1963, 134.
587. Pezold, F. A., *Aerztl. Wochenschr.,* 13(6), 129, 1958.
588. Pezold, F. A., *Clin. Chim. Acta,* 3, 40, 1958.
588a. Pezold, F. A., *Clin. Chim. Acta,* 2, 43, 1957.
589. Pezold, F. A., *Klin. Wochenschr.,* 36, 560, 1958.
590. Pezold, F. A., *Naturwissenschaften,* 43, 280, 1956.
591. Pezold, F. A., in *Protides of the Biological Fluids,* Elsevier, New York, 1966, 247.
592. Pezold, F. A., Nahrungsfett und Atherosklerose, Symposium Klinische Biologie der Fette, Med. Univ. Klinik Erlangen 1960; Med. u. Ernährung 1, 244, 1960.
593. Pezold, F. A. and Thomas, H., *Z. Ges. Exp. Med.,* 129, 412, 1957.
594. Phillips, G. B., *Proc. Soc. Exper. Biol. Med.,* 100, 19, 1959.
595. Phillips, G. B., *J. Clin. Invest.,* 38, 489, 1959.
596. Pierce, F. T. and Gofman, J. W., *Circulation,* 4, 25, 1951.
597. Pierce, F. T. and Bloom, B., *Metabolism,* 1, 163, 1952.
598. Pieri, J. et al., *Presse Med.,* 74, 1933, 17, 1966.
599. Pignataro, L., Bolognesi, G., and Marinoni, G. F., *Boll. Soc. Ital. Biol. Sper.,* 38, 883, 1962.
600. Popják, G., *Br. Med. Bull.,* 14, 197, 1958.
601. Pomeranze, J. and Gadek, R. J., in *Diabetes Mellitus,* Monogr., 3rd Congr. Int. Diabetes Fed. Düsseldorf, 1958; Georg Thieme Verlag, Stuttgart, 1959, 440.
602. Popják, G. and McCarthy, E. F., *Biochem. J.,* 37, 702, 1943.
602a. Popják, G. and McCarthy, E. F., *Biochem. J.,* 40, 608, 789, 1946.
603. Popják, G. and Muir, H., *Biochem. J.,* 46, 103, 1950.
604. Popják, J. and Beeckmans, M. L., *Biochem. J.,* 47, 233, 1950.
605. Popják, G. and Tietz, A., *Biochem. J.,* 56, 46, 1957.
606. Popják, G. and McCarthy, E. F., *Discuss. Faraday Soc.,* 6, 97, 1949.
607. Porath, J., *Adv. Protein Chem.,* 135, 223, 1963.
608. Porath, J. and Flodin, P., *Nature (London),* 183, 1657, 1959.
609. Porath, J. and Flodin, P., in *Protides of the Biological Fluids,* Monogr., Peeterse, H., Ed., Elsevier, Amsterdam, 1963, 290.
610. Poulik, M. D., *J. Immunol.,* 82, 502, 1959.
611. Poulik, M. D., *Nature (London),* 180, 1477, 1957.
612. Predergast, J. J. and Teagus, D. M., *Circulation,* 4, 23, 1951.
613. Raacke, I. D., *Arch. Biochem. (New York),* 62(1), 184, 1956.
614. Randle, P. J. and Smith, G. H., in *Diabetes Mellitus,* 3rd Congr. Int. Diabetes Fed., Düsseldorf, 1958, Georg Thieme Verlag, Stuttgart, 1959, 50.
615. Randle, P. J. et al., *Ann. N. Y. Acad. Sci.,* 131, 324, 1965.
616. Randle, P. J., *Colloquia on Endocrinology,* Ciba Foundation, Vol. 15, London, 1964.
617. Randle, P. J., in *On the nature and treatment of diabetes, 5th Congr. Int. Diabetes Fed., Toronto, 1964,* Ed. Exc. Med. Found., 1965, 361.
618. Ray, B. R., Davisson, E. O., and Crespi, P. O., *J. Am. Chem. Soc.,* 74, 5807, 1952.
619. Ray, R. B., Davidson, E. O., and Crespi, H. L., *J. Phys. Chem.,* 58, 841, 1954.
620. Raymond, S. et al., *Science,* 151, 346, 1966.
621. Raynaud, R., d Eshougues, J. R., Pasquet, R., and di Giovanni, S., *Ann. Biol. Clin.,* 11, 377, 1953.
622. Reaven, G. M., Hill, D. B., Gross, R. C., Farquhar, J. E., *J. Clin. Invest.,* 44, 1826, 1965.
623. Rebeyrotte, P. et al., *Ann. Inst. Pasteur.,* 87, 697, 1954.
624. Renninger, W., *Dtsch. Med. Wochenschr.,* 89, 2249, 1964.
625. Renold, A. E., *Colloquia on Endocrinology,* Ciba Foundation, London, Vol. 15, 1964.
626. Renold, A. E., Crofford, O. B., Burgi, H., and Froesch, E. R., in *On the Nature and Treatment of Diabetes,* Monogr., 5th Congr. Int. Diabetes Fed., Toronto, 1964, Ed. Exc. Med. Found., 1965, 146.
627. Renold, A. E., *Union Med. Can.,* 94, 292, 1965.
628. Ribeiro, L. P. and McDonald, H. J., *J. Chromatogr.,* 10, 443, 1963.
629. Ricketts, H. T., in *On the Nature and Treatment of Diabetes,* 5th Congr. Int. Diabetes Fed., Toronto, 1964, Ed. Exc. Med. Found., 1965, 588.
630. Rizza, O., Pignataro, L., and Marinoni, G. F., *Boll. Soc. Ital. Biol. Sper.,* 36, 894, 1960.
631. Rodbell, M., *Science,* 127, 701, 1958.

632. Roheim, P. S., Miller, L., and Eder, H. A., *J. Biol. Chem.*, 240, 2994, 1965.
633. Roheim, P. S. et al., *J. Clin. Invest.*, 42, 1277, 1963.
634. Root, H. F., *J. Clin. Endocrinol. Metab.*, 12, 458, 1952.
635. Root, H. F., in *Diabetes Mellitus*, 3rd Congr. Int. Diabetes Fed., Dusseldorf, 1958, Georg Thieme Verlag, Stuttgart, 1959, 185.
636. Rosenberg, I. N., Young, E., and Proger, S., *Am. J. Med.*, 16, 818, 1954.
637. Rosenberg, I. N., *J. Clin. Invest.*, 31, 657, 1952.
638. Rosenberg, I. N., *Proc. Soc. Exp. Biol., (New York)*, 80, 751, 1952.
639. Rozynkowa, D. et al., *Pol. Arch. Med. Wewn.*, 36, 697, 1966.
640. Rubin, L. and Aladjem, F., *Am. J. Physiol.*, 178, 263, 1954.
641. Rukovin, J. G., Block, W. D., and Curtis, A. C., *J. Lab. Clin. Med.*, 47, 365, 1956.
642. Russ, E. M., Eder, H. A. and Barr, D. P., *Am. J. Med.*, 11, 468, 1951.
643. Salem, L., *J. Chemical Phys.*, 37, 2100, 1962.
644. Salem, L., *Can. J. Biochem. Physiol.*, 40, 1287, 1962.
645. Salt, H. B. et al., *Lancet*, 2, 325, 1960.
646. Salt, H. B. and Wolff, O. H., *Arch. Dis. Child.*, 32, 404, 1957.
647. Sanbar, S. S. and Alupovic, P., *Biochim. Biophys. Acta*, 71, 235, 1963.
648. Sarbar, N., *Nature (London)*, 189, 929, 1961.
649. Scanu, A., Lewis, L. A., and Bumpus, F. M., *Arch. Biochem. Biophys.*, 74, 390, 1958.
650. Scanu, A. M., *Adv. Lipid Res.*, 3, 63, 1965.
651. Scanu, A. and Grauda, J. L., *Biochemistry*, 5, 446, 1966.
652. Scanu, A. and Hughes, W. L., *J. Biol. Chem.*, 235, 2876, 1960.
653. Scanu, A. and Hughes, W. L., *J. Clin. Invest.*, 41, 1681, 1962.
654. Scanu, A. and Page, J. H., *J. Exp. Med.*, 109, 239, 1959.
655. Scanu, A. and Grauda, J. L., *J. Lab. Clin. Med.*, 64, 1002, 1964.
656. Scanu, A., Page, I. H., *J. Lipid Res.*, 2, 161, 1961.
657. Scanu, A., *Nature (London)*, 207, 528, 1965.
658. Scanu, A., *Proc. Natl. Acad. Sci.*, 54, 1699, 1965.
659. Scücs, S. and Csapo, G., Internationales Symposium über die Diab. Angiopathie, Karlsburg, 1962.
660. Searcy, R. L. et al., *Biochim. Biophys. Acta*, 106, 603, 1965.
661. Searcy, R. L., Asher, T. M., and Bergquist, L. M., *Clin. Chim. Acta*, 8, 148, 1963.
662. Seegers, W., Hirschhorn, K., Burnett, L., Robson, E., and Harris, H., *Science*, 149, 303, 1965.
663. Seliger, V. and Podroužek, V., *Cs. Fysiol.*, 4, 365, 1955.
664. Sharfir, E., *J. Clin. Invest.*, 37, 1775, 1958.
665. Shore, B., *Arch. Biochem. Biophys.*, 71, 1, 1957.
666. Shore, V. et al., *Biochemistry*, 6, 1962, 1967.
667. Shore, V. and Shore, B., *Biochem. Biophys. Res. Commun.*, 9, 455, 1962.
667a. Shore, V. and Shore, B., *Biochem. Biophys. Res. Commun.*, 28, 1003, 1967.
668. Shore, V. and Shore, B., *Plasma*, 2, 621, 1954.
669. Shore, B., Nichols, A. V., and Freeman, N. K., *Proc. Soc. Exp. Biol. (New York)*, 83, 216, 1953.
670. Shubarov, K. et al., *Suvr. Med. (Sofia)*, 16, 607, 1965.
671. Schachman, H. K., *Ultracentrifugation in Biochemistry*, Academic Press, New York, 1959.
672. Schjeide, O. A., Rivin, A. V., and Yoshino, J., *Am. J. Clin. Pathol.*, 39, 329, 1963.
673. Schjeide, O. A., *J. Biol. Chem.*, 210, 355, 1954.
674. Schjeide, O. A., *Growth*, 20(3), 195, 1956.
675. Schjeide, O. A., *UCLA*, 373, 1, 1956.
676. Scheidegger, J. J., *Int. Arch. Allergy Appl. Immunol.*, 7, 103, 1955.
677. Schettler, G., Eggstein, M., and Dietrich, F., *Klin. Wochenschr.*, 34, 684, 1956.
678. Schmidt, H. and Zerlett, G., *Med. Klin.*, 51, 1742, 1956.
679. Scholtan, W., *Z. Ges. Exp. Med.*, 121, 574, 1953.
680. Schultze, H. E. and Schwick, G., *Clin. Chim. Acta*, 4, 15, 1959.
681. Simon, K. H., *Med. Wochenschr.*, 20, 234, 1966.
682. Silberman, H. J., *Biochim. Biophys. Acta*, 24, 647, 1957.
683. Singer, P., *Med. Klin.*, 61, 1210, 1966.
684. Skipski, V. P. et al., *Biochem. J.*, 104, 340, 1967.
685. Slizewicz, P., Rebeyrotte, P., and Lepine, P., *Ann. Inst. Pasteur. Paris*, 89(4), 428, 1955.
686. Smith, C., Sauls, H. C., and Balew, J., *Ann. Intern. Med.*, 17, 681, 1942.
687. Smith, E. P., *Lancet*, II, 530, 1962.
687a. Smith, E. P., *Lancet*, II, 910, 1957.
688. Smithies, O., *Biochem. J.*, 61, 626, 1955.
688a. Smithies, O., *Biochem. J.*, 71, 585, 1959.

689. Smithies, O., *Adv. Protein Chem.*, 14, 65, 1959.
690. Sobotka, H., in Biochemical Problems of Lipids, Popjak, G. and LeBreton, E., Eds., *Proc. 2nd Int. Conf. Biochemical Problems*, Butterworths, London, 1956.
691. Sonnino, F. R. and Gassaniga, P. P., *Clin. Chim. Acta*, 6, 295, 1961.
692. Soothill, J. F., *J. Lab. Clin. Med.*, 59, 859, 1962.
693. Sotgiu, G. and Pellegrini, R., in *Diabetes mellitus*, 3rd Congr. Int. Diabetes Fed., Düsseldorf, 1958, Georg Thieme Verlag, Stuttgart, 1959, 197.
694. Speiser, P. and Pansch, V., *Ann. Paediat.*, 205, 193, 1965.
695. Sperry, W. M. and Webb, M., *J. Biol. Chem.*, 187, 97, 1950.
696. Sperry, W. M., *Methods of Biochemical Analysis*, Interscience, New York, 1955, 83.
697. Spitz, J. M., Rubenstein, A. H., Berson, I., Lowy, C., Wright, D., and Fraser, T. R., 6th Congr. Int. Diabetes Fed., Stockholm, Sweden, 1967.
698. Spritz, N., *J. Clin. Invest.*, 44, 339, 1965.
699. Stadie, W. C., *Diabetes*, 5, 263, 1956.
700. Stajner, A., *Vnitr. Lek.*, 12, 707, 1966.
701. Steigerwaldt, F., in *Diabetes mellitus*, 3rd Congr. Int. Diabetes Fed., Düsseldorf, 1958, Georg Thieme Verlag, Stuttgart, 1959, 105.
702. Steinberg, D., in *Control of Lipid Metabolism*, Grant, J. K., Ed., University Press, New York, 1963.
703. Stern, R. and Suchantke, G., *Arch. Exp. Pathol. Pharmakol.*, 115, 221, 1926.
704. Stern, J. and Lewis, W. H. P., *J. Ment. Defic. Res.*, 2(2), 59, 1958.
705. Stokes, R. P., *J. Med. Lab. Technol.*, 23, 33, 1966.
706. Storiko, K., *Blut*, XVI, 200, 1968.
707. Storiko, K. and Fisher, G. B., in *Protides of the Biological Fluids*, Sect. B., Elsevier, Amsterdam, 1966, 291.
708. Strisower, B., Gofman, J. W., Galioni, E., Rubinger, J. H., O'Brien, G. W., and Simon, A., *J. Clin. Endocrinol.*, 15, 73, 1955.
709. Strisower, E. H. et al., *J. Lab. Clin. Med.*, 65, 748, 1965.
710. Strisower, B., Gofman, J. W., Galioni, E. F., Almanda, A. A., and Simon, A., *Metabolism*, 3, 218, 1954.
711. Strisower, B., Gofman, J. W., Galioni, E. F., Rubinger, J. H., Ponteau, J., and Guzvich, P., *Lancet*, 272, 120, 1957.
712. Strzydlewski, Z. et al., *J. Atheroscler. Res.*, 6, 273, 1966.
713. Sukhareva, B. S., *Cor Vasa*, 3, 259, 1961.
714. Sullivan, J. F. et al., *Curr. Ther. Res.*, 7, 28, 1965.
715. Surgenor, D. M., in *Symposium on Atherosclerosis*, National Research Council Publication Number 338, 1955, 203.
716. Svedberg, T. and Pedersen, K. O., *Handbuch der Kolloidwissenschaften*, Verlag Theodor Steinkopf, Dresden, 1940.
717. Svensson, H., *Kolloid. Z.*, 87, 181, 1939.
718. Svensson, H., *Kolloid. Z.*, 90, 141, 1940.
719. Swahn, B., *Scand. J. Clin. Lab. Invest.*, 4, 98, 1952.
719a. Swahn, B., *Scand. J. Clin. Lab. Invest., Suppl.*, 9, 5, 1953.
720. Swahn, B., *Studies on Blood Lipids*, Akademisk Avhandling, Lund, 1953.
721. Swynghedanw, B. and Beaumont, J. L., *Pathol. Biol. (Paris)*, 10, 1531, 1962.
722. Syllaba, J. and Opplt, J., in *Diabetes Mellitus*, 3rd Congr. Int. Diabetes Fed., Düsseldorf, 1958, Georg Thieme Verlag, Stuttgart, 1959, 200.
723. Szücs, S. and Csapo, G., Int. Symp. über die diab. Angiopathie, Karlsburg, 1962.
724. Štastný, M., Michalec, C., and Nováková, E., *Acta Univ. Carol.*, 14, 271, 1961.
725. Tamplin, A. R., Strisower, B., DeLalla, O. F., Gofman, J. W., and Glazier, F. W., *J. Gerontol.*, 9, 404, 1954.
726. Tayeau, F., *Arch. Mal. Appar. Dig.*, 44, 740, 1955.
727. Tayeau, F., *C. R. Soc. Biol. (Paris)*, 137, 239, 1943.
728. Tencks, W. P. and Durrum, E. L., *J. Clin. Invest.*, 34, 1437, 1955.
729. Theorell, H., *Biochem. Z.*, 223, 1, 1930.
730. Tiselius, A. and Kabat, E. A., *J. Exp. Med.*, 69, 119, 1939.
731. Tiselius, A., Flodin, P., *Adv. Protein Chemistry*, Vol. VIII, 461, 1953.
732. Tomita, S., *Jpn. J. Clin. Pathol.*, 13, 529, 1965.
733. Tracy, R. E., Merchant, E. B., and Kao, V. C., *Circulation Res.*, 9, 472, 1961.
734. Tracy, R. E. et al., *Proc. Soc. Exp. Biol. Med.*, 118, 1095, 1965.
735. Tristham, G. R., *Adv. Prot. Chem.*, 5, 83, 1949.
736. Truman, D. E. S. and Löw, H., *Exp. Cell Res.*, 31, 230, 1963.

737. Turner, R. H., Snavely, J. R., Goldwater, W. H., Randolph, M. L., Sprague, C. C., and Unglaub, W. C., *J. Clin. Invest.*, 30, 1071, 1951.
738. Turner, R. H., Goldwater, W. H., Randolph, M. L., Sprague, C. C., Snavely, R., and Unglaub, W. G., *Trans. Assoc. Am. Physicians*, 63, 230, 1950.
739. Turner, R. H., Snavely, J. R., Goldwater, W. H., and Randolph, M. L., *Yale J. Biol. Med.*, 24, 450, 1952.
740. Uriel, J. and Grabar, P., *Ann. Inst. Pasteur.*, 90, 427, 1956.
741. Uriel, J. and Grabar, P., *Bull. Soc. Chim. Biol.*, 38, 1253, 1956.
742. Vallance-Owen, J., in *On the Nature and Treatment of Diabetes*, 5th Congr. Int. Diabetes Fed., Toronto, 1964, Ed. Exc. Med. Found., 1965, 340.
743. Vandenheuvel, F. A., *Can. J. Biochem. Physiol.*, 40, 1299, 1962.
744. Van Handel, E. and Zilversmit, D. B., *J. Lab. Clin. Med.*, 50, 152, 1957.
745. Vierucci, A. et al., *Nature (London)*, 216, 1231, 23, 1967.
746. Voigt, K. D., *Ber. Deutch. Opthalmol. Ges.*, 61, 192, 1957.
747. Voigt, K. D. and Schrader, E. A., *Z. Kreislauforsch.*, 43, 2, 1954.
748. Wachter, H. et al., *Z. Biol.*, 115, 156, 1965.
749. Walton, K. W., *Immunochemistry*, 1, 267, 1964.
750. Walton, K. W., *J. Clin. Pathol.*, 17, 627, 1964.
751. Warembourg, H. et al., *Presse Med.*, 72, 2439, 1964.
752. Waris, E., *Duodecim*, 76, 25, 1960 /Fin/.
753. Watkin, D. M., Lawry, E. Y., Mann, G. V., and Halperin, M., *J. Clin. Invest.*, 33, 874, 1954.
754. Weiss, W. A., *Klin. Wochenschr.*, 43, 273, 1965.
755. Weller, H., *Klin. Wochenschr.*, 36, 563, 1958.
756. Wiedemann, D., *Vnitř. Lék.*, 4, 623, 1958.
757. Wiedemann, E., *Helv. Chim. Acta*, 40, 2074, 1957.
758. Wiedemann, E., *Chem. Ing. Tech.*, 28, 263, 1956.
759. Wieland, O. and Weiss, L., *Biochem. Biophys. Res. Commun.*, 13, 26, 1963.
760. Wieland, O. and Löffler, G., *Biochem. Z.*, 339, 204, 1963.
761. Wieland, O., *Klin. Wochenschr.*, 385, 1954.
762. Wieland, O., in *On the Nature and Treatment of Diabetes*, Monogr., 5th Congr. Int. Diabetes Fed., Toronto, 1964, Ed. Exc. Med. Found., 1965, 533.
763. Wieland, O., in *Diabetes Mellitus*, 3rd Congr. Int. Diabetes Fed., Düsseldorf, 1958, Georg Thieme Verlag, Stuttgart, 1959, 20.
764. Wieland, O., 4th Congr. Fed. Int. Diabete, Geneve, Juillet, 1961, C. R., 131, Edition médecine et hygiene, Geneve.
765. Wieme, R. J., *Arscia Nitgaven*, Brussel, 1959.
766. Williams, F. G., Pickels, E. G., and Durrum, E., *Science*, 121, 829, 1955.
767. Winegrad, A. J., Yalcin, S., and Mulsahy, P. D., in *On the Nature and Treatment of Diabetes*, 5th Congr. Int. Diabetes Fed., Toronto, 1964, Ed. Exc. Med. Found., 1965, 452.
768. Wolff, R., Brigton, J. J., and Nicolas, J. P., *Ann. Biol. Clin. (Paris)*, 21, 15, 1963.
768a. Wolff, R., Brigton, J. J., and Nicolas, J. P., *Ann. Biol. Clin. (Paris)*, 22, 1139, 1964.
769. Wolff, O. H., *Ergeb. Inn. Med. Kinderheilkd.*, 23, 191, 1965.
770. Wolff, O. H. and Salt, H. B., *Lancet*, I, 707, 1958.
771. Wolff, F., Wales, J., and Viktora, J., 6th Congr. Int. Diabetes Fed., Stockholm, Sweden, 1967.
772. Woolf, N. et al., *J. Pathol. Bacteriol.*, 90, 459, 1965.
773. Wool, I. G., *Am. J. Physiol.*, 199, 719, 1960.
774. Wool, I. G., *Biochim. Biophys. Acta Prev.*, 68, 411, 1963.
775. Wool, I. G. and Krahl, M. E., *Nature (London)*, 183, 1399, 1959.
776. Wool, I. G. and Munro, A. J., *Proc. Natl. Acad. Sci.*, 50, 918, 1963.
777. Wright, P. H., in *On the Nature and Treatment of Diabetes*, Monogr., 5th Congr. Int. Diabetes Fed., Toronto, 1964; Ed. Exc. Med. Found., 1965, 354.
778. Wright, P. H. and Malaisse, W. J., 6th Congr. Int. Diabetes Fed., Stockholm, Sweden, 1967.
779. Wunderly, Ch., Pezold, F. A., *Clin. Chim. Acta*, 3, 40, 1958.
780. Wunderly, Ch. and Pillner, S., *Klin. Wochenschr.*, 425, 1954.
781. Wunderly, Ch. and Pezold, F. A., *Naturwissenschaften*, 39, 493, 1952.
781a. Wunderly, Ch. and Pezold, F. A., *Naturwissenschaften*, 42, 579, 1955.
782. Wunderly, Ch. and Pezold, F. A., *Z. Ges. Exp. Med.*, 120, 613, 1953.
783. Wurm, M., Kositschek, R., and Strauss, R., *Circulation*, 4, 526, 1960.
784. Wurmser, S. F. and Hartman, L., *Bull. Soc. Chim. Biol.*, 44, 919, 1962.
785. Zakelj, A. and Gros, M., *Clin. Chim. Acta*, 5, 947, 1960.
786. Zambrowicz, K., *Pol. Arch. Med. Wewn.*, 35, 383, 1965.

787. Zayala, A. V., Groppa, S., and Baliarda, R. L., 6th Congr. Int. Diabetes Fed., Stockholm, Sweden, 1967.

788. Zilversmit, D. B., J. Lipid Res., 5, 300, 1964.

789. Zinn, W. J. and Griffith, G. C., Am. J. Med. Sci., 220, 597, 1950.

790. Zinn, W. J., Field, J. B., and Griffith, G. C., Proc. Soc. Exp. Biol. Med., 80, 276, 1952.

791. Zöllner, N., Dtsch. Med. Wochenschr., 80(26), 999, 1955.

791a. Zöllner, N., Dtsch. Med. Wochenschr., 448, 609, 1958.

792. Zöllner, N., Verh. Dtsch. Ges. Inn. Med., 63, 631, 1957.

793. Zöllner, N., in Thannhausers Lehrb. d. Stoffwechsels und d. Stoffwechselkrankheiten, Georg Thieme Verlag, Stuttgart, 1957, 581.

793a. Kien, G. and Wetzler-Ligeti, C., Wien. Klin. Wochenschr., 48, 871, 1935.

793b. Stern, L. and Saphiro, B., Br. J. Clin. Pathol., 6, 158, 1953.

793c. Pearson, S., Stern, S., and McGavack, J. H., Anal. Chem., 25, 813, 1953.

793d. Schoenheimer, R. and Sperry, W., J. Biol. Chem., 106, 745, 1934.

793e. Bloor, W. R., J. Biol. Chem., 24, 227, 1916.

793f. Carlson, L. A. and Waldstrom, L. B., Clin. Chem. Acta, 4, 197, 1959.

794. Bacon, G. E. and Shafeck, S., Serum lipids and lipoproteins in diabetic children, Univ. Mich. Med. Cen. J., 34, 84, 1968.

795. Bazex, A., Dupre, A., Christol, B., and Bazex, J., Type IV hyperlipoproteinemia in a female diabetic, Bull. Soc. Fr. Dermatol. Syphiligr., 78, 74, 1971.

796. Bhan, C. K., Kumar, V., and Ahuja, M. M., Studies on neutral fat, lipoproteins and lipoprotein lipase in relation to vascular disease in young Indian diabetics, Acta Diabetol. Lat., 8, 638, 1971.

797. Bhu, N., Sogani, I. C., and Sarin, L. R., Serum lipoproteins and glycoproteins in diabetic subjects, J. Assoc. Physicians India, 17, 573, 1967.

798. Bricker, L. A., et al., The hyperlipoproteinemias: mechanisms, managements and implications. Acta Diabetol. Lat., (Suppl.) 9, 53, 1972.

799. Brown, D. F., The dyslipoproteinemias, Acta Diabetol. Lat. (Suppl.) 9, 18, 1972.

800. Chance, G. W. and Albutt, E. C., Serum lipids and lipoproteins in untreated diabetic children, Lancet, 1, 1126, 1969.

801. Chance, G. W., Albutt, E. C., and Edkins, S. M., Lipids and lipoproteins in untreated diabetes, Lancet, 2, 544, 1969.

802. Davidsen, O., Immunoelectrophoretic determination of serum globulins in newborn infants of diabetic mothers, Acta Paediatr. Scand., 63, 833, 1974.

803. Gligore, V., Hincu, N., and Tecuceanu, R., Significance of hyperlipoproteinemia in diabetes mellitus, Oeff. Gesundheitswes, 36, 1259, 1974.

804. Haller, H., Hanefeld, H., Kretchmar, R., Unger, E., Arnold, F., and Wagner, B., Flotation analysis studies of lipoproteins in correlation with the lipid pattern in essential hyperlipemia and diabetic metabolism status, Schweiz. Med. Wochenschr., 99, 813, 1969.

805. Hart, A., Cohen, H., and Thorp, J. M., Lipoprotein and fibrinogen studies in diabetes, Postgrad. Med. J., 47 (Suppl.), 435, 1971.

806. Hase, M., Abnormalities of serum lipids and lipoproteins in diabetes mellitus, Saishin Igaku, 27, 510, 1972.

807. Howard, C. J., Jr., Correlations of serum triglyceride and pre-β lipoprotein levels severity of spontaneous diabetes in Macaca nigra, J. Clin. Endocrinol. Metab., 38, 356, 1974.

808. Kobierska-Szczepanska, A., Proteinogram, lipoproteinogram and glycoproteinogram of serum protein in diabetic children, Pediatr. Pol., 41, 1151, 1966.

809. Kremer, G. J., Distribution of various hyperlipoproteinemia patterns in patients with diabetes mellitus or peripheral circulation disorders, Verh. Dtsch. Ges. Inn. Med., 78, 1346, 1972.

810. Kumar, D. and Gupta, N. N., Serum cholesterol, phospholipids and β-lipoproteins in untreated diabetics, J. Assoc. Physicians India, 15, 357, 1967.

811. Mazovetskii, A. G., Blood serum lipoproteins in patients with diabetic angiopathy and obliterating atherosclerosis, Ter. Arkh., 41, 63, 1969.

812. Meduri, D. and Polito, R., Changes of the blood lipid pattern in diabetics under treatment with chlorpropamide and phenethyl-1-beguanide, Minerva Med., 60, 2439, 1969.

813. Murthy, D. Y. N., Guthrie, R. A., Womack, W. N., and Jackson, R. L., β-lipoproteins in children with diabetes mellitus: a study, Mo. Med., 66, 273, 1969.

814. Nakano, E., et al., Serum lipoprotein fractions in diabetes mellitus, Jpn. J. Clin. Pathol., 19 (Suppl.), 171, 1971.

815. Naruszewicz, M., Phenotyping of secondary hyperlipoproteinemia in diabetics, Pol. Tyg. Lek, 29, 653, 1974.

816. Opplt, J. J., Makroglobulare Proteins bei der Diabetischen Neuropathie., Abh. Dtsch. Akad. Wiss. Berlin Kl. Med., 3, 221, 1964.

817. **Opplt, J. J., Marcan, K., Spatny, E., and Vakocova, H.,** Atypical proteins in nephrotic syndrome, *Acta Univ. Carol. Med. Suppl.,* 19, 119, 1964.

818. **Opplt, J. J.,** The evaluation of serum-lipoprotein analyses in atherosclerosis, *Acta Univ. Carol. Med.,* 12, 425, 1966.

819. **Opplt, J. J.,** Serum proteins in diabetic angiopathy, *Cas. Lek. Cesk.,* 105, 799, 1966.

820. **Opplt, J. J. and Syllaba, J.,** L'etude des proteines urinaires au cours de la glomerulosclerose diabetigue, *Excerpta Med. Int. Congr. Ser.,* 67, 130, 1963.

821. **Opplt, J. J.,** Lipoproteide in Blutplasma und die Semiologie ihrer Fractionen, *Lipid Res. Suppl.,* 16, 194, 1966.

822. **Opplt, J. J.,** A combined method for determination and evaluation of blood lipoproteins, Section 28, Abstracts of Communications, Vth International Congress of Biochemistry, Moscow, 1961, 541.

823. **Opplt, J. J. and Opplt, M. A.,** Separation of plasma lipoproteins according to molecular size, *Clin. Chem.,* 20, 906, 1974.

824. **Opplt, J. J., Glavaski, B. S., Bahler, C. R., and Opplt, M. A.,** Molecular distribution of plasma lipoproteins in Type IV hyperlipoproteinemia, *Clin. Chem.,* 21, 990, 1975.

825. **Bahler, R. C. and Opplt, J. J.,** Dietary influence on molecular distribution of serum lipoproteins in subjects with proven coronary artery disease; atherosclerosis, *Adv. Exp. Med. Biol.,* 82, 195, 1977.

826. **Opplt, J. J.,** Changes in the vascular wall., Synopsis of metabolic disturbances., and Synopsis of personal investigations, in *Diabetic Nephropathy,* Syllaba, J., Ed., Universitas Carolina Pragensis, 1968.

827. **Opplt, J. J.,** Biochemistry of diabetes mellitus., and Biochemical diagnostic methods, currently used for diagnosis of diabetes and for control of its course and therapy, in *Diabetes Mellitus,* Monogr. Foit, R. and Syllaba, J., Eds., Avicenum, Praha, 1970.

828. **Opplt, J. J., Bahler, R. C., and Opplt, M. A.,** Molecular pathology of serum lipoproteins, in *Atherosclerosis IV,* Monogr. Schettler, G., Goto, Y., Hata, Y., and Klose, G., Ed., Springer-Verlag, Berlin, 1977, 545.

829. **Opplt, J. J.,** Agarose-gel electrophoresis of serum lipoproteins, in *CRC Handbook of Electrophoresis,* Vol. 1, Lewis, L. A. and Opplt, J. J., Eds., CRC Press, Boca Raton, 1979.

830. **Opplt, J. J., Bahler, R. C., and Opplt, M. A.,** Lipoproteins in proven coronary artery disease, in *CRC Handbook of Electrophoresis,* Vol. 1, Lewis, L. A. and Opplt, J. J., Eds., CRC Press, Boca Raton, 1979.

831. **Petrides, R. and Schrapler, P.,** Behavior of lipoproteins in diabetic coma, *Verh. Dtsch. Ges. Inn. Med.,* 77, 603, 1971.

832. **Richletti, A., Scherrer, J. R., Micheli, H., and Pometta, D.,** Etude clinique des relations entre le diabete les hyperlipoproteinemies et l'artheromatose, *Schweiz. Med. Wochenschr.,* 103, 668, 1973.

833. **Roe, R. L. and Walsh, E. J.,** Lipids and lipoproteins in untreated diabetes, *Lancet,* 2, 496, 1969.

834. **Sanwald, R. and Wahl, P.,** On the action of lipoproteins in diabetics under treatment with biguanides, *Schweiz. Med. Wochenschr.,* 94, 1459, 1964.

835. **Schonfeld, G., Birge, C., Miller, P., Kessler, G., and Santiago, J.,** Apolipoprotein levels and altered lipoprotein composition in diabetes, *Diabetes,* 23, 827, 1974.

836. **Schwandt, P.,** Hyperlipoproteinemia and diabetes mellitus, *Hippokrates,* 44, 457, 1973.

837. **Shestakova, S. A.,** Effect of "diabetic" serum on respiration and carbohydrate metabolism in leukocytes of rabbit exudate in experimental diabetes, *Vop. Med. Khim,* 13, 461, 1967.

838. **Silnitskii, P. A.,** Dynamics of blood serum lipid and lipoprotein indices in diabetes mellitus, treated with chlorpropamide and butamide, *Probl. Endokrinol.,* 14, 9, 1968.

839. **Skopichenko, N. F.,** Study of blood proteins, lipids and lipoproteins in patients with diabetic glomerulosclerosis, *Vrach. Delo,* 3, 22, 1969.

840. **Steiner, G.,** Biosynthesis, physiologic role and normal patterns of lipoproteins in normal human plasma, *Acta Diabetol. Lat.,* Suppl. 9, 3, 1972.

841. **Strat, C.,** Study of serum lipoproteins in diabetes mellitus. Clinicometabolic significance, *Rev. Med. Chir. Soc. Med. Nat. Jasi,* 78 (3), 565, 1974.

842. **Streda, M., Kucerova, L., and Stork, A.,** Serum lipoproteins in diabetes mellitus, *Cas. Lek. Cesk.,* 111, 1026, 1972.

843. **Syllaba, J. and Opplt, J. J.,** On the so-called "hepatic diabetes" *Cas. Lek. Cesk.,* 95, 1247, 1956.

844. **Syllaba, J. and Opplt, J. J.,** Des resultats de laboratoire clinique dans 168 subjects avec diabete sucre complique par nephropathie diabetique, Monogr., *Diabetes Mellitus,* C. R. 3rd Kongr. Int. Diabetes Fed., Georg Thieme Verlag, Stuttgart, 1959, 200.

845. **Syllaba, J. and Opplt, J. J.,** Classification du syndrome Kimmelstiel-Wilson base sur des experiences cliniques et des recherches biochimiques, *C. R. 4th Congr. Fed. Int. Diabetes,* Medicine et Hygiene, Geneve, 1962, 513.

846. **Syllaba, J. and Opplt, J. J.,** Nephropathy in diabetic subjects. On the nature and treatment of diabetes, Monogr., Excerpta Medica Foundation, Amsterdam, 1965, 447.

847. **Syllaba, J. and Opplt, J. J.,** Serum proteins in diabetes, C. R. Congr. biochemists, physicists, pharmacologists, Czechoslovakia, 1956.
848. **Syllaba, J.,** Therapy of diabetic complications, *Cas. Lek. Cesk.,* 56, 157, 1967.
849. **Syllaba, J., Ireland, I. T., Opplt, J. J., and Zrůstová, M.,** Diabetic Nephropathy, Monogr., Universitas Carol. Pragensis, 1969.
850. **Syllaba, J.,** Diabetes and Arteriosclerosis, in *Arteriosclerosis,* SZN, Praha, 1957, 74.
851. **Tiengo, A., Muggeo, M., Fedele, D., Marchiori, E., and Crepaldi, G.,** Insulin secretion in hyperlipoproteinemias, *Acta Diabetol. Lat.,* 11, 149, 1974.
852. **Tishenina, R. S.,** Influence of incubating the serum of donors and diabetes mellitus patients with insulin on its effect with regard to several lipid metabolism indices, *Probl. Endokrinol.,* 17, 19, 1971.
853. **Vogelberg, K. A., Gries, F. A., and Dietel, J.,** Clinical picture and treatment of insulin resistance in primary hyperlipoproteinemia. *Dtsch. Med. Wochenschr.,* 98, 1751, 1973.
854. **Wahl, P., Hasslacher, Ch., and Vollmar, J.,** Diabetes and hyperlipoproteinaemias *Dtsch. Med. Wochenschr.,* 99, 2158, 1974.
855. **Wille, L. E.,** Lipoproteins and diabetes mellitus, *J. Nor. Laegeforen,* 90, 313, 1970.
856. **Wille, L. E. and Aarseth, S.,** Lipoproteins and lipids in diabetics with and without coronary heart disease, *J. Oslo City Hosp.,* 24, 113, 1974.
857. **Wille, L. E. and Aarseth, S.,** Demonstration of hyper-α-lipoproteinemia in three diabetic patients, *Clin. Genet.,* 4, 281, 1973.
858. **Zorrilla, E., Hernandez, A., Serrano, P. A., and Magos, C.,** Diabetes and hyperlipoproteinemia in coronary atherosclerosis, *Acta Diabetol. Lat.,* 8, 629, 1971.

HYPERLIPIDEMIA AND HYPERLIPOPROTEINEMIA IN PATIENTS WITH GOUT

H.K. Naito, A.H. Mackenzie, C.E. Willis, and M. Olynyk

The high incidence of atherosclerosis (coronary, cerebrovascular, peripheral vascular, and nephrosclerosis) as a complication of gout has been recognized for about a century; however, the precise relationships of gout or hyperuricemia to coronary artery disease have not been clear. The relationship of primary gout to dysmetabolism of lipids and lipoproteins is also unclear.

Since there is a high degree of correlation between elevated serum lipid and lipoprotein levels to cardiovascular disease and that it is the leading cause of mortality in the U.S., it is exceedingly important that we understand the reason(s) for the high incidence of this pathogenic phenomenon in gouty patients.

One main problem in understanding the effects of gout on lipid or lipoprotein metabolism is failure to detect any direct effect of primary gout or of elevated serum uric acid (SUA). Is hyperlipidemia or hyperlipoproteinemia a secondary manifestation of gout or of elevated SUA levels, or is the occurrence of the lipid abnormalities in gouty subjects no higher than the age-sex match group that is considered healthy for the geographical area? Gout has been associated with obesity, hypertension, renal stone formation and failure, impaired glucose tolerance, and elevated uric acid levels — all of which may have a direct or indirect effect on lipid and lipoprotein metabolism. Another problem associated with past studies is that patients with clinical gout (0.5% prevalence) and with symptomless hyperuricemia (10% prevalence) were combined in the same category which probably obscured any effect that gout itself may have exerted. The relationship might also depend upon the degree of hyperuricemia.

Table 1 summarizes studies (in chronological order) that have reported patients with gout or hyperuricemia and its effect on serum total cholesterol or triglyceride concentrations. While there is some discrepancy in the data on the relationship of elevated SUA and blood-lipid concentrations, there is high correlation between the two parameters, particularly between hypertriglyceridemia with hyperuricemia. It should be pointed out that some investigators[10,29] have not been able to demonstrate hyperlipidemia in patients with gout. Studies done during the 1950s and early 1960s give the impression that elevated serum cholesterol level is consistently found in gouty individuals or in individuals with elevated SUA concentrations, while the correlations with serum triglycerides are unremarkable. This initial impression reverses itself as one reviews papers in the late 1960s and 1970s. The correlation becomes more consistent and distinct with serum triglycerides and not with serum cholesterol, with some notable exceptions.[26,29] Some of the discrepancies in the literature may be due to inadequate or improper diagnosis and classification of patients with primary gout, inclusion of patients with secondary gout; or lipid measurements being made at different stages in the clinical course of gout, rendering comparison impossible.

The mechanism for this hyperlipidemia is still unknown. Nishida et al.,[1] Gibson and Grahame,[2] Frank,[45] and Lang[46] suggested that excessive alcohol intake may play a role in inducing hyperlipidemia in primary gout. It has long been known that alcohol can induce transitory hyperlipidemia in man.[36-41] Excessive alcohol intake appears to elevate pre-β-lipoprotein levels.[40-42] Avogaro and Cazzolato[41] reported that a large intake of ethanol induced acute hypertriglyceridemia, which varied in intensity among individuals, with increases ranging from 26 to 377% above basal values. On the other hand, no significant changes occurred in plasma cholesterol. Results from the agarose[43] and polyacrylamide gel (PAG)[44] electrophoresis showed that the serum pre-β-lipopro-

Table 1
STUDIES ON SERUM LIPID LEVELS IN PATIENTS WITH GOUT[a]

Author(s) and ref. no.	Year	TC[b]	TG[b]
Gertler et al.[15]	1951	↑	
Harris-Jones[10]	1957	↑	
Kramer et al.[16]	1958	↑	
Kohn and Prozan[17]	1959	↑	
Becker[18]	1960	↑	↑
Feldman and Wallace[9]	1964	↑	↔
Berkowitz[19]	1964	↔	↑
Blachos and Reisenauser[20]	1965	↑	
Hall[21]	1965	↑	
Dunn and Moses[35]	1965	↔	↔
Berkowitz[6]	1966	↔	↑
Barlow[22]	1966	↑	↑
Gunther and Knapp[27]	1966	↑	↑
Gunther et al.[23]	1967	↔	↑
Hollister el al.[5]	1967		↑
Benedek[4]	1967	↔	↑
Knick et al.[24]	1968	↔	↑
Barlow[25]	1968	↑	↑
Bernheim et al.[26]	1968	↑	↑
Bluestone et al.[28]	1971		↑
Emmerson and Knowles[29]	1971		↑
Darlington and Scott[30]	1972	↔	↑
Wiedemann et al.[8]	1972	↔	↑
Mielants et al.[31]	1973	↑	↑
Mielants et al.[32]	1973	↑	↑
Klein et al.[33]	1973	↑	↔
Pereira Miguel et al.[64]	1973		↑
Frank[47]	1973	↑	↑
Gibson and Grahame[2]	1974	↔	↔
Frank[45]	1974	↔	↑
Halpern and Pereira Miguel[34]	1974		↑
Nishida et al.[1]	1975	↔	↑

[a]   These abbreviations will be used in the table: TC = total cholesterol; TG = triglycerides.
[b]   ↔ means no change; ↑ means increased.

tein and "chylomicron-like" bands increased in stain intensity in several patients who consumed ethanol. This study demonstrated that the relative increase in triglyceride concentration is not confined to the very low-density lipoprotein (VLDL) fraction; in addition, it is also manifested, although to a lesser degree, in low-density lipoprotein (LDL) and high-density lipoprotein (HDL) fractions. It also suggested that the appearance of intermediate-density lipoprotein (IDL), d 1.006 to 1.019 may result from ethanol ingestion.

Yano et al.[3] studied 8000 Japanese men (age, 45 to 69 years) in Japan, Hawaii, and California and suggested that while alcohol intake and elevated triglycerides were correlated with high SUA levels, obesity seemed to be responsible for the hyperuricemia. However, it is unlikely that obesity per se is the cause of hyperuricemia. Obesity might be associated, to some degree, with elevated SUA levels but is not responsible for or the cause of the actual rise in SUA concentration. Benedek[4] and Hollister et al.[5] found high correlation between SUA concentrations and obesity in patients with hypertriglyceridemia but were careful not to imply that obesity was the etiological factor for hyperuricemia.

The relationship between SUA and triglyceride concentrations is, however, not always predictable.[9] Gunther and Knapp[14] showed the absence of a concomitant reduction of triglyceride levels after allopurinol (an analogue of hypoxanthine which reduces SUA levels) treatment had lowered elevated SUA levels.[55,56] This result suggests that circulating SUA levels may not have a direct influence on serum triglyceride levels.

While hypertriglyceridemia alone[1,4,6,8,19,23,24,30] has been associated with gout, hypercholesterolemia combined with hypertriglyceridemia[18,22,25-27,31,32,36,47] has also been reported in patients with gout (Table 1). Fulton,[65] Kohn and Prozan,[17] and Mielants et al.[31] suggested that since genetic transmission of both hyperuricemia and hypercholesterolemia occurs, both disorders may be linked, inborn metabolic errors. However, the majority of studies on gouty patients fail to reveal a clear mode of inheritance, suggesting that perhaps many genes are involved in the modification of serum urate levels.[48-56] For example, Benedek and Sunder[12] reported that while dietary differences were mainly responsible for differences in serum lipid levels in white and Negro men, dietary effect did not influence SUA concentrations. While not fully acknowledged at the present time, the mechanism of hyperuricemia may also be mediated through the endocrine system.[25]

Studies such as that of Berkowitz,[6] Halpern and Pereira Miguel,[34] and Wiedemann et al.[8] suggest that hypertriglyceridemia in patients with gout is related to carbohydrate intolerance or diabetes; however, genetic studies suggest that familial or primary hypertriglyceridemia is inherited independently from diabetes.[13] Others believe that the hyperlipidemia in gout may be related to genetic factors.[9,11] Alternatively, some believe that hyperuricemia and gout, undoubtedly, represent a common phenotypic expression of a number of different genetic disorders[55] and can be influenced by environmental conditions.

Since lipids are solubilized in serum by binding to proteins in the form of lipoproteins, it is logical to examine the effects of hyperuricemia on lipoprotein metabolism. A review of the literature illustrates that little work has been done in this area, particularly with respect to lipoprotein electrophoresis (Table 2).

Barlow[25] was one of the earlier investigators who studied the relationship between primary gout and lipoprotein levels. He quantitated the β- and α-lipoproteins by polyanion precipitation techniques. While there was little relationship between the serum α-lipoprotein level, the β-lipoprotein concentration was higher in gouty subjects as compared to that of the control.

Using the cellulose acetate lipoprotein electrohporesis method of Chin and Blankenhorn,[66] Bluestone et at.[28] reported on 12 hypertriglyceridemic, gouty patients who showed an excess of pre-β-lipoprotein. The most common pattern was the type IV hyperlipoproteinemia pattern. Using a similar method of electrophoresis,[66] Mielants et al.[31] found a statistically significant increase in pre-β-lipoprotein and a concomitant decrease of both α- and β-lipoproteins in patients with gout. It is interesting that they found a decrease in β-lipoprotein level when, in fact, serum total cholesterol and triglycerides were elevated in these subjects. In their second report,[32] they indicated a log to log correlation between SUA levels and α- and β-lipoprotein values. There was also a log to log relationship between all lipids, lipoproteins, and SUA levels, with the exception of cholesterol. They suggested that the SUA vs. α- and β-lipoprotein regression lines decrease in a parallel way, which led them to conclude that the percent rise of pre-β-lipoprotein is caused by an equal percent fall of α- and β-lipoproteins. This interpretation is questionable, because what was actually measured densitometrically, i.e., relative percent distribution of lipoprotein fractions, may misrepresent the actual concentration of each lipoprotein fraction.

Table 2
HYPERLIPOPROTEINEMIA AND GOUT[a]

Author	Year	TC[b]	TG[b]	Type of hyperlipoprotein-emia	Method of quantitation
Barlow[25]	1968	↑	↑	Hyper-β-lipoproteinemia	Polyanion precipitation
Bluestone et al.[28]	1971		↑	Mainly type IV (one type V and IIa)	Cellulose acetate electro-phoresis
Wiedemann et al.[8]	1972	↔	↑	Type IV	Paper electrophoresis
Mielants et al.[31]	1973	↑	↑	Hyperpre-β-lipoproteinemia	Cellulose acetate electro-phoresis
Mielants et al.[32]	1973	↑	↑	Hyperpre-β-lipoproteinemia	Cellulose acetate electro-phoresis
Halpern and Per-eira Miguel[34]	1974		↑	Type IV, IIb	Not described
Frank[45]	1974	↔	↑	Type IV	Cellulose acetate electro-phoresis

[a] These abbreviations will be used in the table: TC = total cholesterol; TG = triglycerides.
[b] ↔ means no change; ↑ means increased.

Wiedemann et al.[8] studied lipoprotein patterns of gouty subjects using the paper electrophoretic method of Lees and Hatch.[57] There were no significant differences between mean plasma cholesterol of gouty patients and nongouty control subjects, but plasma triglyceride was elevated in gouty subjects (P<.05). Elevated triglyceride level was associated with an increase in pre-β-lipoproteins only (type IV pattern). Similar findings were reported by Halpern and Pereira Miguel,[34] who studied 104 patients and found a high degree of association between hyperuricemia and hypertriglyceridemia in type IV and IIb patients.

Frank[45] reported that the most common type of hyperlipoproteinemia in patients with primary gout was the type IV pattern (60% of gouty patients). In contrast, patients with symptomless hyperuricemia had neither type II nor type IV hyperlipoproteinemia exceeding the frequency found in control subjects.

The authors carefully selected 17 patients at the Cleveland Clinic who were diagnosed as having primary gout. Only patients classified as primary gout, either classical or definite, were included in this study. Patients meeting criteria for probable or possible gout were excluded, including patients with gout secondary to blood dyscrasias, chronic nephritis, drug administration, or glycogen storage disease. Sufficient clinical study to exclude other episodic arthropathies were performed prior to this study. The criteria used were as follows:

1. Asymptomatic hyperuricemia (over 7 mg SUA per 100 ml) was not regarded as gout.
2. Possible gout was defined as one to three gout-like episodes of dramatic, sudden, painful joint inflammation, lasting 3 to 10 days, involving appropriate sites, and separated by symptom-free intervals in a hyperuricemic individual.
3. Probable gout was defined as four or more attacks conforming to possible gout in a chronically hyperuricemic individual (two or more determinations) whose attacks seem to respond to appropriate anti-inflammatory drugs.
4. Three varieties of definite gout were recognized:
a. Probable gout as above, in which attacks declined markedly or ceased upon therapeutic reduction of serum urate

b. Acute monoarthritis, yielding synovial fluid containing crystals of sodium urate monohydrate identified either by specific chemical methods or by polarized light microscopy (strong negative birefringence and other typical characteristics)

c. Possible or probable clinical gout with proven urate nephrolithiasis (No examples of an alternate possible type were included, i.e., urate nephrolithiasis without clinical gout but with chronic hyperuricosuria, over 600 mg/24 hr.)

5. Classical gout was defined as one of the above patterns with tophaceous deposits of sodium urate monohydrate in bone or cartilage, subcutaneous sites, or kidney. (Some of these patients lose symptom-free intervals of gout.)

At the time of this study, gouty attacks in these adult, male patients (mean age = 59.8 ± 2.5, range = 34 to 73 years old) were under control by a combination of drugs and dietary regimen (Table 3). Although their attacks were under control, these individuals should still be considered to have gout.

Of these patients, 41% were classified with definite gout and 59% had classical gout (Table 4). Since it is well known that liver dysfunction can contribute to secondary dyslipoproteinemia, liver function profile were examined by the Technicon SMA 12/60 method.[63] Included in the tests was total bilirubin, alkaline phosphatase, creatine phosphokinase (CPK), lactate dehydrogenase (LDH), serum glutamic oxalacetic transaminase (SGOT), total protein, and albumin.

It is also known that renal dysfunction can alter serum lipoprotein patterns.[68,69] Since gouty patients are known to have urate crystal formation in the kidneys or renal stone formation leading to kidney impairment, we also monitored renal function by the Technicon SMA 6/60 method.[63] Included in the tests were sodium, potassium, chloride, carbon dioxide, blood urea nitrogen (BUN), and creatinine.

Using the method of Sobrinho-Simoes,[67] the SMA 12/60 uric acid test is a colorimetric assay based on the reduction of a phosphotungstate complex to a phosphotungstide complex.

Since it is well established that serum lipid values are age and sex dependent, a group of apparently healthy male individuals (n = 61) were randomly selected to obtain an age-matched group to compare with the gouty patients. The healthy individuals were selected on the basis of having no abnormal clinical history or biochemical data according to their annual physical examination. It is from this group of people that the lipid and lipoprotein data were obtained for "baseline" values for comparison with the gouty subjects. Serum total cholesterol and triglycerides were done by the Auto-Analyzer-II method[59] with slight modifications. The procedure for standardization and certification of total cholesterol and triglyceride determinations was performed according to the method outlined by the Lipid Standardization Laboratory at the Center for Disease Control, Atlanta, Ga. Serum total phospholipids[59] were also determined. In addition to using the agarose lipoprotein electrophoretic method,[62] paper,[60] and PAG[61] methods were used on certain selected samples that were unusual or difficult to interpret.

The data on serum lipids and lipoproteins are shown in Tables 5 to 7.* This study[70] reveals that although these gout patients were asymptomatic due to treatment with diet and drugs, they were still hyperlipemic. The gout patients had a serum cholesterol concentration of 261 ± 12 which was not statistically different from the apparently normal males in fifth decade of life (Table 5).

Using arbitrary cut-off points of >250 mg/dl as being abnormal levels, about 38% of the apparently normal and 35% of the gouty patients can be considered hypercho-

---

* Patients were fasted (>12 hr) before obtaining blood.

Table 3

CLINICAL DATA ON ADULT PATIENTS WITH PRIMARY GOUT IN THE CLEVELAND CLINIC STUDY

Patient	Age	Sex	Height (in.)	Weight (lb)	Classification of gout	Drugs in use	Alcohol consumption
G.C.	67	M	72	208	Classical gout	Allopurinol, Esidrix, ® Sorbitrate, ® Coumadin®	1 drink occasionally
W.W.	62	M	70	172	Classical gout	Allopurinol	14 oz/week
W.G.	61	M	71	217	Definite gout	Benemid,® Deltasone®	56 oz/week
R.V.	46	M	73	182	Definite gout	Butazolidin, Atromid-S,® Benemid	6 oz/week
A.J.	66	M	67	190	Definite gout	Benemid, Proloid,® Nitro-Bid,® Atromid-S	4—6 oz/week
D.W.	63	M	69	184	Classical gout	Diabinese,® Apresoline,® Benemid	Only occasionally
D.E.	44	M	67	173	Definite gout	Benemid, colchicine	6 drinks bimonthly
D.K.	34	M	74	219	Definite gout	Allopurinol, colchicine	None
B.K.	57	M	67	154	Classical gout	Allopurinol, Motrin®	14 oz/week
R.L.	63	M	69	233	Definite gout	Allopurinol, Catapres,® Hydrodiuril,® Inderal®	Only occasionally
R.S.	66	M	63	204	Classical gout	Allopurinol, Aldoril®	14 oz/week
F.D.	70	M	65	163	Classical gout	Allopurinol, Tandearil®	28 oz/week
O.G.	71	M	69	208	Definite gout	Allopurinol, colchicine, Aldomet®	2—3 beers daily
M.R.	63	M	68	218	Classical gout	Allopurinol, colchicine, Diuril®	None
E.S.	59	M	65	162	Classical gout	Allopurinol, colchicine, Isordil®, Or-nade®	14 oz/week
J.H.	71	M	66	144	Classical gout	Anturane®	14 oz/week
H.C.	53	M	69	229	Classical gout	Allopurinol	10 oz/week

Table 4

SUMMARY OF THE PHYSICAL
CHARACTERISTICS OF THE GOUTY PATIENTS

Parameters	Definite Gout	Classical Gout
Classifications (% of total)	41%	59%
Mean age (years)[a]	55	63
Mean height (cm)	180	170
Mean weight(kg)	91.4	82.7

[a]   Mean age of both groups = 59.8± 2.5.

Table 5

SERUM LIPID VALUES

Lipid	"Normal" n = 61	Gout n = 17
Age	53.9 ± 0.7	59.8± 2.5
Total Cholesterol (mg/dl)	283 ± 6	261 ±12
Triglycerides (mg/dl)	194 ± 13	244 ±22
Phospholipids (mg/dl)	—	277 ±11

Table 6

PERCENT OF PATIENTS WITH HYPERLIPIDEMIA

Patients	Number (n)	Elevated blood lipids (%)[a]		
		Total cholesterol (>250 mg/dl)	Triglycerides (>150 mg/dl)	Phospholipids (>300 mg/dl)
Normal[b]	61	38%	45%	—
Gouty[c]	17	35%	65%	41%

[a]   Represents percent of total subjects.
[b]   Mean age = 53.9 ± 0.7.
[c]   Mean age = 59.8 ±2.5.

lesterolemic (Table 6). So the incidence between the two populations is not different. However, the serum triglyceride levels of the gouty subjects were significantly different from the apparently normal subjects (Table 5). Furthermore, using an arbitrary cut-off point of >150 mg/dl, 65% of the gout patients were considered hypertriglyceri-demic as compared to 45% of the "control" subjects (Table 6). Using an arbitrary cut-off point of >300 mg/dl, 41% of the gout patients were classified as being hyper-phospholipidemic.

Using these arbitrary cut-off points, 95% of the gout patients had at least one of the three lipid classes elevated. However, the more meaningful data is the fact that there is a greater occurrence of elevated triglyceride concentrations in gouty subjects as compared to their apparently healthy sex-age matched counterparts.

It cannot be overemphasized that this secondary lipid disorder does not seem to be due to elevated SUA levels since the majority of the SUA values were within normal limits, irrespective of the hyperlipidemic condition (Table 8). This is in agreement with Gunther and Knapp,[14] who found that controlling SUA levels with allopurinol did not

Table 7

LIPOPROTEIN PATTERNS
OF GOUT PATIENTS (%)

Types	"Control" (n = 61)	Gouty (n = 17)
Normal	46%	6%
Type I	0	0
Type IIa	19%	18%
Type IIb	15%	29%
Type III	0	0
Type IV	19%	47%
Type V	1%	0

Table 8

SERUM VALUES IN NORMAL AND
GOUTY SUBJECTS

	Uric acid (mg/dl)	Glucose (mg/dl)
Gout patients (Mean± SEM)	7.7 ± 0.4	111 ± 11
Normal range	2.5—8.0	65—110

lower serum-lipid levels. It does not appear that carbohydrate imbalance is a factor as suggested by other investigators because fasting glucose levels appear normal in our gouty subjects (Table 8). Of the patients, 55% were considered nondrinkers (<6 oz. of alcohol per week), 17% were classified as moderate drinkers (6 to 14 oz/week), and 38% were heavy drinkers (>14 oz/week); thus, only about half of the patients used alcohol. It does not appear that alcohol intake has played a major role in the abnormal lipid and lipoprotein concentration and patterns of this selected group of gouty patients.

Since the liver profile (Table 9) and kidney profile (Table 10) of the gouty subjects were considered normal, it does not appear that the secondary hypertriglyceridemic and hyperlipoproteinemic conditions can be attributed to the dysfunction of these two organs that play a major role in lipid and lipoprotein metabolism.

However, 70% of the patients were considered >25% over ideal weight for their body structure and 65% of the patients had hypertriglyceridemia; thus, being overweight might be one of the more important factors associated with hyperlipidemia and hyperlipoproteinemia in these gouty patients.

It is probable that the association of obesity and hypertriglyceridemia observed in gouty patients is a reflection, in part, of high caloric intake and excessively rich diets. The possibility also exists that gout and its associated lipid and lipoprotein abnormalities may represent a common phenotypic expression which is influenced by dietary intake and obesity.

The data on serum lipoproteins is shown in Table 7. Only 6% of the gouty subjects had normal lipoprotein profile as compared to 46% in the apparently healthy group. Eighteen percent of the gout patients had a type IIa lipoprotein pattern, while 29 and 47% of the patients had type IIb and IV patterns, respectively. The high frequency of type IV lipoprotein patterns is in agreement with the large percentage of patients with hypertriglyceridemia.[28] The mechanism of this lipid and lipoprotein disorder is still not

Table 9
LIVER PROFILE

	Total bilirubin (mg/dl)	Alkaline phosphatase (mU/ml)	CPK (mU/ml)	LDH (mU/ml)	SGOT (mU/ml)	Protein (g/dl)	Albumin (g/dl)
Mean ± SEM (n = 17)	0.65±0.6	76±7	116±12	176±7	24±2	7.3±0.9	4.4±0.4
Normal range	0.15—8.0	30—85	25—145	100—225	8—40	6—8	3.5—5.0

Table 10
KIDNEY PROFILE

	Na⁺ (meq/l)	K⁺ (meq/l)	Cl⁻ (meq/l)	CO₂ (meq/l)	BUN (mg/dl)	Creatinine (mg/dl)
Gout patients[a]	143 ± 0.8	4.5 ± 0.1	104 ± 0.8	29 ± 0.7	20 ± 1.3	1.4 ± 0.68
Normal range	135—145	3.5—5.0	95—105	24—32	10—20	0.7—1.4

[a] Mean ± SEM (n = 17)

clear; however, it does appear that even when gouty patients are under treatment and SUA levels are within normal concentration ranges, hyperlipidemia and hyperlipoproteinemia are still evident and that it is not a result of liver or kidney dysfunction, nor to excessive alcohol intake. The effect of primary, uncontrolled gout on serum lipid and lipoprotein metabolism should also be observed.

As suggested by some investigators[17,31,65] it is possible that both gout and lipid and lipoprotein dysmetabolism may be inborn metabolic errors, the expression of which is influenced by dietary intake and obesity. To elucidate the precise mechanism of abnormal blood lipids and lipoproteins in patients having gout, more detailed and careful studies are needed. Until this is accomplished, treatment for lipid and lipoprotein disorders will be entirely empirical.

Furthermore, there are numerous factors that will elicit an elevation of serum lipid and lipoprotein concentrations, some of which are primary causes and others, secondary. The use of electrophoretic techniques help identify the probable type of dyslipoproteinemia. It is the responsibility of the person who conducts these analytical tests to help determine whether it is a familial or acquired defect. In this instance, a high incidence of type IV and II hyperlipoproteinemias was associated with primary gout.

CONCLUSION

1. About 95% of the definite and classical gout patients have hyperlipidemia. More specifically and accurately, 65% had markedly elevated serum triglyceride levels and at a greater incidence than the age- and sex-matched "control" group. Nintyfour percent had hyperlipoproteinemia, most of which were type IV.
2. Neither alcohol use nor hyperuricemia per se seems to cause the hyperlipidemia.
3. The elevated serum lipids and lipoproteins cannot be related to liver or kidney dysfunction.
4. Obesity was the major underlying factor associated with the hyperlipidemic and hyperlipoproteinemic conditions. Diet may be the etiologic factor that causes these abnormalities. It is highly likely that the association of obesity and hypertriglyc-

eridemia and hyper pre-β-lipoproteinemia in gouty patients is a reflection, in part, of the high caloric intake and excessively rich diets. Under these circumstances the observed lipid and lipoprotein abnormalities in asymptomatic gouty patients is a secondary disorder.

ACKNOWLEDGMENTS

This study was supported in part by Grant No. HL 6835 from the National Heart and Lung Institute, Grant No. 3084R from the American Heart Association, Northeast Ohio Affiliate, Inc., and Grant No. CRP 400 from the Clinical Research Projects Committee of the Cleveland Clinic Foundation. The authors wish to thank Ms. Joan Ann David, Ms. Jeanette H. Repansky, and Mr. Joseph Paksi for their technical assistance and Ms. Penny K. Porter for typing this manuscript.

REFERENCES

1. **Nishida, Y., Akaka, I., Nishizawa, T., and Yoshimura, T.,** Hyperlipidemia in gout, *Clin. Chim. Acta,* 62, 103, 1975.
2. **Gibson, T. J. and Grahame, R.,** Gout and hyperlipidemia, *Ann. Rheum. Dis.,* 33, 298, 1974.
3. **Yano, K., Rhoads, G. G., and Kagan, A.,** Epidemiology of serum uric acid among 8000 men of Japanese ancestry in Hawaii, *Circulation,* 52, 2, 259, 1975, abstr.
4. **Benedek, T. G.,** Correlations of serum uric acid and lipid concentrations in normal, gouty, and atherosclerotic men, *Ann. Intern. Med.,* 66, 851, 1967.
5. **Hollister, L. E., Overall, J. E., and Snow, H. L.,** Relationship of obesity to serum triglyceride, cholesterol, and uric acid and to plasma-glucose levels, *Am. J. Clin. Nutr.,* 20, 777, 1967.
6. **Berkowitz, D.,** Gout, hyperlipidemia and diabetes interrelationship, *J. Am. Med. Assoc.,* 197, 117, 1966.
7. **Wyngaarden, J.B. and Kelley, W. N.,** Gout, in *The Metabolic Basis of Inherited Disease,* Stanbury, J. B., Wyngaarden, J. B., and Fredrickson, D. S., Eds., McGraw-Hill, New York, 1972, 900.
8. **Wiedemann, E., Rose, H. G., and Schwartz, E.,** Plasma lipoproteins, glucose tolerance and insulin response in primary gout, *Am. J. Med.,* 53, 299, 1972.
9. **Feldman, E. B. and Wallace, S. L.,** Hypertriglyceridemia in gout, *Circulation,* 29, 508, 1964.
10. **Harris-Jones, J. N.,** Hyperuricemia and essential hypercholesterolemia, *Lancet,* 1, 957, 1957.
11. **Jensen, J., Blankenhorn, H. P., Chin, H. P., Sturgeon, P., and Ware, A. G.,** Serum lipids and serum uric acid in human twins, *J. Lipid Res.,* 6, 196, 1965.
12. **Benedek, T. G. and Sunder, J. H.,** Comparison of serum lipid and uric acid content in white and Negro men, *Am. J. Med. Sci.,* 260, 331, 1970.
13. **Motulsky, A. G.,** The genetic hyperlipidemias, *N. Engl. J. Med.,* 294, 823, 1976.
14. **Gunther, R. and Knapp, E.,** Der Einslub von Allopurinol (Zyloric) auf Harnsaure Kreatinin Nuchter-blutzucher und Plasma Lipide be: Gichtkranken, *Wien. Klin. Wochenschr,* 82, 78, 1970.
15. **Gertler, M. M., Garn, S.M., and Levine, S. A.,** Serum uric acid in relation to age and physique in health and in coronary heart disease, *Ann. Intern. Med.,* 34, 1421, 1951.
16. **Kramer, D. W., Perilstein, P. K., and DeMedeiros, A.,** Metabolic influences on vascular disorders with particular reference to cholesterol determination in comparison with uric acid levels, *Angiology,* 9, 162, 1958.
17. **Kohn, P. M. and Prozan, G. B.,** Hyperuricemia — relationship to hypercholesterolemia and acute myocardial infarction, *J. Am. Med. Assoc.,* 170, 1909, 1959.
18. **Becker, J. H.,** Gout and serum cholesterol, *Wis. Med. J.,* 59, 735, 1960.
19. **Berkowitz, D.,** Blood lipid and uric acid interrelationship, *J. Am. Med. Assoc.,* 190, 856, 1964.
20. **Blachos, J. and Reisenauser, R.,** Levels of serum uric acid and serum cholesterol in various population groups in Ethiopia, *Am. J. Med. Sci.,* 250, 308, 1965.

21. Hall, A. P., Correlations among hyperuricemia, hypercholesterolemia, coronary disease and hypertension, *Arthritis Rheum.*, 8, 846, 1965.
22. Barlow, K. A., Lipid metabolism in gout, *Proc. R. Soc. Med.*, 59, 325, 1966.
23. Gunther, R., Herbst, M., and Knapp, E., Gicht and Hyperlipamie, *Wien. Klin. Wochenschr.*, 79, 218, 1967.
24. Knick, B., Lang, H. J., Ritter, U., and Schilling, F., Adipositas: Hypertriglyzeridamie, Fettleber und latenter Diabetes bei Gicht-patienten, *Therapiewoche*, 18, 2071, 1968.
25. Barlow, K. A., Hyperlipidemia in primary gout, *Metabolism*, 17, 289, 1968.
26. Bernheim, C., Ott, H., Zahnd, G., and Martin, E., Goutte et diabete, *Schweiz. Med. Wochenschr.*, 98, 33, 1968.
27. Gunther, R. and Knapp, E., Zuk Klinik und Therapie der Gicht unter besonderer Berucksichtigung der Staffwechselwirkungen von Sulfinpyrazon (Anturan) und Allopurinol (Zyloric), *Wien. Klin. Wochenschr.*, 81, 817, 1966.
28. Bluestone, R., Lewis, B., and Mervart, I., Hyperlipoproteinaemia in Gout, *Ann. Rheum. Dis.*, 30, 134, 1971.
29. Emmerson, B. T. and Knowles, B. R., Triglyceride concentrations in primary gout and gout of chronic lead nephropathy, *Metabolism*, 20, 721, 1971.
30. Darlington, L. G. and Scott, J. T., Plasma lipid levels in gout, *Ann. Rheum. Dis.*, 31, 487, 1972.
31. Mielants, H., Veys, E. M., and Deweergt, A., Gout and its relation to lipid metabolism. I. Serum uric acid, lipid, and lipoprotein levels in gout, *Ann. Rheum. Dis.*, 32, 50, 1973.
32. Mielants, H., Veys, E. M., and DeWeerdt, A., Gout and its relation to lipid metabolism. II. Correlations between uric acid, lipid, and lipoprotein levels in gout, *Ann. Rheum. Dis.*, 32, 506, 1973.
33. Klein, R., Klein, B., Cornoni, J. C., Maready, J., Cassel, J. C., and Tyroler, H. A., Serum uric acid, *Arch. Intern. Med.*, 132, 401, 1973.
34. Halpern, M. J. and Pereira Miguel, M. S., Uric acid and coronary heart disease, *J. Am. Geriatr. Soc.*, 22, 86, 1974.
35. Dunn, J. P. and Moses, C., Correlation of serum lipids with uric acid and blood sugar in normal males, *Metabolism*, 14, 788, 1965.
36. Lieber, C. S., Leevy, C. M., Stein, S. W., George, W. S., Cherrick, C. R., Abelmann, W. M., and Davidson, C. S., Effect of ethanol on plasma free fatty acids in man, *J. Lab. Clin. Med.*, 59, 826, 1962.
37. Jones, P. D., and Losowsky, M. S., Davidson, C. S., and Lieber, C. S., Effects of ethanol on plasma lipids in man, *J. Lab. Clin. Med.*, 62, 675, 1963.
38. Losowsky, W. S., Jones, D. P., Davidson, C. S., and Lieber, C. S., Studies of alcoholic hyperlipemia and its mechanism, *Am. J. Med.*, 35, 794, 1963.
39. Schapiro, R. H., Scheig, R. L., Drummey, G. D., Manderson, J. H., and Isselbacher, K. J., Effect of prolonged ethanol ingestion on the transport and metabolism of lipids in man, *N. Engl. J. Med.*, 25, 610, 1965.
40. Nestel, P. J. and Hirsch, E. Z., Mechanism of alcohol-induced hypertriglyceridemia, *J. Lab. Clin. Med.*, 3, 357, 1965.
41. Avogaro, P. and Cazzolato, G., Changes in the composition and physico-chemical characteristics of serum lipoproteins during ethanol-induced lipaemia in alcoholic subjects, *Metabolism*, 24, 1231, 1975.
42. Wilson, D. E., Schreibman, P. H., Bresten, A. C., and Arky, R. A., The enhancement of alimentary lipemia by ethanol in man, *J. Lab. Clin. Med.*, 75, 264, 1970.
43. Seidel, D., Improved techniques for assessment of plasma lipoprotein patterns. I. Precipitation in gels after electrophoresis with polyanionic compounds, *Clin. Chem.*, 19, 737, 1973.
44. Mead, M. G. and Dangerfield, W. G., The investigation of "mid-band" lipoproteins using polyacrylamide gel electrophoresis, *Clin. Chim. Acta*, 51, 173, 1974.
45. Frank, V. O., Untersuchungen über die Häujigkeit von Störugen des Lipid und Kohlenhydratstoffwechsels bei primarer Gicht und symptomloser Hyperurikamie, *Wien. Klin. Wochenschr.*, 86, 252, 1974.
46. Lang, P. D., Fettstoffwechselstörungen bei Gicht, *Muench. Med. Wochenschr.*, 116, 909, 1974.
47. Frank, V. O., Observations concerning the incidence of disturbance of lipid and carbohydrate metabolism in gout, *Isr. J. Med. Sci.*, 9, 16, 1973.
48. Houge, M. and Harvald, B., Heredity in gout and hyperuricemia, *Acta Med. Scand.*, 152, 247, 1955.
49. Emmerson, B. T., Heredity in primary gout, *Australas. Ann. Med.*, 9, 168, 1960.
50. Neel, J. V., Rakic, M. T., Davidson, R. T., Volkenberg, H. A., and Mikkelsen, W. M., Studies on hyperuricemia. II. A reconsideration of the distribution of serum uric acid values in the families of Smyth, Cotterman and Freyberg, *Am. J. Hum. Genet.*, 17, 14, 1965.
51. Howell, R. R., The interrelationships of glycogen storage disease in gout, *Arthritis Rheum.*, 8, 780, 1965.

52. **Mikkelsen, W. M., Dodge, H. J., and Valkenburg, H.,** The distribution of serum uric acid values in a population unselected as to gout or hyperuricemia, *Am. J. Med.,* 39, 242, 1965.
53. **Archeson, R. M. and O'Brian, W. M.,** Dependence of serum uric acid on hemoglobin and other factors in the general population, *Lancet,* 2, 777, 1966.
54. **Stetten, D. W., Jr.,** Gout, *Perspect. Biol. Med.,* 2, 185, 1959.
55. **Kelley, W. N., Rosenbloom, F. M., Miller, J., and Seegmiller, J. E.,** An enzymatic basis for variation in response to allopurinol, *N. Engl. J. Med.,* 278, 286, 1968.
56. **Kelley, W. N., Greene, M. L., Rosenbloom, R. M., Henderson, J. F., and Seegmiller, J. E.,** Review: hypoxanthine-guanine phosphoribosy transferase deficiency in gout, *Ann. Intern. Med.,* 70, 155, 1969.
57. **Lees, R. S. and Hatch, F. T.,** Sharper separation of lipoprotein species key paper electrophoresis in albumin-containing buffer, *J. Lab. Clin. Med.,* 61, 518, 1963.
58. Technicon Autoanalyzer II. Clinical Method No. 24: Simultaneous Cholesterol Triglycerides, Technicon Instruments Corp., Tarrytown, N.Y., 1972.
59. **Naito, H. K.,** Modification of the Fiske and Subbarow method for total phospholipid in serum, *Clin. Chem.,* 21, 1454, 1975.
60. **Naito, H. K. and Lewis, L. A.,** Rapid lipid-staining procedures for paper electrophoretograms, *Clin. Chem.,* 19, 106, 1973.
61. **Naito, H. K., Wada, M., Ehrhart, L. A., and Lewis, L. A.,** Polyacrylamide gel disc-electrophoresis as a screening procedure for serum lipoprotein abnormalities, *Clin. Chem.,* 19, 228, 1973.
62. Pfizer Electrophoresis Operating Manual. V. Serum Lipoprotein Phenotyping, Pfizer Diagnostic Division, New York, N.Y., 1972.
63. Technicon AutoAnalyzer 12/60 and 6/60, Technicon Instruments Corp., Tarrytown, N.Y., 1974.
64. **Pereira Miguel, M. J., Halpern, M., Pereira Miguel, J. M., Amador, M. G., da Silva, J. F., Lima, M., and de Padua, F.,** Hyperlipémies dans la population rurale portugaise, *Ann. Biol. Clin.,* 31, 147, 1973.
65. **Fulton, J. K.,** Essential lipemia, acute gout, peripheral neuritis, and myocardial disease in a Negro man, *Arch. Intern. Med.,* 89, 303, 1952.
66. **Chin, H. P. and Blankenhorn, D. H.,** Separation and quantitative analysis of serum lipoproteins by means of electrophoresis on cellulose acetate, *Clin. Chim. Acta,* 20, 305, 1968.
67. **Sobrinho-Simoes, M.,** A sensitive method for the measurement of uric acid using hydroxamine, *J. Lab. Clin. Med.,* 65, 665, 1965.
68. **Lewis, L. A. and Page, I. H.,** Electrophoretic and ultracentrifugal analysis of serum lipoproteins of normal, nephrotic and hypertensive persons, *Circulation,* 7, 707, 1953.
69. **Wada, M., Akamatsu, A., Naito, H. K., Minamisono, T., Ehrhart, L. A., Nakamoto, S., and Lewis, L. A.,** Unusual lipoprotein (Lp) profiles in patients with various types of renal diseases, *Fed. Proc. Fed. Am. Soc. Exp. Biol.,* 36(3), 574, 1977.
70. **Naito, H. K. and Mackenzie, A. H.,** Secondary hypertriglyceridemia and hyperlipoproteinemia in primary asymtomatic gout patients, *Clin. Chem.,* 25, 371, 1979.

SERUM LIPIDS AND LIPOPROTEINS AND THEIR RELATIONSHIP WITH THYROID FUNCTION

H.K. Naito and M.S. Kumar

It has long been known both from clinical observations and animal experiments that the metabolism of lipids is influenced by hormones, particularly thyroxine. Among the lipids, serum cholesterol levels show the greatest regularity in response to the circulating level of thyroid hormone, but other lipid components may be affected.[1-20,148]

THYROXINE AND SERUM CHOLESTEROL

In 1895, Magnus-Levy made the fundamental observation on the influence of the thyroid on lipid metabolism.[21] He also noted thyroid insufficiency in all cases of myxedema and in many cases of endogenous adiposity. Subsequent canine experiments by Martinez[22] in 1917 demonstrated a decrease in blood cholesterol within 24 hrs of a thyroidectomy, followed by a progressive increase thereafter. In 1926, Baumann and Holly[25] reported that thyroidectomy of rabbits resulted in a rise in serum cholesterol, an effect that was lost when the animals became pregnant. Goldbloom and Gottlieb in 1927[31] and Bronstein in 1933[32] found an increase in the blood cholesterol of certain subjects and cretin rabbits. The feeding of dessicated thyroid caused a reduction in the serum cholesterol of the cretin rabbits but not in that of normal rabbits.

In 1922, Epstein and Lande[23] called attention to an inverse relationship between blood cholesterol values and metabolic rates in man. Gardner and Gainesborough[24] obtained similar results but refrained from attaching great significance to them because of the many exceptions to their findings. It was believed that the definite inverse relationship that exists between blood cholesterol levels and BMR (basal metabolic rate) is true only in extreme cases.[21] Wade[26] found slightly increased values of blood cholesterol in patients with toxic goiter and even higher values after total thyroidectomy.

Mason et al.[21] and Hurxthal[27-30] consistently found hypercholesterolemia in myxedemic patients and a less regular, but significant, decrease in blood cholesterol in thyrotoxicosis. Their exhaustive study of over 500 patients led to the following conclusions:

1. Postoperative myxedema is accompanied by hypercholesterolemia.
2. Subtotal thyroidectomy may be followed by hypercholesterolemia without clinical myxedema, which is interpreted as transient thyroid deficiency.
3. Subtotal thyroidectomy may be followed by low metabolic rates without hypercholesterolemia; myxedema was seldom found in these cases.
4. Thyroxine deficiency produces myxedema and hypercholesterolemia, but at times, myxedema may be clinically imperceptible.
5. Hypercholesterolemia, when not explainable on any other basis, may be considered as possibly of thyroid origin and a rational indication for thyroid administration.
6. Finding hypercholesterolemia in the absence of its few other common causes points more specifically to thyroid deficiency than does the finding of a low metabolic rate.
7. The relationship between blood cholesterol and basal metabolism is usually reciprocal, when they undergo change as a result of variations in the activity of the thyroid gland or thyroid compounds in the body.

8.   Blood cholesterol provides another variable which may be used as a guide in the treatment of thyroid disease.

Observing hypercholesterolemia in patients with low metabolism and no clinical evidence of hypothyroidism, Gilligan and collaborators[34] reported that many patients failed to maintain a consistent relationship between basal metabolism and blood cholesterol after thyroidectomy. That same year, experimental investigations of Cutting and co-workers[35] and Grant and Schube[36] supported that observation. They found no correlation between blood cholesterol levels and BMR elevations induced by dinitrophenol, thus suggesting that cholesterol level is more precisely related to thyroid activity than to BMR. Abundant data have since supported that principle.[3,37-90,100,105,112,139,140,142-149,153,154,156] Similarly, evidence for the relationship between hypocholesterolemia and hyperthyroidism is well documented.[3,37,40,41,44,48,49,51,54,58,59,61,62,64,66,67,75,77,80,82-100,148]

It should be stated that the inverse relationship of thyroid activity to serum cholesterol concentration is not invariable; thus, hypocholesterolemia is rarely of diagnostic value in thyrotoxicosis.[2,24,111,128,138] McGee[164] in 1935 and Peters and Man[2,3] as late as 1950 did not find any relationship between serum cholesterol, BMR, and PBI (Protein-bound iodine). Lamberg et al.[146] found a correlation between cholesterol, PBI, T_4 (L-thyroxine) and T_3 (L-triiodothyronine). In 1973, Jacobson[12] studied 30 untreated hyperthyroid patients and found that serum total cholesterol and phospholipid values correlated inversely with the BMR status of the individual. Serum lipid fractions showed a correlation with T_3 values, but not with serum T_4 or PBI. When these patients were treated with 20 mg thiamazole daily, the initially elevated free fatty acid and glycerol levels decreased, while the initially low cholesterol levels rose. The observed, low triglyceride level did not change with treatment. It is interesting that the T_3 resin uptake test used by Jacobsen was a better correlate of serum lipid values than the other thyroid function tests. This test, an indirect expression of thyroid function, is based upon the content of thyroxine-binding proteins in plasma.[150] Along with other indirect measurements of peripheral response to thyroid function, such as BMR, electrocardiographic changes, and ankle tendon reflex duration, the T_3 resin uptake test can verify mainly marked forms of thyroid dysfunction. These indirect methods have been gradually replaced by more refined tests, most notably the measurement of total or free T_4, T_3, and TSH (thyroid-stimulating hormone) by radioimmunossay.

The correlation of the relationship of thyroid status to serum cholesterol concentration became more meaningful when studies were directed to its relationship at the biochemical level. It was not until 1949 that the study of Karp and Stetten[106] suggested that thyroxine had a direct effect on cholesterol metabolism. The turnover of plasma cholesterol in vivo was studied in intact and thyroidectomized rats using acetate as a precursor. Assuming that the endogenous cholesterol of plasma, both in its free and esterified forms, is synthesized largely by the liver,[107-110] a change in the rate of replacement of plasma cholesterol should signify a change in the rate of hepatic cholesterol biosynthesis. They studied the turnover of plasma cholesterol in normal and thyroidectomized rats and determined that it was significantly decreased in the thiouracil-treated group. Using deuterium as a marker, they also showed increased incorporation of the isotope into the sterols of hyperthyroid rats. Byers and co-workers [112] used tritiated water and found that in hyperthyroid rats rates of cholesterol synthesis and turnover were greater than in normal rats, while in hypothyroid rats both synthesis and turnover rate were lower. These findings on the direct correlation between the amount of circulating plasma thyroxine or its analogues and the cholesterogenesis and turnover rates have been confirmed using labeled acetate,[79,113,114] deuterium,[84] and cholesterol-^{14}C.

In experiments using liver slices, Fletcher and Myant[115] determined that thyroidectomized rats had an 80% lower rate of cholesterol biosynthesis from acetate, while thyroxine injection caused a two- to three-fold increase in synthesis rate. There was no effect on the rate of biosynthesis from mevalonic acid; thus, the affected biosynthetic step would appear to lie between acetate and mevalonate. Similar results were obtained the following year by Boyd.[71]

It is now well known that in the biochemical synthesis of cholesterol, there are regulatory enzymes that determine the rate of cholesterol formation. 3-Hydroxy-3-methylglutaryl- CoA reductase (HMG-CoA), one of the major enzymes known to be subjected to the end-product negative-feedback mechanism, is involved in the major enzymatic step which determines whether or not a steroid is formed.

A decrease in the activity of HMG-CoA reductase has been found in thyroidectomized animals; while administration of thyroxine increases the activities of this enzyme, resulting in increased cholesterogenesis.[116,117] The change in the rate of cholesterol synthesis which occurs under the influence of thyroid-active hormones are opposite in direction from the observed changes in serum cholesterol concentrations. In order for blood cholesterol levels to vary as they do, the rate of cholesterol catabolism and excretion must be higher than the synthetic rate in the hyperthyroid state and be lower than the synthetic rate in the hypothyroid state. This is not to say that other physiological processes, i.e., intestinal absorption and shifting of cholesterol from one pool to another may not be occurring simultaneously.[118] One of the authors and co-workers have found that when 250- to 300-g female rats are made hyperthyroid by injecting 10 μg L-T$_4$ per 100 g body weight for 6 weeks, plasma volume increases about 15% .* Thyroidectomized rats do not have plasma volumes different from those of control rats with intact thyroids. Actual plasma volume (ml/100 g body weight) for adult, male hyper-, eu-, and athyroid rats were 3.78 ± 0.18, 3.17 ± 0.12 and 3.19 ± 0.04 ml/ 100 g body weight (mean $\pm$ S.E.M.), respectively. Therefore, the change in plasma volume in hyperthyroid states could be an additional means of altering the concentration of cholesterol in the blood.

Extensive studies of Rosenman and co-workers[61,62] revealed a decreased output of biliary cholesterol in the hypothyroid rat and an increase in the hyperthyroid rat. Moreover, rates of catabolism as demonstrated by the half life ($t_{1/2}$) of the labeled cholesterol were 17.3 days for controls, 43.0 days for hypothyroid, and 7.3 days for hyperthyroid rats. These results were confirmed by Thompson and Vars,[119-121] who in addition, found a decrease in excretion of cholic acid both in hyper- and hypothyroid states; however, chenodeoxycholic acid was not mentioned and is quantitatively the second most important bile acid in the rat.[122] Eriksson,[123,124] in a detailed study of biliary excretion of both cholic and chenodeoxycholic acid in eu-, hyper-, and hypothyroid, bile-fistulated rats, found decreased excretion of cholate in both hyper- and hypothyroidism. Also, there was a substantial increase in the amount of chenodeoxycholic acid in hyperthyroidism so that total excretion of bile acids was at least the same as in the normal. In hypothyroidism, the amount of chenodeoxycholic acid excreted was less than in the euthyroid state. These results suggest the possibility of a direct, inhibitory effect of thyroxine on 12α-hydroxylase or a stimulatory one on side-chain oxidation, since there was almost a reversal of the normal 4:1 ratio of cholate to chenodeoxycholate to 1:4 in the hyperthyroid state. Similar results were obtained by Van Zye,[125] Strand,[126] Lin et al.,[127] Lepp and co-workers,[78] and Miettinen,[80] but not by Gans.[129] Strand[130] demonstrated effects of thyroid activity on the excretion of bile acids similar to those observed by Eriksson,[123,124] although the proportionate change of the

* Naito, H.K. and Ferreri, L.F., unpublished data.

two major bile acids differed. The normal 3:1 taurochenodeoxycholate to taurocholate ratio changed to 1:3 after D-T_3 administration.

In earlier studies by Rosenman et al.,[62] the hyperthyroid rat cleared intravenously injected cholesterol at a faster rate than did normal or hypothyroid rats. The hyper-cholesterolemic rat plasma injected into hypothyroid, normal, and hyperthyroid rats resulted in 59.7, 71.2, and 95.6% excretion, respectively. Weiss and Marx [63] injected cholesterol-4-^{14}C into hyper- and hypothyroid rats and found that conversion of cho-lesterol to acidic products as well as excretion of both acidic products and neutral sterol was stimulated by thyroid hormones. Calculations of bile acid synthesis and pools in the intact rat indicated that when D- and L-T_3 are administered, total biliary acid synthesis increases as a result of increased chenodeoxycholate synthesis[126] without alteration in rate of cholate synthesis. These results can be interpreted to indicate that thyroid hormone can stimulate degradation or modification of cholesterol, a prereq-uisite step for the formation of bile acids. The mechanism by which this occurs is not clearly understood.

It has been shown that liver mitochondria can oxidize the end-carbon atoms of the cholesterol side chain to CO_2[131] with the production of substances closely related to naturally occurring bile acids.[132] In vitro studies on liver mitochondria from thyroxine-treated rats have demonstrated that four times as much CO_2 is produced in treated rats as in untreated rats.[133] This suggests the possibility of a mechanism in which thy-roxine influences cholesterol degradation by stimulating a rate-limiting reaction at one of the later biochemical steps involved in the modification of the cholesterol side chain. This step could be the cleavage reaction resulting in the formation of a C-24 bile acid and propionyl-CoA.[134] It has also been suggested that thyroxine could stimulate the enzyme catalyzing the formation of di- and trihydroxycoprostanic acids from their respective precursors, di- and trihydroxycoprostane. This could explain why thyroid-active compounds stimulate production of chenodeoxycholate to a greater extent than cholate, since 1,7-α-dihydroxycoprostane is the precursor of chenodeoxycholate and the immediate precursor of 3,7,12-α-trihydroxycoprostane, an intermediate product in the formation of cholate. However, Kritchevsky and associates[135,136] have not been able to demonstrate in vitro the dependence of cholesterol side-chain oxidation by thyroxine.

Story et al.[118] reported that radioactive, acidic fecal steroid excretion changed very little in the propylthiouracil-fed, D-T_4- and L-T_4-treated, and control rats (all fasted and killed in the morning). However, excretion of radioactive, neutral fecal steroids was much higher in D-T_4-and L-T_4-treated rats and much lower in propylthiouracil-treated animals. They suggested that thyroid hormones mainly affected neutral steroid excretion. On the other hand, Balasurbramaniam and Mitropoulos[137] showed that 7α-hydroxylase activity (the rate-limiting enzyme for the conversion of cholesterol to bile acids) was lower in thyroidectomized rats and higher in L-T_4-treated rats as compared to control animals (all nonfasted and killed at night). The differences in the results on the effects of thyroxine on 7α-hydroxylase activity[137] and the little change in acidic fecal steroid excretion between the hypothyroid and control rats are not clear.

One of the authors and co-worker, in a study of 7α-hydroxylase activity of adult female Wistar rats treated with 17β-estradiol-3-benzoate demonstrated that the hor-mone can stimulate hepatic enzyme activity but only under certain conditions of feed-ing, fasting, and light/dark photocycles.[138] Estradiol treatment (25 or 100 μg/day for 20 days) induced a threefold increase in 7α-hydroxylase activity in fasted rats and a twofold increase in nonfasted rats (all killed in the morning) as compared to the re-spective nonestrogen-treated control rats killed at the same time and under the same conditions of fasting. Estrogen-treated rats fasted and killed at midnight had 7α-hydroxylase activity no different from control rats killed at midnight. Estrogen-treated, nonfasted rats killed at midnight had only a 50% higher hepatic enzyme activity as compared to the respective, nonfasted control rats killed at midnight. The point is that the 7α-hydroxylase activity will be different, depending on the time at which the rat is killed and whether it is fasted or nonfasted.

Despite changes in enzyme activity observed under these various conditions, results did not correlate with observed increases in serum cholesterol levels of all estrogen-treated rats. Serum cholesterol concentrations not only failed to decline but actually increased about 25 to 50%. This study emphasizes the complexity of blood cholesterol homeostasis and suggests that in addition to its effect on 7α-hydroxylase, estradiol simultaneously alters one or more physiological processes related to steroid metabolism. For example, it is possible that estradiol stimulated cholesterol synthesis to a greater extent than catabolism, yielding an overall increase in serum cholesterol concentrations. In hyperthyroidism, it appears that the observed lower cholesterol concentration is due to both an increase in cholesterogenesis and an increase in cholesterol degradation rate which leads to a shorter half life of cholesterol in blood. It is certain that the hypocholesterolemic effect of hyperthyroidism is due to the rate of cholesterol catabolism exceeding the rate of cholesterol synthesis. The net effect is a lowering of cholesterol in the blood. What is still not clear is the precise catabolic pathway by which cholesterol breakdown occurs. Some investigators have indicated that thyroxine may increase the formation and secretion of bile acids via cholesterol catabolism, while others have demonstrated that thyroxine mainly affects cholesterol excretion via the bile as neutral steroids. It is possible that both catabolic pathways are important, but that the increased neutral steroid excretion is the primary effect that occurs during hyperthyroidism.

THYROXINE AND SERUM TRIGLYCERIDES

The mechanism by which thyroid hormones mediate the regulation of plasma triglyceride concentration is less well understood than that of cholesterol metabolism.[111,128] Increased fasting plasma triglyceride levels commonly accompany myxedema.[1-20,140,141] The increase may be due to a low lipoprotein lipase activity since it has been shown that plasma PHLA (post-heparin lipolytic activity) is inversely correlated with plasma triglyceride level in these patients.[163]

Plasma PHLA is also depressed in hypothroidism, but the elevated triglyceride levels associated with the hypothyroid state are not related to the fat-induced type. Nikkila and Kekki[10] showed that both hypo- and hyper-thyroidism are characterized by specific patterns of disturbed triglyceride transport kinetics. Hypothyroid patients showed normal synthesis of plasma triglycerides, but the fractional removal of both endogenous and exogenous triglycerides was markedly reduced. This change seems to account for the elevated triglyceride levels associated with thyroid hypofunction.

In thyrotoxicosis, however, the mean PHLA and mean plasma triglyceride levels were both elevated. The mean fractional removal rate was not different from normal, so the slight elevation of plasma triglyceride was associated with augmented production of triglycerides.

In hyperthyroidism, triglyceride concentration might be decreased or in the low "normal" range. There was a significant linear correlation between triglyceride concentration and turnover rate in both eu- and hyperthyroid patients, but the concentration turnover rate was less in the hyperthyroid group, suggesting that efficiency of triglyceride removal from circulation was improved in thyroid hyperfunction. Further evidence to support this is that untreated hyperthyroid patients eliminated intravenously administered particulate fat (Intralipid®) more rapidly than euthyroid control subjects. Upon adequate treatment of the hyperthyroid state, the plasma triglyceride concentration increased, a finding which would be consistent with a reduced triglyceride removal rate.

The turnover rate of endogenous triglycerides was significantly correlated with serum PBI and T_3 uptake in thyrotoxicosis but not in hypothyroidism. A mixed group of hypo- and hyperthyroid patients showed a relationship between PHLA and fractional endogenous triglyceride transport, but the lipase activity was not related to exogenous fat removal. This study suggests that thyroid hormones control both production and removal of plasma triglycerides in man. In thyrotoxicosis, the mechanism for hypertriglyceridemia is probably increased synthesis of endogenous triglycerides since the mean fractional removal rate remains unchanged.[10] Tulloch, however, reported high or normal clearance rates in a series of seven thyrotoxic patients, resulting in low or low normal serum triglycerides.[11] In thyroid hypofunction, the rate of triglyceride synthesis remains unchanged, but moderate hypertriglyceridemia develops as a result of impaired removal of endogenous and exogenous triglycerides.[10] In hyperthyroidism, triglycerides are decreased or in the low normal range, findings consistent with the increased triglyceride removal rate and elevated plasma PHLA. The mechanisms, responsible for these kinetic changes, are still not clearly defined. There is evidence to support the belief that T_4 affects both triglyceride synthesis and clearance rates. Further research is necessary to clearly define which of the two metabolic processes predominates in hypo- and hyperthyroidism.

THYROXINE AND SERUM LIPOPROTEINS

While there have been numerous reports on the effects of thyroid dysfunction on serum lipid levels, very little has been done on serum lipoprotein metabolism and lipoprotein electrophoresis (Table 1). Walton et al.[139] and Ellefson and Mason[75] showed increased synthesis and turnover of LDL, sterols, and phospholipids in hyperthyroid patients.

According to the study of Strisower et al.,[144] daily administration of 3 grains of desiccated thyroid to 50 schizophrenic patients caused a lowering of the S_f 0—12* and 12—20 lipoprotein levels without significant alterations of the S_f 70—100 and 100—400 lipoprotein levels. The effect upon concentration of S_f 0—12 lipoproteins was maximal for 3 weeks and subsequently decreased until 24 weeks, at which time the concentration returned to the pretreatment level in spite of the continuation of thyroid treatment. The magnitude of the drop in concentration of the S_f 0—12 fraction was directly related to the prethyroid treatment value. This data is in agreement with their earlier study in which the S_f 0—12 and 12—20 fractions decreased in 19 schizophrenic patients given 10 gr of desiccated thyroid daily for 9 weeks.[145] With 3 grains of thyroid, the S_f 0—12 fraction decreased, but not the S_f 20—400. Similar findings were reported by Jones et al.[148]

No reduction in α-lipoprotein concentration was found in hypothyroid patients. Malmros and Swahn[149] found an increase in β-lipoprotein concentration in hypothyroidism but a diminution in β-lipoprotein cholesterol and an increase in α-lipoprotein cholesterol concentrations in thyroid treatment patients. Balasurbramaniam and Mitropoulos[137] found that myxoedematous patients show a decreased turnover of [131]I-

---

* S_f 0—12 = (d 1.019 to 1.063) = LDL_2 (low-density lipoprotein); S_f 12—20 = (d 1.006 to 1.019) = LDL_1 or IDL (intermediate-density lipoprotein); S_f 70—400 = (d < 1.006) = VLDL (very low-density lipoprotein).

Table 1
REVIEW OF PAST WORK ON THE EFFECTS OF THYROID DYSFUNCTION OF SERUM LIPOPROTEIN PATTERNS[a]

Authors and	Year	Serum lipid conc		Serum lipoprotein pattern	Method of electrophoresis	Thyroid Status
		TC	TG			
Kunkel, H.G. and Slater, R.G.[152]	1952	452	—	Decreased α-lipoprotein; little β-lipoprotein in usual position but high amounts of a fast-migrating β-lipoprotein	Paper and starch gel	Hypothyroidism
Malmros, H. and Swahn, B.[149]	1953	330—769	—	Hyper-β-lipoproteinemia	Paper	Myxedema
Feldman, E.B. and Carter, A.C.[151]	1963	384[b] 269[c]	451[b] 221[c]	β-lipoprotein/α-lipoprotein ratio decreased with D-T_4 treatment in all euthyroid patients, and only one out of four hypothyroid patients	Paper	
Hazzard, W. R. and Bierman, E. L.[16]	1972	533—584[b] 129—383[c]	478—665 115—343[c]	Type III Type III or IV	Agarose, paper, and polyacrylamide gel	Hypothyroid
Jacobsen, B. B.[12]	1973	166±31[b] 181±40[c]	102±36[b] 108±36[c]	No definite abnormalities in untreated patients; treated patients also showed no change in pattern	Paper	Hyperthyroid; diagnosis based on PBI, T_3 and T_4 resin uptake, I^{131} uptake, and triiodothyronine suppression tests
Lasser, N L. et al.[17]	1974	805 283	2040 760	Type III Type IV	Agarose Agarose	Hypothyroidism T_3 treatment

Table 1 (continued)
REVIEW OF PAST WORK ON THE EFFECTS OF THYROID DYSFUNCTION OF SERUM LIPOPROTEIN PATTERNS[a]

Authors and	Year	Serum lipid conc		Serum lipoprotein pattern	Method of electrophoresis	Thyroid Status
		TC	TG			
Vessby, B. and Wide, L.[15]	1975	—	—	Type III	Agarose	TSH was elevated in Type III patients as compared to normal or Type IIa individuals
Mishkel, M.A. and Crowther, S. M.[179]	1977	386.5 ±119.9	327.8 ±246.6	50% Type IIb, 23% Type IIa, 13% Type III, and 14% Type IV	Agarose	Primary hypo-thyroidism

[a] These abbreviations will be used in the table: TC = total cholesterol; TG = triglycerides; PBI = protein-bound iodine; TSH = thyroid-stimulating hormone.

[b] Serum lipid concentration before treatment.

[c] Serum lipid concentration after treatment.

labeled LDL, the principle carrier of plasma cholesterol. Treatment with T_4 reverted the turnover of labeled LDL normal.

Furman et al.[6] found that the hypercholesterolemia associated with hypothyroidism is of two types: 1. one in which serum triglyceride levels are elevated and 2. that in which triglyceride levels are normal. In the nonhypertriglyceridemic type, a "hyper-α-lipoproteinemia" is noted similar to that seen in primary hypercholesterolemia. They also noted that when normolipidemic individuals are given T_3, a decrease in β-lipoprotein levels occurs,[19] and when T_3 is given to hyperlipidemic subjects, the LDL (d 1.006 to 1.019) concentration decreases. This effect appeared to be independent of the presence of severe hypothyroidism.

The reasons for the inconsistency in the levels of fasting blood triglyceride observed in patients with thyroid dysfunction is not clear. Rössner and Rosenqvist[154] studied nine hypothyroid female patients with both elevated serum triglycerides and cholesterol, which they believed was primarily transported in serum LDL. The triglycerides in the VLDL fraction did not increase, while its cholesterol content was higher than normal. After giving L-thyroxine, serum triglyceride and cholesterol concentrations decreased. The triglyceride content in the LDL was also lower, and the cholesterol content in both LDL and VLDL was decreased. Using the intravenous fat tolerance test, the fractional removal rate of the exogenous triglyceride was estimated to be lower in hypothyroid patients.

Feldman and Carter[151] indicated that the response to D-T_4 therapy in hyperlipemic patients varied according to whether the patients were eu- or hypothyroid. While β-lipoprotein cholesterol and triglycerides decreased in all patients, α-lipoprotein cholesterol and triglycerides increased in euthyroid patients and decreased in hypothyroid patients. The β to α lipoprotein ratio decreased with treatment in all euthyroid patients, but only one of four hypothyroid patients exhibited this change.

Recently, Mishkel and Crowther[179] reported that primary hypothyroidism was the cause of hyperlipidemia in 22 patients. Of the 22 patients, 50% had Type IIb patterns; 23% had Type IIa patterns; 13% had Type III patterns; and 14% had Type IV patterns. With L-T_4 treatment (0.05 to 0.2 mg/day), normal serum lipid concentrations were obtained in 82% of the patients. In two patients, who were originally diagnosed as having Type III hyperlipoproteinemia and yielded completely normal serum lipid concentrations with L-T_4 therapy, the Type III pattern still persisted; however, another Type III hyperlipoproteinemic patient had a normal lipoprotein pattern after L-T_4 treatment. The majority of the Type IV hyperlipoproteinemic patients also had persistent Type IV patterns even after L-T_4 treatment.

Fredrickson et al.[180] have discussed the fact that hypothyroidism may be the primary cause of Types I, II, III, IV, and V. The occasional occurrence of a Type III hyperlipoproteinemic pattern is an interesting phenomenon. Type III patterns may occur transiently in diabetic acidosis and occasionally in patients with dysglobulinemia. Patients on chronic hemodialysis have Type III-like patterns.[180] This is probably due to an increase in the S_f 12—20 fraction[181] or the accumulation of IDL (d 1.006 to 1.019). The β-VLDL observed in Type III patients have an unusually high affinity for triglycerides, and the cholesterol to triglyceride ratio is usually twice as high as compared to other phenotypes.[180] It is believed that hypothyroidism markedly aggravates Type III hyperlipoproteinemia and may conceivably produce it. Conversely, hyperthyroidism may mask Type III by suppressing cholesterol and triglyceride concentrations, which rise rapidly upon return to a euthyroid state. More studies are necessary to elucidate the effects of T_4 on IDL metabolism and to understand why there is an occasional occurrence of Type III hyperlipoproteinemic pattern in hypothyroid individuals.

Because little work has been done on serum lipoprotein electrophoresis on individuals with thyroid dysfunction, a study of the lipid concentrations and lipoprotein distribution in 66 patients with hypothyroidism and 26 patients with hyperthyroidism was conducted. Thirty subjects served as euthyroid controls. The thyroid status of each individual was carefully determined by assaying the serum TSH and total T_4 by double-antibody radioimmunoassay (R.I.A.) techniques[165-167]

R.I.A. for TSH

Human TSH for labeling and rabbit anti-TSH were obtained from (Calbiochem, La Jolla, Cal.) TSH research standard B was obtained from Medical Research Council, National Institute for Medical Research, Mill Hill, London. TSH was labeled with ^{125}I by the chloramine-T method as described by Greenwood et al.[168] Sensitivity of the assay was 2 $\mu U/ml$.* Interassay variation at the level of 3 $\mu U/ml$ was 8%. The mean serum TSH concentration in 45 normal subjects was 2.8 ± 1.0 $\mu U/ml$ (range, 2.0 to 6.0 $\mu U/ml$).

R.I.A. for T_4

Antiserum for T_4 was made in rabbits by immunizing with T_4 conjugated to bovine serum albumin. It showed 4.8% cross reaction with T_3. T_4 ^{125}I was purchased from Industrial Nuclear, St. Louis, Mo. The interassay variation at the level of 0.6 $\mu g/100$ ml was 8%. In 65 normal subjects, the mean T_4 level was 6.4 ± 1.2 $\mu g/100$ ml (range, 4 to 9 $\mu g/100$ ml). Sensitivity of the assay was 0.25 $\mu g/100$ ml.

It is generally accepted that serum TSH concentrations are elevated in patients with primary hypothyroidism.[165, 169-171] In fact, elevation of TSH is now considered *Sine qua non* for diagnosis of hypothyroidism.[172] Some studies have associated the high serum TSH levels in hypothyroid individuals with the low serum T_4 levels.[165,171] The diagnostic value of serum T_3 determinations in hypothyroidism is not clear. The relationship between the concentration of serum TSH, T_4, T_3 and effective thyroxine ratio (ETR) in untreated patients with primary thyroid failure was studied. Next to TSH, T_4 determinations were found to have more diagnostic value than serum T_3 or ETR in hypothyroid patients.[167] Patients with TSH values > 25 $\mu U/ml$ were either definite hypothyroids or suspected to have hypothyroidism by clinical diagnosis. Since it is accepted that elevated serum T_4 values are characteristic of hyperthyroid patients, the R.I.A.-T_4 method was used to diagnose hyperthyroidism. In some patients, ETR, T_3, free thyroxine index, and other clinical studies were done to confirm the questionable thyroid status.

In addition, fasting serum samples were analyzed for total cholesterol,[173] triglycerides,[173] and phospholipids.[174] Serum was electrophoresed using agarose-gel as support medium for lipoprotein studies.[175] On some samples that were difficult to interpret, paper[176] and polyacrylamide-gel disc[177] electrophoretic methods were also used. Results indicated that the serum total cholesterol concentration of the hypothyroid patients was higher than the euthyroid group (Table 2). On the other hand the hyperthyroid patients had slightly lower levels of total cholesterol. The serum triglyceride and phospholipid levels of the hypothyroid group were slightly higher than the eu- or hyperthyroid groups. These findings are in agreement with past reports.

The serum lipoprotein electrophoresis study (Table 2) indicates that there is a higher percentage of β-lipoprotein in the hypothyroid group when compared to the euthyroid patients. This agrees with the elevated serum total cholesterol content seen in the hypothyroid individuals. While there was a reduction in the total cholesterol in the hyperthyroid group, there was no reduction in the distribution of lipoproteins in the β-lipoprotein class. There was, however, less cholesterol (45 ± 3 mg/dl) in the HDL of

---

* μU = micro unit.

Table 2

SERUM LIPID AND LIPOPROTEIN DATA ON EU-, HYPER-, AND HYPOTHYROID PATIENTS

Thyroid status	No. in study	Serum T$_4$ levels (µg/100 ml)	Serum TSH (µU/ml)	Serum lipid levels (mg/dl)			Serum lipoprotein distribution (%)[a]			
				Total cholesterol	Triglycerides	Total phospholipids	Chylomicron	pre-β	β	α
Euthyroid	30	6.06±0.14	—	237±57	150±75	251±63	0	11±6	52±9	37±8
Hyperthyroid	26	12.84±0.58	—	184±40	147±71	231±24	1.0±0.9	10±5	50±8	39±7
Hypothyroid	66	—	73.07±6.03	316±6.03	197±83	287±56	1.0±1.0	9±3	62±6	28±6

Note. All values are given as X±S.E M.

[a] Serum lipoprotein electrophoresis is based on the agarose method.[175]

the hyperthyroid group when compared to the hypothyroid group (53 ± 4 mg/dl). HDL cholesterol concentration for the euthyroid group was 50 ± 4 mg/dl. It is interesting that the slight elevation of serum triglycerides seen in the hypothyroid group did not result in a concomitant rise in the pre-β-lipoprotein area. Whether the increased proportion of β-lipoprotein in the hypothyroid group resulted in more cholesterol and triglycerides carried in that moiety is not clear. Only when serum triglycerides were markedly elevated (>200 mg/dl) did we see a concomitant rise in the pre-β-lipoprotein fraction.

It should be stated that the inverse relationship of thyroid activity to serum lipid levels did not apply to all individuals (Table 3). Euthyroid patients were not all normolipemic. About 57% had normal lipid values (total cholesterol, triglycerides, and phospholipids). Thirteen percent had elevated cholesterol alone, while 27% had elevated triglycerides alone and 10%, elevated phopholipid alone. Seven percent of the euthyroids had all three lipid classes elevated above normal levels. The hyperthyroid followed a similar phenomenon with slight differences in percentage values. On the other hand, 55% of the hypothyroid patients had elevated cholesterol, while 47% and 30% had elevated triglycerides and phospholipid concentrations, respectively. Twenty percent had all three lipid moieties elevated, while only 26% had normal lipid values.

CONCLUSIONS

The thyroid hormone has an important and decisive influence on concentration, distribution, and pattern of the serum lipids and lipoproteins. There is an inverse relationship between the activity of the thyroid gland and serum lipid concentrations, with cholesterol showing the most consistent responsiveness. Serum triglycerides and, to a lesser extent, serum total phospholipid concentrations are also affected by the activity or inactivity of the thyroid gland. The precise mechanisms by which the thyroid gland mediates these serum lipid abnormalities is not clear. Since there are many factors that regulate lipid metabolism, it is not unexpected to find abnormal lipid concentrations in euthyroid subjects as seen in our study. It appears that in these subjects there are other factors that have a greater influence on lipid metabolism, i.e., genetic, dietary and/or environmental.

Serum lipoproteins are also affected by thyroid dysfunction. β-lipoprotein levels vary according to serum cholesterol level in the thyroid condition of the individual. The slight serum triglyceride elevation seen in many hypothyroid patients does not manifest itself by increased pre-β-lipoprotein levels. Only when serum triglycerides are moderately to markedly elevated, does one see a concomitant rise in pre-β-lipoprotein. In hyperthyroid patients, there was no change in distribution of the serum lipoproteins. However, the amount of HDL cholesterol was lower when compared to the hypothyroid group. The mechanism and significance of this is not understood at this time.

While the influence of the thyroid gland on serum lipid and lipoprotein metabolism has been reviewed it should be emphasized that there are numerous other hormones that can also affect lipid and lipoprotein metabolism, either independently of thyroid activity or in a synergistic, agonistic, or antagonistic manner. The use of serum lipoprotein electrophoretic techniques has been a convenient guide for categorizing basic forms of dyslipoproteinemias, but it is an over simplified approach for characterizing abnormal lipoprotein patterns due to inherent limitations of existing methods. Many unusual and complex lipoprotein electrophoretic patterns are frequently observed in secondary or acquired forms of dyslipoproteinemia. These patterns must be interpreted with care since they do not always correlate well with observed changes in serum lipid concentrations. The occasional occurrence of a Type III-like pattern in hypothyroidism is just one example. Many other instances of secondary abnormal lipoprotein patterns

Table 3
PERCENTAGE OF PATIENTS WITH ABNORMAL
SERUM LIPID CONCENTRATIONS[a]

Serum lipids[b]	Thyroid status		
	Euthyroid	Hyperthyroid	Hypothyroid
TC + TG + PL = N	57	65	26
TC + TG + PL > N	7	4	20
Either TC, TG or PL alone > N	36	31	54
TC alone > N	9	6	23
TG alone > N	19	22	19
PL alone > N	8	3	12

[a] Serum lipoprotein electrophoresis is based on the agarose method.[175].

[b] These abbreviations will be used in the table: TC = total cholesterol; TG = triglycerides; PL = phospholipid; N = normal.

related to endocrine dysfunction remain to be clarified, but it will be difficult to do so until the precise role of hormones in normal lipid and lipoprotein metabolism is understood.

ACKNOWLEDGMENTS

This study was supported in part by Grant No. HL-6835 from the National Heart and Lung Institute, Grant No. 3084R from the American Heart Association, Northeast Ohio Affiliate, Inc., and Grant No. CRP-400 from the Clinical Research Projects Committee, Cleveland Clinic Foundation. We wish to thank Ms. Maryann Olynyk, Ms. Joanne R. Bratush, Ms. Jeanette H. Repanszky, Ms. Joan Ann David, and Mr. Joseph Paksi for their technical assistance and Ms. Penny Porter for typing this manuscript.

REFERENCES

1. **Pitt-Rivers, R. and Tato, J. R.**, in *The Thyroid Hormones I*, Pergamon Press, New York, 1959, 350.
2. **Peters, J. P. and Man, E. B.**, The interrelation of serum lipides in patients with thyroid disease, *J. Clin. Invest.*, 22, 715, 1943.
3. **Peters, J. P. and Man, E. B.**, The significance of serum cholesterol in thyroid disease, *J. Clin. Invest.*, 29, 1, 1950.
4. **O'Hara, D. D., Porte, D., Jr., and Williams, R. H.**, The effect of diet and thyroxine on plasma lipids in myxedema, *Metabolism*, 15, 123, 1966.
5. **Barclay, M.**, Lipoprotein class distribution in normal and diseased states, in *Blood Lipids and Lipoproteins: Quantitation, Composition, and Metabolism*, Nelson, G. J., Ed., Interscience, New York, 1972, 585.
6. **Furman, R. H., Howard, R. P., Lakshmi, K., and Norcia, L.N.**, The serum lipids and lipoproteins in normal and hyperlipidemic subjects as determined by preparative ultracentrifugation. Effects of dietary and therapeutic measures. Changes induced by in vitro exposure of serum to sonic forces, *Am. J. Clin. Nutr.*, 9, 73, 1961.
7. **Eisalo, A., Ahrenberg, P., and Nikkilä, E.**, Treatment of hyperlipidemia with d-thyroxine, *Acta. Med. Scand.*, 173, 639, 1963.

8. **Kirkeby, K.,** Post heparin plasma lipoprotein lipase activity in thyroid disease, *Acta Endocrinol.* Copenhagen, 59, 555, 1968.
9. **Calay, R., Kocheleff, P., Johnlaux, G., Sohet, L., and Bastenie, P.A.,** Dextrothyroxine therapy for the disordered lipid metabolism of preclinical hypothyroidism, *Lancet,* 1, 205, 1971.
10. **Nikkilä, E. and Kekki, M.,** Plasma triglyceride metabolism in thyroid disease, *J. Clin. Invest.,* 51, 2103, 1972.
11. **Tulloch, B., Lewis, B., and Fraser, T. R.,** Triglyceride metabolism in thyroid disease, *Lancet,* 1, 391, 1973.
12. **Jacobsen, B. B.,** Blood lipids during treatment of hyperthyroidism. A statistical evaluation of the relationships between lipids and thyroid variables, *Acta Endocrinol.* (Copenhagen), 72, 443, 1973.
13. **Tulloch, B. R.,** Lipid changes in thyroid disease: The effect of thyroid and analogues, *Proc. R. Soc. Med.,* 67, 670, 1974.
14. **Fredrickson, D. S., Levy, R. I., and Lees, R. S.,** Fat transport in lipoproteins: an integrated approach to mechanisms and disorders, *N. Engl. J. Med.,* 276, 32, 94, 148, 215, 273, 1967.
15. **Vessby, B. and Wide, L.,** Serum levels of thyroid-stimulating hormone in hyperlipoproteinemia, *Clin. Chim. Acta,* 62, 293, 1975.
16. **Hazzard, W. R. and Bierman, E. L.,** Aggravation of broad-β-disease (type 3 hyperlipoproteinemia) by hypothyroidism, *Arch. Intern. Med.,* 130, 822, 1972.
17. **Lasser, N. L., Burns, J., and Solar, S.,** Type III hyperlipoproteinemia secondary to hypothyroidism, in *Atherosclerosis III,* Schettler, G. and Weizel, A., Eds., Springer-Verlag, New York, 1974, 621.
18. **Hazzard, W. R., Porte, D., Jr., and Bierman, E. L.,** Abnormal lipid composition of very low density lipoproteins in diagnosis of broad-beta disease (type III hyperlipoproteinemia), *Metabolism,* 21, 1009, 1976.
19. **Furman, R. H., Howard, R. P., and Conrad, L. L.,** Effects of androsterone and triiodothyronine on serum lipids and lipoproteins, nitrogen balance and related metabolic phenomena in subjects with normal and decreased thyroid function, with hyperglyceridemia and/or hypercholesterolemia, *Metabolism,* 11, 76, 1962.
20. **Rossner, S. and Rosenqvist, U.,** Serum lipoproteins and the intravenous fat tolerance test in hypothyroid patients before and during substitution therapy, *Atherosclerosis,* 20, 365, 1974.
21. **Mason, R. L., Hunt, H. M., and Hurxthal, L.,** Blood cholesterol values in hyperthyroidism and hypothyroidism-their significance, *N. Engl. J. Med.,* 203, 1273, 1930.
22. **Martinez, B. D.,** Travauz du laboratorie de physiology de la faculte de medicine de Buenos Aires, *Endocrinology,* 1, 357, 1917.
23. **Epstein, A. A. and Lande, H.,** Studies on blood lipids. I. The relation of cholesterol and protein deficiency to basal metabolism, *Arch. Intern. Med.,* 30, 563, 1922.
24. **Gardner, J. A. and Gainsborough, H.,** The relation of plasma cholesterol on basal metabolism, *Br. Med. J.,* 2, 935, 1928.
25. **Baumann, E. J. and Holly, O. M.,** Cholesterol and phosphatide metabolism in pregnancy, *Am. J. Physiol,* 75, 618, 1926.
26. **Wade, P. A.,** Clinical and experimental studies on calcium and cholesterol in relation to the thyroid-parathyroid apparatus, *Am. J. Med. Sci.,* 177, 790, 1929.
27. **Hurxthal, L. M.,** Blood cholesterol in thyroid disease. I. Analyses of findings in toxic and nontoxic goiter before treatment, *Arch. Intern. Med.,* 51, 22, 1933.
28. **Hurxthal, L. M.,** Blood cholesterol in thyroid disease. II. Effect of treatment, *Arch. Intern. Med.,* 52, 86, 1933.
29. **Hurxthal, L. M.,** Blood and thyroid disease. III. Myxedema and hypercholesterolemia, *Arch. Intern. Med.,* 53, 762, 1934.
30. **Hurxthal, L. M.,** Blood cholesterol and hypometabolism, *Arch. Intern. Med.,* 53, 825, 1934.
31. **Goldbloom, A. and Gottlieb, R.,** The cholesterol content of the blood of infants and children, *Can. Med. Assoc. J.,* 17, 1333, 1927.
32. **Bronstein, I. P.,** Studies in cretinism and hypothyroidism in childhood. I. Blood cholesterol, *J. Am. Med. Assoc.,* 100, 1661, 1933.
33. **Westra, J. J. and Kunde, M. M.,** Blood cholesterol in experimental hypo-and hyperthyroidism (rabbit), *Am. J. Physiol.,* 103, 1, 1933.
34. **Gilligan, D. R., Volk, M. C., Davis, D., and Blumgart, H. L.,** Therapeutic effect of total ablation of normal thyroid on congestive heart failure and angina pectoris. VIII. Relationship between serum cholesterol values, basal metabolic rate and clinical aspects of hypothyroidism, *Arch. Intern. Med.,* 54, 746, 1934.
35. **Cutting, W. C., Rytand, D. A., and Tainter, M. L.,** Relationship between blood cholesterol and increased metabolism from dinitrophenol and thyroid, *J. Clin. Invest.,* 13, 547, 1934.
36. **Grant, L. F. and Schube, P. G.,** The effect of alpha dinitrophenol 1-2-4 on blood cholesterol in man, *J. Lab. Clin. Med.,* 20, 56, 1934.

37. Kunde, M. M., Green, M. F., and Burns, G., Blood changes in experimental hypo- and hyperthyroidism (rabbit), *Am. J. Physiol.*, 99, 469, 1932.

38. Boyd, E M., The effects of thyroidectomy on blood lipids, *Trans. R. Soc. Can. Sect.* 5, 30, 11, 1936.

39. Boyd, E. M. and Connell, W. F., Thyroid disease and blood lipids. *J. Med.*, 5, 455, 1936.

40. Turner, K. B., Present, C. H., and Bidwell, E. H., The role of the thyroid in the regulation of blood cholesterol of rabbits, *J. Exp. Med.*, 67, 111, 1938.

41. Schmidt, L. H. and Hughes, H.B., The free and total cholesterol content of whole blood and plasma as related to experimental variations in thyroid activity, *Endocrinology*, 22, 474, 1938.

42. Gildea, E. F., Man, E. B., and Peters, J. P., Serum lipoids and proteins in hypothyroidism, *J. Clin. Invest.*, 18, 739, 1939.

43. Kendall, E. C., The influence of some of the ductless glands on metabolic processes, *Endocrinology*, 24, 798, 1939.

44. Fleischmann, W., Schumacker, H. B., and Wilkins, L., The effect of thyroidectomy on serum cholesterol and basal metabolic rate in the rabbit, *Am. J. Physiol.*, 131, 317, 1940.

45. Thompson, K. W. and Long, C. N. H., The effect of hypophysectomy upon hypercholesterolemia of dogs, *Endocrinology*, 28, 715, 1941.

46. Chaikoff, I. L., Entenman, C., Changus, G. W., and Reichert, F. L., Influence of thyroidectomy on blood lipids of the dog, *Endocrinology*, 28, 797, 1941.

47. Fleischmann, W. and Wilkins, L., Sterol balance in hypothyroidism, *J. Clin. Endocrinol.*, 1, 799, 1941.

48. Fleischmann, W. and Schumacher, H. B., Jr., The relationship between serum cholesterol and total body cholesterol in experimental hyper- and hypo-thyroidism, *Bull. Johns Hopkins Hosp.*, 71, 175, 1942.

49. Entenman, C., Chaikoff, I. L., and Reichert, F. L., Blood lipids of the hypophysectomized-thyroidectomized dog, *Endocrinology*, 30, 802, 1942.

50. Forbes, J. C., Effect of thyroxine on the neutral fat and cholesterol content of the body and liver of rats, *Endocrinology*, 35, 126, 1944.

51. Fleischmann, W. and Fried, I. A., Studies on the mechanism of the hypercholesterolemia and hypercalcemia induced by estrogen in immature chicks, *Endocrinology*, 36, 406, 1945.

52. Foldes, F. F. and Murphy, A. J., Distribution of cholesterol, cholesterol esters, and phospholipid phosphorus in blood in thyroid disease, *Proc. Soc. Exp. Biol. Med.*, 62, 218, 1946.

53. Steiner, A. and Kendall, R. E., Atherosclerosis and arteriosclerosis in dogs following ingestion of cholesterol and thiouracil, *Arch. Pathol.*, 42, 433, 1946.

54. Chanutin A., Gjessing, E. C., and Ludeqig, S., Alpha napthylthiourea (ANTU) in dogs: electrophoretic and cholesterol studies on blood plasma and pleural effusion, *Proc. Soc. Exp. Biol. Med.*, 64, 174, 1947.

55. Horlick, L. and Havel, R., The effect of feeding prophylthiouracil and cholesterol on blood cholesterol and arterial intima in the rat, *J. Lab. Clin. Med.*, 33, 1029, 1948.

56. Handler, P., The influence of thyroid activity on the liver and plasma lipides of choline- and cystine-deficient rats, *J. Biol. Chem.* 173, 295, 1948.

57. Blumgart, H. L., Freedberg, A. S., and Kurland, G. S., Hypothyroidism produced by radioactive iodine (I^{131}) in treatment of euthyroid patients with angina pectoris and congestive heart failure, *Circulation*, 1, 1105, 1950.

58. Stamler, J., Silber, E. N., Miller, A. J., Akman, K., Bolene, C., and Katz, L. N., Effect of thyroid- and of dinitrophenol-induced hypermetabolism on plasma and tissue lipids and atherosclerosis in the cholesterol fed chick, *J. Lab. Clin. Med.*, 35, 351, 1950.

59. Marx, W., Marx, L. and Shimoda, F., Thyroid hormone and tissue cholesterol distribution, *Proc. Soc. Exp. Biol. Med.*, 73, 599, 1950.

60. Page, I. H. and Brown, H. B., Induced hypercholesterolemia and atherogenesis, *Circulation*, 6, 681, 1952.

61. Rosenman, R. H., Byers, S. O., and Friedman, M., The mechanism reponsible for the altered blood cholesterol content in deranged thyroid states, *J. Clin. Endocrinol. Metab.*, 12, 1287, 1952.

62. Rosenman, R. H., Friedman, M., and Byers, S. O., Observations concerning the metabolism of cholesterol in the hypo- and hyperthyroid rat, *Circulation*, 5, 589, 1952.

63. Weiss, S. B. and Marx, W., The fate of radioactive cholesterol in mice with modified thyroid activities, *J. Biol. Chem.*, 213, 349, 1955.

64. Byers, S. O., The mechanism for changes in blood cholesterol in deranged thyroid states, *Am. J. Clin. Nutr.*, 6, 642, 1958.

65. Deming, Q. B., Mosbach, E. H., Bevans, M., Daly, M. M., Abell, L. L., Martin, E., Brun, L. M., Halpern, E., and Kaplan, R., Blood pressure, cholesterol content of serum and tissues, and atherogenesis in the rat, *J. Exp. Med.*, 107, 581, 1958.

66. **Duncan, C. H. and Best, M. M.,** Effect of thiouracil on serum and liver cholesterol of the athyreotic rat, *Am. J. Physiol.,* 194, 351, 1958.
67. **Duncan, C. H. and Best, M. M.,** Thyroxine-like compounds and cholesterol metabolism: differences in the effects of thyroxine, triiodothyronine, and their formic acid analogues, *Endocrinology,* 63, 169, 1958.
68. **Best, M. M. and Duncan, C. H.,** Effect of thiouracil and sitosterol on diet-induced hypercholesterolemia and lipomatous arterial lesions in the rat, *Am. Heart J.,* 58, 214, 1959.
69. **Thomas, W. A. and Hartroft, W. S.,** Myocardial infarction in rats fed diets containing high fat, cholesterol, thiouracil, and sodium cholate, *Circulation,* 19, 65, 1959.
70. **Gould, R. G.,** The relationship between thyroid hormones and cholesterol biosynthesis and turnover, in *Hormones and Atherosclerosis,* Pincus, G., Ed., Academic Press, New York, 1959, 75.
71. **Boyd, G. S.,** Thyroid function, thyroxine analogs, and cholesterol metabolism in rats and rabbits, in *Hormones and Atherosclerosis,* Pincus, G., Ed., Academic Press, New York, 1959, 49.
72. **Kritchevsky, D., Staple, E., and Whitehouse, M. W.,** Regulation of cholesterol biosynthesis and catabolism, *Am. J. Clin. Nutr.,* 8, 411, 1960.
73. **Kurland, G. S., Lucas, J. L., and Freedberg, A. S.,** The metabolism of intravenously infused C^{14}-labeled cholesterol in euthyroidism and myxedema, *J. Lab. Clin. Med.,* 57, 574, 1961.
74. **Wells, A. F. and Ershoff, B. H.,** Effects of cholesterol feeding on plasma and liver cholesterol levels in the hypophysectomized rat, *Proc. Soc. Exp. Biol. Med.,* 109, 643, 1962.
75. **Ellefson, R. D. and Mason, H. L.,** Effects of thyroid hormone on lipid metabolism in the rat, *Endocrinology,* 71, 425 1962.
76. **Patek, P. R., Bernick, S., Ershoff, B. H., and Wells, A.,** Induction of atherosclerosis by cholesterol feeding in the hypophysectomized rat, *Am. J. Pathol.,* 42, 137, 1963.
77. **Myant, N. B.,** The thyroid and lipid metabolism, in *Lipid Pharmacology,* Paoletti, R., Ed., Academic Press, New York, 1964, 299.
78. **Lepp, A., Wagle, S. R., and Oliver, L.,** Effects of L- and D-thyroxine on cholesterol synthesis and turnover in the chick, *Proc. Soc. Exp. Biol. Med.,* 115, 517, 1964.
79. **Tsung-Chin, H. and Shih-Chen, W.,** Regulation of cholesterol metabolism by thyroid hormone, *Sci. Sin.,* 14, 874, 1965.
80. **Miettinen, T. A.,** Mechanism of serum cholesterol reduction by thyroid-hormones in hypothyroidism, *J. Lab. Clin. Med.,* 71, 537, 1968.
81. **Furman, R. H.,** Endocrine factors in atherogenesis, in *Atherosclerosis: Pathology, Physiology, Aetiology, Diagnosis, and Clinical Management,* Schettler, F. G. and Boyd, G. S., Eds., Elsevier, New York, 1969, 410.
82. **Turner, K. B.,** Studies on the prevention of cholesterol atherosclerosis in rabbits; effects of whole thyroid and of potassium iodide, *J. Exp. Med.,* 58, 115, 1933.
83. **Dauber, D., Horlick, L., and Katz, L. N.,** The role of desiccated thyroid and potassium iodide in cholesterol- induced atherosclerosis of the chicken, *Am. Heart J.,* 38, 25, 1949.
84. **Marx, W., Gustin, S. T., and Levi, C.,** Effects of thyroxine, thyroidectomy and lowered environmental temperature on incorporation of deuterium into cholesterol, *Proc. Soc. Exp. Biol. Med.,* 83, 143, 1953.
85. **Boyd, G. S. and Oliver, M. F.,** Various effects of thyroxine analogues on the heart and serum cholesterol in the rat, *J. Endocrinol.,* 21, 35, 1960.
86. **Boyd, G. S. and Oliver, M. F.,** Thyroid hormones and plasma lipids, *Br. Med. Bull.,* 16, 138, 1960.
87. **Cuthbertson, W. F. J., Elcoate, P. V., Ireland, D. M., Mills, D. C. B., and Shearley, P.,** Effect of compounds related to thyroxine on serum and liver cholesterol and on atherosclerosis and heart weights in rats and mice, *J. Endocrinol.,* 21, 45, 1960.
88. **Best, M. M. and Duncan, C. H.,** Effect of certain thyroxine analogues on liver cholesterol, *Am. J. Physiol.,* 199, 1000, 1960.
89. **Kritchevsky, D., Moynihan, J. L., and Sachs, M. L.,** Influence of thyroactive compounds on serum and liver cholesterol in rats, *Proc. Soc. Exp. Biol. Med.,* 108, 254, 1961.
90. **Greene, R , Pearce, J. F., and Rideout, D. F.,** Effect of D-thyroxine on serum cholesterol, *Br. Med. J.,* 5239, 1572, 1961.
91. **Jepson. E. M.,** Long-term trial of D-thyroxine in hypercholesterolaemia, *Br. Med. J.,* 1, 1446, 1963.
92. **Felt, V.,** Thyroid and steroid hormones and their relation to serum lipid concentrations in man, *J. Clin. Endocrinol. Metab.,* 26, 683, 1966.
93. **Kritchevsky, D. and Tepper, S. A.,** Oxidation of cholesterol by rat liver mitochondria: effect of thyroidectomy, *J. Cell. Comp. Physiol.,* 66, 91, 1967.
94. **Young, J. W.,** Effects of D- and L-thyroxine on enzymes in liver and adipose tissue of rats, *Am. J. Physiol.,* 214, 378, 1968.
95. **Masket, B. H., Levy, R. I., and Fredrickson, D. S.,** The use of polyacrylamide gel electrophoresis in differentiating type III hyperlipoproteinemia, *J. Lab. Clin. Med.,* 81, 794, 1973.

96. Fletcher, K. and Myant, N. B., Effects of thyroxine on the synthesis of cholesterol and fatty acids by cell-free fractions of rat liver, *J. Physiol.* London, 154, 145, 1960.

97. Lewis, L. A., Lipoproteins and their relation to metabolic disease, *Ann. N. Y. Acad. Sci.*, 94, 320, 1961.

98. Naito, H. K., and Lewis, L. A., Lipid (L), lipoprotein (Lp), and protein (P) composition and concentration in six strains of guinea pigs (P), *Fed. Proc., Fed. Am. Soc. Exp. Biol.*, 32, 934, 1973.

99. Leibbrandt, V. D., Naito, H. K., and Lewis, L. A., Influence of hyperthyroidism on ascorbate adequacy for cholesterol catabolism, *Fed. Proc. Fed. Am. Soc. Exp. Biol.*, 34, 466, 1975.

100. Malinow, M. R., McLaughlin, P., Papworth, L., Naito, H. K., Lewis, L., and McNulty, W. P., A model for therapeutic interventions on established coronary atherosclerosis in a nonhuman primate, in *Advances in Experimental Medicine and Biology: Atherosclerosis Drug Discovery*, Vol. 67, Day, C. E. , Ed., Plenum Press, New York, 1976, 3.

101. Kelstrup, J., Free and total thyroxine in serum, *Scand . J. Clin. Lab. Invest.*, 32, 227, 1973.

102. Evered, D. C., Ormston, B. J., Smith, P. A., Hall, R., and Bird, T., Grades of hypothyroidism, *Br. Med. J.*, 1, 657, 1973.

103. Wenzel, K. W., Meinhold, H., Raffenberg, J., Adlkofer, F., and Schleusener, H., Evaluation of hypothyroidism after radioiodine therapy by comparison of total T_4, free-T_4-index, TRH test and T_3 in serum, *Acta Endocrinol. (Copehnagen) Suppl.*, 177, 273, 1973.

104. Wenzel, K. W., Meinhold, H., Raffenburg, M., Adlkofer, F., and Schleusener, H., Classification of hypothyroidism in evaluating patients after radioiodine therapy by serum cholesterol, T_3-uptake, total T_4, FT-$_4$Index, total T_3, basal TSH and TRH-test, *Eur. J. Clin. Invest.*, 4, 141, 1974.

105. Rosenqvist, H., Efendic, S., Jereb, B., and Ostman, J., Influence of the hypothyroid state on lipolysis in human adipose tissue in vitro, *Acta Med. Scand.*, 189, 381, 1971.

106. Karp, A. and Stetten, D., Jr., The effect of thyroid activity on certain anabolic process. Studied with the aid of deuterium, *J. Biol. Chem.*, 179, 819, 1949.

107. Gould, R. G., Lipid metabolism and atherosclerosis, *Am. J. Med.*, 11, 209, 1951.

108. Friedman, M., Byers, S. O., and Michaelis, F., Production and excretion of cholesterol in mammals. IV. Role of liver and restoration of plasma cholesterol after experimentally induced hypocholesterolaemia, *Am. J. Physiol.*, 164, 789, 1951.

109. Friedman, M. and Byers, S. O., Observations concerning the production and excretion of cholesterol in mammals. XVI. The relationship of the liver to the content and control of plasma cholesterol ester, *J. Clin. Invest.*, 34, 1369, 1955.

110. Dietschy, J. M. and Wilson, J. D., Cholesterol synthesis in the squirrel monkey: relative rates of synthesis in various tissues and mechanisms of control, *J. Clin. Invest.*, 47, 166, 1968.

111. Nilsson, G., Nordlander, S., and Levin, K., Studies on subclinical hypothyroidism with special reference to the serum lipid pattern, *Acta Med. Scand.*, 200, (1—2), 63, 1976.

112. Byers, S. O., Rosenman, R. H., Friedman, M., and Biggs, M. W., Rate of cholesterol synthesis in hypo- and hyperthyroid rats, *J. Exp. Med.*, 96, 513, 1952.

113. Dayton, S., Dayton, J., Drimmer, F., and Kendall, F. E., Rates of acetate turnover and lipid synthesis in normal, hypothyroid, and hyperthyroid rats, *Am. J. Physiol*, 199, 71, 1960.

114. Eskelson, C. D., Cazee, C. R., Anthony, W., Towne, J. C., and Walske, B. R., In vitro inhibition of cholesterolgenesis by various thyroid hormone analogs, *J. Med. Chem.*, 13, 215, 1970.

115. Fletcher, K. and Myant, N. B., Influence of the thyroid on the synthesis of cholesterol by liver and skin in vitro, *J. Physiol*, 144, 361, 1958.

116. Gruder, W., Nolte, I., and Wieland, O., The influence of thyroid hormones on β-hydroxy-β-methyl-glutarylcoenzyme A reductase of rat liver, *Eur. J. Biochem.*, 4, 273, 1968.

117. Gries, F. A., Matschinsky, F., and Wieland, O., Induktion der β-Hydroxy-β-methylglutaryl-reductase durch Schilddrusen Hormone, *Biochim. Biophys. Acta*, 56, 615, 1962.

118. Story, J. A., Tepper, S. A., and Kritchevsky, D., Influence of thyroid state on cholesterol hydroxylation and absorption in the rat, *Biochem. Med.*, 10, 615, 1962.

119. Thompson, J. C., Daily cholic acid and cholesterol excretion in the bile of hypo-, hyper-, and euthyroid rats, *Fed. Proc., Fed. Am. Soc. Exp. Biol.*, 12, 404, 1953.

120. Thompson, J. C. and Vars, H. M., Biliary excretion of cholic acid and cholesterol in and euthyroid rats, *Proc. Soc. Exp. Biol. Med.*, 83, 246, 1953.

121. Thompson, J. C. and Vars, H. M., Influence of thyroid activity on the hepatic excretion of cholic acid and cholesterol, *Am J. Physiol.*, 179, 405, 1954.

122. Bergström, S. and Sjövall, J., Occurrence and metabolism of chenodesoxycholic acid in the rat. Bile acids and steroid 13, *Acta. Chem. Scand.*, 8(1), 611, 1954.

123. Eriksson, S., Biliary excretion of bile acids and cholesterol in bile fistula rats. Bile acids and steroids, *Proc. Soc. Exp. Biol. Med.*, 94, 578, 1957.

124. Eriksson, S., Influence of thyroid activity on excretion of bile acids and cholesterol in the rat, *Proc. Soc. Exp. Biol. Med.*, 94, 582, 1957.

125. **Van Zye, A.,** Note on the effects of thyroidectomy and thyroid hormone administration on the concentration of bile cholesterol and cholic acid, *J. Endocrinol.,* 16, 213, 1957.

126. **Strand, O.,** Effects of D- and L-triiodothryonine and propylthiouracil on the production of bile acids in the rat, *J. Lipid Res.,* 4, 305, 1963.

127. **Lin, T H., Rubinstein, R., and Holmes, W. L.,** A study of the effect of D- and L-triiodothyronine on bile acid excretion of rats, *J Lipid Res.,* 4, 63, 1963.

128. **Nordoy, A., Yik-Mo, H., and Berntsen, H.,** Haemostatic and lipid abnormalities in hypothyroidism, *Scand. J. Haematol.,* 16(2), 154, 1976.

129. **Gans, J. H.,** Bile secretion during experimental hyperthyroidism in the dog, *Am. J. Physiol.,* 199, 893, 1960.

130. **Strand, O.,** Influence of propylthiouracil and D- and L-triiodothyronine on excretion of bile acids in bile fistula rats. Bile acids and steroids 118, *Proc. Soc. Exp. Biol. Med.,* 109, 668, 1962.

131. **Anfinsen, C. B. and Horning, M. G.,** Enzymatic degradation of the cholesterol side chain in cell-free preparations, *J. Am. Chem. Soc.,* 75, 1511, 1953.

132. **Fredrickson, D. S.,** The conversion of cholesterol-4-C[14] to acids and other products by liver mitochondria, *J. Biol. Chem.,* 222, 109, 1956.

133. **Mitropoulos, K. A. and Myant, N. B.,** Effect of thyroid hormones upon the metabolism of cholesterol by isolated liver mitochondria, *Biochem. J.,* 91, 20p, 1964.

134. **Suld, H. M., Staple, E., and Gurin, S.,** Mechanism of formation of bile acids from cholesterol: oxidation of 5β-cholestane-3α, 12α-triol and formation of propionic acid from the side chain by rat liver mitochondria, *J. Biol. Chem.,* 237, 338, 1962.

135. **Kritchevsky, D., Cottrell, M. C., and Tepper, S. A.,** Oxidation of cholesterol by rat liver mitochondria: effect of thyroactive compounds, *J. Cell. Comp. Physiol.,* 60, 105, 1962.

136. **Kritchevsky, D. and Tepper, S. A.,** Oxidation of cholesterol by rat liver mitochondria: effect of thyroidectomy, *J. Cell. Comp. Physiol.,* 66, 91, 1965.

137. **Balasurbramaniam, S. and Mitropoulos, K. A.,** The role of cytochrome P-450 in the 7-α-hydroxylation of cholesterol, *Biochem. Soc. Trans.,* 3, 964, 1975.

138. **Ferreri, L. F. and Naito, H. K.,** Stimulation of hepatic cholesterol 7α-hydroxylase activity by administration of an estrogen to female rats, *Steroids,* 29, 229, 1977.

139. **Walton, K. W., Scott, P., Dykes, P., and Davis, J.,** The significance of alterations in serum lipids in thyroid dysfunction. I. The relation between serum lipoproteins, carotenoids, and vitamin A in hypothyroidism and thyrotoxicosis, *Clin. Sci.,* 29, 216, 1965.

140. **Naito, H K.,** Effects of physical activity on serum cholesterol metabolism: a review, *Cleveland Clin. Q.,* 43, 21, 1976.

141. **Naito, H. K. and Griffith, D. R.,** Changes in rats' serum triglyceride concentration with graded levels of thyroxine and exercise, *Proc. Soc. Exp. Med. Biol.,* 154, 372, 1977.

142. **de Paula e Silva, P., Diament, J., Forti, H., Luthold, W. W. and Diogo Giannini, S. D.,** Aspectos do metabolismo lipidico no mixedema, antes e apos tratamento, *Rev. Hosp. Clin. Fac. Med. Univ. Sao Paulo,* 28, 3, 1973.

143. **Mitchell, W. D.,** A comparison of the effect of clofibrate and thyroxine on serum lipids in three hypothyroid subjects, *Clin. Chim. Acta,* 35, 429, 1971.

144. **Strisower, E H., Adamson, G., and Strisower, B.,** Treatment of hyperlipidemias, *Am. J. Med.,* 45, 488, 1968.

145. **Strisower, B., Gofman, J. W., Giacioni, E., Rubinger, J. H., O'Brien, G. W., and Simon, A.,** Effect of long-term administration of desiccated thyroid on serum lipoprotein and cholesterol levels, *J. Clin. Endocrinol. Metab.* 15, 73, 1955.

146. **Lamberg, B. A., Heinonen, O. P., Viherkoski, M., Aro, A., and Liewendahl, K.,** Diagnosis of hyperthyroidism. Statistical evaluation of laboratory and clinical criteria, *Acta Endocrinol. (Copenhagen) Suppl.* 146, 7, 1970.

147. **Bastenie, P. A.,** Thyroide et lipides plasmatiques, *Acta Cardiol.,* Suppl. 15, 49, 1972.

148. **Jones, R., Cohen, L., and Corbus, H.,** The serum lipid pattern in hyperthyroidism, hypothyroidism and coronary atherosclerosis, *Am. J. Med.,* 19, 71, 1955.

149. **Malmros, H. and Swahn, B.,** Lipid metabolism in myxedema, *Acta Med. Scand.,* 145, 361, 1953.

150. **Kiewendahl, K. and Helenius, T.,** Comparison of serum free thyroxine indices and "corrected" thyroxine tests, *Clin. Chim. Acta,* 64, 3, 263, 1975.

151. **Feldman, E. B., and Carter, A. C.,** Reduction of serum triglycerides in patients treated wih sodium D-thyroxine, *Metabolism,* 12, 1132, 1963.

152. **Kunkel, H. G. and Slater, R. J.** Lipoprotein patterns of serum obtained by zone electrophoresis, *J. Clin. Invest.,* 31, 677, 1952.

153. **Strisower, B., Gofman, J. W., Galioni, E. F., Rubinger, J. H., Pouteau, J., and Guzvich, P.,** Long-term effect of dried thyroid on serum-lipoprotein and serum cholesterol levels, *Lancet,* 1, 120, 1957.

154. **Rössner, S. and Rosenqvist, U.,** Serum lipoproteins and the intravenous fat tolerance test in hypothyroid patients before and during substitution therapy, *Atherosclerosis,* 20, 365, 1974.

155. Wieland, H. and Seidel, D., Improved techniques for assessment of plasma lipoprotein patterns. II. Rapid method for diagnosis of type III hyperlipoproteinemia without ultracentrifugation, *Clin. Chem.*, 19, 1139, 1973.

156. Koppers, L. E. and Palumbo, P.J. Lipid disturbances in endocrine disorders, *Med. Clin. North Am.*, 56, 1013, 1972.

157. Fredrickson, D. S. and Levy, R.I., Familial hyperlipoproteinemia, in *The Metabolic Basis of Inherited Disease*, 3rd ed., Stanbury, J. B., Wyngaarden, J. B., and Fredrickson, D. S., Eds., McGraw-Hill, New York, 1972, 545.

158. Shore, V. G. and Shore, B., Heterogeneity of human plasma very low density lipoproteins. Separation of species differing in protein components, *Biochemistry*, 12, 502, 1973.

159. Bilheimer, D. W., Eisenberg, S., and Levy, R. I., Abnormal metabolism of very low density lipoproteins in type III hyperlipoproteinemia, *Circulation*, Suppl. 2, 56, 1971.

160. Bilheimer, D. W., Eisenberg, S., and Levy, R. I.,The mechanisms of very low density lipoprotein proteins. I. Preliminary in vitro and in vivo observations, *Biochim. Biophys. Acta*, 260, 212, 1972.

161. Walton, K. W., The significance of alterations in serum lipids in thyroid disease. II. Alterations of the metabolism of turnover of I^{131}-low-density lipoproteins in myxoedema and thyrotoxicosis, *Clin. Sci.*, 29, 217, 1965.

162. Vessby, B., Studies on the serum lipoprotein composition in a 50-year-old man. A suggestion of chemical criteria for diagnosis of hyperlipoproteinaemia type III (broad-β-disease), *Clin. Chim. Acta*, 69, 29, 1976.

163. Porte, D , Jr., O'Hara, D. D., and Williams, R. H., The relation between postheparin lipolytic activity and plasma triglyceride in myxedema, *Metabolism*, 15, 107, 1966.

164. McGee, L. C., Blood cholesterol and disturbances of the basal metabolic rate, *Ann. Intern. Med.*, 9, 728, 1935.

165. Mayberry, W. E., Gharib, M., and Bilstad, J. M., Radioimmunoassay for human thyrotrophin. Clinical value in patients with normal and abnormal thyroid function, *Ann. Intern. Med.*, 74, 471, 1971.

166. Chopra, I. J., A radioimmunoassay for measurement of thyroxine in unextracted serum, *J. Clin. Endocrinol., Metab.*, 34, 938, 1972.

167. Kumar, M. S., Safa, A. M., Deodhar, S. D., and Schumacher, O. P., Evaluation of T_4 radioimmunoassay as a screening test for thyroid function: comparison with effective thyroxine ratio, *Cleveland Clin. Q.*, 44, 1, 1977.

168. Greenwood, F. C., Hunter, W. M., and Glover, J. S., The preparation of I^{131} labelled human growth hormone of high specific radioactivity, *J. Biochem.*, 89, 114, 1963.

169. Hershman, J. M. and Pittman, J. A., Utility of the radioimmunoassay of serum thyrotrophin in man, *Ann. Intern. Med.*, 74, 481, 1971.

170. Odell, W. D., Wilber, J. F., and Utiger, R. D., Studies of thyrotrophin physiology by means of radioimmunoassay, *Recent Prog. Horm. Res.*, 23, 47, 1967.

171. Reichlin, S. and Utiger, R. D., Regulation of the pituitary-thyroid axis in man: relationship of TSH concentration to concentration of free and total thyroxine in plasma, *J. Clin. Endocrinol. Metab.*, 27, 251, 1967.

172. Wehner, H. W., T_3 hyperthyroidism, *Mayo Clin. Proc.*, 47, 938, 1972.

173. AutoAnalyzer II Clinical Method File No. 24. Simultaneous Cholesterol Triglycerides, Technicon Instruments Corp., Tarrytown, N.Y. 1972.

174. Naito, H. K., Modification of the Fiske and SubbaRow method for total phospholipid in serum, *Clin. Chem.*, 21, 1454, 1975.

175. Pfizer-Pol-E-Film System. V. Serum Lipoprotein Phenotyping, Pfizer Diagnostics Division, New York, 1971.

176. Naito, H. K. and Lewis, L. A., Rapid lipid-staining procedure for paper electrophoretograms, *Clin. Chem.*, 19, 106, 1973.

177. Naito, H. K., Wada, M., Ehrhart, L. A., and Lewis, L. A., Polyacrylamide-gel disc-electrophoresis as a screening procedure for serum lipoprotein abnormalities, *Clin. Chem.*, 19, 228, 1973.

178. Strisower, B., Gofman, J. W., Galioni, E. F., Almada, A. A., and Simon, A., Effect of thyroid extract on serum lipoproteins and serum cholesterol, *Metabolism*, 3, 218, 1954.

179. Mishkel, M. A. and Crowther, S. M., Hypothyroidism, an important cause of reversible hyperlipidemia, *Clin. Chim. Acta*, 74, 139, 1977.

180. Fredrickson, D. S., Gotto, A. M., and Levy, R. I., Familial lipoprotein deficiency, in *The Metabolic Basis of Inherited Disease*, 3rd ed., Stanbury, J. B., Wyngaarden, J. D., and Fredrickson, D. S., Eds., McGraw-Hill, New York, 1972, 493.

181. Wada, M., Naito, H. K., Unoki, T., Okamatsu, A., Minamisono, T., Handa, Y., Kusukawa, R., and Lewis, L. A., A possible mechanism for intermediate-density lipoprotein accumulation in the serum of uremic patients on chronic hemodialysis, *Clin. Chemm.*, 24(6), 1024, 1978.

LIPOPROTEINS IN AUTOIMMUNE HYPERLIPIDEMIA AND IN MULTIPLE MYELOMA

Lena A. Lewis

BINDING OF SERUM LIPOPROTEINS WITH IMMUNOGLOBULINS

In 1965, Beaumont et al, reported in Nouvelle Revue Francaise d'Hematologie the association of a γ-, i. e., IgA, paraprotein with β-lipoprotein in the serum of a patient with multiple myeloma, hyperlipidemia, and xanthomtosis. [1] The isolated γ A, when used as antibody, formed precipitin arcs against the patient's β-lipoprotein and all other β-lipoproteins of human serum tested. There was no antibody activity against α-lipoprotein. They observed delayed vitamin A catabolism in this patient and suggested that the hyperlipidemia was probably due to the lipoprotein-immunoglobulin association. They felt that their results suggested an antigen-antibody situation with the antigen being the β-lipoprotein and established that the IgA paraprotein, alone of the patient's immunoglobulins, had antibody activity against β-lipoprotein.[2,3,5-7]

The same year Lewis and Page [8] reported "an unusual serum lipoprotein-globulin complex" in the serum of a patient with hyperlipidemia. The patient,who had extensive xanthomata, was known to have hyperlipidemia for 11 years, and the lipoprotein and immunoglobulin were firmly associated as a soluble complex. This association was demonstrated by electrophoretic, immunoelectrophoretic, and ultracentrifugal studies. While the patient's bone marrow showed 18% plasma cells, no clinical evidence of multiple myeloma was found. The IgA at that time showed both x and λ chains. There was indication of interaction between IgA and β- and α-lipoproteins since the immunoelectrophoretic study of the patient's serum and d 1.21 lipoprotein concentrate when reacted against anti-IgA antisera showed a precipitin arc extending from the position of the β-lipoprotein to that of the fast α-lipoprotein.

Since these initial studies involving hyperlipidemia associated with hyper-IgA-globulinemia and antigen-antibody characteristics, autoantibody properties against lipoproteins have been demonstrated in unique IgM [10] and IgG[11-13] immunoglobulins, also. Some of the immunoglobulins react specifically only against β-lipoproteins, while others react against a protein fragment, the Pg antigen, common to both α and β-lipoprotein as was clearly demonstrated by Beaumont et al.[14,15] Beaumont also found that the antilipoprotein activity was in the Fab fragment of the immunoglobulin.[13]

A second type of immunoglobulin specificity was found in the serum IgG-x paraprotein of a patient with multiple myeloma.[16] This paraprotein reacts with the LpAS fragment which is found in both α and β-lipoprotein but occurs in only about 6% of the human serum lipoproteins studied.

An IgG type λ paraprotein found in the serum of a patient with multiple myeloma showed binding activity against HDL (high-density lipoprotein), α-lipoprotein, but no activity against β-lipoprotein.[17,18] The serum lipoprotein pattern showed a high β-lipoprotein concentration similar to that of Fredrickson's type II.[53] The hyperlipidemia was apparently not of genetic origin. The investigators felt that the interaction of HDL and paraprotein had characteristics of an antigen-antibody combination.

Cryoglobulinemia is a frequently observed characteristic of sera of patients with autoimmune hyperlipidemia.[10,19,20] Successful reduction of serum immunoglobulin levels of patients with multiple myeloma and autoimmune hyperlipidemia with immunosuppressive drugs has resulted in restoration of serum lipid levels to near normal. Serum lipoprotein patterns were also greatly improved.[19,21] Antibody activity against lipoproteins was not demonstrable during the remission period when normal immunoglobulin levels were present. With exacerbation of the disease, hyperlipidemia and hyperimmunoglobulinemia both occurred. The immunoglobulin-lipoprotein complex appears to interfere with the feedback control of cholesterol synthesis and its secretion, thus leading to the extreme hyperlipidemia observed in some cases.[22]

Occurrence of atherosclerosis and arterial lipid deposition is not prevented by plasma lipoproteins being bound to immunoglobulins.[23,24] IgA and β-lipoprotein were both demonstrated in atherosclerotic plaques in the aorta of a patient who had hyperlipidemia, lipoprotein-IgA complex in his serum.[25] Occurrence of hyperlipidemia and atherosclerosis is unusual in patients with multiple myeloma. However, in patients with autoimmunehyperlipidemia, multiple myeloma is very often the primary disease and atherosclerosis, a common complication.

Low serum lipid and lipoprotein concentrations are generally observed in sera of patients with multiple myeloma.[26] Low lipid levels, however, do not indicate that there is no lipoprotein-immunoglobulin association in their sera.[27-29] Such binding activity against β-lipoprotein has been found in sera of some patients with multiple myeloma who had low lipoprotein concentration. It was suggested that this association of immunoglobulin and lipoprotein may hasten lipoprotein elimination by action of the reticuloendothelial tissue.[28,29]

Autoantibodies against lipoproteins were demonstrated in the sera of five patients with clinical evidence of rheumatoid arthritis, but their sera were negative for rheumatoid factor. Two of the sera reacted against HDL. The reactive site of the Ig molecule, as in the paraprotein of multiple myeloma patients, was the Fab fragment.[30,31]

POSSIBLE BINDING OF SERUM LIPOPROTEINS WITH IMMUNOGLOBULINS

In addition to the lipoprotein-immunoglobulin associations, in which lipoprotein is bound firmly enough to the immunoglobulin to be demonstrable in the position of the Ig by electrophoresis or in the fraction concentrated by ultracentrifugation containing immunoglobulin as well as lipoprotein, another type of interaction has been observed in the sera of three patients with multiple myeloma.[32,33] When the abnormal globulin was added to normal serum, the electrophoretic pattern showed marked alteration in the mobility of the lipoprotein fractions, which were most clearly demonstrated by starch-gel electrophoresis. The mobility of the fast-moving part of the α-lipoprotein was most affected by addition of the abnormal paraprotein. The mechanism or exact nature of the reaction is not understood.

LIPID IN THE IMMUNOGLOBULIN MOLECULE

An abnormal lipid-like material was found in ''myeloma'' protein fractions after electrophoresis of the sera of five out of seven patients with multiple myeloma.[34] The lipid in the ''M'' protein stained an orange color with oil red 0 rather than the usual red color typical of lipoprotein. The lipid was not readily extractable with Bloor's reagent. Faint lipid staining of a paraprotein, γ-type I, was found in the serum of a patient with multiple myeloma with elevated levels of serum lipids.[35] Most of the lipid-stainable material was found in fractions of electrophoretic mobility, typical of normal α- and β-lipoproteins. Clinical manifestations were unique since the patient experienced no bone pain and showed no osteolytic lesions, despite massive plasma cell infiltration of the spleen. There was widespread atherosclerosis. If there was any true association of immunoglobulin and lipoprotein, it was readily broken.

More detailed studies of the nature of the lipid in paraproteins were reported by

Hartmann et al.[38] who found that 2% of the IgM macroglobulin molecule is lipid. Analysis of the lipid of the purified IgM fraction showed that it consisted of cholesterol ester, phospholipid, and triglyceride. It was stainable with Sudan black stain for lipid. In the sera of patients with Waldenstrom's macroglobulinemia in which total macroglobulin concentration may be very high, lipid transport by IgM may have significance. This is especially likely, since total lipoprotein concentration in patients with macroglobulinemia is frequently low.[39] Similar demonstration of lipid in IgG myeloma protein has been reported.[40]

A certain amount of β- and α-lipoprotein may be adsorbed to the immunoglobulin and is removed during isolation of the immunoglobulin. In contrast, the lipid that forms part of the paraprotein molecule is difficult to extract. In those patients with very high IgM or IgG levels, lipid transport by these fractions may be important; lipids may be involved in some of the immunoglobulin functions.

HYPERLIPIDEMIA-HYPERLIPOPROTEINEMIA INDIRECTLY DUE TO AUTOANTIBODY ACTION

Antiheparin Activity Resulting in Accumulation of Chylomicron

Hyperlipidemia associated with autoantibody activity may be the result of interaction of immunoglobulin with some material other than the lipoprotein. A patient with extremely high serum triglyceride levels, chylomicronemia, and a lipoprotein pattern similar to that of type I according to Frederickson's classification[53] was found to have pancreatitis and lupus hepatitis with elevated serum γ-globulin. Detailed metabolic studies established that the patient's serum inhibited postheparin lipoprotein lipase action, thus resulting in accumulation of a high concentration of chylomicron and triglyceride.[41,42] Another example of hyperlipidemia resulting from a molecular complex of immunoglobulin-heparin in a patient's serum was reported.[43] Interaction of heparin with the IgA of the serum resulted in inhibition of antithrombin activity of heparin and of its precipitating effect on lipoproteins. In vivo hyperlipidemia resulted. A multiple myeloma patient's serum which had marked cryoglobulinemic properties [44] showed similar effects on heparin activity; while no effect of the purified IgG could be demonstrated on postheparin lipoprotein lipase activity, the IgG, "M" protein, diminished the heparin-induced prolongation of prothrombin-thromboplastin time. The serum lipoprotein pattern had characteristics similar to those of type III (W.H.O. classification). The patient survived for 17 years after diagnosis of the multiple myeloma.

ATYPICAL LIPOPROTEIN PATTERNS OF PATIENTS WITH MULTIPLE MYELOMA OR MONOCLONAL GAMMOPATHY

Some lipoprotein studies have reported finding a very significant percentage of the total lipid-stainable material migrating with the "M" protein, but the fraction failed to react against antisera to either α- or β-lipoprotein. The nature of this unusual lipid fraction was not identified.[45]

An early, extensive study of lipoproteins in a patient with multiple myeloma showed a five to sixfold increase in level of β-lipoprotein during the course of the disease.[46] Following administration of ^{14}C-labeled L-glumatic acid, radioactive label was demonstrated in β-lipoprotein and other globulins and in the cytoplasm of the plasma cells. The abnormal lipoprotein of the serum had an antigenic component that was equivalent to a similar component in β-2A (present nomenclature IgA), but its significance was not understood at that time.[48] In view of subsequent studies in numerous centers, it appears probable that this was an example of a β-lipoprotein-IgA complex of an antigen-antibody type.

MULTIPLE MYELOMA AND THE NEPHROTIC SYNDROME

The "complexing" of IgA globulin with lipoprotein in the serum of a patient with multiple myeloma and the nephrotic syndrome was felt to be a factor in the development of unusual glomerular changes seen by electron microscopic study of his renal tissue.[47]

GENERAL CONSIDERATIONS

A review of the antibody activity of human myeloma globulins, of which the activity against lipoproteins is an example, was published in 1973.[50] The number of specific antigens against which antibody activity has been demonstrated is large, and it is therefore not surprising that antilipoprotein specificities occur infrequently.

Appreciation of the existence of autoantibodies involving lipoproteins and the effects that they may have on lipid metabolism presents an additional approach to investigations of lipoproteins and their regulation. Electrophoretic techniques are sensitive and useful means of identifying and studying lipoprotein-immunoglobulin associations.

Table 1

ELECTROPHORETIC TECHNIQUES[a]

Author	Year	Medium, Special conditions	Buffer	Cell and special equipment	Electrical conditions	Time of electrophoresis	Temp	Special handling	Staining	Evaluation	Densitometry	Methodological error
Beaumont, J. L., et al.[1]	1965	Electrophoresis on: Paper Agar Agarose Starch gel	Buffers used for each of supporting media were barbital with and without 0.1 mol urea	—	—	—	—	In addition to electrophoresis of serum, lipoprotein fractions isolated at different densities by ultracentrifugation were studied electrophoretically	Sudan black and amido black	Mobilities of fractions and their relative concentrations evaluated	—	—
		Immunoelectrophoresis[a]	Barbital	—	—	—	—	Precipitin arcs examined	—	Shape, position, and concentration of precipitin arcs noted and photographed	—	—
Jacotot, B. et al.[2]	1965	Agarose, Agar-Antisera were developed against lipoproteins and against IgA of patient; commercially available specific anti-β- and anti-α-lipoprotein antisera used in immunoelectrophoresis and immunodiffusion studies	—	—	—	—	—	Lipoprotein fractions isolated at d 1.063 and d 1.21 in ultracentrifuge; protein fractions in subnatant fraction of d 1.21 ultracentrifuge tube were studied	—	—	—	—
Beaumont, J. L. et al.[3]	1965	Ionagar® L, Immunoelectrophoresis[a]	Veronal, pH 8.6	—	—	—	—	—	—	Precipitin lines developed by patient's serum after electrophoresis against anti-β-lipoprotein antisera were more diffuse than those against pure or normal β-lipoprotein	—	—
Beaumont, J. L. Jacotot B. Vilain, C. Beaumont, V. Halpern, B.	1965	Immunoelectrophoresis, ionagar 1%	Veronal, pH 8.6	—	Patient's serum was source of antibody; if patient's serum after removal of lipoproteins from ultracentrifuge at d 1.21 was used as antisera, better precipitin arcs formed	—	—	Precipitin arcs developed in 12—24 hr	Sudan black	Visual and specific color for lipid or protein	—	—
Beaumont, J. L.[4]	1966	IgA-lipoprotein complex separated by ultracentrifugation; when complex subjected to ultracentrifugation in urea at acid pH, IgA sedimented; IgA, thus separated, was studied by immunoelectrophoresis (Grabar), agarose, 1%; and by starch gel	—	Purified β-lipoprotein contains Pβ-antigen common to α- and to β-lipoprotein	—	—	—	—	—	—	—	—

Table 1 (continued)
ELECTROPHORETIC TECHNIQUES[a]

Author	Year	Medium, Special conditions	Buffer	Cell and special equipment	Electrical conditions	Time of electrophoresis	Temp	Special handling	Staining	Evaluation	Densitometry	Methodological error
Beaumont, J. L. et al.[7]	1967	Immunoelectrophoresis Agarose Starch gel	—	—	—	—	—	Antisera specific to patients IgA prepared by author; also antisera to IgG, IgM and IgA (Behringwerke)	—	IgA migrates to γ-position, by combining chromatography, electrophoresis, and double diffusion techniques, interpretation of nature of lipoprotein-IgA complex in patient's serum was determined; no anti-β- or anti-α lipoprotein activity in IgG or IgM of patient's serum	—	—
Lewis, L. A. and Page, I. H.[8]	1965	Free moving boundary, filter paper, starch gel, immunoelectrophoresis; two-dimensional electrophoresis, first dimension on filter paper, second dimension was migration of fractions from filter-paper into starch gel[b]	Phosphate, pH 7.8; barbital, pH 8.6; Tris-borate, pH 8.4; all three buffers were used and tried with each system	Tiselius — Durrum Λ type cell used for filter paper and starch gel; special cell for starch-gel blocks	—	—	Room temp	—	Oil red O for lipid Amido Schwarz® for protein	Visual, special consideration of mobilities and immunological precipitin bands	—	—
Lewis, L. A. et al.[10]	1966	Electrophoretic studies of: 1. Whole serum 2. Cryoglobulin 3. Supernatant after removal of cryoglobulin studies done on paper, starch gel, agarose, and by immunoelectrophoresis; isoelectric point and electrophoretic mobility of purified macroglobulin by free moving boundary technique of Tiselius; pleural fluid of patient also studied	Barbital, pH 8.6	—	—	—	Removal of cryoglobulin at 4°C; electrophoresis carried out at 37°C to prevent precipitation of cryoglobulins	Whole serum kept in incubator at 37°C to prevent precipitation of cryoglobulin	Oil red O and Amido Schwarz	95% of protein-stainable cryoglobulin had mobility γ-globulin, 3% β-globulin; 90% of lipid-stainable cryoglobulin migrated as β-lipoprotein, a trace of α-lipoprotein, and 10% of lipid-stainable material remained at application point in γ-position Isoelectric point of IgM was at pH 6.0 Pleural fluid of patient showed cryoglobulin with properties of IgM-β-lipoprotein, similar to those of serum	Beckman® analytrol	—

Lewis, L. A., and Lazzarini-Robertson, A., Jr.[12]	1974	Paper Immunoelectrophoresis	Barbital, pH 8.6	Room temp	Ultracentrifugally separated at d 1.21 and d 1.063; fractions of the patient's sera were studied by electrophoresis to determine fractions containing immunoglobulins and lipoproteins	Oil red O Amido Schwarz Ponceau R	Frequent electrophoretic studies for protein and lipoprotein during treatment H with immunosuppressive drugs, showed return to nearly normal lipoprotein pattern during remission when serum lipoprotein levels were near normal	Visual inspection of stained pattern and precipitin arcs; IgA found in top fraction of ultracentrifuge preparation which contained lipoproteins and IgA. In case H, lipoproteins and IgG were dissociated by high salt concentration and ultracentrifugation; IgG at bottom of preparation tube had no lipoprotein in this fraction as shown by electrophoresis; this association of lipoprotein-IgG was not broken by electrophoresis; lipoproteins from top fraction of ultracentrifugation preparation had mobility of normal α- and β-lipoprotein
Beaumont, J. L.[13]	1969	Agar gel Agarose Starch gel			Antibody prepared from sera of patients with autoimmune hyperlipidemia; IgA globulins with antilipoprotein activity	Amido Schwarz and Sudan black used for all gels		Definite proof of autoantibody possible if Ig can be separated and studied, although other studies strongly suggest such activity; mobility varied with degree of polymerization of IgA; number of reactive sites of Fab fragment of IgA and IgG against β-lipoprotein and against α-lipoprotein was determined
Beaumont, J. L. et al.[14]	1967	Immunoelectrophoresis in agarose (Graybar); starch gel electrophoresis	Veronal		Specially prepared high-purity β-lipoprotein and α-lipoprotein for hemagglutination studies; purity checked by immunoelectrophoresis and starch-gel electrophoresis	Amido Schwarz Sudan black		Both cases SE and GER had sera reactive against both α- and β-lipoprotein; IgA of SE migrates on starch between slow α_2-macroglobulin Sα_2

Table 1 (continued)
ELECTROPHORETIC TECHNIQUES[a]

Author	Year	Medium, Special conditions	Buffer	Cell and special equipment	Electrical conditions	Time of electrophoresis	Temp	Special handling	Staining	Evaluation	Densitometry	Methodological error
Beaumont, J. L. et al.[13]			—							and application point while that of GER migrates more rapidly; immunoprecipitation arcs of two patients' IgA show identity	—	—
Beaumont, J. L. and Halpern, B.[15]	1967	Immunoelectrophoresis	—	Special purified concentrated IgA from serum of patient with multiple myeloma used as antibody in some of studies; patient had multiple myeloma, hyperlipoproteinemia, and an unusual serum IgA globulin which had antibody activity to α- and β-lipoprotein	—	—	—	—	Sudan black Schwarz			—
Beaumont, J. L. et al.[16]	1970	Physical chemical characteristics of two types of antibodies to lipoproteins studied by immunoelectrophoresis on agarose and electrophoresis on agarose, cellulose, and starch gel	—	IgA with antilipoprotein activity and IgG with antilipoprotein activity prepared from patients' serum (see Table 2); antiserum prepared in rabbits and horse	—	—	—	—	Oil red O Amido Schwarz for protein	All α- and all β-lipoproteins of human serum reacted with purified IgA; purified IgG reacts with both α- and β-lipoproteins LpAS fragment which is present in about 6% of human lipoproteins, may be the common antigenic component with which the immunoglobulins react	—	—
Marien, K. J., and Smeenk, G.[17]	1973	Paper electrophoresis	—	Detailed studies are reported in 1975; see Marien and Smeenk,[18] 1975	—	—	—	—	—	Visual inspection of lipid-stained electrophoretic strips	—	—
Marien, K. J. C. and Smeenk, G.[18]	1975	Paper (Lees-Hatch) immunoelectrophoresis After separation by electrophoresis, anti-HDL antisera introduced to react with isolated myeloma pattern which had also been able to interact with normal HDL	—	Anti-HDL antiserum (Behringwerke); anti-human antiserum and Red Cross transfusion service (Amsterdam)	—	—	—	Whole serum; d 1.006 supernatant	—	Type IIA hyperlipoproteinemia; no floating β-lipoprotein; results showed that if a protein was found in γ-globulin region, probably the paraprotein which had bound HDL	—	—

Reference	Method	Buffer/gel conditions	Sample/antisera	Temperature	Description	Stain	Comments	Results
Perrualt, M. et al.[19] 1971	Protein and lipoprotein electrophoresis of serum	—	—	—	—	Sudan black for lipids Protein stain	Special consideration of migration rates of lipoprotein and abnormal globulin	—
Kodama, H. et al.[20] 1972	Lipoproteins and proteins studied on cellulose acetate and by agar-gel immunoelectrophoresis; lipoproteins also studied on gelatinized cellulose acetate	—	Antihuman antisera (horse)	Cryoglobulin studied at 39°C / 37°C	Separation of cryoglobulin and study of cryoglobulin, serum minus cryoglobulin, and whole serum	Fat red 7B for lipoprotein Ponceau R for protein	Shape and position and fusion of lactescence evaluated in patient's serum in comparison with normals	—
Lewis, L. A. et al.[21] 1973	Immunoelectrophoresis. Cellulose acetate	—	Serum and lipoprotein-immunoglobulin fractions of serum prepared by ultracentrifugation	—	Serum and fractions of serum containing immunoglobulin-lipoprotein complexes were prepared by ultracentrifugation at d 1.063 and d 1.21; they were electrophoresed and then tested against anti-α- and anti-β-lipoprotein, anti-IgG, anti-IgA, and anti-IgM antibodies	Oil red O Amido Schwarz	—	—
Ho, K. J. et al.[22] 1976.	—	—	Lipoprotein and lipid levels were followed after administration of tracer dose of cholesterol-4-^{14}C was administered to patient G; he had autoimmune hyper-IgA-globulinemia, hyperlipoproteinemia; paper electrophoresis, ultracentrifugal quantification of lipoproteins	—	—	Oil red O for lipid Amido Schwarz for protein	—	—
Beaumont, J. L. et al.[23]	Review type of article; no specific electrophoretic or other procedures detailed							
Beaumont, J. L.[24] 1968	Immunoelectrophoresis. Agarose Starch gel Paper	—	Preparations of lipoprotein of human, rat, guinea pig, and rabbit studied and reacted against purified IgA of case I (multiple myeloma — hyperlipoproteinemia, hyper-IgA-globulinemia with autoantibody properties)	—	—	Sudan black for lipid Amido Schwarz for protein	Visual; special attention to migration rate and shape and position of precipitin arc	—
Lewis, L. A. et al.[25] 1975	Paper Starch gel Immunoelectrophoresis* Two-dimensional immunoelectrophoresis	Barbital, pH 8.6, also albuminated for some of serum studies, TRIS-borate was used with starch gel	Durrum Δ type cell	—	In immunoelectrophoresis and two-dimensional electrophoresis, precipitin curves were stained first for lipid, then counterstained for protein	Oil red O Amido Schwarz	IgA levels 1400—3400 mg/dl	Precipitin lines in immunoelectrophoretic patterns of patient's serum lipoprotein and d 1.21 ultracentrifuge preparation against anti-IgA antisera extended

Table 1 (continued)
ELECTROPHORETIC TECHNIQUES[a]

Author	Year	Medium, Special conditions	Buffer	Cell and special equipment	Electrical conditions	Time of electrophoresis	Temp	Special handling	Staining	Evaluation	Densitometry	Methodological error
Lewis, L. A. et al.[25]											from position of β-lipoprotein to that of fast α2-lipoprotein	—
Noseda, G. et al.[27]	1971	Agarose, 0.5% Serum of patient with IgA myeloma	For lipoprotein add 0.5% lyophilized human albumin; veronal, pH 8.6 ionic strength 0.025 and 0.3 % of EDTA	LKB® apparatus	—	—	—	—	Oil red O	Mobility of lipid-stained fractions and relative amounts evaluated	Zeiss® scanner	—
		Cellulose acetate (Gelman®)	Veronal buffer, pH 8.6,μ 0.05						Oil red O	Mobility of lipid-stained fractions and relative amounts evaluated	Zeiss scanner	
		Immunoelectrophoresis, Difco®-Bacto® agar, special noble	Veronal sodium lactate	Antisera, Behringwerke, Marberg/Lahn, or Swiss Central Laboratory Tumor Research					Oil red O	Mobility of lipid-stained fractions and relative amounts evaluated	Zeiss scanner	
Noseda, G. et al.[28]	(2nd) 1971	Agarose gel		Binding of lipoprotein with immunoglobulin further studied by immunodiffusion and passive hemagglutination	Conditions according to Noble[24c]	—	—	—	Oil red O	—	Scanned patterns on Zeiss integrator	—
Riesen, W. et al.[29]	1972	Lipoprotein electrophoresis on agarose plate	Veronal, pH 8.6, 0.025 mol containing 0.3% EDTA; 0.5% human serum albumin	Gelman electrophoresis cell	—	—	—	—	Oil red O	—	Scanned	—
		Isolation of immunoglobulins, Pevikon polyvinyl chloride support; immunoelectrophoresis for purity of fractions		Zone electrophoresis	—	—	—	—	—	—		—
Noseda, G. et al.[30]	1972	Lipoprotein agarose (Noble[24c]) Immunoelectrophoresis;[124]-LDL prepared and used for special identification of fractions			—	—			Oil red O	Examination of precipitin arcs and mobilities; antibody activity against α-lipoprotein independent of that against β-lipoprotein, since addition of β-lipoprotein did not block anti-α-lipoprotein activity and vice versa	—	—
Cohen L. A., et al.[32]	1966	Agar gel Starch block Starch gel Paper Immunoelectrophoresis Special study with delipidized patient's serum; lipoproteins removed by ultracentrifugation after freeze-thawing patient's serum; one part delipidized patient's serum was			—			Lipid analysis of eluted fractions; anti sera to whole human serum, to IgA, to IgG, and to IgM used for immunoelectrophoresis	Oil red O stain for lipids of resolved lipoprotein fractions	No lipid stain of paraprotein was found; no real association of lipoprotein-paraprotein found, but abnormal globulin could alter electro-		—

Reference	Year	Method	Buffer	Apparatus / Antisera	Applied / Materials	Time / Conditions		Sample	Stain	Results	Comments	
Spikes, J. L., Jr. et al.[33]	1968	mixed with two parts of normal serum; five different normal sera were tested with patient's delipidized serum; electrophoresis was conducted on starch gel, of mixture; resulting patterns showed disappearance of most rapidly migrating α-lipoprotein band — Paper Agar gel Starch gel (vertical)	pH 8.4	Vertical cell for starch gel[33a]	500 μl of d < 1.21 fraction applied for starch gel	22—60 hr for starch gel depending on amount of sample applied and degree of resolution desired	—	Paraprotein recovered from application slot; 2.5 mg pure paraprotein	Oil red O for lipid Amido black for protein	Paraprotein did not migrate into starch gel	Purity checked by immunoelectrophoresis and agar and starch-gel electrophoresis; phoretic mobility of fast α-lipoprotein	—
Sachs, B. et al.[34]	1954	Paper	Veronal, pH 8.6, μ 0.1	—	140 V 3.4—4 mA per 7 strips	—	—	0.0125 ml serum applied	Oil red O for lipids Napthalene blue 121 B200 for protein	Optical densitometry of stained strips	—	—
Levin, W. C.[35]	1964	Paper Ultracentrifuge (analytical), d 1.063 Immunoelectrophoresis	Barbital, Spinco® Procedure B[34] (1958)	Durrum type cell	—	—	—	—	Oil red O; most of lipid-stainable material migrated with mobility of normal lipoproteins	—	—	—
Hartmann, L. et al.[38]	1968	Serum lipoprotein preparations separated by ultracentrifugation at d 1.063 (β-lipoprotein) d 1.21 (α-lipoprotein), and delipidized bottom fraction; purified preparations of M. W. (macroglobulin of Waldenstrom) also prepared from patient's serum; electrophoresis on cellulose acetate and filter paper and immunoelectrophoresis	—	Anti- whole human serum, anti-α-lipoprotein, anti-β-lipoprotein, anti-IgG, anti-IgA, anti-IgM, anti-α₁-M, and anti-Hp	—	—	—	Amido Schwarz for protein; Sudan black for lipid; PAS (polysaccharide)	Lipid analysis of IgM fraction as well as evaluation of dye uptake by this fraction; lipids of IgM consist of cholester, phospholipid, and triglyceride; lipoprotein conc of M. W. sera usually low	IgM molecule has lipid content of approximately 2%; therefore, when IgM level high as in M. W., significant lipid transport occurs in this fraction; this lipid transport is in addition to adsorbed β- and α-lipoprotein	—
Glueck, C. J. et al.[a]	1969	Paper	Barbital, albuminated	—	—	Before and after treatment of patient with special diet	—	—	Oil red O	Visual comparison with normal and hereditary type I serum	—	—
Beaumont, J. L., et al.[43]	1970	Agarose electrophoresis	Barbital, pH 8.6	—	—	—	—	—	—	IgA migrated on agarose at pH 8.6 to γ-globulin; when purified, IgA showed antithrombin activity	—	—
Wilson, D. E. et al.[44]	1975	Blood samples collected after patient had fasted for 12—15 hr; plasma lipoproteins were studied by electrophoresis on: agarose gel	—	Monospecific antisera against purified β-lipoprotein and against IgG of patient I[a] were prepared in rabbits	Materials studied: 1. Whole plasma 2. VLDL supernatant Apolipoproteins	—	—	Mobility calculated by comparison with that of marker dye, and that of apolipo-	—	—	—	—

Table 1 (continued)
ELECTROPHORETIC TECHNIQUES[a]

Author	Year	Medium, Special conditions	Buffer	Cell and special equipment	Electrical conditions	Time of electrophoresis	Temp	Special handling	Staining	Evaluation	Densitometry	Methodological error
Wilson, D. E. et al.[44]		polyacrylamide gel lipoprotein X (LpX), a protein found in obstructive liver disease' was also studied using LpX antisera (Immuno-Diagnostic ® antiserum, Vienna, Austria) and immunoelectrophoresis			after delipidation with tetramethylurea			protein fractions from normals; lipoprotein-free plasma of patient 2' had no cryoglobulin properties, but addition of normal VLDL- or LDL-restored cold precipitable properties to lipoprotein-free preparation of patient's plasma				
Koga, S. et al.[47]	1974	Agarose 1.0% Immunoelectrophoresis Acrylamide 7.5%	Veronal, 0.05M, pH 8.6	—	According to Noble Electrophoresis of serum and ultracentrifugal fractions: d <1.21 / d >1.21 —	—		—	—	No evidence of α or β-lipoprotein (immunologic) in γ-globulin lipid-stained fraction	—	—
Neufeld, A. H., et al.[48]	1964	Immunoelectrophoresis; ultracentrifugation at d 1.063 for evaluation of lipoprotein	—	—	—	—		^{14}C-glutamic acid given to patient at one stage to determine incorporation of label in lipoprotein compared with other globulins		Abnormal β-lipoprotein of patient had antigenic component that was equivalent to similar component in β-2A, in present nomenclature IgA, as well as in γ-globulin; the significance of this was not understood	—	—
Rosen, S. et al.[49]	1967	Immunoelectrophoresis, modification of Scheidegger Zonal electrophoresis 1. Filter paper 2. Agar (Lees-Hatch)[a] 3. Starch	—	—		—		Immunodiffusion	Thiazine red R[?] for protein; oil red O for lipid	Visual inspection; mobilities compared with those of normal lipoproteins	—	—
Seligman, M. and Brouet, J. C.[50]	1973	Study of lipoprotein fraction separated in ultracentrifuge at d 1.063 Review type of article										

Review type of article; no specific electrophoretic or other procedures detailed

| Kayden, H. J.,[d] | 1976 | Starch (granular) Agar (immuno-electrophoresis) | — | Not described | Not described | — | — | — | Immunologic studies showed γ-globulin to be γ-G₁ in both patients and light chains λ; Fab fragments of γ-globulin obtained by papain digestion, bound lipoprotein; similar findings for patients 1 and 2; γ-globulin of serum after electrophoresis reacted with anti-β-lipoprotein antisera, and mobility of lipoprotein fraction was slower than normal, although not identical with γ-globulin mobility | — | Estimation of position of bands and precipitin arcs of immunoelectrophoretic patterns; lipoprotein fractions from ultracentrifuge were resolved using D_2O rather than KBr to increase density, because it yielded a larger proportion of lipid still associated with γ-globulin; thus, a high salt concentration aided dissociation of lipoprotein-IgG complex | — | — |

[a] These abbreviations will be used in the table: HDL = high-density lipoprotein; μ = ionic strength; LDL = low-density lipoprotein; PAS = periodic acid-Schiff stain; chol = cholesterol; VLDL = very low-density lipoprotein.
[b] See chapter "Measurement of lipoprotein in arterial wall by quantitative immunoelectrophoresis directly from the tissue into an antibody- containing gel" by E.B. Smith.
[c] See chapter "Changes in the plasma lipoprotein system due to liver disease" by H. Wieland and D. Seidel.
[d] See Table 2 for details.

Table 2
APPLICATIONS OF ELECTROPHORESIS[a]

Author and ref. no	Year	Pattern	Type of character	Normal values	±SD	Reproducibility	Pathology	±SD	Characteristics of subject, age, sex, race	Summary
Beaumont, J. L., et al.[a]	1965	Protein: IgA-κ type multiple myeloma, lipoprotein: increased β-lipoprotein	Paraprotein migrated as β, on paper and agar, and between β, and β, on agarose; some IgA precipitation on agar gel during electrophoresis and anti-β-lipoprotein antibody activity in IgA paraprotein; some IgA floated with β-lipoprotein at d 1.063 and d 1.21; α-lipoprotein precipitin arc after electrophoresis of d 1.21 supernatant extended from point of application to α-lipoprotein position	—	—	Flotation of β-lipoprotein incomplete at d 1.063 and of α-lipoprotein incomplete at d 1.21	Multiple myeloma with hyperlipidemia, tuberous and disseminated papule xanthoma; angina pectoris during exercise; slow vitamin A turnover	—	Age, 57 years	A patient with multiple myeloma, IgA-κ type proteinemia with hyperlipoproteinemia was reported; IgA globulin showed antibody activity against β-lipoprotein
Jacotot, B., et al.[c]	1965	By immunoelectrophoresis and Ouchterlony double diffusion, it was shown that IgA reacted with both β-lipoprotein and α-lipoprotein	—	—	—	—	IgA multiple myeloma with hyperlipidemia	—	57-year-old male	—
Beaumont, J. L. et al.[d]	1965	—	—	—	—	—	Results suggest that patient's serum contained antibody against β-lipoprotein that cannot be due to previous transfusions	—	57-year-old male, SG	Presence of the lipoprotein-IgA complex in the serum may alter catabolism of lipoprotein and explain the hyperlipidemia
Beaumont, J. L. et al.[e]	1965	—	γ,A paraprotein autoantibody in blood in soluble form — not an antibody resulting from transfusions as patient had received none; when tested by immunoelectrophoresis using normal human serum and patient's serum as antisera, arc of precipitation formed in position of β-lipoprotein which stained strongly with Sudan black	—	—	—	Multiple myeloma with hyperlipidemia and xanthomatosis; when tested by immunodiffusion using patient's serum as antibody source, precipitin arc formed against all human β-lipoprotein tested; no reaction against α-lipoproteins of some subjects	—	57-year-old male patient with multiple myeloma	Study of patient's serum with γ-A paraproteinemia showed association of paraprotein with β-lipoprotein of serum; isolated γ,A protein, when used as antibody, formed precipitin arcs against all human β-lipoproteins but showed no action with α-lipoprotein as antigen
Beaumont, J. L.[b]	1966	Antibody activity of IgA to both α- and β-lipoprotein; verified by inhibition of either α- or β-lipoprotein	By immunoelectrophoresis of IgA from patient's serum, two forms of IgA differing in dimensions of molecule were demonstrated. The myeloma IgA protein had specific antibody activity, anti-Pg	—	—	—	γA myeloma with hyperlipidemia and xanthomatosis	—	Male patient with multiple myeloma	A patient with multiple myelomas serum was demonstrated to have antibody activity in IgA (myeloma) protein which was specific against Pg component; common to both α- and β-lipoprotein, IgA found in two forms: the larger, a polymer of the smaller in the patient's serum
Beaumont, J. L.[f]	1967	—	Protein and lipoprotein patterns same as case reported in 1966[b]	—	—	—	Second patient with multiple myeloma, IgA-type protein	—	—	Antibody activity against Pg similar to that of Beaumont[b] (1966)
Lewis, L. A. and Page, I. H.[g]	1965	The serum lipoprotein pattern showed a very concentrated fraction with mobility of a fast β- or a slow α₂-globulin and an α₁-lipoprotein of α low conc; IgA globulin and lipoproteins were firmly associated and concentrated in the d < 1.21 fraction by ultracentrifugation	Free moving boundary; large peak in fast-β position poorly resolved from α-globulin; paper electrophoretic patterns stained for lipid showed an intensely stained band between α₁- and β-fractions; also stained intensely with protein stain	—	—	—	Severe hyperlipidemia xanthomatosis; hyper-IgA-globulinemia complexed with lipoprotein; increased plasma cell concentration in bone marrow but no clinical evidence of multiple myeloma	—	Male, 48 years, when study started (patient G)	Serum proteins of male patient with severe hyperlipoproteinemia for at least 11 years duration contained 18% of α globulin with sedimentation constant of 12 and immunoelectrophoretic IgA; this globulin was shown by electrophoresis and ultracentrifugation to be complexed with lipoproteins
Lewis, L. A. and Lazzarini-Robertson, A., Jr.[i]	1974	Serum lipoprotein pattern of Mr. G. has remained very concentrated, lipidstained band of slow-α₂- or fast-β mobility; lipid is firmly bound to IgA globulin; IgA and β-lipoprotein in atherosclerotic plaques of arterial tissue taken at time of operation; Patient H had a very concentrated lipid-	—	—	—	—	As summarized by (Lewis and Page)[g] Mr. G. had developed peripheral atherosclerosis and an aortic aneurism which was operated on successfully; patient H had multiple myeloma which had been recognized for 5 years, and also had diffuse xanthomatosis	—	Male patient G, 48 years when first examined; female patient H, 67 years in 1967 when first seen at Cleveland Clinic	Association of immunoglobulin with lipoprotein did not prevent lipid deposition in atherosclerotic plaques or in xanthomas

Table 2 (continued)
APPLICATIONS OF ELECTROPHORESIS[a]

Author and ref. no	Year	Pattern	Type of character	Normal values	±SD	Reproducibility	Pathology	±SD	Characteristics of subject, age, sex, race	Summary
Lewis, L. A. and Lazzarini-Robertson, A., Jr.[62]		stained band migrating in position of IgG, with which it was complexed; β-lipoprotein and igG also found in atherosclerotic plaques								
Lewis, L. A. et al.[10]	1966	Serum lipoprotein pattern showed low conc of β-lipoprotein, very low α-lipoprotein, and spike in position of IgM globulin	β-lipoproteins precipitated with cryoglobulin; these patterns were obtained by doing analyses at 37°C	—	—	—	Patient with lymphomatous-type disease and marked cryoglobulinemia, cryoglobulin precipitated at 32°C from serum and pleural fluid; after removal of lipoprotein from serum or cryoglobulin precipitate by ultracentrifugation, cryoproperties were lost, addition of β-lipoprotein restored cryoprecipitate	—	—	Serum contained a cryoglobulin which precipitated at 32°C; it was composed of IgM, 19S macroglobulin, and β-lipoprotein; property of cold precipitibility was dependent on presence of β-lipoprotein, α-lipoprotein was not found in cryoprecipitate
Beaumont, J. L.,[13]	1969	Two patients' purified IgA globulin was digested with papain and the reactivity of Fab fragment with α- and β-lipoprotein, determined; results indicate that IgA has autoantibody activity against both α- and β-lipoprotein; Fab of purified IgG also after papain digestion showed reactivity against both α- and β-lipoprotein	—	—	—	—	Association of lipoproteins with IgA and IgG globulins has been observed more frequently than with IgM; IgM may fix about 2% of lipid (not lipoprotein)	—	Two patients with hyper-IgA-hyperlipidemia, and one with hyper-IgG-hyperlipidemia were studied	The reactive site against lipoproteins in the immunoglobulin was in the Fab fragment of IgA of the first two patients and of IgG of the one who had hyperlipidemia-hyperglobulinemia, showing immunoglobulin-lipoprotein interaction
Beaumont, J. L. et al.[14]	1967	—	Hyperlipidemia, hyper-IgA-globulinemia	—	—	Studies using immunoelectrophoresis, hemagglutinins, and immunoprecipitation to identify and characterize antigenic reactive site	Sera of two patients with multiple myeloma, IgA-type, were extensively evaluated to determine nature of IgA and its antibody activity to α- and β-lipoproteins	—	—	Antigen in α- and β-lipoprotein to which IgA of patients reacts is given the name Pg; it has been found in 85 human sera studied and in some other species
Beaumont, J. L. and Halpern, B.[15]	1967	—	Purity of lipoprotein preparations checked by immunoelectrophoresis	—	—	—	Autoantibody to both α- and β-lipoprotein; best demonstrated by hemagglutination and inhibition	—	—	Pg antigen in α- and β-lipoproteins of human serum and also in rat, dog, and rabbit; concentrated IgA of patient used as antibody will give precipitin lines to α- and β-lipoprotein of human serum after electrophoresis
Beaumont, J. L., et al.[16]	1970	—	IgA molecules of two patients with autoimmune hyperlipidemia, anti-Pg type, is κ light-chain type; β-lipoprotein has about 64 reactive sites and α-lipoprotein, about 20; the site is Pg	—	—	—	Two patients with IgA-type κ- globulinemia — purified IgA of the two patients will react with lipoproteins of all human serum tested; one patient with multiple myeloma IgG-κ light chain which reacts with LpAS — LpAS is found in both α and β-lipoprotein, but in only 3 of 50 sera tested	—	—	In three patients with multiple myeloma, hyperlipidemia is due to accumulation of soluble antigen-antibody complexes of lipoprotein-immunoglobulin; clearance of the complex is slowed due to direct or indirect inhibition of lipolysis; IgA globulin has anti-LpPg activity while IgG has anti-LpAS activity
Marien, K. J. and Smeenk, G.[17]	1973	IgG-type paraprotein; serum lipoprotein type IIA,[33] chol 485 mg/dl: binding of HDL and IgG demonstrated by double diffusion in agar gel	No cryoglobulin; binding activity of IgG protein against HDL was demonstrated; no synthesis of myeloma globulin by skin in culture	—		—	Multiple myeloma treated for 5 years with intermittent schedules of Alkeran® and prednisone; hypercholesteremia in patient apparently not familial; developed planar xanthomata within 1 month; results suggested that HDL-IgG interaction is autoimmune type of reaction	—	62-year-old female	Patient with multiple myeloma developed planar xanthomata in short period; serum showed a myeloma protein IgG-λ that showed binding activity against HDL
Smeenk, G.[18]	1975	Hyper-β-lipoproteinemia normal α- and pre-β-lipoprotein by paper electrophoresis; total serum chol, 485 mg/dl TG, 228 mg/dl, PL, 457 mg/dl	IgG- λ myeloma type IIA hyperlipoproteinemia; IgG showed binding against HDL	—	—	—	Plane xanthomatosis, multiple myeloma, hyperlipoproteinemia; known multiple myeloma, for 5 years; treated with intermittent schedules of melphalan and prednisone; serum chol, markedly elevated, TG and PL, only	—	62-year-old white female whose family showed normal lipid levels and lipoprotein patterns	Results indicate binding occurred between IgG and HDL lipoprotein; explanation as to how this interaction results in type IIA lipoprotein pattern is not available

Table 2 (continued)
APPLICATIONS OF ELECTROPHORESIS[a]

Author and ref. no	Year	Pattern	Type of character	Normal values	±SD	Reproducibility	Pathology	±SD	Characteristics of subject, age, sex, race	Summary
							slightly increased; no Bence Jones protein in urine; 98% plasma cells contained IgG, cultured skin showed synthesis of IgG-λ; no such synthesis in normal skin			
Perrault, M. et al.[]	1971	Marked increase in IgG cryoglobulin and abnormal globulin, stained with lipid stain (Sudan black)	—	—	—		Plane xanthomatosis, increased β-lipoprotein with trailing from α-lipoprotein position		53-year-old female	Female patient with plane xanthomatosis showed association of lipoprotein with IgG globulin, with immunosuppressive treatment, lesions regressed and abnormal lipid-stained band disappeared
Kodama, H. et al.[]	1972	IgG-type κ myeloma protein cryoglobulin thought to be product of autoimmune reaction between IgG and lipoproteins; reaction occurs in vitro at low temp	Whole serum lipoprotein pattern showed high β-pre-β-, normal α-lipoprotein, high TG, and relatively high chol; serum lipoprotein after cryoglobulin removed showed no pre-β-lipoprotein; α- and β-lipoprotein both reduced in conc; autoantibody not identical with that developed in rabbits or other species to normal β-lipoprotein since there was only partial fusion of precipitin lines; autoantibody was not species specific since it reacted with rat and rabbit sera	—	—	—	Plane xanthomatosis with complicated IgG monoclonal gammopathy, hyperlipidemia, and cryoglobulinemia; no manifestation of multiple myeloma in bone marrow or bone X-rays		41-year-old female Japanese	Japanese female, 41 years old, with plane xanthomatosis and cryoglobulinemia was found to have IgG-type κ monoclonal gammapathy with autoimmune properties against α- and β-lipoprotein; reaction was not species specific, since it also reacted against rat and rabbit sera
Lewis, L. A. et al.[]	1973	Mr. G's IgA globulin was bound to VLDL and LDL fractions; only trace amt. of HDL present; all of Mrs. H's lipoproteins migrated with IgG globulin in γ-position.	Despite attempts to regulate serum lipid levels of Mr. G by diet and diet plus drugs, levels stayed high; IgA levels were consistently high for 20 years, during remission of disease following plasmapheresis and L-phenylalanine mustard therapy, IgG and lipoprotein levels of Mrs. H became normal	—	—	—	Patient G had xanthomatosis, peripheral atherosclerosis, autoimmune hyperlipidemia, and hyper-IgA-globulinemia; patient S had multiple myeloma, diffuse xanthomatosis, and extreme hyperlipoproteinemia with very high levels of serum IgG-type κ		Patient G, male, 48 years old when first studied, 18 years before this report	Reduction of serum lipid and IgG levels to normal occurred in a patient with multiple myeloma autoimmunohyperlipidemia during remission by immunosuppressive therapy failure to achieve desirable lipid or IgA levels in patient with autoimmunohyperlipidemia by attempts to regulate serum lipid levels; findings suggest that in autoimmune type hyperlipidemia regulation of immunoglobulin levels may be a more effective way of controlling hyperlipidemia than using hypolipidemic diet or diet plus hypolipidemic drugs.
Ho, K. J. et al.[]	1976	Lipoprotein and IgA firmly associated and migrated on paper electrophoresis to fast β- slow α₂-glogulin position	—	—	—	—	Cholesterol metabolic study of patient with autoimmune hyper-IgA-globulinemia-hyperlipidemia showed rapidly exchangeable pool, 2563% of normal controls, and a slowly exchangeable pool, 654% of normal	—	—	Extreme hyperlipoproteinemia and expanded body cholesterol pools were primarily due to lack of adequate feedback control of cholesterol synthesis as a consequence of lipoprotein-IgA complexing

Beaumont, J. L. et al.[] 1972

Review of atherosclerosis that emphasizes immune mechanisms which may influence atherosclerotic development and course by:

1. A fixation of soluble immune complexes, rich in lipid, in the avascular area of blood vessels (inner intima and media of larger arteries) as well recognized in some autoimmune hyperlipidemias
2. Possible lesions in artery wall due to anti-artery antibodies
3. In hyperimmunized individual, antibody-lipoprotein complexes developing, along with increased immunoglobulin and hyperlipidemia

Author and ref. no	Year	Pattern	Type of character	Normal values	±SD	Reproducibility	Pathology	±SD	Characteristics of subject, age, sex, race	Summary
Beaumont, J. L.[]	1968	Case 1 — γ- globulin spike, immunoelectrophoresis of IgA; starch gel mobility of slow α₂- globulin when γ-globulin purified and tested by immunoelectrophoresis; hemagglutination and immunodiffusion indicate antibody activity against β-lipoprotein	—	—	—	—	Anti-β-lipo autoantibody and hyperlipoproteinemia in two patients; both had clinical evidence of ischemic heart disease; one had IgA myeloma		Case 1 — 58-year-old male with multiple myeloma; Case 2 — 38-year-old male with peripheral atherosclerosis	Anti-β-lipoprotein autoantibody and hyperlipidemia found in two patients (men); one had IgA myeloma and both had ischemic heart disease; results suggest that "autoimmune" hyperlipidemia is a pathological entity of possibly great significance

Table 2 (continued)
APPLICATIONS OF ELECTROPHORESIS[a]

Author and ref. no	Year	Pattern	Type of character	Normal values	±SD	Reproducibility	Pathology	±SD	Characteristics of subject, age, sex, race	Summary
		Case 2 — lipoprotein migrates poorly on paper and agarose; high β-lipoprotein does not migrate into starch gel; immunoreactive globulin appeared to be IgA								
Lewis, L. A et al.[25]	1975	Lipoprotein pattern continues to show very concentrated fraction of slow α_1-mobility	During treatment with Atromid-S® (clofibrate), α-lipoprotein conc increased	—	—	—	Xanthomatosis, peripheral atherosclerosis; β-lipo and IgA globulin demonstrated by immunoelectrophoresis in atherosclerotic plaque of resected artery; present in both internal and medial layers	—	Male patient, 47 years-old when first examined; studied for 21 years at time of present report (Mr. G)	Association of lipoproteins with IgA globulin in the serum of this patient with hyperlipidemia hyper IgA globulinemia did not prevent development of atherosclerotic lesions, although it might have retarded the process since it became manifested only after many years of known hyperlipidemia
Noseda, G. et al.[27]	1971	Lipoprotein pattern of patient showed well-resolved β_1-, pre-β-, and α-lipoprotein, and, in addition, a lipoprotein band with mobility of IgA (M) globulin; β-lipo slightly decreased; after treatment with alkylating agents, myeloma "M" protein, lipid-stained band disappeared		—	—	—	IgA myeloma with hypolipidemia; paraprotein precipitated β-lipoprotein and VLDL, but not HDL; paraprotein agglutinated RBC's coated with VLDL or β-lipo, but no agglutination of cells coated with HDL hemagglutination and hemagglutination inhibition studies demonstrated that purified IgA acted like antibody against VLDL and LDL (a β-lipoprotein) of pool of lipoprotein from normal serum and also patient's serum	—	Patient with IgA multiple myeloma	Patient had a multiple myeloma with IgA paraprotein having autoantibody activity against VLDL and LDL; lipid-stained band with mobility of IgA was present in serum lipoprotein pattern in addition to β- pre-β- and α-lipoproteins; the lipid paraprotein band disappeared with remission of the disease
Noseda, G. et al.[28]	1971	In patient group hypolipidemia with low β-lipoprotein and α-lipoprotein normal	Binding activity against β-lipo detected in patient's sera by passive hemagglutination of β-lipoprotein-coated human erythrocytes	—	—	—	Patients had either multiple myeloma or macroglobulinemia of Waldenstrom; had not received blood transfusions	—	23 patients and 38 control subjects	Binding activity against β-lipoprotein in sera of 23 patients with multiple myeloma or Waldenstrom's macroglobulinemia was demonstrated by means of hemagglutination of β-lipoprotein-coated human erythrocytes; this occurred despite low β-lipoprotein levels, and it is suggested that this association of immunoglobulin-lipoprotein may hasten its elimination
Riesen, W. et al.[29]	1972	Activity against β-lipoprotein demonstrated in Fab fragment of immunoglobulin molecule of myeloma sera	Antibody-like activity against LDL in about 10% of fresh sera of patients; no reaction against HDL of human or LDL of animals; no specificity shown against isoantigens of Ag system; LDL contains the genetically independent polymorphisms, Lp system and Ag system	—	—	—	Human sera containing monoclonal immunoglobulins; anti-β-lipoprotein activity evaluated by passive hemagglutination of LDL- coated human erythrocytes	—	Sera from patients with monoclonal immunoglobulins	Antibody activity demonstrated in Fab fragment of immunoglobulin against β-lipoprotein in about 10% of myeloma or macroglubulinemia sera tested
Noseda, G. et al.	1972	Lipid-stained band migrating to γ-globulin position in sera of two patients	In five cases autoantibodies against LDL, homologous or autologous, were demonstrated by hemagglutination of LDL-coated erythrocytes; active part of Ig molecule is Fab fragment; in two of five cases, there were autoantibodies against HDL, the reactive site of which was the apoprotein; rheumatoid patients with + RF activity showed no antilipoprotein activity	—	—		Four patients studied had polyclonal gammopathy and one monoclonal; all had absence of rheumatoid factor and hypo-β-lipoprotein; when treated with aspirin, the decrease in Ig resulted in increase of β-lipoprotein observed and in one case, of anti-HDL autoantibody; HDL also increased		27 patients with rheumatoid arthritis were selected; 22 had positive RF (rheumatoid factor activity and five had negative RF activity; RF is an antibody against IgG, known as the rheumatoid factor (RF) and can be demonstrated in the serum of 85 - 90% of cases with rheumatoid arthritis; RF is a 19S IgM globulin	Antibody activity in serum against β-lipo in five rheumatoid patients and against α-lipoprotein in two of five was reported; activity was localized to Fab fragment

Table 2 (continued)
APPLICATIONS OF ELECTROPHORESIS[a]

Author and ref. no	Year	Pattern	Type of character	Normal values	±SD	Reproducibility	Pathology	±SD	Characteristics of subject, age, sex, race	Summary
Cohan, L. et al.[12]	1966	Hyperlipidemia aggravated by high carbohydrate diet	γ-globulin type λ myeloma; total protein of serum, 9 - 12 g/dl	—	—	—	55-year-old female patient with hyperlipoproteinemia and xanthomatosis who subsequently developed multiple myeloma; hyperlipidemia was familial; patient and her sister had lipoprotein characteristics, suggestive of type III (not a described type in 1964)	—	55-year-old white female patient	Despite careful study, no association of paraprotein and lipoprotein found; addition of patient's delipidized serum to normal serum resulted in altered lipoprotein mobility on starch gel
Spikes, J. L. Jr. et al.[33]	1968	Purified myeloma globulin obtained by starch-gel electrophoresis	Type II-IgA myeloma globulin; lactescent serum, TG, 1450 mg/dl; Chol and PL, low to normal; β-lipoprotein and IgA myeloma protein migrated separately on agar gel and on filter paper	—	—	—	63% plasma cells in bone marrow; no Bence Jones protein in urine; multiple osteolytic bone lesions; pattern typical of idiopathic hyperlipidemia with most of lipid in d < 1.019 fraction, presently 1976 called type III; this patient's fraction had similar effect on mobility of lipoproteins as that of Cohen et al.[12]; a third of patients with multiple myeloma, whose serum was provided by J. L. Beaumont, had similar effect	—	59-year-old white farmer	Addition of purified globulin to normal serum resulted in: 1. failure of β- lipoprotein to migrate into starch gel, 2. protein staining of β₁-globulins and some haptoglobin region proteins being reduced, 3. lipid-paraprotein mixture remaining at application point and failing to react with β-lipoprotein antisera but reacting with paraprotein antisera (nature of reaction not clear), and 4. similar activity in d <1.006 chylomicron fraction in hypertriglyceridemic sera
Sachs, B. et al.[34]	1954	Normal lipoprotein; faint lipid-stained area from application to beyond β	—	—	—	—	In five of seven myeloma patients' sera, additional lipid-stained band with mobility of abnormal globulin; stained "M" fraction tinged more orange than normal lipoprotein; abnormal fraction still lipid stainable after two extractions of paper strip with Bloor's detergent; normal lipoproteins were no longer lipid stainable	—	12 patients with multiple myeloma; two subjects had high and three, low chol	Abnormal lipid-like material observed in myeloma protein fraction which was not extractable with Bloor's reagent when studied by paper electrophoresis
Levin, W. C. et al.[35]	1964	Myeloma protein, γ type I, and elevated lipid levels	Part of lipid in γ-paraprotein on paper electrophoresis; faint staining of γ with oil red O; readily separable by immunoelectrophoresis or ultracentrifuge	—	—	—	Widespread, severe atherosclerosis at autopsy; massive plasma-cell infiltration of spleen; pleural xanthomatosis; no osteolytic lesions; clinical manifestations were unique since there was no bone pain or osteolytic lesions; hyperlipidemia, but not on basis of myeloid nephrosis.	—	58-year-old white male, nonfamilial hyperlipidemia	Results suggest either a separate γ-paraprotein and lipoprotein or a complex of both that is very easily broken
Hartmann, L. et al.[36]	1968	Generally low levels of α- and β-lipoprotein in macroglobulinemia of Waldenstrom's sera (M.W.)	IgM fraction stains with Sudan black and if IgM sufficiently concentrated, dye binding is significant; total lipid of IgM, about 2% of molecule	—	—	—	—	—	Six normal blood donors; three patients with M.W.	In hyper-IgM-globulinemic patients, lipid transport by IgM may be significant factor, since lipid is 2% of IgM molecule; since immunoglobulins are involved in lipid transport, likely that lipids may become involved in some of immunoglobulin functions; similar binding of lipid to IgG myeloma protein has been reported[37] as partly a loose bonding and partly a firm incorporation with the protein
Glueck, C. J. et al.[41]	1969	Paper — very concentrated chylomicron band at origin	Typical of type I hyperlipidemia	—	—	—	Lupus erythematous hypertriglyceridemia; hyperlipoproteinemia type I	—	—	A case of postheparin lipoprotein lipase deficiency; hyperchylomicronemia with associated lupus erythematosus is described; suggested that possibly because of dysglobulinemia case LE was associated with abnormal heparin binding that linked the coagulation defect and hyperlipoproteinemia; resistance to prolongation of thrombin time by

Table 2 (continued)
APPLICATIONS OF ELECTROPHORESIS[a]

Author and ref. no	Year	Pattern	Type of character	Normal values	±SD	Reproducibility	Pathology	±SD	Characteristics of subject, age, sex, race	Summary
										heparin could be shown in vivo and in vitro; the 7S globulin of patient's serum selectively bound heparin
Beaumont, J. L. et al.[41]	1970	IgA migrated on agarose at pH 8.6 to γ_1-position; it formed molecular complexes with heparin at a site that prevents hydrolysis, thus preventing anticoagulant power; hyperlipidemia resulted from molecular complex of IgA-heparin in patient's serum	—				Hyperlipidemia with chol 300 - 600 mg/dl, TG, 250 — 2,000 mg/dl; IgA-γ-immunoglobulinemia with antiheparin properties compatible with antibody activity; no evidence of myeloma	—	54-year-old female	IgA, when purified, inhibited anti-thrombin activity of heparin and its precipitation of lipoproteins
Wilson, D. E. et al.[44]	1975	Patient 1 — electrophoresis of normal α- and β-lipoprotein; poorly separated pre-β- and floating β-lipoprotein at d 1.006 on ultracentrifuge; studies show increased S_f 10 — 20 and S_f 20 — 400; no interaction of lipoproteins with myeloma protein was demonstrable; patient showed marked decrease in postheparin lipoprotein lipase activity (PHLA), but no effect of purified myeloma protein on PHLA could be demonstrated; half lives of both LDL and VLDL apoproteins was prolonged; "M" protein diminished heparin-induced prolongation of prothrombin thromboplastic time Patient 2 — normal α- and β-lipoprotein bands; faint pre-β-lipoprotein; no floating β-lipoprotein at d 1.006	Cryoprecipitation at 20°C; redissolved in 0.15 M NaCl at pH 7.4; IgG myeloma protein; lipoprotein characteristics similar to those of type III Cryoprecipitate showed β- and pre-β-lipoprotein bands; supernatant after cryoprecipitation contained α-, pre-β-, and β-lipoprotein; total plasma chol 139 mg/dl	—		—	Patient 1 — monoclonal gammapathy, cryoglobulinemic plasma-cell infiltration of bone marrow; multiple fractures caused paraplegia; 4 years after M.M. was diagnosed, xanthelasma, subcutaneous nodules, and grossly lipemic plasma reported; serum cholesterol, then 570 mg/dl; melphalan (1-phenylalinine mustard) therapy started and continued for 11 years; chol decreased from 461±85 to 361±31 mg/dl and finally became normal Patient 2 — plane xanthomatosis; 18% of plasma cells in bone marrow; patient treated with melphalan and prednisone; cryoprecipitate by immunoelectrophoresis showed three lines, one due to albumin, one to β-lipo, and the third, IgG	—	Two patients with multiple myeloma; patient 1 was studied in detail while on a special diet which contained percent carbohydrate, fat and protein, 40:40:20, respectively; the male patient was 31 years old when multiple myeloma diagnosed; survived 17 years with treatment; appeared to be nonfamilial hyperlipidemia; patient 2 was a 75-year-old male	Two patients with multiple myeloma and hyperlipidemia and cryoglobulinemia were reported; first had 17-year survival and great improvement in Ig and lipid levels with treatment; no Ig-lipoprotein interaction demonstrated; cryoglobulin of case 2 contained albumin, β-lipoprotein, and IgG
Koga, S. et al.[47]	1974	Lipid-stained band in position of IgG myeloma protein approximately 10% of total lipid of serum in IgG fraction position	No evidence that either α- or β-lipoprotein were with the IgG lipid-stained band; approximately 5000 mg IgG/dl	—		—	—	Patient with multiple myeloma; IgG-κ type being treated with prednisone	—	—	Patient with IgG-type κ multiple myeloma is described in whose serum approximately 10% of lipid migrated in position of myeloma protein; no evidence found that α- or β-lipoproteins were in IgG position; nature of lipid binding not understood
Neufeld, A. H. et al.[48]	1964	—	—	—		—	Case of patient with multiple myeloma who during course of disease showed five-to six-fold increase in β-lipoprotein; following ^{14}C-labeled L-glutamic administration, found radioactive label in cytoplasm of plasma cells, in β-lipoprotein and other globulins; plasma cells, consistently 50 — 65% of bone marrow cells	—	—	Authors interpreted results to suggest that elevated β-lipoprotein was synthesized by plasma cells in bone marrow
Rosen, S. et al.[49]	1967	Serum protein electrophoresis showed abnormal peak in β-globulin area, decreased γ-globulin, and albumin; IgA globulin, 1400 — 1850 mg/dl; serum β-lipo, six times normal ultracentrifugal d <1.063 lipoprotein fraction contained IgA as well as β-lipoprotein; as much as 14% of IgA in this fraction	Pre-β greatly increased; no α-lipoprotein; on filter paper pre-β-lipo and myeloma protein migrated to same position; by agarose immunoelectrophoresis, separated into two fractions migrating in usual normal position of pre-β-lipoprotein and IgA globulin	—		—	Multiple myeloma, type IgA; hypertension, severe headaches, hyperlipemia ; Proteinuria, 1.5—4.5 g/24 hr; IgA in plasma cells of bone marrow; plasma cells did not stain positively for lipid; proteinuria greater than 3.5 g/24 hr, edema, and doubly refractive, oval fat bodies in urinary specimens fulfill definition of nephrotic syndrome (Schreiner); with electron microscopy, glomeruli shown to have marked foam cell change, confined to mesangium	—	40-year-old white male; no evidence of familial hyperlipoproteinemia	Unusual "complexing" in serum of IgA and lipo may be factor in development of unusual glomerular changes seen by EM studies in patient with multiple myeloma and nephrotic-like kidney involvement

Table 2 (continued)
APPLICATIONS OF ELECTROPHORESIS[a]

Author and ref. no	Year	Pattern	Type of character	Normal values	±SD	Reproducibility	Pathology	±SD	Characteristics of subject, age, sex, race	Summary
Seligman, M and Brouet, J. C.,[30]	1973	General review of antibody activity of human myeloma globulins; Beaumont's studies on anti-lipoprotein globulins cited; excellent table summarizing antigens against which specific antibody activity demonstrated								
Beaumont, J. L. et al.[31]	1974	Patient Lac — type IV (according to WHO classification), patient LER, type II	After infusion of albumin pattern reverted to near normal	—	—	—	Purified immunoglobulins evaluated by hemagglutination and immunoprecipitins, with purified lipoprotein and other plasma proteins	Patient LMC — nephrotic syndrome with failure to respond to usual therapy and ultimate development of chronic renal failure and hypertension; patient LER — monoclonal gammopathy without signs of myeloma; kidney biopsy showed segmental glomerulosclerosis	—	LAC — 38-year-old male; LER — 45-year old male	Hyperlipidemia, glomerulonephritis, and IgG-γ monoclonal gammopathy found in two patients; active IgG-γ prepared from chylomicron and serum γ-globulin; each purified immunoglobulin had reactivity with specific lipoprotein fractions; results suggest that these patients may have immune- complex type disease
Kayden, H. J.[32]	1976	Markedly concentrated lipid-stained band in γ-globulin; both protein and lipid bands in γ-globulin markedly increased; second patient showed markedly increased γ-globulin and 80% of chol migrated in this position; γ-globulin, intensely stained with lipid as well as protein stain	Hypercholesterolemia and hypertriglyceridemia marked in patient; however, present but less marked in three brothers; all γ-globulin in bottom of preparative ultracentrifuge at d 1.21, while lipoproteins floated; Incubating patients' γ-globulin with isolated d 1.005 — 1.019 fraction of normal or patient's serum resulted in greatly increased lipid in γ-position; electrophoresis of γ-globulin of hypercholesterolemic and hypertriglyceridemic patient's serum plus lipoprotein fractions showed migration of lipid to γ-globulin position; also shown with normal lipoprotein; cryoprecipitate contained lipoprotein and γ-globulin separated by ultracentrifugation at d 1.063; no binding of Fab fraction of γ with lipoprotein demonstrated	—	—	—	Patient 1 — flat xanthomata hepatosplenomegaly — increased plasma cells in bone marrow, increased γ-globulin; by ultracentrifugation, trace of 10—12S; clinical course and autopsy findings compatible with multiple myeloma, although X-ray studies revealed no bony lesions; Patient 2 — xanthoma planum — increased plasma cells in bone marrow, γ-globulin mainly 7S with trace 25—30S; Never developed bony lesions but showed increased number of plasma cells in bone marrow	—	First patient, 39-year-old white male; second patient, 67-year-old white male whose family history was incomplete but one sibling and two children had normal serum lipids; patient 3, 62-year-old white male	In addition to consideration of lipoprotein-Ig interactions reported in literature, this paper describes three patients studied by author; two of them have many characteristics suggesting immunoglobulin-lipoprotein interaction; third failed to show such properties as Fab fragment of γ-globulin interacting with lipoprotein
Lahav, M. et al.	1977	One patient had normal total protein in serum, while second patient had high total protein level; both had normal serum lipid, chol and TG levels; both patients' serum contained cryoglobulin, rich in IgMk and β-lipoprotein; the IgM and β-lipoprotein had different electrophoretic mobilities	—	—	—	—	Two patients with macroglobulinemia whose sera showed cryoprecipitation	Two patients with macroglobulinemia were studied	Results of physical chemical studies of cryoglobulins suggest that β-lipoprotein-IgM combination in cold precipitated globulin is probably due to nonimmunohydrophobic or electrostatic interaction

*   These abbreviations will be used in the table: chol = cholesterol; HDL = high-density lipoprotein; TG = triglyceride; PL = phospholipid; VLDL = very low-density lipoprotein; LDL = low-density lipoprotein.

REFERENCES

1. **Beaumont, J. L., Jacotot, B., Beaumont, V., Warnet, J., and Vilain, C.,** Myelome, hyperlidemie et xanthomatose, *N., Nouv. Rev. Fr. Hematol.*5, 507, 1965.

2. **Jacotot, B., Trong, T., Beaumont, J. L.,** Myelomae, hyperlipidemie et xanthomatose. II. Recherches complementaires sur l'association entre la paraproteine et les lipoproteines legeres, *Nouv. Rev. Fr. Hematol.*, 5, 777, 1965.

3. **Beaumont, J. L., Jacotot, B., Vilain, C., and Beaumont, V.,** Myelome, hyperlipidemie et xanthomatose. III. Une syndrome du a la presence d'un auto-anticorps anti-beta-lipoproteine., *Nouv. Rev. Fr. Hematol.*, 5, 787, 1965.

4. **Grabar, P. and Williams, C. A.,** Methode immunoelectrophoretique d'analyse de melanges de substances antigeniques, *Biochim. Biophys. Acta,*10, 7, 1955.

5. **Beaumont, J. L., Jacotot, B., Vilain, C., Beaumont, V., and Halpern, B.,** Presence d'un auto-anticorps anti-beta-lipoprotein dans un serum de myelome, *C. R. Acad. Sci.,* 260, 5960, 1965.

6. **Beaumont, J. L.,** Une γ-A globuline de myelome douee d'une activite specifique antilipoproteine. L'auto-anticorps anti-Pg., *C. R. Acad. Sci.,*263, 2046, 1966.

7. **Beaumont, J. L.,** L'activite specifique des proteines M vers un nouveau critere de classification des myelomes et des macroglobulinemies, *Rev. Fr. Etud. Clin. Biol.,* 12, 319, 1967.

8. **Lewis, L. A. and Page, I. H.,** An unusual serum lipoprotein-globulin complex in a patient with hyperlipemia, *Am. J. Med.,*38, 286, 1965.

9. **Poulik, M. D.,** Immunoelectrophoresis, in *Electrophoresis in Physiology,* 2nd ed., Lewis, L. A., Charles C. Thomas, Springfield, Ill., 1960, 71.

10. **Lewis, L. A., Van Ommen, R. A., and Page, I. H.,** Associatiation of cold-precipitability with beta-lipoprotein and cryoglobulin, *Am. J. Med.,*40, 785, 1966.

11. **Kayden, H. J. , Franklin, E. C., and Rosenberg, B.,** Interaction of myeloma globulin with human beta lipoprotein, *Circulation,* 26, 659, 1962.

12. **Lewis, L. A. and Lazzarini-Robertson, A., Jr.,** Hyperimmuno-globulinemia-lipoproteinemia and atherogenesis, in *Atherosclerosis III,* Schettler, G. and Weizel, A., Eds., Springer-Verlag, New York, 1974, 595.

13. **Beaumont, J. L.,** Gamma-globuline et hyperlipidemie, l'hyperlipidemie par auto-anticorps, *Ann. Biol. Clin.* (Paris), 27, 10, 1969.

14. **Beaumont, J. L., Poullin, M. F., Jacotot, O., and Beaumont, V.,** Myelome et hyperlidemie, nature de l'activite specifique antilipoproteine, *Nouv. Rev. Fr. Hematol.,* 7, 481, 1967.

15. **Beaumont, J. L. and Halpern, B.,** Une specificite commune aux alpha- et beta- lipoproteines du serum revelee par un auto-anticorps de myelome. L'antigene Pg., *C. R. Acad. Sci.,* 264, 185, 1967.

16. **Beaumont, J. L., Beaumont, V., Antonnucci, M., and Lemort, N.,** Les auto-anticorps antilipoproteines de myelome. Etude comparee de deux types l'IgA anti-Lp P.G. et l'IgG anti Lp A.S, *Ann. Biol. Clin.* (Paris), 28, 387, 1970.

17. **Marien, K. J. and Smeenk, G.,** Generalized planar xanthomata associated with multiple myeloma and hyperlipoproteinemia, *Arch. Belq. Dermatol.,* 29, 317, 1973.

18. **Marien, K. J. C. and Smeenk, G.,** Plane xanthomata associated with multiple myeloma and hyperlipoproteinaemia, *Br. J. Dermatol.,* 93, 407, 1975

19. **Perrault, M., Duperrat, B., Dry. J., and Courmont, J.,** Xanthomas plans normo-lipidemiques et paraproteinemiques, *Ann. Med. Interne,* 1, 2, 799, 1971.

20. **Kodama, H., Nakagawa, S., and Tanioku , K.,** Plane xanthomatosis with antilipoprotein auto-antibody, *Arch. Dermatol.,* 105, 722, 1972.

21. **Lewis, L. A., Page, I. H., Battle, J. D., and de Wolfe, V. G.,** How should hyperlipoproteinemia hyperimmunoglobulinemia-manifestations of an antilipoprotein auto-antibody be treated?, *Int. Res. Commun. Syst.,* 1, 117, 1973.

22. **Ho, K. J., de Wolfe, V. G., Siler, W., and Lewis, L. A.,** Cholesterol dynamics in autoimmune hyperlipidemia, *J. Lab. Clin. Med.,* 88, 769, 1976.

23. **Beaumont, J. L., Jan, B., Jacotot, B., Beaumont, V., and Buxtorf, J. C.,** La pathogenie de l'atherosclerose, *Ann. Biol. Clin.* (Paris), 30, 605, 1972.

24. **Beaumont, J. L.,** Hyperlipidemia with circulating antibeta lipoprotein autoantibody in man. Autoimmune hyperlipidemia, its possible role in atherosclerosis, *Prog. Biochem. Pharmacol.,* 4, 110, 1968.

25. **Lewis, L. A., de Wolfe, V. G., Butkus, A., and Page, I. H.,** Autoimmune hyperlipidemia in a patient, *Am. J. Med.,* 59, 208, 1975.

26. **Lewis, L. A. and Page, I. H.,** Serum protein and lipoproteins in multiple myelomatosis, *Am. J. Med.,* 17, 670, 1954.

27. **Noseda, G., Butler, R., Schlumpf, E., and Riesen, W.,** Binding von β-Lipoprotein durch ein IgA-paraproteine: Eine Antigen-Antikorper-Reaction, *Schweiz. Med. Wochenschr.*, 101, 893, 1971.
28. **Noseda, G., Riessen, W., and Butler, R.,** Das Verhalten der Serumlipide par Paraproteinamien Nachweis von Paraprotein-Lipoprotein-Komplexen, *Schweiz. Med. Wochenschr.*, 101, 1787, 1971.
28a. **Noble, R. P.,** Lipoprotein electrophoresis on agarose, *J. Lipid Res.*, 9, 693, 1968.
29. **Riesen, W., Noseda, G., and Butler, R.,** Anti-β-lipoprotein of human monoclonal immunoglobulins, *Vox Sang.*, 22, 420, 1972.
30. **Noseda, G., Riesen, W., Schlumpf, E., and Morell, A.,** Antikorper gegen Lipoproteine und Hypolipidemie Seronegative primar-chronischer Polyarthritis, *Schweiz. Med. Wochenschr.*, 102, 969, 1972.
31. **Noseda, G., Riesen, W., Schlumpf, E., and Morell, A.,** Hypo beta-lipoproteinemia associated with autoantibodies against beta-lipoproteins, *Eur. J. Clin. Invest.*, 2, 342, 1972.
31a. **Robbins, S. L.,** *Pathologic Basis of Disease*, W. B. Saunders, Philadelphia, 1974, 1467.
32. **Cohen, L., Blaisdell, R. K.,** Djordjevich, L., Ormiete, V., and Dabrilovic, L., Familial xanthomatosis and hyperlipidemia with myelomatosis, *Am. J. Med.*, 40, 299, 1966.
33. **Spikes, J. L., Jr., Cohen, L., and Djordjevich, J.,** The identification of a myeloma serum factor which alters serum beta-lipoproteins, *Clin. Chim. Acta*, 20, 413, 1968.
33a. **Smithies, O.,** An improved procedure for starch-gel electrophoresis: further variations in the serum proteins of normal individuals, *Biochem. J.*, 7, 585, 1959.
34. **Sachs, B., Cady, P., and Ross, G.,** An abnormal lipid-like material and carbohydrate in the sera of patients with multiple myeloma, *Am. J. Med.*, 17, 662, 1954.
35. **Levin, W. C., Aboumrad, M. H., Ritzmann, H., and Brantley, C.,** Gamma-type I myeloma and xanthomatosis, *Arch. Intern. Med.*, 114, 688, 1964.
36. Spinco Procedure B, Technical Bulletin TB6050A, Beckman® Spinco, Palo Alto, Cal., 1958.
37. **Lawrence, M.,** Technique of immunoelectrophoresis, *Am. J. Med. Technol.*, 30, 209, 1964.
38. **Hartmann, L., Filitti-Wurmser, S., Ollier, M. P., and Laudat, P.,** Lipides, lipoproteines et immunoglobulines IgM., *Ann. Biol. Clin.* (Paris), 26, 881, 1968.
39. **Lewis, L. A.,** Electrophoresis of proteins, *CRC Crit. Rev. Clin. Lab. Sci.* 1, 233, 1970.
40. **Valdiguie, P., de la Farge, F., and de la Riviere, R. D.,** Sur les lipides des proteines myelomateuses. Acides gras lies et adsorbes des proteines de Bence-Jones, *C. R. Acad. Sci.*, 268, 865, 1969.
41. **Glueck, C. J., Levy, R. I., Glueck, H. I., Gralnick, H. R., Gretin, H., and Fredrickson, D. S.,** Acquired type I hyperlipoproteinemia with systemic lupus erythematosis dysglobulinemia and heparin resistance, *Am. J. Med.*, 47, 318, 1969.
42. **Glueck, C. J., Levy, R. I., Glueck, H. I., Gralnick, H. R., Kaplan, A. P., Barth, W. F., and Fredrickson, D. S.,** Low post-heparin lipolytic activity and exogenous fat intolerance associated with abnormal heparin resistance and heparin binding globulins, *Circulation*, 38, VI-7, 1968.
43. **Beaumont, J. L., Lemort, N., and Halpern, B.,** Une immunoglobuline anti-heparine dans un serum hyperlipoproteinique, *C. R. Acad. Sci.*, 271, 2452, 1970.
44. **Wilson, D. E., Flowers, C. M., Hershgold, E. J., and Eaton, R. P.,** Multiple myeloma, cryoglobulinemia and xanthomatosis. Distinct clinical and biochemical syndromes in two patients, *Am. J. Med.*, 59, 721, 1975.
45. **Hatch, F. T. and Lees, R. S.,** Practical methods for plasma lipoprotein analysis, *Adv. Lipid Res.*, 6, 1, 1968.
46. **Crowle, A. J.,** *Immunodiffusion*, 2nd ed., Academic Press, New York, 1973.
47. **Koga, S., Kozuru, M., Hirayama, C., and Ibayashi, H.,** An unusual lipid-protein complex observed in an IgG myeloma patient, *Clin. Chim. Acta*, 54, 169, 1974.
48. **Neufeld, A. H., Morton, H. S., and Halpenny, G. W.,** Myelomatosis with xanthomatosis multiforme, *Can. Med. Assoc. J.*, 91, 374, 1964.
49. **Rosen, S., Cortell, S., Adner, M. M., Papidopoulis, N. M., and Barry, K. G.,** Multiple myeloma and the nephrotic syndrome. A biochemical and morphologic study, *Am. J. Clin. Pathol*, 47, 567, 1967.
50. **Seligman, M. and Brouet, J. C.,** Antibody activity of human myeloma globulins, *Semin. Hematol.*, 10, 163, 1973.
51. **Beaumont, J. L., Antonucci, M., Lagrue, G., Guedon, J., and Perol, R.,** Nephrotic syndrome, monoclonal gammapathy and auto-immune hyperlipidemia, *Clin. Exp. Immunol.*, 18, 225, 1974.
52. **Kayden, H. J.,** Protein-lipoprotein interactions in man. A re-examination, *Ann. N. Y. Acad. Sci.*, 275, 145, 1976.
53. **Frederickson, D. S., Levy, R. I., and Lees, R. S.,** Fat transport in lipoproteins: an integrated approach to mechanisms and disorders, *N. Engl. J. Med.*, 276, 32, 94, 148, 215, 273, 1967.
54. **Lahav, M., Zur, S., Zheleznik, M., Teuerstein, I., Gotfryd, Z., Klinger, I., Leiba, I. H., and Cahane, P.,** Cold co-precipitation of homogeneous macroglobulins and β-lipoproteins, *Isr. J. Med. Sci.*, 13, 6, 1977.

LIPOPROTEINS AND NEOPLASTIC DISEASES

M. Barclay and R. K. Barclay

After it was recognized that lipids were transported in the blood solubilized as complexes with proteins, various technical procedures were devised to separate the "lipoproteins". One of the techniques at hand was moving boundary electrophoresis which separated proteins according to their electrical charges in appropriate buffer systems. Other less involved electrophoretic techniques using solid media were soon available for separation of proteins, e.g., starch, paper, agarose, cellulose acetate, etc., but the lipid-laden lipoproteins did not always separate well, or even move at all, on such "zonal" media. However, these were not insurmountable problems, and by using optimum conditions informative separations of the then recognized three main classes of lipoproteins, β-, α_1-, and α_2-lipoproteins, based primarily upon their electrophoretic properties, could be obtained.[1,2] These relatively simple and inexpensive tools for separating groups or classes of lipoproteins attained widespread popularity, especially after the clinical significance of blood lipoprotein levels was recognized.[3,4]

It followed, then, that these techniques would be used to study lipoprotein levels in sera from patients who had malignant diseases. As with other clinical syndromes, electrophoretic techniques were especially useful in cases where certain lipoproteins were present in excessive amounts and would produce pronounced differences. In the cases of more subtle differences or in instances where a decrease in a certain lipoprotein might be indicative of metabolic dyscrasia, electrophoretic techniques, for the most part, were not sufficiently sensitive. A case in point was the "α-lipoproteins" or Cohn fraction IV, which were observed routinely to be lowered in patients with cancer.[5,6] Subsequently, it was shown by ultracentrifugal techniques that it was not the entire α-lipoprotein class that was lowered, but only a part of it, the high-density lipoprotein with a density range of 1.0635 to 1.1250 g/mℓ (HDL$_2$).[7,8] Until recently, electrophoretic techniques to separate HDL$_2$ from the other part of the α-lipoproteins (HDL$_3$, d 1.1250 to 1.210 g/mℓ) were not available, so investigators using these techniques would observe only a decreased combined fraction. From a practical aspect this may have been misleading, since HDL$_3$ usually comprises most of the total high-density lipoproteins or α-lipoprotein fraction.

As a result of the present emphasis upon the roles of the apoproteins, polypeptides, and amino acids as special "polarizers" or "binders" for water-insoluble lipids, the protein moieties of the lipoprotiens need detailed study.[9-11] For this, certain electrophoretic techniques, especially polyacrylamide gels, are inordinately qualified. An important criticism of all electrophoretic techniques for the study of lipoproteins in normal as well as pathological specimens is lack of numerical quantitation, such as is possible with the ultracentrifuge;[2] however, rather good indications or trends can be obtained, and in the studies with serum or plasma, the data obtained by electrophoresis has generally confirmed that obtained with the ultracentrifuge.

In 1956 Miller and Erf[12] published one of the earliest papers dealing with the separation by electrophoresis of serum lipoproteins in patients with cancer. As support medium, they used Whatman® No. 1 paper, 3.8 × 30.5 cm, in veronal buffer, pH 8.6 and ionic strength 0.03. A small quantity of serum, 0.01 ml, was applied to the paper, and electrophoresis was run for 12 hr at 15°C with a potential difference of 150 V, 0.4 mA/cm. After electrophoresis, paper strips were dried at 100°C and each strip cut in half lengthwise. One half of the strip was stained for protein with bromphenol blue and the other half with Sudan III for lipoproteins. Stained areas were scanned at 560 nm, and areas under curves were measured with a planimeter to determine the relative proportions of components separated on the paper strips. Table 1 illustrates their results.

Miller and Erf reported these details: there were two distinct Sudanophilic zones, the first (closest to the origin) was in the region to which β-globulins migrate and "frequently consisted of two well differentiated components." The second zone, the authors suggested, might be α_2-lipoproteins (VLDL?), since the blood was obtained at a blood donor center and it was most unlikely that the subjects had fasted. They also observed two peaks in the α_1-globulin region and apparently added them together as 24.4% of the total Sudanophilic material. Here, they had probably partially separated HDL_2 from HDL_3. Average percentages given, approximately 25% for α-lipoproteins and 75% for β-lipoproteins in normal subjects, are in agreement with the relative quantities for these (and their subcomponents as is now known), which had been obtained by other means.[5,6,13]

When lipoproteins from serum of patients with cancer were separated by paper electrophoresis, Miller and Erf reported significant increases in β-lipoprotein areas and greatly diminished areas of α-lipoproteins, especially in serum from patients with advanced disease. In three patients with carcinoma of the breast, α-lipoproteins were not detectable. Examination of Figures 1 and 2 shows that the two peaks seen in normal subjects were not present on the paper strips. Conversely, with sera from these three patients, β-lipoproteins were greatly increased, and often the mobility on the paper was "unusual". This could occur easily if large amounts of serum very low density lipoproteins with density <1.006 g/mℓ (VLDL) and low density lipoproteins with density <1.0635 g/mℓ (LDL) were present. Patients who survived as long as 30 years after mastectomy (cured) had relatively normal lipoprotein patterns with significantly greater amounts of α-lipoproteins. In general, the method of Miller and Erf using paper electrophoresis as a means to separate two of the main groups of serum lipoproteins (there are at least four main components and perhaps numerous lesser ones) was successful, since they obtained a pattern, suggestive of what can be obtained with lipoproteins in sera from patients with cancer. Their suggestive pattern not only confirmed previous observations made with serum lipoproteins by other techniques but has been, in turn, confirmed with paper electrophoretic techniques since their paper was published. Toyoda, in 1964, also using paper electrophoresis, reported low percentages of α-lipoproteins in sera from 124 patients with advanced, inoperable or metastatic carcinoma.[14]

Falor et al.[15] using the more or less standard conditions (veronal buffer, pH 8.6, ionic strength 0.05) separated lipoproteins in the sera from 38 patients with bronchogenic carcinoma of the lung, 35 patients with acute or chronic pulmonary disease of inflammatory origin, and 7 controls. The 7 "controls" were patients with upper respiratory problems of various kinds. After electrophoresis on Schleicher and Schuell 2043 A mgl, the paper strips were stained with oil red O and scanned with the Beckman® Spinco Analytrol, Model RB with a B-5 cam.

Table 2 is constructed from the results of Falor et al. There are slight trends toward decreases in the percentages of α-lipoproteins and increases in the percentages of β-lipoproteins in serum (and also in chyle) from patients with cancer, as shown previously by Miller and Erf,[12] who used a very similar paper electrophoretic technique. Although the ranges are wide, they do not reflect upon the technique so much as on the diversity of subjects in each category.

Although percentages of total serum proteins (albumin(s), α_1- and α_2-globulins, β-globulins, and γ-globulins) were measured, a "trailing" component was observed in both sera and chyle only when lipoproteins were visualized on strips stained with oil red O. Whether the trail followed the α- or β-lipoproteins is not clear. Since there was more (17%) trailing in normal chyle than in normal serum (5%) and chyle is known

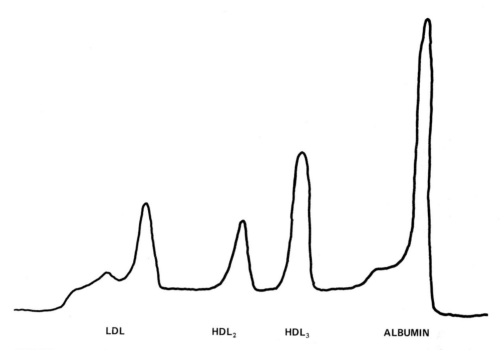

FIGURE 1. Densitometric tracing of a polyacrylamide gel showing electrophoresis of a sample of normal serum using 10-cm tubes. Gel was fixed for 1 hr with 12% trichloroacetic acid and then placed in 7% acetic acid.

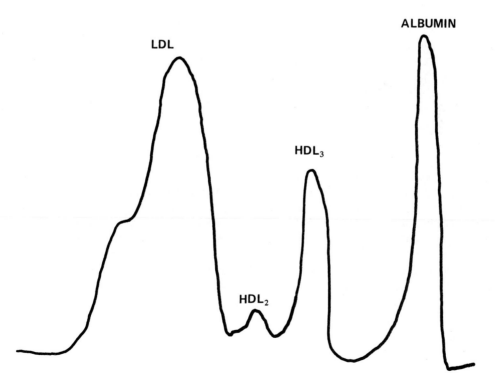

FIGURE 2. Densitometric tracing of a polyacrylamide gel showing electrophoresis of a sample of serum from a patient with cancer. A shorter tube was used; otherwise, the treatment was the same as for Figure 1. For details of instruments employed, see text.

Table 1
SERUM LIPOPROTEINS IN NORMAL SUBJECTS AND IN
PATIENTS WITH CANCER

Subjects	Number	α-Lipoproteins[a]		β-Lipoproteins[a]	
		Avg.	Range	Avg.	Range
Normals	30	24	13—43	76	57—88
Patients, carcinoma of breast	14	8	0—24	92	76—100
Patients, carcinoma of alimentary tract	10	17	15—19	83	81—86
Patients, carcinoma of breast "cured"	8	43	33—59	58	47—67

[a] Averages and ranges given in percent of total lipoproteins.

Data taken from Miller, B. J. and Erf, L., *Surg. Gynecol. Obstet.*, 102, 487, 1956.

to transport more lipid, especially triglycerides, the trail most probably followed the β-lipoproteins and was triglyceride-rich VLDL (α_2-lipoproteins). However, it may also have trailed the α_1-lipoproteins. Since it was described as a "trail", it was not likely to be the pre-β-lipoproteins, unless they were very close to the α_1-lipoproteins. They reported a decrease in the trailing fraction in the chyle of both groups of patients concomitant with an increase in the α_2-globulins in serum, which suggests transport of VLDL from chyle to serum. It is now known that VLDL is elevated in the serum from patients with cancer, even in the fasting state.[7,8]

Ma et al.[16] used similar electrophoretic conditions as reported above except that a vertical tank was used with Whatman No. 1 paper (36 × 3 cm). Frequently, triplicate strips were used. Electrophoresis was run at 110 V for 9 hr (for eight strips). After drying, strips were stained with Sudan black B for 3 hr. These authors also emphasized the necessity for absolute consistency in staining, washing, and drying in order to obtain reproducible results. This is a very important point. They measured the proportional amounts of dye intensities in the bands formed by α- and β-lipoproteins with a Beckman Spinco Analytrol, Model RA, without optical filters.

Table 3 was constructed from the data of Ma et al. Since in patients with hydatidiform mole (associated with choriocarcinoma) gonadotropins are increased, women with normal pregnancies were studied in addition to normal nonpregnant women. The difference in α-lipoprotein levels between Group A with hydatidiform mole and normal nonpregnant women (Group C) was highly significant ($P < 0.001$). There was less spread in values from women without the mole, as indicated by the reasonably low standard deviations, than in patients with the mole. The rather consistent results may have been derived from the use of groups of patients or subjects more homogeneous within themselves, certainly as to sex and probably as to age range or from the care taken in these experiments to standardize the electrophoresis, staining, washing, and drying of the paper strips.

Higazi et al.[17] studied sera from patients with various forms of leukemia in an attempt to clarify the relationships of the different types of this disease to levels of lipoproteins. The electrophoretic procedures used were essentially those of Swahn.[18] Table 4 shows the data from their patients listed according to decreasing amounts of total lipid to show any possible correlations and to help explain the "nonmobile" fraction which was not characterized further. It can probably be assumed that this was a lipid-

Table 2
LIPOPROTEINS IN SERUM AND CHYLE FROM CONTROL SUBJECTS, PATIENTS WITH INFLAMMATORY DISEASES, AND PATIENTS WITH CARCINOMA OF THE LUNGS

Subjects	Number	α-Lipoproteins[a]			β-Lipoproteins[a]			"Trailing"[a]		
		Avg.	Range	±SD	Avg.	Range	±SD	Avg.	Range	±SD
					In serum					
Controls	7	19	4—37	10.58	76	60—95	10.95	5	1—13	4.24
Inflammatory diseases	35	15	2—40	8.94	77	56—96	11.48	7	1—25	6.70
Carcinoma of the lungs	38	14	5—35	7.87	79	62—94	10.40	5 +	1—20	4.10
					In chyle					
Controls	7	9	1—26	7.48	74	66—92	13.15	17	1—25	7.80
Inflammatory diseases	35	10	1—41	11.57	77	44—99	15.32	14	1—34	10.14
Carcinoma of the lungs	38	8	1—25	7.74	83	54—99	12.28	8	1—24	6.48

[a] Averages and ranges given in percent of total lipoprotein.

Data taken from Falor, W. H., Dettling, J. J., Kerkian, A., Bachtel, W., Krismann, E., and Thomas, M. A., *Arch. Surg.*, 91, 671, 1965.

laden fraction which did not migrate on the paper (adsorbed VLDL and LDL?). There was a close positive correlation between the amounts of total lipid and those of non-mobile lipoproteins in milligram percent. Electrophoresis of serum from patients revealed only a suggestive trend in relating the levels of α- and β-lipoproteins to those in normal subjects, values for whom fell, in general, in between those from patients with chronic myeloid and lymphocytic leukemia. Two patient groups had somewhat more α-lipopoteins (and also more β-lipoproteins) than controls. However, all groups of patients, except the one patient with acute lymphocytic leukemia, had more of the nonmobile fraction (VLDL) than controls did. The authors did not explain the expression of their values as milligram percent, nor did the β- to α-lipoprotein ratios always agree with the milligram percent values given, presumably from which they were calculated.

In conjunction with ultracentrifugation to separate lipoproteins with densities less than 1.063 g/mℓ (VLDL + LDL or pre-β- + β-lipoproteins) from those with densities greater than 1.063 g/mℓ (HDL or α-lipoproteins), lipoproteins were also determined by paper electrophoresis. Feldman and Carter[19] used the procedure of Hatch and Lees,[20] in which albumin is added to the buffer.

Table 5 is part of one from Feldman and Carter, and Table 6 is constructed from data in the text of their paper. The patients were 231 women with progressive metastatic breast carcinoma. These patients were more than one year postmenopausal or had had radiation or surgical castration. They ranged in age from 26 to 85 years with a mean age of 57 years. The control group described contained 53 healthy, postmenopausal women, ranging in age from 34 to 90 years with a mean age of 62 years. In Table 5 the lipoproteins listed as percent of total lipoprotein separated on paper electrophoresis do not differ in the two groups of subjects. However, levels of lipoproteins calculated from the ultracentrifuge, at least the α-lipoproteins, regardless of whether the milligrams per 100 mℓ refers to the amount calculated from the analytical ultracentrifuge or from the cholesterol content, show lower quantities in the serum or plasma from patients with cancer. This implies that differences in levels of lipoproteins in samples from two groups of subjects are not always apparent or, at least, were not

Table 3
SERUM LIPOPROTEINS IN NORMAL SUBJECTS, IN PREGNANT WOMEN, AND IN PATIENTS WITH HYDATIDIFORM MOLE

Subjects			α-/β-Lipoprotein ratio[a]	Percent α-lipoproteins		
Condition	No.	Group		Avg.	Range	±SD
Normal, non-pregnant	12	C	26.2/73.8	26.2	23—30	2.9
Pregnant						
Week 5—12	12	B-1	22.1/77.9	22.1	18—27	2.8
Week 13—24	16	B-2	22.9/77.1	22.9	18—25	2.7
Week 25—37	16	B-3	18.2/81.8	18.2	16—21	2.5
Patients with hydatidiform mole	8	A	12.3/87.7	12.3	5—20	4.8

[a] Expressed as a percentage of the α-lipoprotein fraction.

Data taken from Ma, L., Chew, W., Pang, L. S. C., and Wong, C. W., *Clin. Chim. Acta,* 11, 561, 1965.

Table 4
SERUM LIPOPROTEINS IN NORMAL SUBJECTS AND IN PATIENTS WITH LEUKEMIA

Subjects	No.	Total Lipid[a]		α-Lipoproteins[a]		β-Lipoproteins[a]		"Nonmobile"[a]		β/α[b]	
		Avg.	±SD	Avg.	±SD	Avg.	±SD	Avg.	±SD	Avg.	±SD
Controls	35	483	83	154	38	211	43	117	34	1.46	0.49
Patients											
Acute myeloid	6	607[c]	113	180	64	249	40	169[c]	54	1.45	0.53
Monocytic	2	588	153	185	109	230	13	173	32	1.24	6.0
Chronic myeloid	23	491	119	143	42	213	62	136	52	1.63	0.87
Chronic lymphocytic	12	437	154	117	55	199	85	121	25	1.59	0.49
Acute lymphatic	1	212		37		166		11		4.47	

[a] Averages and SD expressed in milligram percent.
[b] Averages and SD expressed as ratios.
[c] Higazi et al.[17] indicated that these values were significantly higher than the corresponding values from controls.

Data taken from Higazi, A. M., Ata, A. A., Abdel-Rahman, Y. M., Malek, A., and Mansour, K., *J. Egypt. Med. Assoc.*, 49, 679, 1966.

Table 5
SERUM LIPOPROTEINS IN POSTMENOPAUSAL WOMEN CONTROLS AND IN WOMEN WITH METASTATIC CARCINOMA OF THE BREAST

Subjects	Lipoproteins by electrophoresis[a]						Lipoproteins by ultracentrifugation[b]			
	α		β		Origin		α		Cholesterol in β + pre-β	
	Mean	±SD	Mean	±SD	Mean	±SD	Mean	±SD	Mean	±SD
Controls	25	1.2	67	1.3	8	0.67	65	—	192	11.2
No. of subjects	44		44		44		11		11	
Patients	29	1.54	66	1.24	10	0.74	57	2.57	161	7.99
No. of subjects	162		109		109		50		50	

[a] Mean and SD expressed as percent of total lipoprotein.
[b] Preparative ultracentrifugation at solution density of 1.063 g/mℓ. Mean and SD are expressed in mg/100 mℓ serum.

Data taken from Feldman, E. B. and Carter, A. C., *J. Clin. Endocrinol. Metab.*, 33, 8, 1971.

Table 6
PLASMA LIPID AND LIPOPROTEIN CONCENTRATIONS IN POSTMENOPAUSAL WOMEN CONTROLS AND IN WOMEN WITH METASTATIC CARCINOMA OF THE BREAST

Subjects[a]	No. of subjects	Lipids[b]			Lipoproteins[b]	
		Chol	PL.	TG.	Chol in α^c	Chol in β + pre-β^c
Controls	5	257	250	83	80	155
Patients	10	227	250	80	63	152

[a] Controls: mean age was 61 years, mean weight was 156 lb; patients: mean age was 55 years, mean weight was 163 lb.

[b] Values are expressed in mg/100 mℓ plasma. Abbreviations used are chol = cholesterol, PL = phospholipid, TG = triglycerides.

[c] Lipoproteins separated by preparative ultracentrifugation at solution density of 1.063 g/ml.

Data taken from Feldman, E. B. and Carter, A. C., *J. Clin. Endocrinol. Metab.*, 33, 8, 1971.

when studied by means of paper electrophoresis. Lipoproteins as separated with preparative ultracentrifugation and measured indirectly by cholesterol content do show a difference, which agrees with most published results comparing α-lipoproteins in serum from normal women and from patients with cancer.[8]

Table 6 refers to different sets of subjects and also shows less α-lipoproteins (as measured by cholesterol) in patients with cancer. There is slightly less triglyceride in the serum of cancer patients, so this may not account for the 10% of lipoproteins which remain at the origin as shown in Table 5. In both tables, however, values for lipoproteins at the origin and for triglycerides measured are very much the same in both groups of subjects.

These experiments strongly suggest that paper electrophoresis does not, and cannot be expected to, give quantitative data that correlates positively to a significant degree with data derived from the ultracentrifuge, whether calculated as such or measured by means of a principal component, such as cholesterol, in the individual lipoproteins. The two techniques cannot be expected to agree exactly, because lipoproteins are separated according to different physical properties in the two systems.

Certain aspects of the technique using polyacrylamide gels were published in the 1960s by Ornstein, [21] Raymond,[22] and Davis.[23] This medium for separating proteins by size, shape (molecular sieving), and charge has had many applications widespread use, and of course, many modifications. An important modification is that of changing or adjusting the pore size (percent of polyacrylamide) to accommodate the substances to be separated. There have also been other changes, such as buffer system(s), pH, the use of gradients, changes in electrical current, etc. Of all the so-called zone electrophoretic procedures, polyacrylamide gel gives the greatest resolution of substances to be separated.

Narayan et al.[24] were the first to publish the utilization of "disc" electrophoresis to separate human serum lipoproteins. They essentially used the procedure of Ornstein and Davis.[25] Many investigators working with serum have thought it desirable to prestain the serum before applying the sample to the gel. For this purpose Narayan and co-workers used the procedure of Ressler et al.[26] In most protein work, bromphenol blue is used as a tracking dye to visualize movement of the albumin and thus enable cessation of electrophoresis before the proteins leave the gels; however, Narayan and fellow workers did not use tracking dye in their early experiments on serum lipoproteins.

The usual percentage of polyacrylamide for most protein separations is 7.5%. The pore size of this concentration of gel is about 50 Å, and most large molecules do not penetrate.[25] Narayan et al. reported that LDL d 1.006 to 1.0635 g/ml, did not enter the separating gel and that both high-density lipoproteins (HDL$_2$, d 1.0635 to 1.125 g/ml, and HDL$_3$, d 1.125 to 1.210 g/ml) were resolved into three components when 30 λ of each were applied. Electrophoresis was carried out at 2.5 mA for 73 to 85 min. One of the great advantages of polyacrylamide gel electrophoresis is that only very small quantities of protein or lipoproteins or other components are required in low volumes of solvent. These lipoproteins had been applied in solutions containing their respective density salts, which ranged from approximately 9 to 24%. Since polyacrylamide gel electrophoresis is usually done at low ionic strengths, these salts could effectively reduce the mobility of lipoproteins.

Additional experiments[24] were carried out at lower polyacrylamide concentrations: 5% and 3.75% with HDL$_2$ and HDL$_3$ dialyzed against 1 *M*NaCl. A volume of 10 λ for each lipoprotein was applied to 3.75% gel, and electrophoresis was carried out at 2.5 mA for only 34 min. With this low concentration of gel and somewhat greater pore size, LDL entered the gel but did not resolve further; whereas, HDL$_2$ and HDL$_3$ were reported to have moved into the gel. HDL$_2$ was also reported to separate into two or three components and HDL$_3$, into five, most of which moved more rapidly down the gel than HDL$_2$.

The fact that possibly HDL$_2$ and HDL$_3$ can be resolved with polyacrylamide gel electrophoresis is of some importance, because previously no zone electrophoretic nor even moving boundary process had accomplished this separation. All high-density lipoproteins were reported as one peak or band, α-lipoproteins. The possible exception is in the report by Miller and Erf,[12] cited above; they claimed to have seen two bands but did not claim to have separated the two high-density lipoproteins.

When 10-λ samples of prestained serum were applied to both 5% and 3.75% gels, Narayan et al. reported that in 5% gel, LDL did not migrate but HDL separated into two components in all serum samples. In 3.75% gel, LDL migrated into the gel and was better resolved, but HDL diffused rapidly at 2.5 mA for 35 min.

It is very difficult to see the components of both LDL and HDL in the figures shown by Narayan et al.,[24] who explain that photography of the gels was difficult. They have, however, claimed initially that it may be possible to separate classes of serum lipoproteins with an electrophoretic technique and thus obtain estimates of differences between levels of lipoproteins in the serum from patients with abnormal or unusual levels of lipoproteins. For example, in the three sera from patients with coronary artery disease whom they studied, it is apparent that the discs representing LDL in the 3.75% gel are wider and perhaps more densely stained than the corresponding control sera.

In a subsequent publication, Narayan[27] used the stain, Sudan black B, to visualize the albumin and subjected the gels to densitometric tracing (Model E microdensitometer from Canalco®). In human serum, he reports two major peaks (components): one, slow moving and close to the gel-spacer gel interface (LDL?), and at least one faster moving component located behind (above) the albumin. These latter would be the high-density lipoproteins. Table 7 illustrates results obtained with different concentrations of gels. The data show the importance of gel concentrations in dye uptake, in width of band, and possibly in the effective separation of the components. One serious, operational problem, not mentioned by Narayan, is the softness and delicacy of the 3.75% gel. These gels break easily and are readily damaged, especially during removal from the tubes, whether or not they have been treated to make removal easier. Small

Table 7
EFFECT OF POLYACRYLAMIDE GEL CONCENTRATIONS UPON THE INTENSITIES OF DISCS CONTAINING SERUM LIPOPROTEINS

Gel concentration (%)	Peak areas[a] (mm²)	HDL/LDL ratio[b]
3.75	408	0.9
5.00	360	1.3
7.50	248	1.9

[a] Peak areas are obtained from densitometry of low-density lipoproteins.

[b] Ratio is not corrected for albumin "contamination."

Data taken from Narayan, K. A., *Lipids*, 2, 282, 1967.

abrasions or chips can give a "peak" with a sensitive densitometer. Fixing or stabilizing gels with 12% trichloroacetic acid is an aid, but still, 3.75% gels, one of the best for lipoprotein separations from whole serum, must be manipulated with great caution.

An additional problem is selection of a suitable stain and whether or not to prestain the serum. Lipid stains do not always lend themselves well to prestaining of serum samples before electrophoresis. Because the quantity of lipid attached to protein in certain diseases can vary greatly, it is difficult to judge when the ideal stain has been used or to remove the background coloring from the gel. Gels with background staining are difficult to evaluate when scanned with a densitometer, and accurate baselines are virtually impossible to attain. As Narayan[27] observed, "satisfactory correlation between lipoprotein concentration and stain intensity . . . was not observed because the dye uptake is apparently dependent upon the lipid composition of the lipoproteins".

Nishioka[28] used the technique for polyacrylamide gel as described by Davis[23] with certain modifications: glass tubes (7.0 × 0.5 mm I.D.) containing 2.4% polyacrylamide in the upper half and 7.0% in the lower half. The buffer used was TRIS-glycine, pH 8.9, and the gels were stained by immersion for 1 hr in 1% amido black in 7% acetic acid. Stained gels were scanned, and relative amounts of the proteins of interest were obtained by planimetry of the scans to obtain areas under the curves. When normal samples were used, many bands separated rather sharply in the lower part of the gel, the region where the α-lipoproteins can be seen as a fairly sharp peak above, and well separated from, albumin.

When sera from patients with primary liver carcinoma were studied, the bands on the gels were certainly less well separated and not nearly so sharp as they appeared in normal samples. Nishioka concluded that the number of components was decreased, but this observation may result from the fact that the densitometric tracing of the gel from the samples of sera from patients reflected the diffusion and less well-spaced bands in these samples. However, gels from patients with cancer had a noticeable band just behind albumin which when traced appeared as a shoulder on albumin. From observation of the photographs of the gels, the α-lipoprotein band appeared less strongly stained and was too near the shoulder band to be traced as an entity as in normal samples. Nishioka concluded, from the immunodiffusion of the cut segment containing this shoulder, that it represents α-fetoprotein located between α-lipoprotein and albumin. He saw the band primarily in sera from patients with primary liver carcinoma, not in sera from normal persons, but it was also found in saline extracts of fetal tissue and hepatoma tissue. It is absent in patients with hepatitis, liver cirrhosis, and carcinoma metastasized from the stomach. Nishioka thus concluded that the superior sensitivity of polyacrylamide gel electrophoresis made possible revelation of this α-fetoprotein, not seen in paper electrophoresis.

Terebus-Kekish et al..[29] by using less concentrated gel and longer tubes, were able to obtain good separation of the lipoproteins as follows: A gel, 3.5% polyacrylamide, was prepared in the usual way,[30] and the only buffer was TRIS-glycine, pH 8.4. Freshly obtained, nonhemolyzed fasting serum is preferred. A small quantity of ethylenediaminetetraacetic acid may be added to a final concentration of $0.001 M$. The volume of serum which produces the most discrete separation is 0.02 ml. The most efficient size for the electrophoresis tubes is 10 cm (length) × 5 mm (I.D.), 7 mm (O.D.) With these tubes there is sufficient length of gel and time of electrophoresis to separate the components from each other, as seen in Figure 1. It was necessary to run the electrophoresis for approximately 2 hr at 3 mA per tube, or until the visible band formed by albumin (or a tracking dye may be used) is less than 1.0 cm from the bottom of the gel. This moves the albumin boundary sufficiently away from HDL_3 (the faster moving of the two α-lipoprotein components) to minimize "trailing" and thus contamination of the lipoprotein.

In all experiments carried out on many samples, HDL_3 has been observed toward the bottom of the gel, i.e., toward the albumin area, with the other component of 79the α-lipoproteins, HDL_2, above it by 5 to 10 mm, as illustrated in Figure 1, which shows a tracing of a typical gel from a 10-cm tube. It may be noted that β-lipoproteins (LDL) are present and well separated, but in this normal, fasting subject, there was essentially no α_2-lipoprotein (VLDL). After electrophoresis the gel was placed in a solution of 12% tricholoroacetic acid (TCA) for 1 hr and then transferred to a tube containing 7% acetic acid. It was then placed in the quartz holder of a Joyce-Loebel UV scanner, scanned at 280 nm, and traced with a Hewlett-Packard® 680 strip chart recorder.

If the investigator wishes to know the relative quantities of lipoproteins, a more precise tracing of lipoprotein "discs" can be made if the gels are removed from the tubes and scanned immediately and directly without acid treatment. This should be done with dispatch so that no diffusion occurs. A batch of 12 samples can be easily and accurately done immediately after removing the gels from the tubes. When 10-cm tubes are used, the discs are separated sufficiently well (Figure 1) so that a reasonably accurate baseline for the different lipoproteins can be drawn. If the sides of the peaks produced by the densitometer do not touch the baseline, the conventional procedure of projecting straight lines, which best describe the descending slopes to the baseline, may be used to enclose the total area. The total areas may be obtained by planimetry, reading areas directly in square inches to 0.01 in.², or simply by counting the squares within the completed areas on the strip chart paper. As emphasized previously, these gels are "tender" and fragile, but with sufficient patience and care, they can be handled and will give reproducible results.

Although this procedure was utilized in experiments concerned mainly with high-density lipoproteins (α-lipoproteins), it separates low-density lipoproteins (β-lipoproteins) from other lipoproteins equally well. In subjects in whose serum β-lipoprotein levels are quite elevated as in Figure 2, a large band or disc is produced from the large amount of component(s). This band may interfere with separation of the α-lipoproteins, especially if at least a 10-cm glass tube is not used, as was the case in the sample illustrated in Figure 2, for which a 7.5-cm tube was used. It can be seen that the bands producing the peaks must have been somewhat closer together than in Figure 1, for which a 10-cm tube was used.

That the two peaks so labeled represent high-density lipoproteins was demonstrated in the following ways: When HDL_2 was isolated by ultracentrifugal means,[31] dialyzed against buffer, and then placed on the gel, it migrated as one band to the position at which it appeared in serum. This also occurred with HDL_3 which moved to its relative position below HDL_2. No other bands appeared. When eight samples of HDL_2 (or HDL_3) ranging in amounts from 33 to 113μg protein were applied to eight gels, bands appeared in increasing intensities but in the same locations. Thus, the location of the bands is not dependent upon concentration of the lipoproteins, if the amounts are not excessive. The same serum samples analyzed over a period of several days produced the same patterns. In addition, when gels were placed in different combinations of lipoprotein precipitants, such as sulfated polysaccharides and metal ions,[32,33] the discs appeared in the same relative positions as they did when precipitated by 12% TCA. However, these precipitants did not produce sharp peaks with clean backgrounds such as is obtainable when proteins are precipitated by 12% TCA, or simply by scanning the untreated gels immediately after electrophoresis. Several lipid-staining dyes proved unsatisfactory, because the destaining process did not remove the dye, which remained in the gel between the bands. This would result in distorted densitometric tracings.

Serum from a healthy male, when titrated over a twofold range (100 to 50% in 5 increments), showed straight-line relationships between fractions obtained by ultracentrifugation and with polyacrylamide gel electrophoresis.

Data in Table 8A, concerned only with high-density lipoproteins, show that women subjects without evidence of cancer had more HDL_2 in their sera than men, when measured with the ultracentrifuge. The women also had more HDL_3, a more unusual observation. Women usually have more HDL_2 in their sera than men have,[34] but differences observed in these hospital patients are not as great as would be encountered in women subjects, more highly selected with regard to age and hormonal status. In the latter instance, women may have an average of 50 mg more HDL_2 per 100 ml serum than men of an equal age range and state of health. Most of the subjects in the present experiments were in the hospital for a variety of reasons and were not fasting. No knowledge of their drug intake was available. Although their ages ranged from 9 to 80 years, the averages for men and women were similar.

HDL values obtained from polyacrylamide gels are, however, different; women have essentially the same amounts of both HDL_2 and HDL_3 as men. Comparing the two techniques within individuals, the data in Table 8 show that with both techniques and for both sexes, there are greater amounts of HDL_3 than HDL_2. This corresponds to most published data regarding the relative amounts of these two high-density lipoproteins or α-lipoproteins.[34]

When coefficients of correlation (r) between calculated values (in milligram percent serum) from the ultracentrifuge and measurements of areas (in in.^{2}) were calculated, they were usually too low to claim any direct quantitative relationships between the two techniques. Also, it may be seen in Table 8 that the standard deviations are high for both techniques. This may be partially a result of the great heterogeneity of the subjects. The techniques, themselves, especially the ultracentrifuge, will give quite reproducible results in a single subject whose serum is analyzed at different time periods.

The few patients with different types of cancer (Table 8) had even more variable amounts of lipoproteins in their sera. Many of the women had higher values for HDL_2 especially and HDL_3 than usually observed in women with no clinical evidence of cancer. This may have resulted from drugs administered, especially estrogens or thyroxin. Although two women patients had high values for HDL_2 on polyacrylamide gels, two had very low values. These values did not agree with the ultracentrifuge data except in one out of five cases. The ultracentrifuge data on the men with cancer agree with

Table 8

COMPARATIVE DATA FROM TWO TECHNIQUES, ANALYTICAL
ULTRACENTRIFUGATION AND POLYACRYLAMIDE GEL, FOR
SEPARATING SERUM LIPOPROTEINS FROM PATIENTS WITH AND
WITHOUT CANCER

	Ultracentrifugation[a]				Polyacrylamide Gel[b]			
	HDL$_2$		HDL$_3$		HDL$_2$		HDL$_3$	
Subjects[c]	Avg.	±SD	Avg.	±SD	Avg.	±SD	Avg.	±SD
A. Subjects without cancer								
Women (63)	84	37.8	100	38.8	0.35	0.20	0.45	0.19
Men (45)	71	35.0	89	40.7	0.32	0.20	0.44	0.16
B. Subjects with cancer								
Women (5)	87	36.8	101	30.8	0.29	0.20	0.45	0.58
Men (4)	36	10.0	77	49.8	0.73	0.36	0.65	0.34

[a]   Average and SD values expressed in milligram percent.
[b]   Average and SD values expressed in inches square.
[c]   Numbers in parentheses are the numbers of subjects. Ages for the women ranged from 11 to
    80 years with 60% over age 40; ages for the men ranged from 9 to 73 years with 63% over
    age 40.

previously published observations[35] but do not correspond to the areas of the two high-density lipoproteins on polyacrylamide gel.

The polyacrylamide gel technique will readily separate lipoproteins, even when excessive quantities of one. e.g., β-lipoproteins, are present. It will reveal any conditions of excessive, moderate, or low amounts of lipoproteins , present in serum. In subjects without cancer, even in a hospital population, the technique will give relative amounts which agree with data from the ultracentrifuge. In women with cancer this was also true, but not in men. Admittedly, data from these few samples cannot be definitive, but they do suggest possible trends in what might be considered a usual sampling as compared to a highly controlled matched sampling. From a pragmatic point of view, the less well-matched sampling is the one met in clinical chemistry and the population in general.

These experiments illustrate again that serum lipoproteins applied to zonal media for electrophoretic separation cannot be expected to be quantitatively the same in value as the results obtained with analytical ultracentrifugation.[35] The chemical and physical bases of the separations are different. But, estimations, trends, and proportional amounts can be obtained with electrophoresis such that these simple, inexpensive, rapid procedures can have significant clinical value in diagnosis and the subsequent treatment of patients. They may also be valuable in evaluating the effects of drugs, diets, or other parameters employed in treating various diseases, in which serum lipoproteins may be altered chemically or in quantity.

SUMMARY

Most reports concerned with the effects of neoplastic diseases upon the levels of lipoproteins in serum of human subjects have shown that the α-lipoproteins or high-density lipoproteins are below normal values. Concomitantly, α_2- and β-lipoproteins

Table 9
ELECTROPHORETIC TECHNIQUES[a]

Author and ref. no.	Year	Medium special conditions	Buffer	Cell and special equipment	Electrical conditions	Time of electrophoresis	Temp	Special handling	Staining	Evaluation	Densitometry	Methodological error
Miller, B. J. and Erf, L.[12]	1956	Paper, Whatman® #1	Veronal, pH 8.6	—	0.4 mA/cm; 150 V	12 hr	15°C	Dried at 100°C, then cut in half lengthwise	Bromphenol blue and Sudan III	Percent of total lipoprotein	Scan at 560 nm and planimetry	—
Falor, W. H. et al.[13]	1965	Paper, Schleicher and Schuell #2043A	Veronal, pH 8.6, μ 0.05	Spinco, Model R kit	Not given	Not given	Not given	—	Protein—bromthymol blue and lipid oil red O	Percent of total lipoprotein	Spinco:Analytrol, Model RB	—
Ma, L. et al.[14]	1965	Paper	Veronal, pH 8.6, μ, 0.05	Vertical tank	110 V	9 hr	Not given	Dried at 105°C for 15 min	Sudan black B for 3 hr	α/β ratio percent, α-lipoprotein in milligram percent	Analytrol, Model RA	—
Narayan, K. A. et al.[24]	1965	Polyacrylamide gel	7.0% page not given	Disc	2.5 mA/tube	73—85 min	Not given	Prestained sera; see Ressler et al.[48]	Sudan black B	Areas of peaks in mm^2	Densitometry	—
Higazi, A. M. et al.[17]	1966	Paper	Not given; see Swahn[18]	—	Not given; see Swahn[18]	See Swahn[18]	Not given	—	See Swahn[18]	—	—	—
Narayan, K. A.[17]	1967	PAGE at 3.75 and 5%	TRIS-glycine, pH 8.9	Disc	2.5 mA/tube	73—85 min	—	—	—	—	—	—
Feldman, E. B. and Carter, A. C.[19]	1971	Paper	Veronal and albumin	—	Not given; see Hatch and Lees[20]	See Hatch and Lees[20]	See Hatch and Lees[20]	—	See Hatch and Lees[20]	Percent of total lipoprotein	See Hatch and Lees[20]	—
Nishioka, M.[28]	1971	PAGE at 2.4 and 7%	TRIS-glycine, pH 8.9	Disc	Not given	Not given	Not given	—	Amido black for 1 hr	areas of peaks	Densitometry and planimetry of peaks or curves	—
Terebus-Kekish, O. et al.[29]	1976	PAGE at 3.5%	TRIS-glycine, pH 8.4	Disc	3.0 mA/tube	1.5 hr	Room temp	10-cm tubes used	TCA	areas of peaks	Densitometry and planimetry of peaks or curves	±0.2

[a] These abbreviations will be used in the table: PAGE = polyacrylamide gel electrophoresis; TCA = trichloroacetic acid.

Table 10
APPLICATIONS OF ELECTROPHORESIS[a]

Author and ref. no.	Year	Type or characteristics	Pattern	Normal values	±SD	Reproducibility	Pathology	±SD	Characteristics of subjects, age, sex, race	Summary of results
Miller, B. J. and Erf, L.[12]	1956	Carcinoma of breast	Lower α and higher β in carcinoma	α = 24.4%; β = 75.6%	Not given	Ranges for α-lipoprotein = 13—43 and β = 57—88	Carcinoma of breast: α = 7.6%, β = 92.4%	Ranges: α = 0—24%; β = 76—100%	Carcinoma of breast; various ages; Caucasian	Lower α-lipoprotein; higher β-lipoprotein in patients
Falor, W. H. et al.[13]	1965	Carcinoma of lungs	Lower α and higher β in carcinoma	α = 19%; β = 76%; "trail" = 5%	α = 11%; β = 20%; "trail" = 4%	Ranges for α-lipoprotein 4—37% β-lipoprotein = 69—95%; "trail" = 1—13%	Carcinoma of lung: α = 14%, β = 79%, "trail" < 5%	± α = ±7, β = ±10.4, "trail" = ±4.1	Caucasian	Lower α-lipoprotein higher β-lipoprotein in patients with carcinoma of lung
Ma, L. et al.[14]	1965	Hydatidiform mole (associated with choriocarcinoma)	Lower α and higher β in patients with mole	α/β = 26.2/73.8;α = 26.2%	Percent of α = 2.9	Range for α-lipoprotein = 23—30	Mole: α/β = 12.3/87.7, Percent of α = 12.3	Percent α = ±4.8	Women; ages, 22—38 years; Oriental	Ratio of lower β to higher β in women with hydatidiform mole (associated with choriocarcinoma)
Higazi, A. M. et al.[15]	1966	Different leukemias	Lower α in three of five types of leukemias; β higher in three	α = 154 mg%; β = 211 mg%	α = ±37.6		Chronic myeloid: α = 143%; β = 213%	±41.6 ±62.3	Egyptian subjects	Generally lower α; More variable β
Feldman, E. B. and Carter, A. C.[16]	1971	Carcinoma of breast	Lower α with ultracentrifugation, not with electrophoresis	α = 25%; β = 67%; "origin" = 8%	β = ±43.0	chronic lymphocytic: α = 117% β = 199%	acute myeloid: α = 180% β = 249% carcinoma of breast: α = 28.5%, β = 65.7%, "origin" = 10.7%	±55.1 ±84.8 ±64.2 ±40.1	Women of various ages; all subjects were postmenopausal	Inconclusive with electrophoresis; α-lipoprotein probably lower with preparative ultracentrifugation experiments
Nishioka, M.[17]	1971	Primary liver carcinoma	Bands on PAGE less sharp and poorly separated in patients	Not given	Not given	Good?	Presence of α-fetoprotein in gel of serum from patients with primary hepatocarcinoma: Separations less sharp.		Japanese subjects	—
Terebus-Kekish, O. et al.[18]	1976	General hospital population with a variety of causes for hospitalization and of both sexes; most patients not reported to have cancer; groups studied separately	Studied HDL with both ultracentrifugation and PAGE; HDL_3 and HDL_2 higher in women; both women and men had more HDL_3 than HDL_2	Hospital patients with no clinical evidence of cancer; values obtained by ultracentrifugation in mg%: female — HDL_3, 84, and HDL_2, 100, male — HDL_3, 71, and HDL_2, 89; PAGE, areas in in.²: female — HDL_3, 0.35, and HDL_2, 0.45, male — HDL_3, 0.32, and HDL_2, 0.44	SD for ultracentrifugation: female — HDL_3, ±37.8, and HDL_2, ± 38.8; male — HDL_3, ±35.0, and HDL_2, ±40.7; PAGE, areas in in.²: female — HDL_3, ±0.20, and HDL_2, ±0.19; male — HDL_3, ±0.20, and HDL_2, ±0.16	Techniques are excellent for ultracentrifuge and good for PAGE	Ultracentrifuge values in mg%: female — HDL_3, 87, and HDL_2, 77; PAGE, areas in in.²: female — HDL_3, 0.29, and HDL_2, 0.45, male — HDL_3, 0.73, and HDL_2, 0.65	SD for ultracentrifugation: female — HDL_3, ±36.8, and HDL_2, ±30.8, male — HDL_3, ±10.0, and HDL_2, ±49.8; PAGE, areas in in.²: female — HDL_3, ±0.20, and HDL_2, ±0.20, and HDL_3, ±0.58, male — HDL_2, ±0.36, and HDL_3, ±0.34	Mixed, typical general hospital sampling; women's ages, 11—80 years; men 9—73 years; 60% > 40 years; 63 women, 45 men with no cancer; 5 women, 4 men with cancer; none were fasted; avg. American population	According to ultracentrifuge, women in both groups had more HDL_3 and HDL_2 than men; only men with cancer had lower HDL_3 and HDL_2 than "control" men; according to PAGE, both groups had more HDL_3 than HDL_2; women with cancer had slightly less HDL_3

[a]   These abbreviations will be used in the table: HDL = high-density lipoprotein; PAGE = polyacrylamide gel electrophoresis.

or very low- and low-density lipoproteins may be elevated. These elevations can be significant even in fasting subjects. These observations hold, for the most part, regardless of the techniques used, providing, of course, that there has been some control over the selection of patients (essentially no treatment with drugs) and suitable control subjects and over the handling of electrophoretic equipment.

It can be stated, however, that there is sufficient concordance of the trends and relationships of the lipoprotein levels observed by electrophoresis and the ultracentrifuge technique to justify employment of electrophoresis. The latter technique is simpler, more rapid, and reasonably reproducible and sensitive, but not as quantitative as analytical ultracentrifugation. The equipment, even including a recording densitometer, is less expensive and less complicated to operate than the analytical ultracentrifuge. For surveys of clinical material, electrophoresis should give acceptably reliable indications of serum lipoprotein patterns in both normal subjects and patients with neoplastic diseases uncomplicated by liver abnormalities and drugs.

ACKNOWLEDGMENT

Part of M. Barclay's contribution was supported by grant CA08748 from the National Cancer Institute, and the Elsa U. Pardee Foundation.

REFERENCES

1. Kunkel, H. G. and Slater, R. J., Lipoprotein patterns of serum obtained by zone electrophoresis, *J. Clin. Invest.*, 31, 677, 1952.
2. Lewis, L. A. and Page, I. H., Electrophoretic and ultracentrifugal analysis of serum lipoproteins of normal, nephrotic and hypertensive persons, *Circulation*, 7, 707, 1953.
3. Gofman, J. W., Hanig, M., Jones, H. B., Lauffer, M. A., Lawry, E. Y., Lewis, L. A., Mann, G. W., Moore, F. E., Olmsted, F., and Yeager, J. F., Evaluation of serum lipoproteins and cholesterol measurements as predictors of clinical complications of atherosclerosis, *Circulation*, 14, 691, 1956.
4. Fredrickson, D. S., Levy, R. I., and Lees, R. S., Fat transport in lipoproteins: an integrated approach to mechanisms and disorders, *N. Engl. J. Med.*, 276, 34, 1967.
5. Barclay, M., Cogin, G. E., Escher, G. C., Kaufman, R. J., Kidder, E. D., and Petermann, M. L., Human plasma lipoproteins. I. In normal women and in women with advanced carcinoma of the breast, *Cancer*, 8, 253, 1955.
6. Barclay, M., Calathes, D. N., DiLorenzo, J. C., Helper, A., and Kaufman, R. J., The relation between plasma lipoproteins and breast carcinoma: effect of degrees of breast disease on plasma lipoproteins and the possible role of lipid metabolic aberrations, *Cancer*, 12, 1163, 1959.
7. Barclay, M., Escher, G. C., Kaufman, R. J., Terebus-Kekish, O., Greene, E. M., and Skipski, V. P., Serum lipoproteins and human neoplastic disease, *Clin. Chim. Acta*, 10, 39, 1964.
8. Barclay, M., Skipski, V. P., Terebus-Kekish, O., Greene, E. M., Kaufman, R. J., and Stock, C. C., Effects of cancer upon high-density and other lipoproteins, *Cancer Res.*, 27, 2420, 1970.
9. Shore, B. and Shore, V., Isolation and characterization of polypeptides of human serum lipoproteins, *Biochemistry*, 8, 4510, 1969.
10. Scanu, A. and Granda, J. L., Human serum lipoproteins: advances in physical and chemical methods of isolation and analysis, in *Progress in Clinical Pathology*, Vol. 1, Stefanini, M., Ed., 1966, 398.
11. Alaupovic, P., Conceptual development of the classification systems of plasma lipoproteins, *Protides Biol. Fluids Proc. Colloq.*, 19, 9, 1972.
12. Miller, B. J. and Erf, L., The serum proteins and lipoproteins in patients with carcinoma and in subjects free of recurrence, *Surg. Gynecol. Obstet.*, 102, 487, 1956.
13. Russ, E. M., Eder, H. A., and Barr, D. P., Protein-lipid relationships in human plasma, *Am. J. Med.*, 11, 468, 1951.
14. Toyoda, N., Serum lipoproteins in cancer patients; the influence of the grade of cancer and its surgical treatment, *Excerpta Med. Sect. 16*, 12, 5, 1964, abstr.

15. **Falor, W. H., Dettling, J. J., Kerkian, A., Bachtel, W., Krismann, E., and Thomas, M. A.,** Electrophoresis of chyle in bronchogenic carcinoma, *Arch. Surg.,* 91, 671, 1965.
16. **Ma, L., Chew, W., Pang, L. S. C., and Wong, C. W.,** Serum lipoproteins in hydatidiform mole, *Clin. Chim. Acta,* 11, 561, 1965.
17. **Higazi, A. M., Ata, A. A., Abdel-Rahman, Y. M., Malek, A., and Mansour, K.,** Electrophoretic pattern of serum proteins and lipids in leukaemias, *J. Egypt. Med. Assoc.,* 49, 679, 1966.
18. **Swahn, B.,** Studies on blood lipids, *Scand. J. Clin. Lab. Invest., Suppl.* 5, 9, 1953.
19. **Feldman, E. B. and Carter, A. C.,** Circulating lipids and lipoproteins in women with metastatic breast carcinoma, *J. Clin. Endodrinol. Metab.,* 33, 8, 1971.
20. **Hatch, F. T. and Lees, R. S.,** Practical methods for plasma lipoprotein analysis, *Adv. Lipid Res.,* 6, 1, 1968.
21. **Ornstein, L.,** Disc electrophoresis. I. Background and theory, *Ann. N. Y. Acad. Sci.,* 121, 321, 1964.
22. **Raymond, S.,** Acrylamide gel electrophoresis, *Am. N. Y. Acad. Sci.,* 121, 350, 1964.
23. **Davis, B. J.,** Disc electrophoresis. II. Method and application to human serum proteins, *Ann. N. Y. Acad. Sci.,* 121, 404, 1964.
24. **Narayan, K. A., Narayan, S., and Kummerow, F. A.,** Disk electrophoresis of human serum lipoproteins, *Nature,* 205, 246, 1965.
25. **Ornstein, L. and Davis, B. J.,** Disc electrophoresis, Distillation Products Industries, Rochester, New York, 1962.
26. **Ressler, N., Springgate, R., and Kaufman, J.,** A method of lipoprotein electrophoresis, *J. Chromatogr.,* 6, 409, 1961.
27. **Narayan, K. A.,** Disc electrophoresis of human and animal serum lipoproteins, *Lipids,* 2, 282, 1967.
28. **Nishioka, M.,** Polyacrylamide gel electrophoresis of sera from patients with primary liver carcinoma, *Clin. Chim. Acta,* 31, 439, 1971.
29. **Terebus-Kekish, O., Barclay, M., and Stock, C. C.,** Discrete separation of HDL$_2$ from HDL$_3$ of human serum by means of polyacrylamide gel, *Clin. Chim. Acta,* 88, 9, 1978.
30. **Instruction Manual,** Canalco,® Model 1200, *Scientific Products,* Edison, New Jersey.
31. **Barclay, M., Barclay, R. K., Terebus-Kekish, O., Shah, E. B., and Skipski, V. P.,** Disclosure and characterization of new high-density lipoproteins in human serum, *Clin. Chim. Acta,* 8, 721, 1963.
32. **Burstein, M. and Scholnick, H. R.,** Isolation of lipoproteins from human serum by precipitation with polyanions and divalent cations *Protides Biol. Fluids Proc. Colloq.,* 19, 21, 1971.
33. **Seidel, D., Wieland, H., and Ruppert, C.,** Improved techniques for assessment of plasma lipoprotein patterns. I. Precipitation in gels after electrophoresis with polyanionic compounds, *Clin. Chem.,* 19, 737, 1973.
34. **Barclay, M.,** Lipoprotein class distribution in normal and diseased states, in *Blood Lipids and Lipoproteins, Quantitation, Composition and Metabolism,* Nelson, G., Ed., John Wiley and Sons, New York, 1972, 585.
35. **Barclay, M. and Skipski, V. P.,** Lipoproteins in relation to cancer, *Prog. Biochem. Pharmacol.,* 10, 76, 1975.
36. **Lewis, L. A.,** Screening for serum lipoprotein abnormalities: comparison of ultracentrifugal, paper and thin-layer starch-gel electrophoresis techniques, *Lipids,* 4, 60, 1969.

Lipoprotein Literature

BIBLIOGRAPHY OF LIPOPROTEIN LITERATURE

Marie A. Opplt

TABLE OF CONTENTS

The author presents references concerning electrophoretic analyses of lipoproteins in humans, performed in years 1964—1976. The more complete lipoprotein electrophoretic literature begins in 1952—1953 with Swahn's publication of the first successful staining of two electrophoretic fractions of serum lipoproteins, separated by electrophoresis on paper as a support medium. When the work on the "Handbook of Electrophoresis of Lipoproteins" Volume I, started in 1977, the existing bibliography published in 24 years, was divided into two parts: the older (1952—1963) and the newer (1964—1976). This second part is presented here. Volume II, expected in 1980, will contain the recent bibliography (1977—1979).

The publications covering different problems are arranged according to the years in which they were published. This method allows the reader to follow the development of different methodologies and problems studied during a certain period of time and gives him the opportunity to survey the published literature. The bibliography is subject and problem oriented. The author recognizes shortcomings in the citation of only one author mentioned in every publication.

LIPOPROTEINS IN GENERAL

Lipidoproteins and Lipoproteins in General

Bosch, V. et al., Serum lipoproteins of Amazonian Indians and inhabitants of an urban area of Venezuela fractionated by preparative ultracentrifugation, *Metabolism*, 13, 1456, 1964.

Chudnova, I. M., The effects of saluzide and streptomycin on the blood serum proteins and lipoproteins *(in Russian)*, *Farmakol. Toksikol.*, 27, 349, 1964.

Freeman, N. K. et al., Serum lipid-lipoprotein interrelationships between males and females, *U.S. Atomic Energy Commission*,University of California Radiation Laboratory, 134, 1964.

Rehnborg, C. S. et al.,Changes in lipoprotein composition following prolonged incubation of human serum, *U.S. Atomic Energy Commission, University of California Radiation Laboratory*, 61, 1964.

Ewing, A. M. et al., The analysis of human serum lipoprotein distributions, *Adv. Lipid Res.*, 3, 25, 1965.

Memeo, S. A., Lipoproteins and their importance in clinical diagnosis (in Italian), *Minerva Med.*, 56, 2533, 1965.

Raczynski, G. et al., Lipo-amino acids and lipo-peptides (in Polish), *Postepy Biochem.*, 11, 499, 1965.

Albutt, E. C., A study of serum lipoproteins, *J. Med. Lab. Technol.*, 23, 61, 1966.

Bufardeci, F. et al., On changes of lipoprotein content of the blood in drivers of motor vehicles (in Italian), *Boll. Soc. Ital. Biol. Sper.*, 42, 361, 1966.

Chong, Y. H. et al., A comparison of plasma lipoprotein levels in Asians, Africans and Europeans, *Med. J. Malaya*, 20, 284, 1966.

Parkins, R. A. et al., A preliminary study of factors affecting blood lipid levels in three groups of Yemenite Jews, *Am. J. Clin. Nutr.*, 18, 134, 1966.

Turner, J. D., Lipoproteins revisited, *Cardiovasc. Res. Cent. Bull. Houston*, 4, 99, 1966.

Wehr, H., Serum lipoproteins (in Polish), *Postepy Biochem.*, 12, 383, 1966.

Bosch, V., Lipoproteins in ultracentrifugal fractions of sera of a nomadic South American Indian Tribe, *Acta Cient. Venez.*, 18, 159, 1967.

Harlan, W. R., Jr. et al., Constitutional and environmental factors related to serum lipid and lipoprotein levels, *Ann. Intern. Med.*, 66, 540, 1967.

Ito, H., Lipoproteins (in Japanese), *Jpn. J. Clin. Pathol.*, Suppl. 12, 28, 1967.

Nichols, A. V., Human serum lipoproteins and their interrelationships, *Adv. Biol. Med. Phys.*, 11, 109, 1967.

Strasser, H. et al., On the clinical significance of serum lipid and serum lipoprotein examinations (in German), *Schweiz. Med. Wochenschr.*, 97, 1397, 1967.

Watson, D., Binding of dye anions to plasma albumin and lipoproteins, *Clin. Chim. Acta*, 15, 121, 1967.

Heimberg, M., Lipids and lipoproteins of human serum, *J. Tenn. Med. Assoc.*, 61, 167, 1968.

Lloyd, J. K., Disorders of the serum lipoproteins. I. Lipoprotein deficiency states, *Arch. Dis. Child.*, 43, 393, 1968.

Lloyd, J. K., Disorders of the serum lipoproteins. II. Hyperlipoproteinaemia states, *Arch. Dis. Child.*, 43, 505, 1968.

Wurm, M. et al., Characterization of human serum proteins and lipoproteins, *Am. J. Clin. Pathol.*, 50, 175, 1968.

Griel, L. C., Jr. et al., Blood serum lipoproteins: a review, *J. Dairy Sci.*, 52, 1233, 1969.

Ellis, J. B. et al., Lipoproteins in clinical medicine, *V. Med. Mon.*, 96, 672, 1969.

Lloyd, J. K., Lipoprotein deficiency disorders, *Bristol Med. Chir. J.*, 84, 159, 1969.

Masopust, J., Lipoproteins — their chemistry, physiology and pathology (in Czech), *Cesk. Pediatr.*, 24, 65, 1969.

Pache, W. et al., Metabolic products of microorganisms. Binding of antibiotics to lipoproteins (in German), *Arch. Mikrobiol.*, 66, 281, 1969.

Schumaker, V. N. et al., Circulating lipoproteins, *Ann. Rev. Biochem.*, 38, 113, 1969.

Botre, C. et al., Alkali cation selectivity in a lipoprotein complex, *Farmaco (Sci).*, 25, 939, 1970.

Dyerberg, J., Lipid and lipoprotein analysis-clinical chemical overview-indication for lipid analysis (in Danish), *Nord. Med.*, 84, 1538, 1970.

Pozner, H. et al., Effect of smoking on blood-clotting and lipid and lipoprotein levels, *Lancet*, 760, 1318, 1970.

Soukupova, K. et al., Serum lipids, glucose tolerance tests and clinical signs in healthy men with a raised β - α lipoprotein index, *Gerontol. Clin.*, 12, 1, 1970.

Vogelberg, K. H. et al., Tri-di-and monoglycerides in human serum and their relation to the lipoprotein spectrum, *Klin. Wochenschr.*, 48, 227, 1970.

Winkelman, J. et al., Correlation of laboratory tests and clinical evaluation in phenotyping of lipoproteinemias, *Clin. Chem.*, 16, 594, 1970.

Worowski, K. et al., The role of lipids and lipoproteins in hemostatic processes (in Polish), *Postepy Biochem.*, 16, 51, 1970.

Yasugi, T. et al., Lipoprotein deficiency, *Jpn. J. Clin. Med.*, 28, 2105, 1970.

Bang, H. O. et al., Plasma lipid and lipoprotein pattern in Greenlandic Westcoast Eskimos, *Lancet*, 1, 1143, 1971.

Berg, K., Compositional relatedness between histocompatibility antigens and human serum lipoproteins, *Science*, 172, 1136, 1971.

Dietrich, R. A., Lipoproteins-clinical application, *Med. Ann. D. C.*, 40, 157, 1971.

Folch-Pi, J., The Folch-Pi-Lees proteolipid (FLPL), *Neurosci. Res. Program Bull.*, 9, 544, 1971.

Folch-Pi, J., Wolfgram proteolipid (WPRL), *Neurosci. Res. Program Bull.*, 9, 545, 1971.

Isselbacher, K. J., The intestine lipid transport and lipoproteins, *Klin. Sci.*, 40, 16P, 1971.

Koren, E. et al., Classification of lipoproteinemias (in Croation with English abstract), *Acta Med. Iugosl.*, 25, 227, 1971.

Lopez, A., Serum lipid transport systems: recent advances, *Lipids*, 6, 369, 1971.

Myant, N. B., Turnover of plasma lipoprotein cholesterol, *Biochem. J.*, 123, 19P, 1971.

Quarfordt, S. H. et al., Transfer of triglyceride between isolated human lipoproteins, *Biochim. Biophys. Acta*, 231, 290, 1971.

Scanu, A., Plasma lipoproteins: concepts and trends, *Clin. Sci.*, 40, 15P, 1971.

Schuman, J. M., Lipoprotein analysis: a comparative study, *Am. J. Med. Technol.*, 37, 400, 1971.

Schwandt, E., Protides of biological fluids. Report on the 19th colloquium, Brugge, Belgium, *Fortschr. Med.*, Suppl. 89, 8, 1971.

Seidel, D., Plasma lipoproteins: characterization and clinical importance (in German), *Dtsch. Med. J.*, 22, 445, 1971.

Strunge, P. et al., The lipoprotein pattern in a Danish family. Preliminary report, *Acta Med. Scand.*, 189, 73, 1971.

Yasugi, T. et al., Lipoprotein — how to read its figures (in Japanese), *Jpn. J. Clin. Med.*, Suppl. 29, 305, 1971.

Bang, H. O. et al., Plasma lipids and lipoproteins in Greenlandic west coast Eskimos, *Acta Med. Scand.*, 192, 85, 1972.

Bermes, E. W., Jr. et al., The stability of human serum lipoproteins in vitro, *Ann. Clin. Lab. Sci.*, 2, 226, 1972.

Dyerberg, J. et al., Plasma lipid and lipoprotein levels in Danish population, *Acta Med. Scand.*, 191, 413, 1972.

Folch-Pi, J., Proteolipids, *Adv. Exp. Med. Biol.*, 32, 171, 1972.

Illingworth, D. R. et al., Independence of phospholipid and protein exchange between plasma lipoproteins in vivo and in vitro, *Biochim. Biophys. Acta*, 280, 281, 1972.

Kanno, T. et al., Function and disorders of lipoproteins (in Japanese), *Jpn. J. Clin. Pathol.*, 20, 252, 1972.

Kawai, T., Past and present of the study of lipoprotein analysis (in Japanese), *Saishin Igaku*, 27, 412, 1972.

Kretschmar, R. et al., On the determination of the distribution functions of the lipoproteins in human sera, *Biochim. Biophys. Acta*, 280, 105, 1972.

Lewis, L. A., Serum lipids and lipoproteins. Clinical relevance, *Cleveland Clin. Q.*, 39, 9, 1972.

Mammel, B. T. et al., The stability of cholesterol protein complexes in disease, *J. Am. Osteopath. Assoc.*, 71, 973, 1972.

Matsumoto, H., Hereditary traits concerning lipoprotein (in Japanese), *Saishin Igaku*, 27, 460, 1972.

Métais, P., Lipemia and lipoproteinemia. Detection and classification of their anomalies, *Rev. Med. Liege*, 27, 767, 1972.

Otani, H., Congenital lipoprotein deficiency (in Japanese), *Saishin Igaku*, 27, 491, 1972.

Shuster, M. et al., Lipid and lipoprotein screening tests, *J. Med. Soc. N.J.*, 69, 121, 1972.

Simons, L. et al., A study of hyper- β -and hyper pre- β -lipoproteinaemia, *Br. Heart J.*, 34, 960, 1972.

Strunge, P. et al., The lipoprotein pattern in a Danish family. Children and adolescents, *Acta Med. Scand.*, 192, 331, 1972.

Uzawa, H., Metabolism of serum lipoprotein — with special reference to metabolism of endogenous triglycerides (in Japanese), *Saishin Igaku*, 27, 476, 1972.

Wood, P. D. et al., Prevalence of plasma lipoprotein abnormalities in a free-living population of the Central Valley, California, *Circulation*, 45, 114, 1972.

Asmal, A. C. et al., Is lipoprotein estimation really necessary?, *Lancet*, 1, 609, 1973.

Bilheimer, D. W. et al., Origin and fate of lipoproteins, *Adv. Exp. Med. Biol.,* 38, 39, 1973.

Dito, W. R., A simple time-saving method for interpretive report generation. II. Lipoprotein phenotyping, *Am. J. Clin. Pathol.,* 59, 448, 1973.

Fasoli, A., Biological significance of serum lipoproteins, *Adv. Exp. Med. Biol.,* 38, 23, 1973.

Guardiola, J. et al., Lipoproteins (in Spanish with English abstract), *Rev. Clin. Esp.,* 130, 229, 1973.

Keith, A. D. et al., Spin labeled lipid probes in serum lipoproteins, *Chem. Phys. Lipids,* 10, 223, 1973.

Lees, R. S., Editorial: a progress report on lipoprotein phenotyping, *J. Lab. Clin. Med.,* 82, 529, 1973.

Lewis, B., Classification of lipoproteins and lipoprotein disorders, *J. Clin. Pathol.,* Suppl. 5, 26, 1973.

Lorimer, A. R. et al., Prevalence of lipoprotein abnormalities in the west of Scotland, *Br. Heart J.,* 35, 862, 1973.

Scanu, A. M., Research on plasma lipoproteins: theoretical aspects and relevance to clinical medicine, *J. Chronic Dis.,* 26, 325, 1973.

Schettler, G. et al., Advances in lipoprotein research: biochemical and clinical aspects, *Med. J. Aust.,* 1, 942, 1973.

Schettler, G. et al., Advances in lipoprotein research. Biochemical and clinical aspects, *Arch. Inst. Cardiol. Mex.,* 43(3), 474, 1973.

Schwertner, H. A. et al., Changes in lipid values and lipoprotein patterns of serum samples contaminated with bacteria, *Am. J. Clin. Pathol.,* 59, 829, 1973.

Seidel, D., Advances in lipoprotein research. Biochemical and clinical aspects, *Nutr. Metab.,* 15, 9, 1973.

Seidel, D. et al., Biochemical and clinical aspects of various lipoprotein patterns, *Ann. Biol. Clin. (Paris),* 31, 87, 1973.

Sekimoto, H. et al., Lipoproteins — interpretation of the test results, *Jpn. J. Clin. Med.,* 31, 1371, 1973.

Skrzydlewski, Z. et al., Blood serum lipoproteins and their role in the etiology of some pathological states, *Ginekol. Pol.,* 44, 455, 1973.

Avogaro, P. et al., Serum lipids levels, lipoprotein phenotypes and analysis of lipoprotein in Italian population group (in Italian with English abstract), *G. Ital. Cardiol.,* 4, 237, 1974.

Balta, M. et al., Serum lipoprotein changes in certain internal diseases, *Rev. Roum. Med.,* 12(2), 91, 1974.

Brunet, M. R. et al., Techniques in study of serum lipoproteins, Minerva, *Pediatrics,* 26, 288, 1974.

Byers, S. O. et al., Neurogenic hypercholesterolemia: influence upon lipoproteins, *Proc. Soc. Exp. Biol. Med.,* 145, 442, 1974.

Descovich, G. C. et al., Studies of serum lipid and lipoprotein composition in healthy subjects in the province of Bologna, *Boll. Soc. Ital. Biol. Sper.,* 50(13), 970, 1974.

Ellafson, R. D., Letter: blood lipoproteins, *Arthritis Rheum.,* 17(6), 1057, 1974.

Heiberg, A., The heritability of serum lipoprotein and lipid concentrations. A twin study, *Clin. Genet.,* 6(4), 307, 1974.

Heiberg, A. et al., Serum lipid and lipoprotein concentrations in a Norwegian population sample, *Acta Med. Scand.,* 196(3), 155, 1974.

Ho, W. K. et al., Serum lipid and lipoprotein levels in normal Chinese, *Clin. Chim. Acta,* 57(2), 149, 1974.

Jones, J. J. et al., Fasting serum lipoproteins in rural Africans in Rhodesia, measured by membrane filtration and nephelometry, *Clin. Chim. Acta,* 57(2), 13, 1974.

Polychronopoulou, A. et al., Lipoprotein types, serum cholesterol, and ABO blood groups, *Br. J. Prev. Soc. Med.,* 28, 60, 1974.

Schettler, G. et al., Advances in lipoprotein research: biochemical and clinical aspects, *Folia Clin. Int.,* 24(5), 357, 1974.

Schlierf, G., Diurnal fluctuations of plasma lipids and lipoproteins: the influence of corticosteroids and growth hormone, in *Blood and Arterial Wall in Atherogenesis and Arterial Thrombosis,* Hautvast, J. G. A. J., et al., Eds., Leiden, Brill, 1975, 23.

Carlson, L. A. et al., Quantitative and qualitative serum lipoprotein analysis. Part I. Studies in healthy men and women, *Atherosclerosis,* 21(3), 417, 1975.

Fredrickson, D. S., It's time to be practical, *Circulation,* 51(2), 209, 1975.

Gulbrandsen, C. L. et al., Serum lipoprotein patterns in Puerto Rican men, *Bol. Assoc. Med. P.R.,* 67(6), 148, 1975.

Ho, W. K. et al., Evaluation of serum lipid and lipoprotein levels in normal Chinese. The influence of dietary habit, body weight, exercise and familial record of coronary heart disease, *Clin. Chim. Acta,* 61(1), 19, 1975.

Jones, J. J. et al., Effects of twenty-four hours starvation and of heparin on the serum lipoproteins of rural Africans in Rhodesia, *Clin. Chim. Acta,* 58(3), 299, 1975.

Kimura, M. et al., Prevalence of plasma lipoprotein abnormalities in the farming village of Tanushimaru and the fishing village of Ushibuka. From the epidemiological aspects, *Jpn. Circ. J.,* 39(3), 299, 1975.

Lindgren, F. T. et al., Lipid and lipoprotein measurements in a normal adult American population, *Lipids,* 10(12), 750, 1975.

Pilcher, J., Lipid abnormalities in adolescence and later life, *Practitioner,* 214(1280), 213, 1975.

McCallum, H., Cholesterol, triglyceride and lipoprotein fraction levels in 122 healthy Malawians, *Cent. Afr. J. Med.*, 21(8), 171, 1975.

Greten, H., Lipoprotein analysis and typing system. Problems in diagnosis of lipid metabolism disorders in practice, *Dtsch. Med. Wochenschr.*, 100(9), 439, 1975.

Kimura, N. et al., Classification and metabolic patterns of hyperlipoproteinemia with special reference to serum lipoprotein metabolism (in Japanese), *Jpn. J. Clin. Med.*, 33(11), 3147, 1975.

Levy, R. I., Pathophysiology of lipid transport disorders, *Postgrad. Med. J.*, 51(8), 16, 1975.

Cichocki, T. et al., Lipoprotein metabolism — physiological and clinical aspects. Occurrence, distribution and composition of lipids in the human body, *Przegl. Lek.*, 33(4), 468, 1976.

Eisenberg, S., Metabolism of very low density lipoproteins, in *Lipoprotein Metabolism*, Greten, H., Ed., Springer-Verlag, Berlin, 1976, 32.

Jackson, R. L. et al., Lipoprotein structure and metabolism, *Physiol. Rev.*, 56(2), 259, 1976.

Jonado, M. et al., Dynamic organization of lipoprotein (in Japanese), *Protein Nucleic Acid Enzyme*, 21(3), 184, 1976.

Kostner, G. M., Apo β-deficiency (a-β-lipoproteinaemia): a model for studying the lipoprotein metabolism, in *Lipid Absorption: Biochemical and Clinical Aspects*, Rommel, K. and Bohmer, R., Eds., University Park Press, Baltimore, 1976, 203.

Ostrowski, W., Postgraduate teaching. Lipoprotein metabolism — physiological and clinical aspects. Lipids: classification, structure, metabolism (in Polish), *Przegl. Lek.*, 33(5), 534, 1976.

Polonovski, J., Some aspects of lipoprotein metabolism (in French), *Biochemie*, 58(8), 971, 1976.

Portman, O. W. et al., Influence of lysophosphatidylcholine on the metabolism of plasma lipoproteins, *Biochim. Biophys. Acta*, 450(3), 322, 1976.

Schettler, G., New results in the clinical investigation of the fat metabolism (in Spanish), *Folia Clin. Int.*, 26(10), 391, 1976.

Sznajd, J., Postgraduate courses: lipoprotein metabolism — its physiological and clinical aspects (in Polish), *Przegl. Lek.*, 33(3), 398, 1976.

Sznajderman, H., Disorders of triglyceride metabolism and their role in pathology (in Polish), *Przegl. Lek.*, 33(8), 745, 1976.

Kizaki, H. et al., Analysis of serum lipoproteins and its chemical significance (in Japanese), *Bull. Seishin Igaku Inst.*, 17, 25, 1977.

Metabolism

Ayrault-Jarrier, M. et al., Current data on the metabolism of circulating lipoproteins (in French), *Rev. Prat.*, 15, 4035, 1965.

Scanu, A. H., Factors affecting lipoprotein metabolism, *Adv. Lipid Res.*, 3, 63, 1965.

Roheim, P. S. et al., Alterations of lipoprotein metabolism in orotic acid, *Nutr. Rev.*, 24, 27, 1966.

Mokrasch, L. C. et al., Non-enzymic incorporation of amines into proteolipid protein, *J. Neurochem.*, 15, 1207, 1968.

Skořepa, J. et al., A study on the plasmatic lipoproteins metabolism. II. The preparation of subtrates (in Czech), *Sb. Lek.*, 70, 282, 1968.

Dyerberg, J., Lipids and lipoproteins in plasma. Metabolism and pathophysiology — a clinical-chemical survey (in Danish), *Ugeskr. Laeg.*, 132, 2267, 1970.

Greten, H., Clinical aspects of lipid metabolism, *Naunyn Schmiedebergs Arch. Pharmacol.*, 269, 379, 1971.

Levy, R. I. et al., The structure and metabolism of chylomicrons and very low density lipoproteins (VLDL), *Biochem. Soc. Symp.*, 33, 3, 1971.

Marsh, J. B., Biosynthesis of plasma lipoproteins, *Biochem. Soc. Symp.*, 33, 89, 1971.

Norum, K. R., Composition and metabolism of circulating lipoproteins (in Norwegian with English abstract), *Nord. Med.*, 86, 1173, 1972.

Eder, H. A. et al., Mechanism of lipoprotein production, *Expo. Annu. Biochim. Med.*, 31, 47, 1972.

Hamilton, R. L., Synthesis and secretion of plasma lipoproteins, *Adv. Exp. Med. Biol.*, 26(0), 7, 1972.

Infante, R., Plasma lipoprotein biosynthesis and regulation, *Expo. Annu. Biochim. Med.*, 31, 55, 1972.

Levy, R. I., Turnover of plasma lipoproteins, *Verh. Dtsch. Ges. Inn. Med.*, 78, 1293, 1972.

Lewis, B., Metabolism of the plasma lipoproteins, *Sci. Basis Med.*, 118-44, 1972.

Stein, Y., Biosynthesis and degradation of plasma lipoproteins, *Verh. Dtsch. Ges. Inn. Med.*, 78, 1293, 1972.

Wille, L. E., Some new aspects on serum lipoproteins — physiology and metabolism, *Tidsskr. Nor. Laegeforen.*, 92, 2087-089, 1972.

Biserte, G. et al., Lipoprotein metabolism, *Ann. Biol. Clin. (Paris)*, 31, 119, 1973.

Davis, M. A. et al., Protein hydrophobicity and lipid-protein interaction, *Biochim. Biophys. Acta*, 317, 214, 1973.

Eisenberg, S. et al., On the metabolic conversion of human plasma very low density lipoprotein to low density lipoprotein, *Biochim. Biophys. Acta*, 326, 361, 1973.

Greten, H., Catabolism of human plasma lipoproteins, *Verh. Dtsch. Ges. Inn. Med.,* 79, 1275, 1973.

Myant, N. B., The metabolism of LDL, *Adv. Exp. Med. Biol.,* 38, 53, 1973.

Polonovski, J., Metabolism of plasma lipoproteins, *Ann. Biol. Clin. (Paris),* 31, 101, 1973.

Rey, J. et al., Idiopathic disorders of intestinal fat transport (Anderson's disease). A further case (in French), *Arch. Fr. Pediatr.,* 30, 564, 1973.

Eisenberg, S. et al., Lipoprotein metabolism, *Adv. Lipid. Res.,* 13, 1, 1975.

Eisenberg, S. et al., Pathways of lipoprotein metabolism: integration of structure, function and metabolism, *Adv. Exp. Med. Biol.,* 63, 61, 1975.

Felts, J. M. et al., The mechanism of assimilation of constituents of chylomicrons, very low density lipoproteins and remnants — a new theory, *Biochem. Biophys. Res. Commun.,* 66(4), 1467, 1975.

Gangl, A. et al., Intestinal metabolism of lipids and lipoproteins, *Gastroenterology,* 68(1), 167, 1975.

Glomset, J. A., Recent studies of the role of the lecithin-cholesterol acyltransferase reaction in plasma lipoprotein metabolism, in *Lipoprotein Metabolism,* Greten, H. Ed., Springer-Verlag, Berlin, 1976, 28.

Goldstein, J. L. et al., Role of lysosomal acid lipase in the metabolism of plasma low density lipoprotein. Observations in cultured fibroblasts from a patient with cholesteryl ester storage disease, *J. Biol. Chem.,* 250(21), 8487, 1975.

Greten, H., Interconversion and catabolism of plasma lipoproteins (in German), *Verh. Dtsch. Ges. Inn. Med.,* 81, 868, 1975.

Tanaka, H. et al., Serum total cholesterol, triglyceride and lipoprotein levels in a Japanese rural population, *Osaka City Med. J.,* 21(1), 1, 1975.

Tyroler, H. A. et al., Black-white differences in serum lipids and lipoproteins in Evans County, *Prev. Med.,* 4(4), 541, 1975.

Hedstrand, H. et al., Serum lipoprotein concentration and composition in healthy 50-year old men, *Ups. J. Med. Sci.,* 81(1), 37, 1976.

Jovanovic, S. et al., Studies on the correlation of the selectivity of proteinuria and serum lipoproteins (in Serbian), *Med. Pregl.,* 29(1-2), 33, 1976.

Langelier, M. et al., Plasma lipoprotein profile and composition in White Carnean and Show Racer breeds of pigeons, *Can. J. Biochem.,* 54(1), 27, 1976.

Levy, R. I., The plasma lipoproteins: an overview, *Prog. Clin. Biol. Res.,* 5, 25, 1976.

Composition
Serum lipids and lipoproteins

Pazzanese, D. et al., Serum lipid levels in a Brazilian Indian population, *Lancet,* 2, 615, 1964.

Cancio, M., The serum lipid picture. A review, *Bol. Assoc. Med. P.R.,* 58, 563, 1966.

Farstad, M., Determination of total lipids and the lipids of lipoprotein fractions of serum, *Clin. Chim. Acta,* 14, 341, 1966.

Clark, D. A. et al., Longitudinal study of serum lipids. 12 year report, *Am. J. Clin. Nutr.,* 20, 743, 1967.

Klein, E. et al., In vitro formation of lipid-protein complexes, *Life Sci.,* 6, 1309, 1967.

Sakagami, T., Total lipids and phospholipids (in Japanese), *Jpn. J. Clin. Med.,* 25, 1807, 1967.

Sakagami, T. et al., Lipid metabolism experiment-lipoproteins (in Japanese), *Protein,* Suppl., 219, 1967.

Skaternikov, V. A. et al., Study of initial routes of metabolism of exogenous lipids in the human organism (in Russian), *Lab. Delo,* 12, 715, 1967.

Skipski, V. P. et al., Lipid composition of human serum lipoproteins, *Biochem. J.,* 104, 340, 1967.

Allard, C. et al., Seromucoids and serum lipids: an epidemiological study of an active Montreal population, *Can. Med. Assoc. J.,* 99, 650, 1968.

Canal, J. et al., Mechanism of the fixation of sulfated polysaccharide of lipoproteins: the role of phospholipids (in French), *Bull. Soc. Chim. Biol.,* 50, 1523, 1968.

Chahud Isee, A. et al., Total lipids and lipoproteins in indigenous Indians who do or do not ingest coca (in Spanish), *Rev. Clin. Esp.,* 111, 161, 1968.

Davignon, J., Lipoproteins and lipid transport (in French), *Union Med. Can.,* 97, 420, 1968.

Jones, R. J. et al., Poor predictability of lipoprotein cholesterol from whole serum lipids, *J. Atheroscler. Res.,* 8, 463, 1968.

Noma, A. et al., Dynamics of blood lipids (in Japanese), *Naika,* 22, 804, 1968.

Postma, T. et al., Lipid screening in clinical chemistry, *Clin. Chim. Acta,* 22, 569, 1968.

Clark, D. A. et al., Effects of hydrogen peroxide on lipoproteins and associated lipids, *Lipids,* 4, 1, 1969.

Deliamure, L. L., A study into the strength of the bond in protein-lipid complexes of donors' blood serum (in Russian), *Sov. Med.,* 32, 124, 1969.

Levy, R. I., The diagnosis and treatment of lipid transport disorders, *J. Kans. Med. Soc.,* 70, 367, 1969.

Levy, R. I., Diagnosis and treatment of lipid transport disorders, *Med. Ann. D.C.,* 38, 656, 1969.

Nunn, S. L. et al., A lipid clinic. Preliminary observations, *Minn. Med.,* 52, 1253, 1969.

Dalderup, L. M. et al., Serum lipids, typing, fibrinolysis, and smoking, *Br. Med. J.,* 1, 223, 1970.

Debuch, H., Lipido-composition of lipoproteins (summary), *Ann. Biol. Clin. (Paris),* 31, 65, 1973.

Girard, M. L. et al., Differential diagnosis of hyperlipoproteinemia using simultaneous determination of the lipids of the β-pre β-lipoprotein group and of the lipid(β and pre-β)-cholesterol ratio, *Ann. Biol. Clin. (Paris),* 31, 91, 1973.

Ilinov, P. et al., Lipid fractions in low density lipoproteins, *Vntr. Boles,* 12, 80, 1973.

Schneider, H. et al., The lipid core model of lipoproteins, *Chem. Phys. Lipids,* 10, 328, 1973.

Yoshida, E., Determination of lipid content of individual lipoprotein fractions by means of "lipid coefficients" — distribution of total lipid, total cholesterol, triglyceride and phospholipid among three lipoprotein fractions, *Jpn. J. Clin. Pathol.,* 21, 785, 1973.

Appelbaum, J. J. et al., Altering lipid patterns in adult males, *J. Occup. Med.,* 16, 539, 1974.

Canal, J. et al., Autoanalysis of serum lipids of the β- and pre-β group of lipoproteins, *Pathol. Biol.,* 22, 173, 1974.

Hamilton, J. A. et al., Rotational and segmental motions in the lipids of human plasma lipoproteins, *J. Biol. Chem.,* 249, 4872, 1974.

Ricci, G. et al., The quantitative evaluation of lipid spectrum in population studies, *G. Ital. Cardiol.,* 4, 341, 1974.

Shorey, R. L. et al., Alteration of serum lipids in a group of free-living adult males, *Am. J. Clin. Nutr.,* 27, 268, 1974.

Simon, K. H., Serum lipids. I. Biochemistry of serum lipids, *Med. Monatsschr.,* 28(12), 550, 1974.

Fox, J. M., A glossary of essential phospholipids, lipids and lipoproteins, in *Phosphatidylcholine,* Peeters, H., Ed., Springer-Verlag, Berlin, 1976, 2.

Lipids as parts of lipoproteins

D'Andrea, L. et al., Lipoproteins, total cholesterol and lipoprotein fractions of the blood serum in a group of engineers and employees of the P.L. railways (in Italian), *Boll. Soc. Ital. Biol. Sper.,* 40, 633, 1964.

Giongo, F. et al., Clinical evaluation of the opacimetric method of determining α- and β-lipoprotein cholesterol (in Italian), *Acta Gerontol.,* 14, 158, 1964.

Goodman, D. S., The in vivo turnover of individual cholesterol esters in human plasma lipoproteins, *J. Clin. Invest.,* 43, 2026, 1964.

Goodman, D. S., Fatty acid composition of human plasma lipoprotein fractions, *J. Lipid Res.,* 5, 307, 1964.

Rehngorg, C. S. et al., The fate of cholesteryl esters in human serum incubated in vitro at 38 degrees, *Biochim. Biophys. Acta,* 84, 596, 1964.

Zilversmit, D. B., Extraction of cholesterol from human serum lipoprotein films, *J. Lipid Res.,* 5, 300, 1964.

Baraud, J. et al., Fatty acids of lipoproteins in corn grain (in French), *C.R. Acad. Sci.,* 261, 4272, 1965.

Enev, N. et al., Determination of cholesterol in serum α- and β-lipoproteins, *Suvr. Med.,* 16, 159, 1965.

Engelbrecht, F. H., β-lipoproteins and their physiological relationship to cholesterol and phospholipid in serum, *S. Afr. Med. J.,* 39, 644, 1965.

Lohmann, D., On the fat transport of serum lipoproteins, *Dtsch. Z. Verdau. Stoffwechselkr.,* 25, 264, 1965.

Shafrir, E. et al., Partition of fatty acids of 20-24 carbon atoms between serum albumin and lipoproteins, *Biochim. Biophys. Acta,* 98, 365, 1965.

Glomset, J. A. et al., Role of plasma lecithin: cholesterol acyltransferase in the metabolism of high density lipoproteins, *J. Lipid Res.,* 7, 638, 1966.

Gustafson, A. et al., Studies of the composition and structure of serum lipoproteins. Separations and characterization of phospholipid-protein residues obtained by partial delipidization of very low density lipoproteins of human serum, *Biochemistry,* 5, 632, 1966.

Isselbacher, K. J., Biochemical aspects of fat absorption, *Gastroenterology,* 50, 78, 1966.

Mitchell, F. L. et al., The relationship between serum total cholesterol and β-lipoprotein, *Clin. Chim. Acta,* 14, 1, 1966.

O'Hara, D. D. et al., Use of constant composition polyvinylpyrrolidone columns to study the interaction of fat particles with plasma, *J. Lipid Res.,* 7, 264, 1966.

Scott, P. J. et al., Diet and the fatty acids in cholesterol esters of plasma lipoproteins, *Nutr. Rev.,* 24, 14, 1966.

Shapiro, T. L. et al., Exchange of free and esterified cholesterol between serum α- and β-lipoproteins: preliminary studies, *Life Sci.,* 5, 1423, 1966.

Andac, S. O., The correlation between low density lipoprotein and serum total cholesterol, *Turk. J. Pediatr.,* 9, 41, 1967.

Fredrickson, D. S. et al., Fat transport in lipoproteins — an integrated approach to mechanisms and disorders, *N. Engl. J. Med.,* 276, 34, 1967.

Fredrickson, D. S. et al., Fat transport in lipoproteins — an integrated approach to mechanisms and disorders, *N. Engl. J. Med.,* 276, 94, 1967.

Rossi, F. et al., Activity of a phospholipid association on serum lipoproteins and on cholesteremia (in Italian), *Minerva Med.,* 58, 2608, 1967.

Searcy, R. L., Determination of electrophoretic distribution of cholesterol-bearing proteins in serum, *Clin. Chim. Acta,* 15, 73, 1967.

Walton, K. W., Fat transport by lipoprotein in health and disease, *J. Atheroscler. Res.,* 7, 533, 1967.

Warburton, F. G. et al., The significance of disagreement between serum cholesterol levels and serum β-lipoprotein as determined by immunocrit, *Clin. Chim. Acta,* 18, 75, 1967.

Adlkofer, F. et al., Demonstration, concentration and characterization of a lysolecithin-liberating enzyme from human serum (in German), *Hoppe Seylers Z. Physiol. Chem.,* 349, 417, 1968.

Akanuma, Y. et al., In vitro incorporation of cholesterol-14C into very low density lipoprotein cholesteryl esters, *J. Lipid Res.,* 9, 620, 1968.

Babel, E. et al., Simple micromethod of determination of cholesterol in serum α- and β-lipoproteins (in Rumanian), *Rev. Medicochir. Iasi.,* 72, 981, 1968.

Demerdash, H. et al., Cholesterol esters and lipoproteins, *Lancet,* 1, 902, 1968.

Gjone, E. et al., Familial serum cholesterol ester deficiency. Clinical study of a patient with a new syndrome, *Acta Med. Scand.,* 183, 107, 1968.

Lindall, A. W., Biochemical factors related to the serum cholesterol levels, *Minn. Med.,* 51, 537, 1968.

Rose, H. G., Studies on the equilibration of radioisotopic cholesterol with human serum lipoprotein cholesterol in vitro, *Biochim. Biophys. Acta,* 152, 728, 1968.

Sidorenkov, W. et al., Effect of glycerophosphate and iodoacetate on the lipid and phosphatide contents of blood and tissues, and β-lipoprotein content of the blood (in Russian), *Vopr. Med. Khim.,* 14, 200, 1968.

Antonini, F. M. et al., Statistical study on the ratio between β- and α-lipoproteins, triglycerides and cholesterol, *Clin. Chim. Acta,* 24, 19, 1969.

Boyd, G. S. et al., Control of cholesterol biosynthesis by a plasma apo-lipoprotein, *Nature (London),* 221, 574, 1969.

Brown, D. F. et al., Studies with orally administered H_3 palmitate in two types of hyperlipoproteinemia, *Am. J. Med. Sci.,* 258, 121, 1969.

Raz, A. et al., Various factors affecting cholesterol esterification in plasma lipoproteins by lecithin-cholesterol acyltransferase, *Biochim. Biophys. Acta,* 176, 591, 1969.

Grundy, S. M. et al., The effects of unsaturated dietary fats on absorption, excretion, synthesis, and distribution of cholesterol in man, *J. Clin. Invest.,* 49, 1135, 1970.

Robertson, G. et al., An evaluation of cholesterol determinations in serum lipoprotein fractions by a semi-automated fluorimetric method, *J. Clin. Pathol.,* 23, 243, 1970.

Thiele, O. W., The chemistry of lipoprotein, *Dtsch. Med. J.,* 21, 264, 1970.

Myant, N. B. et al., Turnover of cholesteryl esters in plasma low-density and high-density lipoproteins in familial hyper-β-lipoproteinaemia, *Clin. Sci. Mol. Med.,* 45, 551, 1973.

Rössner, S., The intravenous fat tolerance test with intralipid in various types of hyperlipidaemias and comparison between metabolism of intralipid and VLDL, *Adv. Exp. Med. Biol.,* 38, 69, 1973.

Badzio, T., Isolation of exogenous cholesterol and lecithin by the system of serum lipoproteins (in Polish), *Pol. Tyg. Lek.,* 29(40), 1711, 1974.

Brown, M. S. et al., Expression of the familial hypercholesterolemia gene in heterozygotes: mechanism for a dominant disorder in man, *Science,* 185, 61, 1974.

Chait, A. et al., Proceedings: unsaturated fat and plasma triglyceride metabolism in man, *Clin, Sci. Mol. Med.,* 46, 12P, 1974.

Kartsova, S. V. et al., Phospholipid makeup of the blood plasma α- and β-lipoproteins in man and some animals (with English abstract), *Zh. Evol. Biokhim. Fiziol.,* 10(5), 508, 1974.

Naito, C. et al., A possible role of circulating lipoprotein-triglycerides in the increase in concentration of free fatty acids and in insulin resistance in "total" lipodystrophy, *J. Clin. Endocrinol. Metab.,* 39(6), 1030, 1974.

Traill, M. A. et al., Separate estimation of serum β-lipoprotein cholesterol and pre-β-lipoprotein cholesterol: clinical aspects, *Med. J. Aust.,* 1, 987, 1974.

Bilheimer, D. W. et al., Reduction in cholesterol and low density lipoprotein synthesis after portacaval shunt surgery in a patient with homozygous familial hypercholesterolemia, *J. Clin. Invest.,* 56(6), 1420, 1975.

Castelli, W. P., Mortality significance of hypercholesterolemia and hypertriglyceridemia, *Trans. Assoc. Life. Ins. Med. Dir. Am.,* 58, 272, 1975.

Higgins, M. J. et al., A new type of familial hypercholesterolaemia, *Lancet,* 2(7938), 737, 1975.

Havel, R. J., Lipoproteins and lipid transport, *Adv. Exp. Med. Biol.,* 63, 37, 1975.

Nanbu, S., Significance of plasma lipids — from aspects of plasma lipoprotein, *Jpn. J. Clin. Pathol.,* 23(3), 168, 1975.

Brown, M. S. et al., Receptor-mediated control of cholesterol metabolism, *Science,* 191(4223), 150, 1976.

Chatterjee, S. et al., Glycosphingolipids of human plasma lipoproteins, *Lipids,* 11(6), 462, 1976.

Dawson, G. et al., Distribution of glycosphingolipids in the serum lipoproteins of normal human subjects and patients with hypo- and hyperlipidemias, *Lipid Res.,* 17(2), 125, 1976.

Goldstein, J. L. et al., Heterozygous familial hypercholesterolemia: failure of normal allele to compensate for mutant allele at a regulated genetic locus, *Cell,* 9(2), 195, 1976.

Kupke, I. R., Enzymatic determination of cholesterol in serum lipoproteins, *J. Clin. Chem. Clin. Biochem.,* 14(5), 217, 1976.

Miller, N. E. et al., Relationships between plasma lipoprotein cholesterol concentrations and the pool size and metabolism of cholesterol in man, *Atherosclerosis,* 23(3), 535, 1976.

Rabega, C. et al., Diffusible cholesterol as a risk factor in the biochemical stage of atherosclerosis (with English abstract), *Med. Interne,* 14(4), 265, 1976.

Henck, C. C. et al., β-lipoprotein cholesterol quantitation with polycations, *Clin. Chem.,* 23(3), 536, 1977.

Apolipoproteins

Eder, H. A. et al., An apoprotein of the lipoproteins, *Trans. Assoc. Am. Physicians,* 77, 259, 1964.

Roheim, P. S. et al., The formation of plasma lipoproteins from apoprotein in plasma, *J. Biol. Chem.,* 240, 2994, 1965.

Granda, J. L. et al., Solubilization and properties of the apoproteins of the very low- and low-density lipoproteins of human serum, *Biochemistry,* 5, 3301, 1966.

Levy, R. S. et al., Amino acid composition of the proteins from chylomicrons and human serum lipoproteins, *J. Lipid Res.,* 8, 463, 1967.

Scanu, A., Binding of human high density lipoprotein apoprotein with aqueous dispersions of phospholipids, *J. Biol. Chem.,* 242, 711, 1967.

Shore, B. et al., The protein moiety of human serum β-lipoproteins, *Biochem. Biophys. Res. Commun.,* 28, 1003, 1967.

Camejo, G. et al., Lipid monolayers: interactions with the apoprotein of high density plasma lipoprotein, *J. Lipid Res.,* 9, 562, 1968.

Gotto, A. M. et al., The structure and properties of human β-lipoprotein and β-apoprotein, *Biochem. Biophys. Res. Commun.,* 31, 699, 1968.

Gotto, A. M., β-apoprotein sufficiency and function, *N. Engl. J. Med.,* 280, 1297, 1969.

Pinon, J. C. et al., Blood β-apoprotein: amino acid composition in Type II familial cholesterolemia (in French), *Biochim. Biophys. Acta,* 187, 144, 1969.

Shore, B. et al., Isolation and characterisation of polypeptides of human serum lipoproteins, *Biochemistry,* 8, 4510, 1969.

Berger, K. U. et al., A spin label study of the recombined lipid and apoprotein of human erythrocyte membranes, *Biochem. Biophys. Res. Commun.,* 40, 1273, 1970.

Brown, W. V. et al., Further separation of the apoproteins of the human plasma very low density lipoproteins, *Biochim. Biophys. Acta,* 200, 573, 1970.

Mahley, R. W. et al., Identity of very low density lipoprotein apoproteins of plasma and liver Golgi apparatus, *Science,* 168, 380, 1970.

Onajobi, F. D. et al., Accumulation of squalene during hepatic cholesterol synthesis in vitro. Role of a plasma apolipoprotein, *Eur. J. Biochem.,* 13, 203, 1970.

Sodhi, H. S. et al., Interaction of apo-HDL with HDL and with other lipoproteins, *Atherosclerosis,* 12, 439, 1970.

Alaupovic, P., Apolipoproteins and lipoproteins, *Atherosclerosis,* 13, 141, 1971.

Forte, T. M. et al., Electron microscopic study on reassembly of plasma high density apoprotein with various lipids, *Biochim. Biophys. Acta,* 248, 381, 1971.

Alaupovic, P. et al., Peptide composition of human plasma apolipoproteins A, B and C, *Expo. Annu. Biochim. Med.,* 31, 145, 1972.

Brown, W. V., Some functional aspects of the plasma apolipoproteins, *Verh. Dtsch. Ges. Inn. Med.,* 78, 1292, 1972.

Eisenberg, S. et al., The metabolism of very low density lipoprotein proteins. II. Studies on the transfer of apoproteins between plasma lipoproteins, *Biochim. Biophys. Acta,* 280, 94, 1972.

Fredrickson, D. S., Tonsils, apolipoproteins and enzymes, *N. Engl. J. Med.,* 286, 601, 1972.

Fredrickson, D. S. et al., The apolipoproteins, *Adv. Exp. Med. Biol.,* 26(0), 25, 1972.

Hendrickson, H. et al., Isolation of the Folch-Lees proteolipid apoprotein fraction from bovine brain myelin by a procedure involving rapid water partitioning using Sephadex LH-20, *J. Neurochem.,* 19, 2233, 1972.

Jackson, R. L. et al., A study of the cystine-containing apolipoprotein of human plasma high density lipoproteins: characterization of cyanogen bromide and tryptic fragments, *Biochim. Biophys. Acta,* 285, 36, 1972.

Lees, M. B. et al., Modification of the Lowry procedure for the analysis of proteolipid protein, *Anal. Biochem.,* 47, 184, 1972.

Lux, S. E. et al., The influence of lipid on the conformation of human plasma high density apolipoproteins, *J. Biol. Chem.,* 247, 2598, 1972.

Lux, S. E. et al., Isolation and characterization of apo-Lp-Gln-II (apo A-II) a plasma high density apolipoprotein containing two identical polypeptide chains, *J. Biol. Chem.,* 247, 7510, 1972.

Lux, S. E. et al., Isolation and characterization of the tryptic and cyanogen bromide peptides of apo-Lp-Gln-II (apo A-II), plasma high density apolipoprotein, *J. Biol. Chem.,* 247, 7519, 1972.

Scanu, A. M., The polypeptides of human serum lipoproteins and their clinical significance, *Ann. Clin. Lab. Sci.,* 2(2), 147, 1972.

Ayrault-Jarrier, M. et al., Heterogeneity of apolipoprotein A- of the HDL of human serum, *Ann. Biol. Clin. (Paris),* 31, 73, 1973.

Baker, H. N. et al., Isolation and characterization of the cyanogen bromide fragments from the high-density apolipoprotein glutamine, *Biochemistry,* 12, 3866, 1973.

Etienne, J. et al., Serum α-lipoprotein deficiency: study of an abnormal apo-HDL, *Pathol. Biol.,* 21, 385, 1973.

Harvie, N. R., A method of measuring protein dry weight in solutions of unknown salt concentration and an application to lipoprotein, *Anal. Biochem.,* 53, 252, 1973.

Havel, R. J. et al., Interchange of apolipoproteins between chylomicrons and high density lipoproteins during alimentary lipemia in man, *J. Clin. Invest.,* 52, 32, 1973.

Havel, R. J. et al., Primary dysbetalipoproteinemia: predominance of a specific apoprotein species in triglyceride-rich lipoproteins, *Proc. Natl. Acad. Sci. U.S.A.,* 70, 2015, 1973.

Jackson, R. L. et al., Human high density lipoprotein, apolipoprotein glutamine II. The immunochemical and lipid-binding properties of apolipoprotein glutamine II derivatives, *J. Biol. Chem.,* 248, 5218, 1973.

Jackson, R. L. et al., Human plasma high density lipoprotein. Interaction of the cyanogen bromide fragments from apolipoprotein glutamine II. (A-II) with phosphatidyl-choline, *J. Biol. Chem.,* 248, 8449, 1973.

Kane, J. P., A rapid electrophoretic technique for identification of subunit species of apoproteins in serum lipoproteins, *Anal. Biochem.,* 53, 350, 1973.

McConathy, W. J. et al., Isolation and partial characterization of apolipoprotein D: a new protein moiety of the human plasma lipoprotein system, *FEBS Lett.,* 37, 178, 1973.

Miller, A. L. et al., Activation of lipoprotein lipase by apolipoprotein glutamic acid. Formation of a stable surface film, *J. Biol. Chem.,* 248, 3359, 1973.

Morrisett, J. D. et al., Interaction of an apolipoprotein (apo Lp-alanine) with phosphatidylcholine, *Biochemistry,* 12, 1290, 1973.

Morrisett, J. D. et al., Methods for studying lipid-protein interactions, *Cardiovasc. Res. Cent. Bull. (Houston),* 12(2), 39, 1973.

Nicot, C. et al., Study of Folch-Pi apoprotein. I. Isolation of two components, aggregation during delipidation, *Biochim. Biophys. Acta,* 322, 109, 1973.

Scanu, A. M. et al., The protein of plasma lipoproteins: properties and significance, *Adv. Clin. Chem.,* 16, 111, 1973.

Sparrow, J. T. et al., Chemical synthesis and biochemical properties of peptide fragments of apolipoprotein-alanine, *Proc. Natl. Acad. Sci. U.S.A.,* 70, 2124, 1973.

Whikehart, D. R. et al., Amino- and carboxyl-terminal amino acids of proteolipid proteins, *J. Neurochem.,* 20, 1303, 1973.

Alaupovic, P. et al., Apolipoproteins and lipoprotein families in familial lecithin: cholesterol acyltransferase deficiency, *Scand. J. Clin. Lab. Invest.,* Suppl. 33, 83, 1974.

Bachorik, P. S. et al., Resolubilization of certain apoprotein components of human plasma high density lipoproteins in TCA-fixed polyacrylamide gels during destaining in acetic acid solutions, *Anal. Biochem.,* 60, 631, 1974.

Brewer, H. B., Jr. et al., The complete amino acid sequence of alanine apolipoprotein (apo C-3), and apolipoprotein from human plasma very low density lipoproteins, *J. Biol. Chem.,* 249, 4975, 1974.

Brown, W. V. et al., Some functional aspects of apolipoproteins: apo Lp-α inhibition of lipoprotein lipase and deinhibition by monoolein, *Ann. Otolaryngol. Chir. Cervicofac.,* 91(4-5), 11, 1974.

Chen, C. H. et al., Subunit structure of the apoprotein of human serum low density lipoproteins, *Biochem. Biophys. Res. Commun.,* 60(2), 549, 1974.

Forte, T. et al., Interaction by sonication of C-apolipoproteins with lipid: an electron microscopid study, *Biochim. Biophys. Acta,* 337(2), 169, 1974.

Fredrickson, D. S., Plasma lipoproteins and apolipoproteins, *Harvey Lect,* 68, 185, 1974.

Glickman, R. M. et al., The apoproteins of various size classes of human chylous fluid lipoproteins, *Biochim. Biophys. Acta,* 371(1), 255, 1974.

Gwyne, J. et al., The molecular properties of Apo-I from high density lipoprotein, *J. Biol. Chem.,* 249, 2411, 1974.

Jackson, R. L. et al., Effects of maleylation on the lipid-binding and immunochemical properties of human plasma high density apolipoprotein-A-II, *Biochem. Biophys. Res. Commun.,* 61(4), 1317, 1974.

Jackson, R. L. et al., The primary structure of apolipoprotein-serine, *J. Biol. Chem.*, 249, 5308, 1974.

Kruski, A. W. and Scanu, A. M., Interaction of human serum high density lipoprotein apoprotein with phospholipids, *Chem. Phys. Lipids*, 13(1), 27, 1974.

Lee, D. M. et al., Composition and concentration of apolipoproteins in very-low- and low-density lipoproteins of normal human plasma, *Atherosclerosis*, 19, 501, 1974.

Levy, R. I. et al., The lipoprotein apoproteins. Their role in normal lipid transport and dyslipoproteinemia, *Ann. Biol. Clin. (Paris)*, 32, 1, 1974.

Morrisett, J. D. et al., Structure of the major complex formed by interaction of phosphatidylcholine bilamellar vesicles and apolipoproteinalanine (APO-C-III), *Biochemistry*, 13(23), 4765, 1974.

Nussbaum, J. L. et al., Amino acid analysis and N-terminal sequence determination of P7 proteolipid apoprotein from human myelin, *FEBS Lett.*, 45(1), 295, 1974.

Pownall, H. J. et al., The requirement for lipid fluidity in the formation and structure of lipoproteins: thermotropic analysis of apolipoproteinalanine binding to dimyristoyl phosphatidylcholine, *Biochem. Biophys. Res. Commun.*, 60(2), 779, 1974.

Reynolds, J. A. et al., The interaction of polypeptide components of human high density lipoprotein with sodium dodecyl sulfate, *J. Biol. Chem.*, 249, 3937, 1974.

Schonfeld, G. et al., Assay of total plasma apolipoprotein B concentration in human subjects, *J. Clin. Invest.*, 53, 1458, 1974.

Schonfeld, G. et al., The structure of human high density lipoprotein and the levels of apolipoprotein A-I in plasma as determined by radioimmunoassay, *J. Clin. Invest.*, 54, 236, 1974.

Shore, B. et al., An apolipoprotein preferentially enriched in cholesteryl ester-rich very low density lipoproteins, *Biochem. Biophys. Res. Commun.*, 58, 1, 1974.

Shulman, R. S. et al., Isolation and alignment of the tryptic peptides of alanine apolipoprotein, an apolipoprotein from human plasma very low density lipoproteins, *J. Biol. Chem.*, 249, 4969, 1974.

Träuble, H. et al., Interaction of a serum apo-lipoprotein with ordered and fluid lipid bilayers. Correlation between lipid and protein structure, *FEBS Lett.*, 49(2), 269, 1974.

Verdery, R. B. et al., Interaction of lysolecithin micelles and lecithin vesicles with apo-lipoprotein Gln I from serum high density lipoproteins, *Biochem. Biophys. Res. Commun.*, 57, 1271, 1974.

Badley, R. A., The location of protein in serum lipoproteins: a fluorescence quenching study, *Biochim. Biophys. Acta*, 379(2), 517, 1975.

Albers, J. J. et al., Immunoassay of human plasma apolipoprotein B, *Metabolism*, 24(12), 1339, 1975.

Baker, H. N. et al., The primary structure of human plasma high density apolipoprotein glutamine I (Apo A-I). II. The amino acid sequence and alignment of cyanogen bromide fragments IV, III, and I, *J. Biol. Chem.*, 250(7), 2725, 1975.

Bautovich, G. J. et al., Radioimmunoassay of human plasma apoprotein. Part I. Assay of apolipoprotein-B, *Atherosclerosis*, 21(2), 217, 1975.

Blum, C. B. et al., Interconversions of apolipoprotein fragments, *Annu. Rev. Med.*, 26, 345, 1975.

Chapman, M. J. et al., Stability of the apoprotein of human serum low density lipoprotein: absence of endogenous endopeptidase activity, *Biochem. Biophys. Res. Commun.*, 66(3), 1030, 1975.

David, J. S. K. et al., Interaction of human plasma apolipoproteins and phospholipids. Semi-quantitative studies using polacrylamide gel electrophoresis, *Biochim. Biophys. Acta*, 398(1), 72, 1975.

Delahunty, T. et al., The primary structure of human plasma high density apolipoprotein glutamine I (Apo A-I). I. The amino acid sequence of cyanogen bromide fragment II, *J. Biol. Chem.*, 250(7), 2718, 1975.

Ekman, R. et al., Effects of apolipoproteins on lipoprotein lipase activity of human adipose tissue, *Clin. Chim. Acta*, 63(1), 29, 1975.

Fainaru, M. et al., Radioimmunoassay of human high density lipoprotein apoprotein A-I, *Biochim. Biophys. Acta*, 386(2), 432, 1975.

Farish, E. et al., Plasma apolipoprotein A levels in healthy human adults, *Clin. Chim. Acta*, 62(1), 97, 1975.

Fontaine, M. et al., Carbohydrate content of human VLDL, IDL, LDL, and HDL plasma apoproteins from fasting normal and hyperlipemic patients, *Clin. Chim. Acta*, 64(1), 91, 1975.

Grow, T. E. et al., Lipoprotein geometry. I. Spatial relationship of human HDL apoproteins studied with a bi-functional reagent, *Biochem. Biophys. Res. Commun.*, 66(1), 352, 1975.

Gwynne, J. et al., The interaction of apo A-I from human high density lipoprotein with lysolecithin, *J. Biol. Chem.*, 250(18), 7300, 1975.

Gwynne, J. et al., The self-association of apo A-II an apoprotein of the human high density lipoprotein complex, *Arch. Biochem. Biophys.*, 170(1), 204, 1975.

Hoff, H. F. et al., Apo-low density lipoprotein localization. Intracranial and extracranial atherosclerotic lesions from human normolipoproteinemias and hyperlipoproteinemias, *Arch. Neurol.*, 32(9), 600, 1975.

Hoff, H. F. et al., Localization of apo-lipoproteins in human carotid artery plaques, *Stroke*, 6(5), 531, 1975.

Jackson, R. L. et al., A comparative study on the removal of cellular lipids from Landschütz ascites cells by human plasma apolipoproteins, *J. Biol. Chem.*, 250(18), 7204, 1975.

Kane, J. P. et al., Apoprotein composition of very low density lipoproteins of human serum, *J. Clin. Invest.*, 56(6), 1622, 1975.

Lee, D. M., Identification of apolipoproteins in lipoprotein density classes of hypercholesterolemia (type II.a), *FEBS Lett.*, 51(1), 116, 1975.

Miller, J. et al., Changes in the apoprotein composition of VLD lipoproteins in man following eating, *Experientia*, 31(10), 1132, 1975.

Morrisett, J. D. et al., The interaction of apolipoprotein-alanine (apo C-III) with lipids: study of structural features required for binding, *Adv. Exp. Med. Biol.*, 63, 1, 1975.

Murthy, V. K. et al., In vitro labeling of β-apolipoprotein with 3H or 14C and preliminary application to turnover studies, *J. Lipid Res.*, 16(1), 1, 1975.

Osborne, J. C., Jr. et al., The self-association of the reduced Apo A-II apoprotein from the human high density lipoprotein complex, *Biochemistry*, 14(17), 3741, 1975.

Phillips, M. C. et al., A comparison of the interfacial interaction of the apoprotein from high density lipoprotein and β-casein with phospholipids, *Biochim. Biophys. Acta*, 406(3), 402, 1975.

Reichl, D. et al., Observations on the passage of apoproteins from plasma lipoproteins into peripheral lymph in two men, *Clin. Sci. Mol. Med.*, 49(5), 419, 1975.

Rubenstein, B. et al., A comparison of the peptide composition of human serum low and very low density lipoproteins, *Can. J. Biochem.*, 53(2), 128, 1975.

Shulman, R. S. et al., The complete amino acid sequence of C-I (apo Lp-Ser), an apolipoprotein from human very low density lipoproteins, *J. Biol. Chem.*, 250(1), 182, 1975.

Sigurdsson, G. et al., Conversion of very low density lipoprotein to low density lipoprotein. A metabolic study of apolipoprotein B kinetics in human subjects, *J. Clin. Invest.*, 56(6), 1481, 1975.

Simons, L. A. et al., The metabolism of the apoprotein of plasma low density lipoprotein in familial hyper-β-lipoproteinemia in the homozygous form, *Atherosclerosis*, 21(2), 283, 1975.

Sneiderman, A. et al., Determination of B protein of low density lipoprotein directly in plasma, *J. Lipid Res.*, 16(6), 465, 1975.

Sparrow, J. T., The mechanism of lipid binding by the plasma lipoproteins: synthesis of model peptides, in *Peptides: Chemistry, Structure and Biology*, Walter, R. and Meinhofer, J., Ed., Ann Arbor, Ann Arbor Science Publishers, 1975, 597.

Stein, Y. et al., The removal of cholesterol from aortic smooth muscle cells in culture and Landschütz ascites cells by fractions of human high-density apolipoprotein, *Biochim. Biophys. Acta*, 380(1), 106, 1975.

Stone, W. L. et al., The self-association of the apo-Gln-I and apo-Gln-II polypeptides of human high density serum lipoproteins, *J. Biol. Chem.*, 250(20), 8045, 1975.

Tall, A. R. et al., Apoprotein stability and lipid-protein interactions in human plasma high density lipoproteins, *Proc. Natl. Acad. Sci. U.S.A.*, 72(12), 4940, 1975.

Utermann, G., Isolation and partial characterization of an arginine-rich apolipoprotein from human plasma very-low-density lipoproteins: apolipoprotein E, *Hoppe Seylers Z. Physiol. Chem.*, 356(7), 1113, 1975.

van der Bijl, P. et al., Determination of human low density lipoprotein apoproteins (apo-LDL) by the Lowry procedure: standardization and effect of sodium chloride, *Clin. Chim. Acta*, 61(3), 407, 1975.

Albers, J. J. et al., Quantitation of apolipoprotein A-I of human plasma high density lipoprotein, *Metabolism*, 25(6), 633, 1976.

Carlson, L. A. et al., Changing relative proportions of apolipoproteins C II and C III of very low density lipoproteins in hypertriglyceridaemia, *Atherosclerosis*, 23(3), 563, 1976.

Cham, B. E. et al., In vitro partial relipidation of apolipoproteins in plasma, *J. Biol. Chem.*, 251(20), 6367, 1976.

Chapman, M. J. et al., Comparison of the serum low density lipoprotein and of its apoprotein in the pig, rhesus monkey and baboon with that in man, *Atherosclerosis*, 25(2-3), 267, 1976.

Curry, M. D. et al., Determination of apolipoprotein A and its constitutive A-I and A-II polypeptides by separate electroimmunoassays, *Clin. Chem.*, 22(3), 315, 1976.

Curry, M. D. et al., Determination of human apolipoprotein E by electroimmunoassay, *Biochim. Biophys. Acta*, 439(2), 413, 1976.

Durrington, P. N. et al., A comparison of methods for the immunoassay of serum apolipoprotein-B in man, *Clin. Chim. Acta*, 71(1), 95, 1976.

Eaton, R. P., Incorporation of 75 Se-selenomethionine into human apoproteins. I. Characterization of specificity in VLD and LDLs, *Diabetes*, 25(1), 32, 1976.

Eaton, R. P. et al., Incorporation of 75 Se-selenomethionine into human apoproteins. II. Characterization of metabolism of VLD and LD lipoproteins in vivo and in vitro, *Diabetes*, 25(1), 44, 1976.

Eaton, R. P. et al., Incorporation of 75 Se-selenomethionine into human apoproteins. III. Kinetic behavior of isotopically labeled plasma apoprotein in man, *Diabetes*, 25(8), 679, 1976.

Eder, H. A. et al., Plasma lipoproteins and apolipoproteins, *Ann. N.Y. Acad. Sci.*, 275, 169, 1976.

Ganesan, D. et al., Is decreased activity of C-II activated lipoprotein lipase in type III hyperlipoproteinemia (broad-β-disease) a cause or an effect of increased apolipoprotein E levels?, *Metabolism,* 25(11), 1189, 1976.

Kayden, H. J., Protein-lipoprotein interactions in man: a reexamination, *Ann. N.Y. Acad. Sci.,* 275, 145, 1976.

Glangeand, M. C. et al., Very low density lipoprotein. Dissociation of apolipoprotein-C during lipoprotein-lipase induced lipolysis, *Biochim. Biophys. Acta,* 486(1), 23, 1976.

Harding, D. R. et al., Letter: synthesis of a protein with the properties of the apolipoprotein C-I (Apo Lp-Ser), *J. Am. Chem. Soc.,* 98(9), 2664, 1976.

Le, T. N. et al., A study of Folch-Pi apoprotein. II. Relation between polymerization state and conformation, *Biochim. Biophys. Acta,* 427(1), 44, 1976.

Lee, P. et al., The carbohydrate composition of human apo-low-density-lipoprotein from normal and type II hyperlipoproteinemic subjects, *Can. J. Biochem.,* 54(1), 42, 1976.

Lim, C. T. et al., Apoproteins of human serum high density lipoproteins. Isolation and characterization of the peptides of Sephadex fraction V from normal subjects and patients with a-β-lipoproteinemia, *Biochim. Biophys. Acta,* 420(2), 1332, 1976.

McConathy, W. J. et al., Studies on the isolation and partial characterization of apolipoprotein D and lipoprotein D of human plasma, *Biochemistry,* 15(3), 515, 1976.

Malmendier, C. L. et al., Apoprotein profile of plasma and chylous ascites lipoproteins, *Clin. Chim. Acta,* 68(3), 259, 1976.

Middelhoff, G. et al., Study of the lipid binding characteristics of the apolipoproteins from human high density lipoprotein. I. Electron microscopic and gel filtration studies with synthetic phosphatidylcholines, *Biochim. Biophys. Acta,* 441(1), 57, 1976.

Mohiuddin, G. et al., The location of apoprotein in plasma high-density lipoproteins: photochemical labelling studies, *FEBS Lett.,* 70(1), 85, 1976.

Morrisett, J. D. et al., Interaction of apolipoprotein C-III with phosphatidylcholine vesicles. Dependence of apoprotein-phospholipid complex formation on vesicle structure, *Biochim. Biophys. Acta,* 486(1), 36, 1976.

Nakai, T. et al., The amino- and carboxyl-terminal sequences of canine apolipoprotein A-I, *FEBS Lett.,* 64(2), 409, 1976.

Novosad, Z. et al., Structure of an apolipoprotein-phospholipid complex: apo C-III induced changes in the physical properties of dimyristoylphosphatidyl-choline, *Biochemistry,* 15(15), 3176, 1976.

Osborne, J. C., Jr. et al., The thermodynamics of the self-association of the reduced and carboxymethylated form of apo-a-II from the human high density lipoprotein complex, *Biochemistry,* 15(2), 317, 1976.

Pinon, J. C. et al., Apolipoprotein of the low density lipoprotein of human plasma: structural study in familial hyper-β-lipoproteinemia, *Clin. Chim. Acta,* 70(2), 259, 1976.

Rosseneu, M. et al., Interaction of the apoproteins of very low density and high density lipoproteins with synthetic phospholipids, *Eur. J. Biochim.,* 70(1), 285, 1976.

Rosseneu, M. et al., Studies on the lipid binding characteristics of the apolipoproteins from human high density lipoprotein. II. Calorimetry of the binding of apo AI and apo AII with phospholipids, *Biochim. Biophys. Acta,* 441(1), 68, 1976.

Schonfeld, G. et al., Structure of high density lipoprotein. The immunologic reactivities of the COOH- and NH_2-terminal regions of apolipoprotein A-I, *J. Biol. Chem.,* 251(13), 3921, 1976.

Sigurdsson, G. et al., Metabolism of very low density lipoproteins in hyperlipidaemia: studies of apolipoprotein B kinetics in man, *Eur. J. Clin. Invest.,* 6(2), 167, 1976.

Swaney, J. B. et al., Cross-linking studies on the state of association of apo A-I and apo A-II from human HDL, *Biochem. Biophys. Res. Commun.,* 71(2), 636, 1976.

Tall, A. R. et al., Conformational and thermodynamic properties of apo A-I of human plasma high density lipoproteins, *J. Biol. Chem.,* 251(12), 3749, 1976.

Vitello, L. B. et al., Studies on human serum high density lipoproteins. Self-association of apolipoprotein A-I in aqueous solutions, *J. Biol. Chem.,* 251(4), 1131, 1976.

Vitello, L. B. et al., Studies on human serum high-density lipoproteins. Self-association of human serum apolipoprotein A-II in aqueous solutions, *Biochemistry,* 15(5), 1161, 1976.

Jackson, R. L. et al., Amino acid sequence of a major apoprotein from hen plasma very low density lipoproteins, *J. Biol. Chem.,* 252(1), 250, 1977.

Nicoll, A. et al., Intravenous fat tolerance. Correlation with very low density lipoprotein apoprotein-B kinetics in man, *Atherosclerosis,* 26(1), 17, 1977.

Ritter, M. C. et al., Role of apolipoprotein A-I in the structure of human serum high density lipoproteins. Reconstitution studies, *J. Biol. Chem.,* 252(4), 1208, 1977.

Sigurdsson, G. et al., Turnover of apolipoprotein-B in two subjects with familial hypo-β-lipoproteinemia, *Metabolism,* 26(1), 25, 1977.

Lipoprotein Classes
HDL

Barclay, M. et al., Additional evidence for the existence of "new" high-density lipoproteins in human serum, *Clin. Chim. Acta*, 10, 470, 1964.

Nichols, A. V. et al., Lipid transfer by high density lipoproteins of human serum in vitro, *Biochem. Biophys. Res. Commun.*, 17, 512, 1964.

Levy, R. I. et al., Heterogeneity of plasma high density lipoproteins, *J. Clin. Invest.*, 44, 426, 1965.

Scanu, A., Heterogeneity of the protein moiety of the human serum high-density lipoproteins, *Nature (London)*, 207, 528, 1965.

Scanu, A., Studies on the conformation of human serum high-density lipoproteins HDL-2 and HDL-3, *Proc. Natl. Acad. Sci. U.S.A.*, 54, 1699, 1965.

Strisower, E. H. et al., The effect of Sf 20-10-5 concentration changes induced by ethyl chlorophenoxyiso-butyrate on high-density lipoprotein lipid composition, *J. Lab. Clin. Med.*, 65, 748, 1965.

Alaupovic, P. et al., Studies on the composition and structure of serum lipoproteins. Isolation and characterization of very high density lipoproteins of human serum, *Biochemistry*, 5, 4044, 1966.

Fujii, T. et al., Aggregation of high-density lipoproteins from egg yolk, *Chem. Pharm. Bull.*, 14, 1430, 1966.

Nichols, A. V. et al., Lipid transfer between human serum high density lipoproteins and egg yolk lipoproteins in incubation mixtures, *J. Lipid Res.*, 7, 215, 1966.

Scanu, A. et al., Effects of ultracentrifugation on the human serum high-density (1.063 less than p less than 1.21 g/ml) lipoprotein, *Biochemistry*, 5, 446, 1966.

Scanu, A. et al., Forms of human serum high density lipoprotein protein, *J. Lipid Res.*, 7, 295, 1966.

Scanu, A., Serum high-density lipoprotein: effects of change in structure on activity of chicken adipose tissue lipase, *Science*, 153, 640, 1966.

Shore, V. et al., Some physical and chemical studies on the protein moiety of a high-density (1.126-1.195g/ml) lipoprotein fraction of human serum, *Biochemistry*, 6, 1962, 1967.

Sodhi, H. S. et al., Combination of delipidized high density lipoprotein with lipids, *J Biol. Chem.*, 242, 1205, 1967.

Akanuma, Y. et al., A method for studying the interaction between lecithin: cholesterol acyltransferase and high density lipoproteins, *Biochem. Biophys. Res. Commun.*, 32, 639, 1968.

Cohen, L. et al., Changes in human serum high-density lipoproteins induced by disulfide-exchange reagents, *Proc. Soc. Exp. Biol. Med.*, 129, 788, 1968.

Shore, B. et al., Heterogeneity in protein subunits of human serum high-density lipoproteins, *Biochemistry*, 7, 2773, 1968.

Shore, V. et al., Some physical and chemical studies on two polypeptide components of high density lipoproteins of human serum, *Biochemistry*, 7, 3396, 1968.

Scanu, A. et al., On the structure of human serum high-density lipoprotein: studies by the technique of circular dichroism, *Proc. Natl. Acad. Sci. U.S.A.*, 59, 890, 1968.

Gotto, A. M. et al., Conformation of human serum high density lipoprotein and its peptide components, *Nature (London)*, 224, 69, 1969.

Gotto, A. M., Jr., Recent studies on the structure of human serum low- and high-density lipoproteins, *Proc. Natl. Acad. Sci. U.S.A.*, 64, 1119, 1969.

Scanu, A. et al., Fractionation of human serum high density lipoprotein in urea solutions. Evidence for polypeptide heterogeneity, *Biochemistry*, 8, 3309, 1969.

Scanu, A. M., On the temperature dependence of the conformation of human serum high density lipoprotein, *Biochim. Biophys. Acta*, 181, 268, 1969.

Camejo, G. et al., The apo-lipoproteins of human plasma high density lipoprotein: a study of their lipid binding capacity and interaction with lipid monolayers, *Biochim. Biophys. Acta*, 218, 155, 1970.

Hirz, R. et al., Reassembly in vitro of a serum high-density lipoprotein, *Biochim. Biophys. Acta*, 207, 364, 1970.

Scanu, A. et al., Degradation and reassembly of a human serum high-density lipoprotein. Evidence for differences in lipid affinity among three classes of polypeptide chains, *Biochemistry*, 9, 1327, 1970.

Albers, J. J. et al., Isoelectric heterogeneity of the major polypeptide of human serum high density lipoprotein, *Biochem. Med.*, 5, 48, 1971.

Albers, J. J. et al., Precipitation of I[125]-labeled lipoproteins with specific polypeptide antisera. Evidence for two populations with differing polypeptide compositions in human high density lipoproteins, *Biochemistry*, 10, 3436, 1971.

Barratt, M. D. et al., Protein-protein and protein-lipid interactions in human serum high-density lipoprotein: an analysis by a spin label method, *Chem. Phys. Lipids*, 7, 345, 1971.

Bornt, J. C. et al., Immunochemical heterogeneity of human high density serum lipoproteins, *Immunochemistry*, 8, 865, 1971.

Camejo, G. et al., The size and chemical characteristics of six fractions obtained by differential centrifugation from human high density lipoprotein, *Acta Cient. Venez.*, 22, 45, 1971.

Ellefson, R. D. et al., Pre-β (or α-2) lipoprotein of high density in human blood, *Mayo Clin. Proc.*, 46, 328, 1971.

Gallango, M. L., Precipitating antigen-antibody system of high density lipoproteins, *Nature (New Biol.)*, 234, 111, 1971.

Leslie, R. B., Some physical and physico-chemical approaches to the structure of serum high density lipoproteins (HDL), *Biochem. Soc. Symp.*, 33, 47, 1971.

Leslie, R. B., Some physical approaches to the structure of serum high-density lipoproteins, *Biochem. J.*, 123, 17P, 1971.

Muesing, R. A. et al., Disruption of low- and high-density human plasma lipoproteins and phospholipid dispersions by 1-anilino-napthalene-8-sulfonate, *Biochemistry*, 10, 2952, 1971.

Pearlstein, E. et al., The human serum high density lipoprotein peptides of very low density lipoproteins and chylomicrons — an appendix, *Immunochemistry*, 8, 865, 1971.

Scanu, A., Human plasma high-density lipoproteins: past, present and future, *Biochem. J.*, 123, 17P, 1971.

Scanu, A. M., Human plasma high density lipoproteins, *Biochem. Soc. Symp.*, 33, 29, 1971.

Scanu, A. M. et al., Solubility in aqueous solutions of ethanol of the small molecular weight peptides of the serum very low density and high density lipoproteins: relevance to the recovery problem during delipidation of serum lipoproteins, *Anal. Biochem.*, 44, 576, 1971.

Scanu, A. M., Temperature transitions of lipid mixtures containing cholesterol esters. Relevance to the structural problem of serum high density lipoprotein, *Biochim. Biophys. Acta*, 231, 170, 1971.

Sodhi, H. S. et al., Evaluation of methods for preparing proteins from plasma high density lipoproteins, *Can. J. Biochem.*, 49, 1076, 1971.

Torsvik, H. et al., Amino acid composition of serum high density lipoprotein in patients with familial lecithin: cholesterol acyltransferase deficiency, *Clin. Genet.*, 2, 91, 1971.

Kang, K. Y. et al., Relationship between high density lipoprotein and alkaline phosphatase in the sera from patients with liver metastasis of cancer, *Tohoku J. Exp. Med.*, 105, 141, 1971.

Ayrault-Jarrier, M. et al., Demonstration and separation of the protein components of human serum α-lipoproteins, *Biochimie*, 54, 973, 1972.

Camejo, G., The structure of the high density lipoproteins of the plasma: various properties of its protein component, *Acta Cient. Venez.*, Suppl. 23, 5, 1972.

Ditschuneit, H. H. et al., Demonstration of lipoproteins of high density (HDL) from the plasma with the aid of zonal centrifugation, *Med. Welt*, 23, 1425, 1972.

Edelstein, C. et al., On the subunit structure of the protein of human serum high density lipoprotein. I. A study of its major polypeptide component (Sephadex fraction 3), *J. Biol. Chem.*, 247, 5842, 1972.

Eggena, P. et al., Isoelectric heterogeneity of human serum high-density of lipoproteins, *Biochem. Med.*, 6, 184, 1972.

Jackson, R. L. et al., Isolation of a helical, lipid-binding fragment from the human plasma high density lipoprotein, apo-LP, GLN-I, *Biochem. Biophys. Res. Commun.*, 49, 1444, 1972.

Kaminski, M., Do serum lipoproteins have esterase activity?. II. Study of high density lipoprotein of human serum, *Biochimie*, 54, 1223, 1972.

Kaminski, M. et al., Are serum lipoproteins esterases? Esterase activity of the immunoprecipitates of duck-serum high density lipoprotein in agar immunoelectrophoresis and immunodiffusion, *Eur. J. Biochem.*, 29, 175, 1972.

Kostner, G. et al., Studies of the composition and structure of plasma lipoproteins. Separation and quantification of the lipoprotein families occurring in the high density lipoproteins of human plasma, *Biochemistry*, 11, 3419, 1972.

Lux, S. E. et al., Identification of the lipid-binding cyanogen bromide fragment from the cystine-containing high density apolipoprotein, APOLP-GLN-II, *Biochem. Biophys. Res. Commun.*, 49, 23, 1972.

Nichols, A. V. et al., Degradation products from human serum high density lipoproteins following dehydration by rotary evaporation and solubilization, *Biochim. Biophys. Acta*, 270, 132, 1972.

Scanu. A. M. et al., On the subunit structure of the protein of human serum high density lipoprotein. II. A study of Sephadex fraction IV, *J. Biol. Chem.*, 247, 5850, 1972.

Shipley, G. G. et al., Small-angle X-ray scattering of human serum high-density lipoproteins, *J. Supramol. Struct.*, 1, 98, 1972.

Sundarman, G. S. et al., Heterogeneity of human plasma high density lipoproteins, *Proc. Soc. Exp. Biol. Med.*, 141, 842, 1972.

Torsvik, H., Studies on the protein moiety of serum high density lipoprotein from patients with familial lecithin: cholesterol acyltransferase deficiency, *Clin. Genet.*, 3, 188, 1972.

Aster, R. H. et al., Histocompatibility antigens of human plasma. Localization to the HDL-3 lipoprotein fraction, *Transplantation*, 16, 205, 1973.

Jackson, R. L. et al., A comparison of the major apolipoprotein from pig and human high density lipoproteins, *J. Biol. Chem.*, 248, 2639, 1973.

Mackenzie, S. L. et al., Heterogeneity of human serum high-density lipoprotein (HDL$_2$), *Clin. Chim. Acta*, 43, 223, 1973.

Nakagawa, M. et al., Role of high density lipoproteins in the lecithin-cholesterol acyltransferase activity with sonicated lecithin-cholesterol dispersions as substrate, *Biochim. Biophys. Acta*, 296, 577, 1973.

Rittner, C. et al., Demonstration of HL-A inhibitors in human serum. I. Does an association exist between HL-A and serum lipoproteins?, *Z. Immunitaetsforsch.*, 146(2), 123, 1973.

Assmann, G. et al., Lipid-protein interactions in high density lipoproteins, *Proc. Natl. Acad. Sci. U.S.A.*, 71, 989, 1974.

Blomhoff, J. P., High density lipoproteins in cholestasis, *Scand. J. Gastroenterol.*, 9(6), 591, 1974.

Davis, M. A. et al., Comparative studies on porcine and human high density lipoproteins, *Comp. Biochem. Physiol. B*, 47, 831, 1974.

Glamgeaud, M. C. et al., Study of sub-units of high density lipoproteins of human serum after partial delipidation, *Biochimie*, 56, 245, 1974.

Koren, E. et al., Interaction of human serum high density lipoproteins with triglycerides in vitro, *FEBS Lett.*, 44, 43, 1974.

Lutmer, R. F. et al., Analysis of α-lipoprotein cholesterol in 50 microliters of plasma, *J. Lipid Res.*, 15(6), 611, 1974.

Makino, S. et al., The interaction of polypeptide components of human high density serum lipoprotein with detergents, *J. Biol. Chem.*, 249(23), 7379, 1974.

Olofsson, S. O. et al., Degradation of high-density lipoproteins (HDL) in vitro, *Scand. J. Clin. Lab. Invest.*, Suppl. 33, 57, 1974.

Sodhi, H. S. et al., Isoelectric fractionation of plasma high-density lipoproteins, *Scand. J. Clin. Lab. Invest.*, Suppl. 33, 71, 1974.

Sundaram, G. S. et al., Preparative isoelectric focusing of human serum high-density lipoprotein (HDL$_3$), *Biochim. Biophys. Acta*, 337(2), 196, 1974.

Avogaro, P. et al., Familial hyper-HDL-(α)-cholesterolemia, *Atherosclerosis*, 22(1), 63, 1975.

Barter, P. J. et al., The transport of triglyceride in the high-density lipoproteins of human plasma, *J. Lab. Clin. Med.*, 85(2), 260, 1975.

Basu, M. K. et al., Studies on the binding of 1-anilino-8-naphthalene sulfonate to very low density and high density human serum lipoproteins, *Biochim. Biophys. Acta*, 398(3), 385, 1975.

Danielsson, B. et al., An abnormal high density lipoprotein in cholestatic plasma isolated by zonal ultracentrifugation, *FEBS. Lett.*, 50(2), 180, 1975.

Gwynne, J. et al., The molecular behavior of apo A-I in human high density lipoproteins, *J. Biol. Chem.*, 250(6), 2269, 1975.

Harberland, M. E. et al., Interaction of L-α-palmitoyl lysophospatidylcholine with the A-I polypeptide of high density lipoprotein, *J. Biol. Chem.*, 250(17), 6636, 1975.

Hauser, H., Lipid-protein interaction in porcine high-density (HDL$_3$) lipoprotein, *FEBS Lett.*, 60(1), 71, 1975.

Henderson, T. O. et al., 31-P nuclear magnetic resonance studies on serum low and high density lipoproteins: effect of paramagnetic ion, *Biochemistry*, 14(9), 1915, 1975.

Kruski, A. V. et al., Properties of rooster serum high density lipoproteins, *Biochim. Biophys. Acta*, 409(1), 26, 1975.

Lutmer, R. F. et al., Letter: high-density lipoprotein and atherosclerosis, *Lancet*, 1(7908), 691, 1975.

Miller, N. E. et al., Letter: high-density lipoprotein and atherosclerosis, *Lancet*, 1(7914), 1033, 1975.

Schonfeld, G. et al., Structure of human high density lipoprotein reassembled in vitro. Radioimmunoassay studies, *J. Biol. Chem.*, 250(19), 7934, 1975.

Stone, W. L. et al., Hydrophobic interactions of the apo-Gln-I polypeptide component of human high density serum lipoprotein, *J. Biol. Chem.*, 250(10), 3584, 1975.

Sundaram, G. S. et al., Isoelectric focusing of human serum high density lipoprotein in the presence of radioactive ampholines, *Biochim. Biophys. Acta*, 388(3), 349, 1975.

Utermann, G. et al., Plasma lipoprotein abnormalities in a case of primary high-density lipoprotein (HDL) deficiency, *Clin. Genet.*, 8(4), 258, 1975.

Verdery, R. B. et al., Arrangement of lipid and protein in human serum high density lipoproteins: a proposed model, *Chem. Phys. Lipids*, 14(2), 123, 1975.

Bachorik, P. S. et al., Plasma high-density lipoprotein cholesterol concentrations determined after removal of other lipoproteins by heparin/manganese precipitation or by ultracentrifugation, *Clin. Chem.*, 22(11), 1828, 1976.

Barter, P. J. et al., The transport of esterified cholesterol in plasma high density lipoproteins of human subjects: a mathematical model, *J. Lab. Clin. Med.*, 88(4), 627, 1976.

Carew, T. E. et al., A mechanism by which high-density lipoproteins may slow the atherogenic process, *Lancet*, 1(7973), 1315, 1976.

Charlton, S. C. et al., Stopped flow kinetics of pyrene transfer between human high density lipoproteins, *J. Biol. Chem.*, 251(24), 7952, 1976.

Editorial: HDL and CHD, *Lancet*, 2(7977), 131, 1976.

Friedberg, S. J. et al., The molar ratio of the two major polypeptide components of human high density lipoprotein, *J. Biol. Chem.*, 251(13), 4005, 1976.

Gerassimova, E. N., Steroid hormones and cholesterol of α-lipoproteins of blood plasma (in Russian with English abstract), *Ter. Arkh.*, 48(6), 40, 1976.

Keenan, R. W. et al., The binding of (3H) dolichol by plasma high density lipoproteins, *Biochim. Biophys. Acta*, 486(1), 1, 1976.

Levy, R. I. et al., The composition, structure and metabolism of high density lipoprotein, in *Lipoprotein Metabolism*, Greten, H., Ed., Springer-Verlag, Berlin, 1976, 56.

Nichols, A. V. et al., Effects of quanidine hydrochloride on human· plasma high density lipoproteins, *Biochim. Biophys. Acta*, 446(1), 226, 1976.

Olofsson, S. O. et al., Studies on human serum HDLs. V. Isolation and characterization of a cholesterol ester-rich lipoprotein after in vitro incubation, *Scand. J. Clin. Lab. Invest.*, 36(1), 67, 1976.

Olofsson, S. O. et al., Studies on human serum high-density lipoproteins. VI. Studies on a cholesterol ester-releasing reaction in vitro, *Scand. J. Clin. Lab. Invest.*, 36(5), 481, 1976.

Pattnaik, N. M. et al., Kinetic study of the action of snake venom phospholipase A_2 on human serum high density lipoprotein, *J. Biol. Chem.*, 241(7), 1984, 1976.

Reynolds, T. A., Conformational stability of the polypeptide components of human high density serum lipoprotein, *J. Biol. Chem.*, 251(19), 6013, 1976.

Steele, B. W. et al., Enzymatic determinations of cholesterol in high-density-lipoprotein fractions prepared by a precipitation technique, *Clin. Chem.*, 22(1), 98, 1976.

Segrest, J. P., Molecular packing of high density lipoproteins: a postulated functional role, *FEBS Lett.*, 69(1), 111, 1976.

LDL

Walton, K. W., Apparent variation in the characteristics of human low-density lipoproteins during immunoelectrophoresis in agar, *Immunochemistry*, 1, 279, 1964.

Walton, K. W. et al., Estimation of the low-density (β) lipoproteins of serum in health and disease using large molecular weight dextran sulphate, *J. Clin. Pathol.*, 17, 627, 1964.

Walton, K. W. et al., Immunological characteristics of human low-density lipoproteins, *Immunochemistry*, 1, 267, 1964.

Janado, M. et al., Interaction of dextran sulfate with low-density lipoproteins of plasma, *J. Lipid Res.*, 6, 331, 1965.

Nishida, T. et al., Interaction of ferrihemoglobin with peroxidized serum low density lipoproteins, *J. Biol. Chem.*, 240, 225, 1965.

Scott, P. J. et al., Role of the gut in synthesizing the protein component of low-density (β) lipoprotein, *Nature (London)*, 208, 494, 1965.

Tracy, R. E. et al., Sequestration of serum low-density lipoproteins in the arterial intima by complex formation, *Proc. Soc. Exp. Biol. Med.*, 118, 1095, 1965.

Aladjem, F., Immunoelectrophoretic properties of low density human serum lipoproteins, *Nature (London)*, 209, 1003, 1966.

Netto, R. F. et al., Comparative study of serum lipoproteins of low density (LDL) in inhabitants of the city vagrants from the north shore of the state of Sao Paulo and natives of the region of the Rio Xinges (in Portugese), *Arq. Brazil Cardiol.*, 19, 331, 1966

Stokes, R. P., A quantitative small scale method for the separation of low density (β) lipoproteins from serum, *J. Med. Lab. Technol.*, 23, 33, 1966.

Scott, P. J., Serum low-density lipoprotein levels in a New Zealand population sample, *Clin. Chim. Acta*, 15, 449, 1967.

Stokes, R. P. et al., The isolation of low-density (β)-lipoprotein from small volumes of human serum, *J. Atheroscler. Res.*, 7, 186, 1967.

Walton, K. W., Immunochemistry of the low-density lipoproteins, *Rev. Atheroscler.*, Suppl. 9(1), 193, 1967.

Mills, G. L. et al., The distribution of cholesterol in the low-density lipoproteins of human serum, *Clin. Chim. Acta*, 22, 251, 1968.

Nishida, T., Effect of phospholipase A treatment of low density lipoproteins on the dextran sulfate-lipoprotein interaction, *J. Lipid Res.*, 9, 627, 1968.

Roelcke, D. et al., The subdivisibility of human lipoproteins of low density with regard to their Lp (a)-property with ion-exchange-column chromatography, *Blut*, 18, 160, 1968.

Rudman, D. et al., Observations on the protein components of human high- and low-density lipoproteins, *Biochemistry*, 7, 3136, 1968.

Scanu, A. et al., Human serum low-density lipoprotein protein: its conformation studied by circular dichroism, *Nature (London)*, 218, 200, 1968.

Adams, G. H. et al., Polydyspersity of human low density lipoproteins, *Ann. N.Y. Acad. Sci.,* 164, 130, 1969.

Adams, G. H. et al., Rapid molecular weight estimates for low-density lipoproteins, *Anal. Biochem.,* 29, 117, 1969.

Albers, J. J. et al., Identification and genetic control of two new low-density lipoprotein allotypes: phenogroups at the Lpq locus, *J. Immunol.,* 103, 155, 1969.

Bütler, R. et al., Further observations on human anti-LDL antibodies: serological behaviour, frequency and mode of formation, *Vox Sang.,* 17, 230, 1969.

Bütler, R. et al., On the genetics of the low density lipoprotein factors (Ag(c) and Ag(e)), *Hum. Hered.,* 19, 174, 1969.

Cogan, U. et al., Effect of various short and long chain fatty acids on the dextran sulfate-low density lipoprotein interaction, *Biochim. Biophys. Acta,* 187, 444, 1969.

Dearborn, D. G. et al., Reversible thermal conformation changes in human serum low-density lipoprotein, *Proc. Natl. Acad. Sci. U.S.A.,* 62, 179, 1969.

Leslie, R. B. et al., Nuclear magnetic resonance studies on serum low-density lipoproteins (LDL$_2$), *Chem. Phys. Lipids,* 3, 152, 1969.

Lindgren, F. T. et al., Flotation rates, molecular weights and hydrated densities of the low-density lipoproteins, *Lipids,* 4, 337, 1969.

Mills, G. L., The protein content and flotation rate of human low density lipoproteins, *Biochim. Biophys. Acta,* 194, 222, 1969.

Pollard, H. et al., On the geometrical arrangement of the protein subunits of human serum low-density lipoprotein evidence for a dodecahedral model, *Proc. Natl. Acad. Sci. U.S.A.,* 64, 304, 1969.

Sato, J. et al., Immunochemical properties of lipid protein complex. II. Antigenicity of human serum low density lipoprotein and its derivatives, *Jpn. J. Exp. Med.,* 39, 621, 1969.

Scanu, A. et al., On the conformational instability of human serum low-density lipoprotein: effect of temperature, *Proc. Natl. Acad. Sci. U.S.A.,* 62, 171, 1969.

Simons, K. et al., Heterogeneity of maleylated and partially delipidated low-density lipoproteins from human plasma, *Ann. Med. Exp. Biol. Fenn.,* 47, 48, 1969.

Adams, G. H. et al., Equilibrium banding of low-density lipoproteins. I. Determination of gradients, *Biochim. Biophys. Acta,* 202, 305, 1970.

Adams, G. H. et al., Equilibrium banding of low-density lipoproteins. II. Analysis of banding patterns, *Biochim. Biophys. Acta,* 202, 315, 1970.

Adams, G. H. et al., Equilibrium banding of low density lipoproteins. III. Studies on normal individuals and effects of diet and heparin-induced lipase, *Biochim. Biophys. Acta,* 210, 462, 1970.

Charlton, R. K. et al., Soluble HL-A$_7$ antigen: localization in the β-lipoprotein fraction of human serum, *Science,* 170, 636, 1970.

Fielding, C. J., Human lipoprotein lipase. II. Inhibition of enzyme activity by plasma low density lipoproteins, *Biochim. Biophys. Acta,* 206, 118, 1970.

Fisher, W. R., The characterization and occurrence of an Sf 20 serum lipoprotein, *J. Biol. Chem.,* 245, 877, 1970.

Goldstein, S. et al., Study of the antigenicity of human serum β-lipoprotein (in French), *Ann. Biol. Clin. (Paris),* 28, 217, 1970.

Goldstein, S. et al., Study of the antigenicity of human serum β-lipoprotein. II. Immunoprecipitation graphs (in French), *Ann. Biol. Clin. (Paris),* 28, 223, 1970.

Hanefeld, M. et al., Investigations on the role of low density lipoproteins in the pathogenesis of the essential carbohydrate-induced hypertriglyceridemia, *Experientia,* 26, 481, 1970.

Hurley, P. J. et al., Plasma turnover of Sf 0-9 low-density lipoprotein in normal men and women, *Atherosclerosis,* 11, 51, 1970.

Knüchel, F., Determination of total serum cholesterol and serum β-lipoprotein (in German), *Med. Welt.,* 27, 1265, 1970.

Lee, D. M. et al., Studies of the composition and structure of plasma lipoproteins. Isolation, composition and immuno-chemical characterization of low density lipoprotein subfractions of human plasma, *Biochemistry,* 9, 2244, 1970.

Lees, R. S., Immunoassay of plasma low-density lipoproteins, *Science,* 169, 493, 1970.

Mauldin, J. et al., pH and ionic strength dependent aggregation of serum low-density lipoproteins, *Biochemistry,* 9, 2015, 1970.

Nishida, T. et al., Nature of the interaction of dextran sulfate with low density lipoproteins of plasma, *J. Biol. Chem.,* 245, 4689, 1970.

Roelcke, D. et al., Physicochemical, immunologic and biochemical characterization of the moiety of the low-density lipoprotein (in German), *Z. Klin. Chem. Klin. Biochem.,* 7, 467, 1970.

Seidel, D. et al., An abnormal low-density lipoprotein in cholestases. I. Isolation and characterization, *Dtsch. Med. Wochenschr.,* 95, 1774, 1970.

Seidel, D. et al., An abnormal low density lipoprotein in cholestasis. II. Significance for the differential diagnosis of jaundice (in German), *Dtsch. Med. Wochenschr.*, 95, 1805, 1970.

Slack, J. et al., Anomalous low density lipoproteins in familial hyper-β-lipoproteinaemia, *Clin. Chim. Acta*, 29, 15, 1970.

Smith, E. B. et al., The chemical and immunological assay of low density lipoproteins extracted from human aortic intima, *Atherosclerosis*, 11, 417, 1970.

Stone, M. C. et al., Comparison of membrane filtration and nephelometry with analytical ultracentrifugation, for the quantitative analysis of low-density lipoprotein fractions, *Clin. Chim. Acta*, 30, 809, 1970.

Chong, J. H. et al., Serum low density lipoproteins, triglycerides and cholesterol levels in Malaysia, *Clin. Chim. Acta*, 34, 85, 1971.

Fisher, W. R. et al., Hydrodynamic studies of human low density lipoproteins. Evaluation of the diffusion coefficient and the preferential hydration, *Biochemistry*, 10, 1622, 1971.

Hammond, M. G. et al., The characterization of a discrete series of low density lipoproteins in the disease, hyper-pre-β-lipoproteinemia. Implications relating to the structure of plasma lipoproteins, *J. Biol. Chem.*, 246, 5454, 1971.

Helenius, A. et al., Removal of lipids from human plasma low-density lipoprotein by detergents, *Biochemistry*, 10, 2542, 1971.

Jones, R. J. et al., Distribution of β-lipoprotein between low-density and very low-density lipoproteins, *J. Lab. Clin. Med.*, 78, 994, 1971.

Lewis, B., Low-density lipoproteins, *Biochem. J.*, 123, 15P, 1971.

Lewis, B., Low-density lipoproteins, *Biochem. Soc. Symp.*, 33, 19, 1971.

Pinon, J. C. et al., Low density lipoprotein of human plasma: N-terminal amino-acids in familial hyper-β-lipoproteinemia (type II), *Clin. Chim. Acta*, 32, 131, 1971.

Pollard, H. B. et al., Construction of a three-dimensional iso-density map of the low-density lipoprotein particle from human serum, *Biochem. Biophys. Res. Commun.*, 44, 593, 1971.

Albers, J. J. et al., Human serum lipoproteins. Evidence for three classes of lipoproteins in Sf 0-12, *Biochemistry*, 11, 57, 1972.

Burstein, M., Antisera of rabbits immunized against human low density lipoproteins (1,006-1,063), *Nouv. Rev. Fr. Hematol.*, 12, 251, 1972.

Finatti, A. A. et al., Correlation between low density serum lipoproteins (LDL), cholesterol and triglycerides. Regressive equations, *Arq. Bras. Cardiol.*, 25, 461, 1972.

Fisher, W. R. et al., Measurements of the molecular weight variability of plasma low density lipoproteins among normals and subjects with hyper-lipoproteinemia. Demonstration of macromolecular heterogeneity, *Biochemistry*, 11, 519, 1972.

Friedewald, W. T. et al., Estimation of the concentration of low-density lipoprotein cholesterol in plasma, without use of the preparative ultracentrifuge, *Clin. Chem.*, 18, 499, 1972.

Gotto, A. M. et al., Evidence for the identity of the major apoprotein in low density and very low density lipoproteins in normal subjects and patients with familial hyperlipoproteinemia, *J. Clin. Invest.*, 51, 1486, 1972.

Grant, E. H. et al., A dielectric investigation of the water of hydration of low density lipoproteins in familial hyper-β-lipoproteinaemia, *Lancet*, 1, 1159, 1972.

Maitrot, B. et al., Demonstration of several antigens in the LDL fraction of human serum lipoproteins, *Biochimie*, 54, 381, 1972.

Mateu, L. et al., On the structure of human serum low density lipoprotein, *J. Mol. Biol.*, 70, 105, 1972.

Smith, R. et al., The size and number of polypeptide chains in human serum low density lipoproteins, *J. Biol. Chem.*, 247, 3376, 1972.

Wilson, D. E. et al., Metabolic relationships among the plasma lipoproteins. Reciprocal changes in the concentrations of very low and low density lipoproteins in man, *J. Clin. Invest.*, 51, 1051, 1972.

Ghosh, S. et al., Charge heterogeneity of human low density lipoprotein (LDL), *Proc. Soc. Exp. Biol. Med.*, 142, 1322, 1973.

Gotto, A. M., Jr. et al., A comparative study of the effects of chemical modification on the immunochemical and optical properties of human plasma low-density lipoproteins and apoproteins, *Biochem. J.*, 133, 369, 1973.

Havel, R. J. et al., Cofactor activity of protein components of human very low density lipoproteins in the hydrolysis of triglycerides by lipoproteins lipase from different sources, *Biochemistry*, 12, 1828, 1973.

Herbert, R. J. et al., Effect of low-density lipoprotein preparations on plasmin, *Clin. Sci.*, 45, 129, 1973.

Kocher, D. B. et al., Hyper-low density lipoproteinemia in United States Air Force recruits, *Circulation*, 48, 1304, 1973.

Chen, G. C. et al., Contribution of carotenoids to the optical activity of human serum low-density lipoprotein, *Biochemistry*, 13, 3330, 1974.

Hornal, D. K. et al., Preparation and properties of an antibody to human low density lipoprotein, *Z. Klin. Chem. Klin. Biochem.*, 12(5), 246, 1974.

Jubb, J. S. et al., A comparison of three methods of human plasma low density lipoprotein analysis, *Z. Klin. Chem. Klin. Biochem.,* 12(5), 244, 1974.

Krishnaiah, K. V. et al., Demonstration of protease-like activity in human serum low density lipoprotein, *FEBS Lett.,* 40, 265, 1974.

Lee, D. M. et al., Physicochemical properties of low density lipoproteins of normal human plasma. Evidence for the occurrence of lipoprotein-β in associated and free forms, *Biochem. J.,* 137, 155, 1974.

Reichl, D. et al., The metabolism of low-density lipoprotein in a patient with familial hyper-β-lipoproteinaemia, *Clin. Sci. Mol. Med.,* 47(6), 635, 1974.

Brown, M. S. et al., Receptor-dependent hydrolysis of cholesteryl esters contained in plasma low density lipoprotein, *Proc. Natl. Acad. Sci. U.S.A.,* 72(8), 2925, 1975.

Chen, G. C. et al., Temperature dependence of the optical activity of human serum low density lipoprotein. The role of lipids, *Biochemistry,* 14(15), 3357, 1975.

Deckelbaum, R. J. et al., Thermal transitions in human plasma low density lipoproteins, *Science,* 190(4212), 392, 1975.

Goldstein, J. L. et al., Inhibition of proteolytic degradation of low density lipoprotein in human fibroblasts by chloroquine, concanavalin A and Triton WR 1339, *J. Biol. Chem.,* 250(19), 7854, 1975.

Monahan, L. K. et al., Low density lipoprotein cholesterol (CLDL) and lipoprotein phenotyping, *Am. J. Med. Technol.,* 41(8), 307, 1975.

Aggerbeck, L. P. et al., Enzymatic probes of lipoprotein structure. Hydrolysis of human serum low density lipoprotein-2 by phospholipase A_2, *J. Biol. Chem.,* 251(12), 3823, 1976.

Basu, S. K. et al., Degradation of cationized low density lipoprotein and regulation of cholesterol metabolism in homozygous familial hypercholesterolemia fibroblasts, *Proc. Natl. Acad. Sci. U.S.A.,* 73(9), 3178, 1976.

Berg, K. et al., Genetic variation in serum LDLs and lipid levels in man, *Proc. Natl. Acad. Sci. U.S.A.,* 73(3), 937, 1976.

Brown, M. S. et al., Role of the LDL receptor in the regulation of cholesterol and lipoprotein metabolism, in *Lipoprotein Metabolism,* Greten, H., Ed., Springer-Verlag, Berlin, 1976, 82.

Camejo, G. et al., Differences in the structure of plasma low-density lipoproteins and their relationship to the extent of interaction with arterial wall-components, *Ann. N.Y. Acad. Sci.,* 275, 153, 1976.

Cham, B. E., Nature of the interaction between low-density lipoproteins and polyanions and metal ions, as exemplified by heparin and Ca_2^+, *Clin. Chem.,* 22(11), 1812, 1976.

Fischer-Dzoga, K. et al., Stimulation of proliferation in stationary primary cultures of monkey aortic smooth muscle cells. Part 2. Effect of varying concentrations of hyperlipemic serum and low density lipoproteins of varying dietary fat origins, *Atherosclerosis,* 24(3), 515, 1976.

Ikai, A., Serum low density lipoprotein, *Protein Nucleic Acid Enzyme,* 21(2), 981, 1976.

Jackson, R. L. et al., Comparative studies on plasma LDLs from pig and man, *Comp. Biochem. Physiol. B,* 53(2), 245, 1976.

Mills, G. L. et al., Low-density lipoproteins in patients hemozygous for familial hyper-β-lipoproteinemia, *Clin. Sci. Mol. Med.,* 51(3), 221, 1976.

Myant, N. B. et al., The metabolism in vivo and in vitro of plasma low-density lipoprotein from a subject with inherited hypercholesterolaemia, *Clin. Sci. Mol. Med.,* 51(5), 463, 1976.

Packard, C. J. et al., Low density lipoprotein metabolism in a family of familial hypercholesterolemia patients, *Metabolism,* 25(9), 995, 1976.

Packard, C. J. et al., Low-density lipoprotein metabolism in type II hyperlipoproteinaemia, *Biochem. Soc. Trans.,* 4(1), 105, 1976.

Salmon, S. et al., Immunologic study of human low density lipoproteins (in French with English abstract), *Paroi Arterielle,* 3(3), 115, 1976.

Sears, B. et al., Temperature-dependent 13C nuclear magnetic resonance studies of human serum low density lipoproteins, *Biochemistry,* 15(19), 4151, 1976.

Sigurdsson, G. et al., The metabolism of low density lipoprotein in endogenous hypertriglyceridaemia, *Eur. J. Clin. Invest.,* 6(2), 151, 1976.

Skrzydlewski, Z. et al., Low density lipoprotein (LDL) complexes with protamine, *Bull. Acad. Pol. Sci. Sur. Sci. Biol.,* 24(4), 201, 1976.

Swaminathan, N. et al., The monosaccharide composition and sequence of the carbohydrate moiety of human serum low density lipoproteins, *Biochemistry,* 15(7), 1516, 1976.

Deckelbaum, R. J. et al., Structure and interactions of lipids in human plasma low density lipoproteins, *J. Biol. Chem.,* 252(2), 744, 1977.

VLDL

Augustyniak, J. et al., Characterization of lipovitellin components and their relation to low-density lipoprotein structure, *Biochim. Biophys. Acta,* 84, 721, 1964.

Gustafson, A. et al., Studies on the composition and structure of serum lipoproteins: physical-chemical characterization of phospholipid-protein residues obtained from very-low-density human serum lipoproteins, *Biochim. Biophys. Acta*, 84, 767, 1964.

Nichols, A. V. et al., The effect of very-low-density lipoproteins on lipid transfer in incubated serum, U.S. Atomic Energy Commission, University of California Radiation Laboratory, 145, 1964.

Gustafson, A. et al., Studies of the composition and structure of serum lipoproteins: isolation, purification, and characterization of very-low-density lipoproteins of human serum, *Biochemistry*, 4, 596, 1965.

Reaven, G. M. et al., Kinetics of triglyceride turnover of very low density lipoproteins of human plasma, *J. Clin. Invest.*, 44, 1826, 1965.

Spritz, N., Effect of fatty acid saturation on the distribution of the cholesterol moiety of very low density lipoproteins, *J. Clin. Invest.*, 44, 339, 1965.

Bierman, E. L. et al., Particle-size distribution of very low density plasma lipoproteins during fat absorption in man, *J. Lipid Res.*, 7, 65, 1966.

Gustafson, A., Studies on human serum very-low-density-lipoproteins, *Acta Med. Scand. Suppl.*, 446, 1, 1966.

Levy, R. I. et al., The nature of pre-β (very low density) lipoproteins, *J. Clin. Invest.*, 45, 63, 1966.

Ockner, R. K. et al., Very-low-density lipoprotein in intestinal lymph: evidence for presence of the A protein, *Science*, 162, 1285, 1968.

Splitter, S. D., Relation between dietary fat and fatty acid composition of "endogenous" and "exogenous" very low density lipoprotein triglycerides (D-1.006), *Metabolism*, 17, 544, 1968.

Brown, W. V. et al., Studies on the protein in human plasma very low density lipoproteins, *J. Biol. Chem.*, 244, 5687, 1969.

Menlendijk, P. N., The direct determination of very low density lipoproteins using Thorp's nephelometer (in Dutch), *Ned. Tijdschr. Geneeskd.*, 113, 1348, 1969.

Ockner, R. K. et al., Very low density lipoproteins in intestinal lymph: role in triglyceride and cholesterol transport during fat absorption, *J. Clin. Invest.*, 48, 2367, 1969.

Barter, P. J. et al., The distribution of triglyceride in subclasses of very low density plasma lipoproteins, *J. Lab. Clin. Med.*, 76, 925, 1970.

Brown, W. V. et al., Further characterization of apolipoproteins from the human plasma very low density lipoproteins, *J. Biol. Chem.*, 245, 6588, 1970.

Hanefeld, M. et al., Role of very-low-density lipoproteins in the pathogenesis of primary hypertriglyceridemia (in German with English abstract), *Z. Gesamte Inn. Med.*, 25, 961, 1970.

Havel, R. J. et al., Splanchnic metabolism of free fatty acids and production of triglycerides of very low density lipoproteins in normotriglyceridemic and hypertriglyceridemic humans, *J. Clin. Invest.*, 49, 2017, 1970.

Hazzard, W. R. et al., Very low density lipoprotein subfractions in a subject with broad-β disease (Type 3 hyperlipoproteinemia) and a subject with endogenous lipemia (Type IV). Chemical composition and electrophoretic mobility, *Biochim. Biophys. Acta*, 202, 517, 1970.

Puppione, D. L. et al., Partial characterization of serum lipoproteins in the density interval 1.04-1.06 g/mℓ, *Biochim. Biophys. Acta*, 202, 392, 1970.

Schonfeld, G., Changes in the composition of very low density lipoprotein during carbohydrate induction in man, *J. Lab. Clin. Med.*, 75, 206, 1970.

Schumaker, V. N. et al., Very low density lipoproteins: surface volume changes during metabolism, *J. Theor. Biol.* 26, 89, 1970.

Albers, J. J. et al., Isoelectric fractionation and characterization of polypeptides from human serum very low density lipoproteins, *Biochim. Biophys. Acta*, 236, 29, 1971.

Herbert, P. et al., Correction of COOH-terminal amino-acids of human plasma very low density apolipoproteins, *J. Biol. Chem.*, 246, 7068, 1971.

Ruderman, N. B. et al., A biochemical and morphologic study of very low density lipoproteins in carbohydrate induced hypertriglyceridemia, *J. Clin. Invest.*, 50, 1355, 1971.

Wille, L. E. et al., Demonstration of an atypical pre-β-lipoprotein in human serum, *Clin. Genet.*, 2, 242, 1971.

dePury, G. G. et al., Very low density lipoproteins and lipoprotein lipase in serum of rats deficient in essential fatty acids, *J. Lipid Res.*, 13, 268, 1972.

Eaton, R. P. et al., Incorporation of Se 75 selenomethionine into a protein component of plasma very low density lipoprotein in man, *Diabetes*, 21, 744, 1972.

Eisenberg, S. et al., On the apoprotein composition of human plasma very low density lipoprotein subfractions, *Biochim. Biophys. Acta*, 260, 329, 1972.

Jones, A. L. et al., Intestinal very low density lipoprotein production, *J. Med. Sci.*, 8, 836, 1972.

Nakaya, N., Composition of lipoprotein-influence of high carbohydrate and high fat diet on the composition of very low density lipoproteins, *Jpn. J. Geriatr.*, 9, 29, 1972.

Pearlstein, E. et al., Subpopulations of human serum very low density lipoproteins, *Biochemistry*, 11, 2553, 1972.

Quarfordt, S. H. et al., Heterogeneity of human very low density lipoproteins by gel filtration chromatography, *J. Lipid Res.*, 13, 435, 1972.

Dahlen, G. et al., Study on pre-beta-1 lipoprotein fraction, *Ann. Biol. Clin. (Paris)*, 31, 115, 1973.

Herbert, P. N. et al., Fractionation of the C-apoproteins from human plasma very low density lipoproteins, *J. Biol. Chem.*, 248, 4941, 1973.

Pearlstein, E. et al., Quantitation of three subpopulations of human serum very low density lipoproteins, *Biochem. Med.*, 8, 28, 1973.

Shore, V. G. et al., Heterogeneity of human plasma very low density lipoproteins. Separation of species differing in protein components, *Biochemistry*, 12, 502, 1973.

Avogaro, P., "Sinking" lipoprotein, *G. Ital. Cardiol.*, 4(6), 651, 1974.

Barter, P. J., Origin of esterified cholesterol transported in the very low density lipoproteins of human plasma, *J. Lipid Res.*, 15, 11, 1974.

Cloarec, M. et al., Pre-beta lipoproteins. The so-called slow pre-beta fraction. Practical value of slow pre-beta lipoproteins study, *Ann. Med. Interne*, 125(10), 737, 1974.

Ito, Y. et al., On the selective estimation methods for β- and pre-β lipoproteins (in Japanese), *Jpn. J. Clin. Pathol.*, 22(4), 293, 1974.

Phillips, G. B., Studies on the dense pre-beta-lipoprotein of human serum, *Clin. Chim. Acta*, 53, 127, 1974.

Shelburne, J. A. et al., A new apoprotein of human plasma very low density lipoproteins, *J. Biol. Chem.*, 249, 1428, 1974.

Albers, J. J. et al., Lp (a) lipoprotein: relationship to sinking pre-beta lipoprotein hyperlipoproteinemia and apolipoprotein, *Metabolism*, 24(9), 1047, 1975.

Avogaro, P. A. et al., "Sinking": lipoprotein in normal hyperlipoproteinaemic and atherosclerotic patients, *Clin. Chim. Acta*, 61(3), 239, 1975.

Bijl, V. D. et al., Human very low density lipoproteins: loss of electrophoretic mobility on enzymatic removal of sialic acid residues, *Clin. Chim. Acta*, 60(2), 191, 1975.

Kekki, M. et al., Turnover of plasma total and very low density lipoprotein triglyceride in man, *Scand. J. Clin. Lab. Invest.*, 35(2), 171, 1975.

Rubenstein, B. et al., A comparison of the peptide composition of human serum low and very low density lipoprotein, *Can. J. Biochem.*, 53(2), 128, 1975.

Chylomicrons

Alaupovic, P. et al., Isolation and characterization of human chyle chylomicrons and lipoproteins, *Ann. N.Y. Acad. Sci.*, 149, 791, 1968.

Nichols, A. V., Functions and interrelationships of different classes of plasma lipoproteins, *Proc. Natl. Acad. Sci. U.S.A.*, 64, 1128, 1969.

Hazzard, W. R. et al., Abnormal lipid composition of chylomicrons in broad-beta disease (type 3 hyperlipoproteinemia), *J. Clin. Invest.*, 49, 1853, 1970.

Kostner, G. et al., Characterization and quantitation of the apolipoproteins from human chyle chylomicrons, *Biochemistry*, 11, 1217, 1972.

Brunzell, J. D. et al., Evidence for a common, saturable, triglyceride removal mechanism for chylomicrons and very low density lipoproteins in man, *J. Clin. Invest.*, 52, 1578, 1973.

Burstein, M. et al., Turbidimetric estimation of chylomicrons and very low density lipoproteins in human sera after precipitation by sodium lauryl sulfate, *Biomedicine*, 19, 16, 1973.

Chait, A. et al., Functional overlap between "chylomicra" and "very low density lipoproteins" of human plasma during alimentary lipaemia, *Atherosclerosis*, 17, 455, 1973.

Ferrans, V. T. et al., Chylomicrons and the formation of foam cells in type I hyperlipoproteinemia. A morphologic study, *Am. J. Pathol.*, 70, 253, 1973.

Giardini, O. et al., Case of primary hyperchylomicronemia, *Minerva Pediatr.*, 25, 506, 1973.

Lipoprotein Fractions

α_1-Lipoproteins

Rejnek, J. et al., Study of the changes in α_1-lipoprotein in vitro (in German), *Hoppe Seylers Z. Physiol. Chem.*, 336, 40, 1964.

Schen, R. J. et al., The affirmity of Evans blue for the α_1-lipoprotein of human serum, *Clin. Chim. Acta*, 16, 445, 1967.

Laissue, J. et al., Foam cells in an α_1-lipoproteinemia (in German), *Virchows Arch. A*, 344, 119, 1968.

Beaumont, J. L. et al., Purification of serum α_1-lipoproteins (in French), *Ann. Biol. Clin.* 27, 237, 1969.

Schnitzler, S. et al., The cleavage of human α-lipoproteins by extracts from *Actinia equina* (in German), *Acta Biol. Med. Ger.*, 22, 437, 1969.

Torsvik, H., Presence of α_1-lipoprotein in patients with familial plasma lecithin:cholesterol acyltransferase deficiency, *Scand. J. Clin. Lab. Invest.*, 24, 187, 1969.

Rudmann, D. et al., A new method for isolating the nonidentical protein subunits of human plasma α-lipoproteins, *J. Clin. Invest.*, 49, 365, 1970.

Dahlén, G. et al., Lp(α) lipoprotein/pre-β_1-lipoprotein, serum lipids and atherosclerotic disease, *Clin. Genet.*, 9(6), 558, 1976.

Edwards, J. A. et al., Changes in the immunoelectrophoretic characteristics of serum α_1-lipoprotein induced by various dyes, *Clin. Chim. Acta*, 67(1), 35, 1976.

Klimov, A. N. et al., Identification of Macheboeuf protein as α-lipoprotein (in Russian), *Biochemistry*, 41(1), 110, 1976.

Muckle, T. J., A third form of electrophoretic dye-induced mobility alteration (DIMA), polymorphism of α_1-lipoprotein as revealed by immunoelectrophoresis with Thymol Blue, *Clin. Chim. Acta*, 73(1), 57, 1976.

β-Lipoproteins

Badin, J. et al., Further evidences of the qualitative abnormality of β-lipoprotein in liver diseases (in French), *Clin. Chim. Acta*, 9, 531, 1964.

Berg, K., Precipitation reactions in agar gel between albumin and β-lipoprotein of human serum, *Acta Pathol. Microbiol. Scand.*, 62, 287, 1964.

Berg, K., A new hereditary system concerning human serum β-lipoprotein, *Sangre*, 9, 24, 1964.

Bernfeld, P. et al., Proteolysis of human serum β-lipoprotein, *J. Biol. Chem.*, 239, 3341, 1964.

Franzini, C. et al., Observations on the behavior of total cholesterolemia and β-lipoprotidemia in various morbid conditions (in Italian), *Prog. Med.*, 20, 227, 1964.

Ilca, S. et al., Simple routine methods for determining total lipids, β-lipoprotein cholesterol, β-lipoproteins and cholesterol (in German), *Z. Gesamte Inn. Med. Ihre Grenzgeb.*, 19, 323, 1964.

Ingiulla, A. et al., Probable iso-immunization against hereditary groups of β-lipoproteins (Ag) in a multipara with anti-Rh(D) antibodies (in Italian), *Riv. Clin. Pediatr.*, 74, 474, 1964.

Ledvina, M. et al., Lipolysis of raised serum β-lipoproteins (in German), *Z. Gesamte Inn. Med. Ihre Grenzgeb.*, 19, 597, 1964.

Link, J. et al., Experiences with a single β-lipoprotein determination. I. β-Lipo-proteins in chronic liver diseases (in German), *Z. Gesamte Inn. Med. Ihre Grenzgeb.*, 19, 400, 1964.

Sany, C. et al., Lipid-protein complexes in the blood serum, determination of β-lipoproteins (in French), *Montpellier Med.*, 65, 223, 1964.

Szymanska, H. et al., Determination of serum β-lipoproteins using the Burstein-Prawerman method. A new text for the differentiation of jaundice (in Polish), *Pol. Tyg. Lek.*, 19, 127, 1964.

Varone, D. et al., Behavior of a genetic determinant of β-lipoproteins (Ag-groups) in the serum of mother-child pairs (in Italian), *Riv. Clin. Pediatr.*, 74, 499, 1964.

Vasileiskii, S. S. et al., On the variations in fetal β-proteins in individual fetuses (in Russian), *Biull. Eksp. Biol. Med.*, 7, 52, 1964.

Vierucci, A. et al., Antibodies against β-lipoprotein genetic factors (the Ag groups): a new aspect of plasma iso-immunization in thalassemic children having multiple transfusions (in Italian), *Riv. Clin. Pediatr.*, 74, 451, 1964.

Viggiani, A., Values of blood lipids, blood cholesterol and blood β-lipoproteins in young 20 year olds in the province of Pescara (in Italian), *G. Ig. Med. Prev.*, 5, 349, 1964.

Báliut, S. et al., Photometric determination of serum β-lipoproteins with the Burstein and the Frie-Hoeflmayr methods (in Hungarian), *Orv. Hetil.*, 106, 743, 1965.

Berg, K., Comparative studies on the Lp and Ag serum type systems of human β-lipoprotein, *Bibl. Haematol.*, 23, 385, 1965.

Bihari-Varga, M., Study of the relation of the β-lipoprotein content of human sera to the degree of its combination with aortic mucopolysaccharides in vitro (in Hungarian), *Orv. Hetil.*, 106, 353, 1965.

Blumberg, B. S., Inherited serum protein variants with emphasis on isoprecipitins and the β-lipoproteins, *Vox Sang.*, 10, 366, 1965.

Cabau, N. et al., Value of the determination of antistreptolysin on serum lacking β-lipoprotein (in French), *J. Med. Lyon*, 46, 1299, 1965.

Franzini, C., On the cholesterol-protein ratio of human serum β-lipoprotein in some physiological conditions, *Clin. Chim. Acta*, 12, 33, 1965.

Gosztonyi, J. et al., Labelling of β-lipoproteins with tritium, *Nature*, 208, 381, 1965.

Henry, J. et al., Determination of phosphorus in serum β-lipoproteins (in French), *Ann. Biol. Clin.*, 23, 275, 1965.

Hoeflmayr, J. et al., Simplified clinico-chemical laboratory test (photometric methods). 3. Fatty acid ester and β-lipoproteins (in German), *Ther. Ggw.*, 104, 354, 1965.

Ledvina, M. et al., The lipolytic activity of microorganisms determined by means of serum β-lipoproteins (in Slovak), *Biologia*, 20, 671, 1965.

Ohkawa, K., Histochemical studies on the lipid requirement for the oxidative enzyme activities. II. Studies on the effects upon the oxidative enzyme activities of phospholipids, CoQ and β-lipoprotein, *Acta Tuberc. Jpn.*, 15, 22, 1965.

de Oliveira, J. W., Effects of a new anabolic agent on lipids and β-lipoproteins (in Portugese), *Folia Med.*, 50, 251, 1965.

Shubarov, K. et al., Normal values of β-lipoproteins, *Suvrem. Med.*, 16, 607, 1965.

Tomita, S., Tests for β-lipoproteins (in Japanese), *Jpn. J. Clin. Pathol.*, 13, 529, 1965.

Zambrowicz, K., Attempted determination of β-lipoproteins using the nephelometric method (in Polish), *Pol. Arch. Med. Wewn.*, 35, 383, 1965.

Cabau, N. et al., Influence of the non-specific inhibition of streptolysin O. New methods of determination of the inhibiting power of β-lipoprotein (in French), *Rev. Fr. Etud. Clin. Biol.*, 11, 533, 1966.

Contu, L., A new antigen of human β-lipoproteins (in French), *Nouv. Rev. Fr. Hematol.*, 6, 671, 1966.

Frézal, J. et al., Influence of a linoleic acid rich oil on the fatty acid composition of plasma lipids with congenital absence of β-lipoproteins (in French), *Rev. Fr. Etud. Clin. Biol.*, 11, 69, 1966.

Klimov, A. N. et al., Turbidimetric method of determination of β-lipoproteins and chylomicrons in blood plasma and tissues (in Russian), *Lab. Delo*, 5, 276, 1966.

Lewis, L. A. et al., Association of cold-precipitability with β-lipoprotein and cryoglobulin, *Am. J. Med.*, 40, 785, 1966.

Margolis, S. et al., Studies on human serum β-1-lipoprotein. I. Amino acid composition, *J. Biol. Chem.*, 241, 469, 1966.

Margolis, S., Studies on human serum β-1-lipoprotein. II. Chemical modifications, *J. Biol. Chem.*, 241, 477, 1966.

Mills, G. L. et al., The "normal" human plasma β-lipoprotein pattern, *Clin. Chim. Acta*, 14, 273, 1966.

Skrzydlewski, Z. et al., Inhibition of proteolytic enzymes by β-lipoprotein, *J. Atheroscler. Res.*, 6, 273, 1966.

Trzaski, M. et al., Evaluation of usefulness of determination of serum β-lipoproteins in comparison with cholesterol level determination (in Polish), *Pol. Tyg. Lek.*, 21, 206, 1966.

Allard, C. et al., Usefulness of a simple serum β-lipoprotein assay: an epidemiological study of an active Montreal population, *Can. Med. Assoc. J.*, 97, 1321, 1967.

Jaegermann, K., Studies of the group system of β-lipoprotein (Lp) (in German), *Z. Aerztl. Forbild.*, 61, 778, 1967.

Kellen, J., Immunochemical and turbidimetric studies on β-lipoproteins in epidemic hepatitis (in German), *Acta Hepatosplen.*, 11, 144, 1967.

Lukeš, J. et al., Structure of β-lipoprotein and its significance for the evaluation of lipemia components (in Czech), *Cas. Lek. Cesk.*, 106, 543, 1967.

Zuber, E. et al., Blood serum content of β-lipoprotein in certain diseases (in Polish), *Pol. Tyg. Lek.*, 22, 512, 1967.

Zuber, E., Contents of total lipids, cholesterol and β-lipoproteins in the blood serum (in Polish), *Pol. Tyg. Lek.*, 22, 832, 1967.

Boyle, E., Jr. et al., Serum β-lipoproteins and cholesterol in men. Relationships to smoking, age and body weight, *Geriatrics*, 23, 102, 1968.

Gotto, A. M. et al., Human serum β-lipoprotein., *Nature (London)*, 219, 1157, 1968.

Gotto, A. M. et al., Human serum β-lipoprotein: preparation and properties of a delipidated, soluble derivative, *Biochem. Biophys. Res. Commun.*, 31, 151, 1968.

Gotto, A. M. et al., Observations on the conformation of human β-lipoprotein, evidence for the occurrence of β-structure, *Proc. Natl. Acad. Sci. U.S.A.*, 60, 1436, 1968.

Krauer-Mayer, B. et al., Turbidimetric determination of total β-lipoproteins: correlation between venous blood and capillary blood (in German), *Helv. Paediatr. Acta*, 23, 529, 1968.

Pola, P. et al., Fibrinolytic effect of streptokinase on whole plasma and on plasma deprived of β-lipoproteins (in Italian), *Haematol. Lat.*, 11, 77, 1968.

Reisz, G., Concomitant determination of β-lipoprotein and total cholesterol and of fractional and total lipoproteins by a rapid micromethod, *Rom. Med. Rev.*, 12, 35, 1968.

Schnitzler, S. et al., Precipitation of human β-lipoprotein by extracts of *Actinia equina* (in German), *Z. Gesamte Hyg. Ihre Grenzgeb.*, 14, 726, 1968.

Sone, A. et al., Evaluation of simple measurement of β-lipoprotein with special reference to β-L Kit, the product of H. Haury Co. (in Japanese), *Jpn. J. Clin. Pathol.*, 16, 711, 1968.

Vierucci, A. et al., Synthesis of β-lipoproteins (Ag groups) in the foetus and the newborn, *Vox Sang.*, 14, 151, 1968.

Badin, J. et al., Effects of β-lipoprotein phospholipids on serum inhibitors of streptolysin O (in French), *Pathol. Biol.*, 17, 1107, 1969.

Beaumont, J. L. et al., Method of purification of rabbit antibody against human β-lipoprotein starting with specific precipitates (in French), *Immunochemistry*, 6, 489, 1969.

Burstein, M., Immune serum anti-beta lipoproteins preparation, activity, species specificity, *Rev. Fr. Transfus.*, 12, 271, 1969.

Burstein, M. et al., Precipitation of serum β-lipoproteins by anions with a strong electronegative charge. Role of the pH, *Nouv. Rev. Fr. Hematol.*, 9, 645, 1969.

Fukui, I. et al., New examination method for β-lipoprotein and β-cholesterol (in Japanese), *Naika*, 24, 29, 1969.

Gotto, A. M. et al., Human serum β-lipoprotein and β-apoprotein, *Nature (London)*, 223, 835, 1969.

Houston, R. G., Determination of serum β-lipoprotein (in German), *Med. Lab.*, 22, 104, 1969.

Imanishi, A. et al., Normal levels of serum β-lipoprotein in healthy middle-aged men in a city (in Japanese), *Med. Biol.*, 79, 263, 1969.

Lukeš, J. et al., Isolation of serum β-lipoprotein in patients (in Czech), *Cas. Lek. Cesk.*, 108, 989, 1969.

Kostner, G. et al., A method for the isolation of antibodies to human β-1-lipoprotein, *Biochim. Biophys. Acta*, 188, 157, 1969.

Kuzuya, F. et al., Activation of β-flucuronidase by β-lipoprotein and inhibition of this effect by sulphated polysaccharide, *J. Atheroscler. Res.*, 9, 215, 1969.

Richet, G. et al., Asymptomatic familial hypolipoproteinemia involving mainly beta-lipoprotein revealed during the study of an isolated proteinuria (in French), *Presse Med.*, 77, 2045, 1969.

Tikhonov, V. P. et al., Polarographic studies of serum β-lipoproteins (in Russian), *Lab. Delo*, 9, 571, 1969.

Alcindor, L. G. et al., Induction of the hepatic synthesis of β-lipoproteins by high concentrations of fatty acids. Effect of actinomycin D, *Biochim. Biophys. Acta*, 210, 483, 1970.

Badin, J. et al., Streptolysin O inhibition by serum gamma G-globulin and β-lipoprotein after blocking of nonesterified cholesterol by digitonin, *J. Lab. Clin. Med.*, 75, 975, 1970.

Boman, H., Studies on inherited antigenic variation of human serum beta-lipoprotein by passive hemagglutination, *Int. J. Pept. Protein Res.*, 2(4), 209, 1970.

Buxtorf, J. C. et al., Amino acids in beta-lipoproteins of normal human serum, *Pathol. Biol.*, 18, 648, 1970.

Kane, J. P. et al., Subunit heterogeneity in human serum β-lipoprotein, *Proc. Natl. Acad. Sci. U.S.A.*, 66, 1075, 1970.

Skrzydlewski, Z. et al., Correlation between the concentration of β-lipoproteins in the serum and the activity of the fibrinolytic system in plasma in healthy subjects at various ages (in Polish), *Wiad. Lek.*, 23, 1735, 1970.

Boxer, L. A. et al., Correlation of β-lipoprotein levels and serum cholesterol concentration, *Experientia*, 27, 635, 1971.

Okijio, T. et al., Fluctuation in serum lipoprotein fractions in various types of diseases and its diagnostic significance (in Japanese), *Jpn. J. Clin. Pathol. Suppl.*, 19, 169, 1971.

Baba, T. et al., Effect of β-lipoprotein in the serum on the measurement of ASLO antibody (Japanese with English abstract), *Ryumachi*, 12, 152, 1972.

Badin, J. et al., Progressive inactivation of streptolysin O incubated along or in the presence of beta-lipoprotein. Antagonistic effect of albumin and a reducing agent, *Ann. Biol. Clin.*, 30, 433, 1972.

Boman, H. et al., Studies on inherited antigenic variation of human serum β-lipoprotein by passive hemagglutination. VI. Screening for antibodies in sera from multiparous women and from normal individuals, *Clin. Genet.*, 3, 180, 1972.

Boman, H. et al., Studies on inherited antigenic variation of human serum β-lipoprotein by passive hemagglutination. VII. Isoantibodies in sera from children with thalassaemia, *Clin. Genet.*, 3, 201, 1972.

Freedman, F. et al., Quantitation of β-lipoprotein cholesterol, *Clin. Biochem.*, 5, 83, 1972.

Mermall, H. L. et al., Release of enzymic activity from human beta-lipoproteins by sonic radiation, *Proc. Soc. Exp. Biol. Med.*, 141, 735, 1972.

Nowak, A. et al., Relationships between skin surface lipids and blood betalipoprotein levels, sex and age in man, *Przegl. Dermatol.*, 59, 475, 1972.

Plamieniak, F. et al., Serum β-lipoprotein curve in type II and IV hyperlipoproteinemia after glucose loading (in Polish with English abstract), *Pol. Tyg. Lek.*, 27, 13, 1972.

Canal, J. et al., Properties of heavy (Sf 4—6) carbamylated human beta-lipoproteins, *Ann. Biol. Clin.*, 31, 97, 1973.

Delahunty, T. et al., Glycosylation of beta-lipoprotein and the possible role of choline derivatives, *Can. J. Biochem.*, 52, 359, 1974.

Giacone, G. O. et al., Beta lipoproteins and anti-streptococcal antibodies: nonspecific increases in the antistreptolysin O titer, *Ann. Osp. Maria Vittoria Torino*, 17(1—6), 14, 1974.

van Husen, N. et al., The increased activity of unspecific beta-lipoprotein-bound alkaline phosphatase in human serum, *Clin. Chim. Acta*, 53, 91, 1974.

Ito, Y. et al., On the selective estimation methods for beta- and pre beta-lipoproteins (in Japanese), *Jpn. J. Clin. Pathol.*, 22(4), 293, 1974.

Ito, Y. et al., Determination of β-lipoprotein by immunocrit methods (in Japanese), *Jpn. J. Clin. Pathol.*, 18, 579, 1974.

Kanezashi, K. et al., Problems of beta-lipoprotein determination, *Jpn. J. Clin. Pathol.*, 22(10), 136, 1974.

Kusumi, K. et al., On the determination of serum beta-lipoprotein, *Jpn. J. Clin. Pathol.*, 22(12), 889, 1974.

Monnier G. et al., Effects of factors influencing the activity of the reticuloendothelial system on the half-life of labelled beta-lipoprotein: preliminary report, *J. Med.*, 5(5), 217, 1974.

Kuzuya, F., Determination of β-lipoprotein by partigen method and its normal value (in Japanese), *Jpn. J. Clin. Med.*, 33(8), 2648, 1975.

Vaverková, H. et al., β-lipoprotein cholesterol, calculation without centrifugation and clinical importance (in Czech with English abstract), *Vnitr. Lek.*, 21(11), 1084, 1975.

Morganti, G. et al., Contributions to the genetics of serum β-lipoprotein in man. IV. Evidence for the existence of the Ag-a1-d and Ag-c-g loci, closely linked to the Ag-x-y locus, *Humangenetik*, 10, 244, 1977.

Pre-β-lipoprotein

Loeper, J. et al., Prebetalipoproteinemia (based on 70 patients and 20 controls) (in French), *Rev. Atheroscler.*, 10, 30, 1968.

Naito, C., Determination of pre-β-lipoprotein and its clinical significance (in Japanese), *Naika*, 24, 36, 1969.

Fujikuro, I. et al., Serum lipid pattern in ischemic heart disease — study of fatty acid composition and pre-beta-lipoprotein (in Japanese), *Jpn. Circ. J.*, 34, 479, 1970.

Srinivasan, S. R. et al., A simple technique for semiquantitative, clinical estimation of serum β- and pre-β-lipoproteins, *Angiologica*, 7, 344, 1970.

Srinivasan, S. R. et al., Complexing of serum pre-β and β-lipoproteins and acid mucopolysaccharides, *Atherosclerosis*, 12, 321, 1970.

Fedele, D. et al., Production of growth hormone in primary dyslipidemias with hyper-pre-β-lipoproteinemia (in Italian), *Folia Endocrinol.*, 24, 352, 1971.

Horder, M. et al., Plasma pre-β-lipoprotein in healthy adults, *Atherosclerosis*, 14, 31, 1971.

Loeper, J. et al., Study of mixed hyperlipemia with a distinct excess of β and pre-β-lipoproteins (78 cases) (in French), *Presse Med.*, 79, 797, 1971.

Dahlén, G. et al., Studies on an extra pre-β-lipoprotein fraction, *Acta Med. Scand. Suppl.*, 531, 1, 1972.

Sobra, J. et al., Inborn errors of lipid metabolism. XXVI. 2 subfractions of pre-β-lipoproteins (in Czech with English abstract), *Cas. Lek. Cesk.*, 111, 556, 1972.

Carmena, R. et al., Pre-beta-lipoprotein in the plasma of normal adults, *Rev. Clin. Esp.*, 132, 319, 1974.

Dahlen, G. et al., Lp (alpha) lipoprotein and pre-beta 1-lipoprotein in relation to lipid levels in males, *Acta Med. Scand.*, 198(4), 263, 1975.

Källberg, M. et al., Studies on an additional pre-β-lipoprotein sinking pre-β (SPB). II. Fatty acid composition of lipid moieties, *Scand. J. Clin. Lab. Invest.*, 36(1), 59, 1976.

Lindén, L. et al., Studies on an additional pre-β-lipoprotein, sinking pre-β (SPB). I. Isolation and characterization, *Scand. J. Clin. Lab. Invest.*, 36(1), 51, 1976.

Odrowaz-Sypniewska, G. et al., Occurrence of pre-β_1-lipoprotein fraction in primary hyperlipoproteinemia (in Polish with English abstract), *Przegl. Lek.*, 33(6), 644, 1976.

Abnormal Lipoproteins
Different Abnormal Lipoproteins

Seegers, W. et al., Double β-lipoprotein: a new variant in man, *Science*, 149, 303, 1965.

Chisiu, N. S., A new class of lipoproteins, normal component of human blood serum: the ultra-slow lipoproteins, *Stud. Cercet. Med. Interna*, 10, 445, 1969.

Fredrickson, D. S. et al., A comparison of heritable abnormal liprotein patterns as defined by two different techniques, *J. Clin. Invest.*, 47, 2446, 1969.

Mincu, T. et al., The 3d class of lipoproteins in the blood serum of some diabetics: ultra-slow lipoproteins demonstrated by immunoelectrophoretic studies, *Med. Interna*, 21, 955, 1969.

Sodhi, H. S., New lipoprotein differing in charge and density from known plasma lipoproteins, *Metabolism*, 18, 852, 1969.

Roggenbach, H. J. et al., Abnormal β-lipoprotein with reduced lipid-binding ability in a patient with generalized lymph node swellings, hepatosplenomegaly and hypocholesteremia, *Verh. Dtsch. Ges. Inn. Med.*, 77, 594, 1971.

Sugano, T. et al., Function and disorder of lipoprotein — the structure of abnormal lipoprotein (in Japanese), *Jpn. J. Clin. Pathol.*, 19, Suppl., 173, 1971.

Pagnan, A. et al., Examination of a migrant double lipoprotein band in the pre-beta region, preliminary study in two family groups, *Minerva Med.*, 64, 3065, 1973.

Avogaro, P., "Sinking" lipoprotein, *G. Ital. Cardiol.*, 4(6), 651—55, 1974.

Wille, L. E., Further characterization of the serum pre-alpha-lipoprotein, *Clin. Chim. Acta*, 57(1), 63, 1974.

Albers, J. J. et al., Lp(α) lipoprotein, relationship to sinking pre-β lipoprotein, hyperlipoproteinemia, and apolipoprotein β, *Metabolism*, 24(9), 1047, 1975.

Hewitt, D. et al., Heritability of "sinking" pre-β-lipoprotein level, a twin study, *Clin. Genet.*, 11(3), 224, 1977.

Lipoprotein-X

Seidel, D., A new immunochemical technique for a rapid, semi-quantitative determination of the abnormal lipoprotein (LP-X) characterizing cholestasis, *Clin. Chim. Acta*, 31, 225, 1971.

Poley, J. R. et al., Lipoprotein-X and the double ^{131}I-Rose Bengal test in the diagnosis of prolonged infantile jaundice, *J. Pediatr. Surg.*, 7, 660, 1972.

Seidel, D., Structure of lipoprotein-X, *Expo. Annu. Biochim. Med.*, 31, 17, 1972.

Danielsson, B. et al., Separation of lipoprotein X and beta-lipoprotein by zonal centrifugation or hydroxy-apatite chromatography, *Clin. Chim. Acta*, 47, 365, 1973.

Gjone, E. et al., Studies of lipoprotein-X (LP-X) and bile acids in familial LCAT deficiency. Preliminary report, *Acta Med. Scand.*, 194, 377, 1973.

Neurath, A. R. et al., Letter: Hepatitis-B antigen, immunological relationship to LP-X, the abnormal plasma-lipoprotein of obstructive jaundice, *Lancet*, 2, 1394, 1973.

Petek, W. et al., Investigation of methods for the immunological determination of lipoprotein-X (LP-X) (in German with English abstract), *Z. Klin. Chem. Klin. Biochem.*, 11(10), 415, 1973.

Poley, J. R. et al., Diagnosis of extrahepatic biliary obstruction in infants by immunochemical detection of LP-X and modified ^{131}I-Rose Bengal excretion test, *J. Lab. Clin. Med.*, 81, 325, 1973.

Prexl, H. J. et al., Significance of lipoprotein-X and serum cholinesterase in preoperative diagnosis of obstructive jaundice, *Chirurg*, 44, 310, 1973.

Ritland, S. et al., Lipoprotein-X (LP-X) in liver disease, *Scand. J. Gastroenterol.*, 8, 155, 1973.

Ritland, S., The lipoprotein X-test, pointing out abnormal lipoprotein in cholestasis, *Tidsskr. Nor. Laegeforen.*, 93, 1155, 1973.

Vergani, C. et al., Study of an abnormal lipoprotein (LP-X) associated with cholestasis, *Minerva Med.*, 64, 1461, 1973.

Vergani, C. et al., Study of the abnormal lipoprotein-X in obstructive and non-obstructive jaundice, *Clin. Chim. Acta*, 48, 243, 1973.

Wieland, H. et al., A new and simplified method for the detection of LP-X, a cholestasis specific lipoprotein, *Dtsch. Med. Wochenschr.*, 98, 1474, 1973.

Colalongo, G. et al., The serum lipoprotein X (LP-X) test in the diagnosis of cholestatic hepatopathies (preliminary note), *Boll. Soc. Ital. Biol. Sper.*, 50(18), 1440, 1974.

Kostner, G. M. et al., Immunochemical measurement of lipoprotein-X, *Clin. Chem.*, 20, 676, 1974.

James, J. A. et al., Proceedings: experimental production of lipoprotein X (LP-X) in the absence of obstructive jaundice, *Gut*, 15, 343, 1974.

Jonas, A. et al., Properties of the abnormal human plasma lipoprotein (LP-X) characteristic of cholestasis after chemical modification with succinic anhydride, *Arch. Biochem. Biophys.*, 163, 200, 1974.

Magnani, H. N. et al., Proceedings: obstructive lipoprotein (LP-X) and liver disease, *Clin. Sci. Mol. Med.*, 47(5), 20P, 1974.

Ritland, S., Demonstration of the abnormal lipoprotein of cholestasis, LP-X, by precipitation with polyanion, *Scand. J. Gastroenterol.*, 9, 507, 1974.

Ritland, S., A method for quantitative determination of the abnormal lipoprotein of cholestasis, LP-X, *Clin. Chim. Acta*, 55(3), 359, 1974.

Vierucci, A. et al., Cholestasis lipoprotein (LP-X), characteristics, diagnostic value and distribution in some liver diseases in children, *Minerva Pediatr.*, 26, 131, 1974.

Baldi, A. et al., Behavior of lipoprotein X (LP-X) in viral hepatitis (English abstr.), *Quad. Sclavo. Diagn.*, 11(1), 122, 1975.

Fischer, M. et al., The diagnosis of cholestasis, lipoprotein X (LP-X) (in German), *Wien. Klin. Wochenschr.*, 87(16), 524, 1975.

Fukazawa, T. et al., Obstructive lipoprotein (LP-X) (in Japanese), *Jpn. J. Clin. Pathol.*, 23(12), 926, 1975.

Garijo, J. M. et al., Clinical contribution to the study of X (LP-X) lipoprotein (English abstr.), *Rev. Clin. Esp.*, 137(5), 407, 1975.

Mast, A. et al., Lipoprotein-X (LP-X), a new test for the diagnosis of cholestasis (in Dutch), *Tijdschr. Gastroenterol.*, 18(5), 293, 1975.

Mayr, K., The significance of lipoprotein X in the diagnosis of obstructive jaundice, comparison with other biochemical tests, *Dtsch. Med. Wochenschr.*, 100(43), 2193, 1975.

Milewski, B. et al., Significance of serum lipoprotein-X (LP-X) determination for the diagnosis of cholestasis in chronic liver diseases, *Pol. Med. Sci. Hist. Bull.*, 15(5—6), 551, 1975.

Mordasini, R. C. et al., Lipoprotein X in hepatobiliary diseases (in German), *Schweiz. Med. Wochenschr.*, 105(27), 863, 1975.

Neubeck, W. et al., Direct method for measuring lipoprotein-X in serum, *Clin. Chem.*, 21(7), 853, 1975.

Palynyczko, Z. et al., Diagnostic importance of lipoprotein-X (LP-X) in the diagnosis of cholestasis, *Mater. Med. Pol.,* 7(4), 327, 1975.

Ras, M. R. et al., The effect of heparin on serum lipoprotein-X, *Clin. Chim. Acta,* 61(1), 91, 1975.

Rubies-Prat, J. et al., Letter: lipoprotein-X and gamma glutamyl transpeptidase in cholestasis, comparative study, *Nouv. Presse Med.,* 4(19), 1433, 1975.

Schrandt, P., Cholestasis-lipoprotein LP-X, *Hippokrates,* 46(1), 107, 1975.

Tanno, H. et al., LP-X in cholestasis, *Acta Hepatogastroenterol.,* 22(5), 289, 1975.

Uminska, H. et al., Preliminary evaluation of lipoprotein X (LP-X) detection in the serum in obstructive jaundice, *Pol. Przegl. Chir.,* 47(10), 1195, 1975.

Alegre, B. et al., Semiquantitative determination of lipoprotein X (LP-X), its usefulness in differential diagnosis of jaundice (in Spanish with English abstract), *Rev. Clin. Esp.,* 143(1), 79, 1976.

Brocklehurst, D. et al., Serum alkaline phosphatase, nucleotide pyrophosphatase, 5'-nucleotide and lipoprotein-X in cholestasis, *Clin. Chim. Acta,* 67(3), 269, 1976.

Degenaar, C. P. et al., Electrophoresis of γ-glutamyl transpeptidase on cellogel. The appearance of the α_2-β-band in positive LP-X sera, *Clin. Chim. Acta,* 67(1), 63, 1976.

Frison, J. C. et al., Comparative study of lipoprotein-X and leucine aminopeptidase in the diagnosis of cholestasis (in Spanish), *Rev. Esp. Enferm. Apar. Dig.,* 47(4), 457, 1976.

Lebacq, E. G. et al., LP-X test in sarcoidosis patients with liver involvement, comparison with other liver function tests, *Ann. N.Y. Acad. Sci.,* 278, 439, 1976.

Magnani, H. N. et al., Utilization of the quantitative assay of lipoprotein X in the differential diagnosis of extrahepatic obstructive jaundice and intrahepatic diseases, *Gastroenterology,* 71(1), 87, 1976.

Magnani, H. N., The influence of LP-X and other lipoproteins associated with hepatic dysfunction on the activity of lecithin:cholesterol acyltransferase, *Biochim. Biophys. Acta,* 450(3), 390, 1976.

Manzato, E. et al., Formation of lipoprotein-X. Its relationship to bile compounds, *J. Clin. Invest.,* 57(5), 1248, 1976.

Mayr, K., Lipoprotein-X and other clinico-chemical parameters in cholestasis (in German), *Wien. Med. Wochenschr.,* 126(25—27), 378, 1976.

Michel, B. et al., The diagnostic importance of low-density lipoprotein (LP-X) for the diagnosis of cholestasis (in German with English abstr.), *Z. Gastroenterol.,* 14(5), 556, 1976.

Milosavljevic, J. et al., Lipoprotein-X (LP-X) in sera of patients with liver, gallbladder and bile duct diseases with and without hyperbilirubinemia (in Serbo-Croatian with English Abstr.), *Acta Med. Iugosl.,* 30(5), 533, 1976.

Patsch, J. R. et al., Isolation and partial characterization of two abnormal human plasma lipoproteins: LP-X_1 and LP-X_2, *Biochim. Biophys. Acta,* 434(2), 419, 1976.

Pirola, R. et al., Diagnostic value of serum lipoprotein-X in jaundice, *Med. J. Aust.,* 1(23), 886, 1976.

Ramenghi, M. et al., Lipoprotein X (LP-X) in beginning hepatitis with an-α-lipoproteinemia. Preliminary study (in Italian), *Minerva Pediatr.,* 28(7), 399, 1976.

Ritland, S. et al., Changes in the concentration of lipoprotein-X during incubation of postheparin plasma from patients with familial lecithin:cholesterol acyltransferase (LCAT) deficiency, *Clin. Chim. Acta,* 67(1), 63, 1976.

Rubies-Prat, J. et al., Comparative study of X lipoprotein and γ-glutamyl transpeptidase in cholestasis and in hepatic tumors (in Spanish with English Abstr.), *Rev. Clin. Esp.,* 140(2), 133, 1976.

Samsioe, G. et al., On the occurrence of lipoprotein-X in non-hemolytic jaundice, *Acta Chir. Scand.,* 142(3), 187, 1976.

Seidel, G. et al., On the metabolism of lipoprotein-X (LP-X), *Clin. Chim. Acta,* 66(2), 195, 1976.

Uminska, H. et al., Diagnostic value of lipoprotein-X determination in gallbladder and bile ducts diseases (in Polish with English Abstr.), *Pol. Tyg. Lek.,* 31(18), 765, 1976.

Verkleij, A. J. et al., The fusion of abnormal plasma lipoprotein (LP-X) and the erythrocyte membrane in patients with cholestasis studied by electron microscopy, *Biochim. Biophys. Acta,* 436(2), 366, 1976.

Molecular Characteristics

Bergquist, L. M. et al., Formation of lipoprotein-tetracycline complexes in human serum, *Antimicrob. Agents Chemother.,* 3, 477, 1963.

Folch Pi, J., Some considerations on the structure of proteolipids, *Fed. Proc.,* 23, 630, 1964.

Hauton, J. C. et al., Structure of the bile proteolipid complex, *J. Gastroenterol.,* 7, 381, 1964.

Hauton, J. C. et al., Is a partly peptide "cohesion center" necessary for the transport of lipids? (in French), *Bull. Soc. Chim. Biol.,* 47, 2175, 1965.

Schen, R. J. et al., The affinity of tetracycline for human serum lipoproteins demonstrated by fluoro-immunoelectrophoresis, *Isr. J. Med. Sci.,* 2, 86, 1966.

Travims, E. G. et al., Models for lipoprotein synthesis, *J. Theor. Biol.,* 12, 311, 1966.

Vandenheuvel, F. A., Structural role of water in lipoprotein systems, *Protoplasma,* 63, 188, 1967.

Wolfgram, F., The amino acid compositions of some non-neural proteolipid proteins, *Biochim. Biophys. Acta,* 147, 383, 1967.

Felmeister, A. et al., Interaction of protein and lipoprotein monolayers with nitrogen dioxide-*trans*-2-butene gaseous mixtures, *Arch. Biochem.,* 126, 962, 1968.

Scanu, A. et al., Molecular weight and subunit structure of human serum high density lipoprotein after chemical modifications by succinic anhydride, *Biochim. Biophys. Acta,* 160, 32, 1968.

Scanu, A. et al., Properties of human serum low density lipoproteins after modification by succinic anhydride, *J. Lipid Res.,* 9, 342, 1968.

Camejo, G., The structure of human density lipoprotein: a study of the effect of phospholipase A and trypsin on its components and of the behavior of the lipid and protein moieties at the air-water interphase, *Biochim. Biophys. Acta,* 175, 290, 1969.

Day, C. E. et al., Molecular structure of serum lipoproteins, *J. Theor. Biol.,* 23, 387, 1969.

Fredrickson, D. S., Plasma lipoproteins: micellar models and mutants, *Trans. Assoc. Am. Physicians,* 82, 68, 1969.

Lees, M. B. et al., Carboxymethylation of sulphydryl groups in proteolipids, *J. Neurochem.,* 16, 1025, 1969.

Guerrina, G. et al., On serum lipoproteins: structure, classification and methods of separation, *Pathologica,* 62, 82, 1970.

Alaupovic, P. et al., Studies on the composition and structure of plasma lipoproteins. Distribution of lipo-protein families in major density classes of normal human plasma lipoproteins, *Biochim. Biophys. Acta,* 260, 689, 1972.

Hara, I., Structure of lipoproteins, *Saishin Igaku,* 27, 453, 1972.

Harmison, C. R. et al., Conformational variation in a human plasma lipoprotein, *Biochemistry,* 11, 4985, 1972.

Scanu, A. M. et al., Serum lipoproteins: structure and function, *Annu. Rev. Biochem.,* 4, 703, 1972.

Scanu, A. M., Structure of human serum lipoproteins, *Ann. N. Y. Acad. Sci.,* 195, 390, 1972.

Shore, B. et al., Structure of normal and pathological lipoproteins, *Expos. Annu. Biochim. Med.,* 31, 4, 1972.

Badin, J. et al., Structure of the beta-lipoprotein, used with digitonin and streptolysin as specific markers of free cholesterol, *Ann Biol. Clin.,* 31, 71, 1973.

Douste-Blazy, L., General comments on the structure of normal lipoproteins, *Ann. Biol. Clin.,* 31, 61, 1973.

Erokhin, Iu. E. et al., Characteristics of proteins (number of chains and molecular weights) for pigment lipoprotein complexes from chromatium, *Dokl. Akad. Nauk SSSR,* 212, 495, 1973.

Girard, M., Structure and metabolism of lipoproteins introduction, *Ann. Biol. Clin.,* 31, 59, 1973.

Laggner, P. et al., Studies on the structure of lipoprotein A of human high density lipoprotein HDL_3, the spherically averaged electron density distribution, *FEBS Lett.,* 33, 77, 1973.

Picard, J. et al., Structure of normal and pathologic lipoproteins, *Ann. Biol. Clin.,* 31, 77, 1973.

Sokolovskii, V. V. et al., Participation of sulfhydryl groups in the conjugation of proteins with lipids, *Dokl. Akad. Nauk SSSR,* 209, 738, 1973.

Hamilton, J. A. et al., Rotational and segmental motions in the lipids of human plasma lipoproteins, *J. Biol. Chem.,* 249, 4872, 1974.

Assmann, G. et al., A molecular model of high density lipoproteins, *Proc. Natl. Acad. Sci. U.S.A.,* 71, 1534, 1974.

Scanu, A. M. et al., Application of physical methods to the study of serum lipoproteins, *CRC Crit. Rev. Biochem.,* 2, 175, 1974.

Segrest, J. P. et al., A molecular theory of lipid-protein interactions in the plasma lipoproteins, *FEBS Lett.,* 38, 247, 1974.

Fredrickson, D. S., Function and structure of plasma lipoproteins, *Ann. Otolaryngol. Chir. Cervicofac.,* 91(4—5), 2, 1974.

Opplt, J. J. and Opplt, M. A., Separation of plasma lipoproteins according to molecular size, *Clin. Chem.,* 20, 990, 1975.

Jackson, R. L. et al., The mechanism of lipid-binding by plasma lipoproteins, *Mol. Cell. Biochem.* 6(1), 43, 1975.

Morrisett, J. D. et al., Lipoproteins, structure and function, *Annu. Rev. Biochem.,* 44, 183, 1975.

Wehr, H., Structure and function of human blood lipoproteins, *Postepy Biochem.,* 21(1), 75, 1975.

Gotto, A. M., Jr. et al., Molecular association of lipids and proteins in the plasma lipoproteins: a review, in *Lipoprotein metabolism,* Greten, H., Ed., Berlin, Springer-Verlag, Berlin, 1976, 152—57.

Hedstrand, H. et al., Serum lipoprotein concentration and composition in healthy 50-year-old men, *Uppsala J. Med. Sci.,* 81(1), 37, 48, 1976.

Rosseneu, M. et al., Application of microcalorimetry to the study of lipid-protein interaction, *Chem. Phys. Lipids,* 17(1), 38, 1976.

Wakeham, W. A. et al., Diffusion coefficients for protein molecules in blood serum, *Atherosclerosis,* 25(2—3), 225, 1976.

Smith, E. B., Molecular interactions in human atherosclerotic plagues, *Am. J. Pathol.*, 86(3), 665, 1977.

Opplt, J. J., Bahler, R. C., and Opplt, M. A., Molecular pathology of serum lipoproteins, *Atherosclerosis IV*, Monogr., Springer-Verlag, Berlin, 1977, 247.

Radiolipoproteins

Sisson, J. C., Labeled plasma lipids after ingestion of radioactive fats, *J. Nucl. Med.*, 6, 210, 1965.

Krepsz, I. et al., The effect of internal irradiation with ^{32}P on serum glucoproteins and lipoproteins (in Rumanian), *Fiziol. Norm. Patol.*, 13, 161, 1967.

Sodhi, H. S. et al., Labeling plasma lipoproteins with radioactive cholesterol, *J. Lab. Clin. Med.*, 82, 111, 1973.

Fidge, N., A review of methods and metabolic studies associated with the radioiodination of lipoproteins, *Clin. Chim. Acta*, 52, 5, 1974.

Roberts, M. M. et al., Letter: radioimmunoassay of apo-L.D.L., *Lancet*, 1, 941, 1974.

Sodhi, H. S. et al., A physiological method for labeling plasma lipoproteins with radioactive cholesterol, *Clin. Chim. Acta*, 51, 292, 1974.

Eisenberg, S. et al., Radioiodinated lipoproteins, absorption of ^{125}I radioactivity by high density solutions, *J. Lipid Res.*, 16(6), 468, 1975.

Bedford, D. K. et al., Radioimmunoassay for human plasma apolipoprotein B, *Clin. Chim. Acta*, 70(2), 267, 1976.

Clarke, J. T. et al., Uptake of radiolabeled galactosyl-(α_1, goes to 4)-galactosyl-(β, goes to 4)-glucosyl-ceramide by human serum lipoproteins in vitro, *Biochim. Biophys. Acta*, 441(1), 165, 1976.

Eaton, R. P., Incorporation of ^{75}Se-selenomethionine into human apoproteins. I. Characterization of specificity in VLD and LD-lipoproteins, *Diabetes*, 25(1), 32, 1976.

Eaton, R. P. et al., Incorporation of ^{75}Se-selenomethionine into human apoproteins. II. Characterization of metabolism of very-low-density and low-density-lipoproteins in vivo and in vitro, *Diabetes*, 25(1), 44, 1976.

Eaton, R. P. et al., Incorporation of ^{75}Se-selenomethionine into human apoproteins. III. Kinetic behavior of isotopically labeled plasma apoprotein in man, *Diabetes*, 25(8), 679, 1976.

Shepherd, J. et al., Radioiodination of human low density lipoproteins: a comparison of four methods, *Clin. Chim. Acta*, 66(1), 97, 1976.

Thompson, G. R. et al., Solid phase radioimmunoassay of apolipoprotein B (apo B) in normal human plasma, *Atherosclerosis*, 24(1—2), 107, 1976.

Glycolipoproteins

Bihari-Varga, M. et al., Quantitative studies on the complexes formed between aortic mucopolysaccharides and serum lipoproteins, *Biochim. Biophys. Acta*, 144, 202, 1967.

Mookerjea, S. et al., Studies on the synthesis of plasma glycolipoprotein and hepatic sub-cellular glycoprotein in early choline dificiency, *Can. J. Biochem*, 45, 825, 1967.

Drygin, Iu. F. et al., Glycolipoproteins in bacterial DNA preparations (in Russian), *Biokhimiia*, 35, 1014, 1970.

Lipoproteins and Other Proteins

Elliot, G. B. et al., False-positive C-reactive protein reactions due to serum lipoproteins, *Am. Rev. Respir. Dis.*, 90, 453, 1964.

Delage, J. M., Complement in the perinatal period. Antagonism of ceruloplasmin and α-2-lipoproteins (in French), *C. R. Acad. Sci. Ser. D*, 262, 2113, 1966.

Etienne, J. et al., Influence of ordinary proteins on stability of lipoproteins (in French), *C. R. Acad. Sci. Ser. D*, 265, 2104, 1967.

Lipoproteins and Immunoglobulins

Lewis, L. A. et al., An unusual serum lipoprotein-globulin complex in a patient with hyperlipemia, *Am. J. Med.*, 38, 286, 1965.

Keler-Bacoka, M., A specific property of the plasmocytoma serum paraprotein lipids with respect to lipid precipitation by heparin, *Clin. Chim. Acta*, 16, 365, 1967.

Hartmann, L. et al., Lipids, lipoproteins and IgM immunoglobulins (in French), *Ann. Biol Clin.*, 26, 881, 1968.

Tedeschi, C. G. et al., Fat macroglobulinemia and fat embolism, *Surg. Gynecol. Obstet.*, 126, 83, 1968.

Valdiguié, P. et al., On the lipids of myelomatous proteins. Fatty acids bound and absorbed by Bence-Jones proteins (in French), *C. R. Acad. Sci. Ser. D*, 268, 865, 1969.

Noseda, G. et al., Binding of β-lipoprotein by an IgA paraprotein: an antigen-antibody reaction? (in German with English abstract), *Schweiz. Med. Wochenschr.*, 101, 893, 1971.

Noseda, G. et al., The behavior of serum lipids in paraproteinemias. Demonstration of paraprotein-lipoprotein complexes (in German with English abstract), *Schweiz. Med. Wochenschr.*, 101, 1787, 1971.

Beaumont, J. L., Interactions between lipoproteins and immunoglobulins, *Expos. Annu. Biochim. Med.*, 31, 169, 1972.

Riesen, W. et al., Monoclonal immunoglobulins with the characteristics of antibodies (in German with English abstract), *Schweiz. Med. Wochenschr.*, 102, 1176, 1972.

Beaumont, J. L. et al., Anti-lipoprotein and anti-emulsive activity of monoclonal IgM. "A series of specific anti-emulsive antibodies," *Pathol. Biol.*, 21, 241, 1973.

Pagé, M. et al., Letter: lipophilic paraproteins, *N. Engl. J. Med.*, 291, 475, 1974.

Riesen, W. et al., Paraproteins with antibody properties, with special reference to anti-lipoprotein autoantibodies, *Minerva Med.*, 66(50), 2471, 1975.

Lipoproteins and Enzymes

Lipase

Marinetti, G. V., The action of phospholipase A on lipoproteins, *Biochim. Biophys. Acta*, 98, 554, 1965.

Infante, R. et al., *Rhizopus arrhizus* lipase. I. Action of the lipase on the triglycerides of chylomicrons and lipoproteins in vitro (in French), *C. R. Soc. Biol.*, 162, 50, 1968.

Fried, M. et al., Serum lipoprotein hydrolysis by purified lipases, *Lipids*, 6, 276, 1971.

Uthe, J. F. et al., Phospholipase A2: comparative action of three different enzyme preparations on selected lipoprotein systems, *Can. J. Biochem.*, 49, 785, 1971.

Tutterova, M. et al., Enhanced lipolysis during interaction of serum and adipose tissue: its mechanism and effect of starvation, *Acta Biol. Med. Ger.*, 34, 710, 1651, 1975.

Pancreatic Lipase

Herfort, K. et al., Hyperlipoproteinemia and exocrine pancreas (English abstract), *Cas. Lek. Cesk.*, 111, 233, 1972.

Brodanová, M. et al., Iron resorption from the digestive tract in patients with primary hyperlipoproteinemia and its influencing by pancreatic extract, *Sb. Lek.*, 75, 244, 1973.

Dabels, J., Pancreatitis and hyperlipoproteinemia in reciprocal relations, *Dtsch. Z. Verdau. Stoffwechselkr.*, 33, 213, 1973.

Herfort, K. et al., Familial hyperlipoproteinemia and the exocrine pancreas, *Rev. Czech. Med.*, 19, 230, 1973.

de Gennes, J. L. et al., Pancreatitis in idiopathic hyperlipidemias. Personal series of 40 cases, *Ann. Med. Interne*, 125(4), 333, 1974.

Jedlovský, A., So-called beta-lipoprotein index and acute pancreatitis associated with cholelithiasis (English abstr.), *Cesk. Gastroenterol. Vyz.*, 28(8), 518, 1974.

Barcz, L. et al., Case of acute pancreatitis and cholangitis in a patient with hyperlipoproteinemia (English abstract), *Wiad. Lek.*, 28(7), 571, 1975.

Carmena, R. et al., Hyperlipoproteinemia and acute pancreatitis, *Rev. Exp. Enferm. Apar. Dig.*, 45(3), 269, 1975.

Lipoprotein Lipase

Harlan, W. R., Jr. et al., Tissue lipoprotein lipase in normal individuals and in individuals with exogenous hypertriglyceridemia and the relationship of this enzyme to assimilation of fat, *J. Clin. Invest.*, 46, 239, 1967.

Bier, D. M. et al., Activation of lipoprotein lipase by lipoprotein fractions of human serum, *J. Lipid Res.*, 11, 565, 1970.

Fielding, C. J. et al., A protein component of serum high density lipoprotein with co-factor activity against purified lipoprotein lipase, *Biochem. Biophys. Res. Commun.*, 39, 889, 1970.

Diogo Giannini, J. et al., The relation between lipoproteic lipase and metabolic-lipidic parameters. I. Normal people (in Spanish), *Rev. Hosp. Clin. Fac. Med. Univ. Sao Paulo*, 25, 23, 1970.

Havel, R. J. et al., Role of specific glycopeptides of human serum lipoproteins in the activation of lipoprotein lipase, *Circ. Res.*, 27, 595, 1970.

Boberg, J., Heparin-released blood plasma lipoprotein lipase activity in patients with hyperlipoproteinemia, *Acta Med. Scand.*, 191, 97, 1972.

Paluszak, J. et al., Hyperchylomicronemia and hyper-prebetalipoproteinemia. Study of the lipoprotein lipase system, *Pol. Arch. Med. Wewn.*, 49, 175, 1972.

Fielding, C. J., Kinetics of lipoprotein lipase activity: effects of the substrate apoprotein on reaction velocity, *Biochim. Biophys. Acta*, 316, 66, 1973.

Jansen, H. et al., Lipoprotein lipase from heart and liver, an immunological study, *Biochem. Biophys. Res. Commun.,* 55, 30, 1973.

Krauss, R. M. et al., Further observations on the activation and inhibition of lipoprotein lipase by apolipoproteins, *Circ. Res.,* 33, 403, 1973.

Krauss, R. M. et al., Selective measurement of two lipase activities in postheparin plasma from normal subjects and patients with hyperlipoproteinemia, *J. Clin. Invest.,* 54(5), 1107, 1974.

Kučerová, L. et al., Lipoprotein lipase and post-heparin esterase activity in primary type IV and V hyperlipoproteinemia (English abstr.), *Cas. Lek. Cesk.,* 113(7), 519, 1974.

Higgins, J. M. et al., Lipoprotein lipase. Mechanism of formation of triglyceride-rich remnant particles from very low density lipoproteins and chylomicrons, *Biochemistry,* 14(11), 2288, 1975.

Fielding, C. J. et al., Mechanism of salt-mediated inhibition of lipoprotein lipase, *J. Lipid Res.,* 17(3), 248, 1976.

LCAT

Glomset, J. A. et al., Plasma lipoproteins in familial lecithin:cholesterol acyltransferase deficiency: lipid composition and reactivity in vitro, *J. Clin. Invest.,* 49, 1827, 1970.

Forte, J. et al., Plasma lipoproteins in familial lecithin:cholesterol acyltransferase deficiency: structure of low and high density lipoproteins as revealed by electron microscopy, *J. Clin. Invest.,* 50, 1141, 1971.

Norum, K. R. et al., Plasma lipoproteins in familial lecithin:cholesterol acyltransferase deficiency: physical and chemical studies of low and high density lipoproteins, *J. Clin. Invest.,* 50, 1131, 1971.

Takatori, T., Serum lipoprotein and lecithin:cholesterol acyltransferase (in Japanese), *Jpn. J. Clin. Pathol. Suppl.* 19, 175, 1971.

Seidel, D., Plasma lipoproteins in patients with familial plasma lecithin:cholesterol acyltransferase deficiency: apolipoprotein composition of isolated fractions, *Verh. Dtsch. Ges. Inn. Med.,* 78, 1292, 1972.

Utermann, G. et al., Lipoproteins in LCAT-deficiency, *Humangenetik,* 16, 295, 1972.

Christian, J. C. et al., A stable plasma lipoprotein and lecithin:cholesterol acyltransferase control material, *Clin. Chim. Acta,* 43, 23, 1973.

Glomset, J. A., The metabolic role of lecithin:cholesterol acyltransferase: perspectives from pathology, *Adv. Lipid Res.,* 11(0), 1, 1973.

Glomset, J. A. et al., Plasma lipoproteins in familial lecithin:cholesterol acyltransferase deficiency. Further studies of very low and low density lipoprotein abnormalities, *J. Clin. Invest.,* 52, 1078, 1973.

Marcel, Y. L. et al., Lecithin:cholesterol acyitransferase of human plasma. Role of chylomicrons, very low, and high density lipoproteins in the reaction, *J. Biol. Chem.,* 248, 8254, 1973.

McConathy, W. J. et al., Identification of lipoprotein families in familial lecithin:cholesterol acyltransferase deficiency, *Biochim. Biophys. Acta,* 326, 406, 1973.

Norum, K. R., The role of lecithin:cholesterol acyltransferase in the metabolism of plasma lipoproteins, *Ann Biol. Clin.,* 31, 123, 1973.

Ritland, S. et al., Lecithin:cholesterol acyltransferase and lipoprotein-X in liver disease, *Clin. Chim. Acta,* 49, 251, 1973.

Akanuma, Y., Symposium on hyperlipidemia — epidemiological, clinical and experimental studies. Plasma lipoprotein metabolism and lecithin:cholesterol acyltransferase (in Japanese), *J. Jpn. Soc. Intern. Med.,* 63(10), 1149, 1974.

Alaupovic, P. et al., Apolipoproteins and lipoproteins families in familial lecithin:cholesterol acyltransferase deficiency, *Scand. J. Clin. Lab. Invest.,* Suppl. 33, 83, 1974.

Fielding, C. J., Phospholipid substrate specificity of purified human plasma lecithin:cholesterol acyltransferase, *Scand. J. Clin. Lab. Invest. Suppl.,* 33, 15, 1974.

Forte, T. et al., The ultrastructure of plasma lipoproteins in lecithin:cholesterol acyltransferase deficiency, *Scand. J. Clin. Lab. Invest. Suppl.,* 33, 121, 1974.

Gjone, E. et al., Possible association between an abnormal low density lipoprotein and nephropathy in lecithin:cholesterol acyltransferase deficiency, *Clin. Chim. Acta,* 54, 11, 1974.

Glomset, J. A. et al., Plasma lipoprotein metabolism in familial lecithin:cholesterol acyltransferase deficiency, *Scand. J. Clin. Lab. Invest. Suppl.,* 33, 165, 1974.

Langer, K. H. et al., On the polypeptide composition of an abnormal high density lipoprotein (LP-E) occurring in LCAT-deficient plasma, *FEBS Lett.,* 45(1), 29, 1974.

Marcel, Y. L. et al., Lecithin:cholesterol acyltransferase of human plasma. Role of chylomicrons, very low, and high density lipoproteins in the reaction, *Scand. J. Clin. Lab. Invest. Suppl.,* 33, 45, 1974.

Nichols, A. V. et al., Effect of lysophosphatidyl choline on interaction between phosphatidyl choline and activation protein (apolipoprotein A-I) of lecithin:cholesterol acyltransferase, *Scand. J. Clin. Lab. Invest. Suppl.,* 33, 147, 1974.

Norum, K. R. et al., Lecithin:cholesterol acyltransferase (LCAT) and lipoprotein metabolism. Results of studies of familial LCAT deficiency, in *Blood and arterial wall in atherogenesis and arterial thrombosis,* Hautvast, J. G. A. J. et al., Eds., Brill, Leiden, Netherlands, 1975, 17.

Scherer, R. et al., Investigation of lecithin-cholesterol-acyltransferase (LCAT) activity in the serum of these patients with Tangier disease, *Klin. Wochenschr.,* 21, 1059, 1973.

Wengeler, H. et al., Does lipoprotein-X (LP-X) act as a substrate for the lecithin:cholesterol acyltransferase (LCAT)?, *Clin. Chim. Acta,* 45, 429, 1973.

Seidel, D. et al., Plasma lipoproteins in patients with familial plasma lecithin:cholesterol acyltransferase (LCAT) deficiency — studies on the apolipoprotein composition of isolated fractions with identification of LP-X, *Ann. Otolaryngol. Chir. Cervicofac.,* 91(4—5), 6, 1974.

Solaas, M. H., Structural studies on serum lipoproteins in homozygotes and heterozygotes for the lecithin: cholesterol acyltransferase deficiency gene, *Scand. J. Clin. Lab. Invest. Suppl.,* 33, 133, 1974.

Glomset, J. A. et al., Plasma lipoproteins in familial lecithin:cholesterol acyltransferase deficiency: effects of dietary manipulation, *Scand. J. Clin. Lab. Invest. Suppl.,* 35, 142, 3, 1975.

Norum, K. R. et al., Plasma lipoproteins in familial lecithin:cholesterol acyltransferase deficiency: effects of incubation with lecithin:cholesterol acyltransferase in vitro, *Scand. J. Clin. Lab. Invest. Suppl.,* 35, 142, 31, 1975.

Ritland, S. et al., Quantitative studies of lipoprotein-X in familial lecithin:cholesterol acyltransferase deficiency and during cholesterol esterification, *Clin. Chim. Acta,* 59(2), 109, 1975.

Soutar, A. K. et al., Effect of human plasma apolipoproteins and phosphatidylcholine acyl donor on the activity of lecithin:cholesterol acyltransferase, *Biochemistry,* 14(14), 3057, 1975.

Utermann, G. et al., Lipoproteins in lecithin-cholesterol-acyltransferase (LCAT)-deficiency. II. Further studies on the abnormal high-density-lipoproteins, *Humangenetik,* 27(3), 185, 1975.

Wallentin, L. et al., Lecithin:cholesterol acyl transfer in plasma of normal persons in relation to lipid and lipoprotein concentration, *Scand. J. Clin. Lab. Invest.,* 35(7), 669, 1975.

Yasugi, T. et al., Lipoproteins and hyperlipidemia; special reference on the LCAT and apo-lipoproteins., *Jpn. Circ. J.,* 39(3), 339, 1975.

Miller, J. P. et al., A critical examination of the value of combined determination of lecithin:cholesterol acyltransferase and lipoprotein-X in the differential diagnosis of liver disease, *Clin. Chim. Acta,* 69(1), 81, 1976.

Salmon, S. et al., An immunoelectrophoretic study of plasma lipoproteins in a case of familial deficiency of lecithin-cholesterol acyltransferase (authors' transl.) (in French with English abstract), *Clin. Chim. Acta,* 66(3), 311, 1976.

Sigler, G. F. et al., The solid phase synthesis of a protein activation for lecithin-cholesterol acyltransferase corresponding to human plasma apo-C-I, *Proc. Natl. Acad. Sci. USA,* 73(5), 1422, 1976.

Sabesin, S. M. et al., Abnormal plasma lipoproteins and lecithin-cholesterol acyltransferase deficiency in alcoholic liver disease, *Gastroenterology,* 72(3), 510, 1977.

Other Enzymes

Ahmed, K. et al., Preparation of lipoproteins containing cation-dependent ATPase, *Biochim. Biophys. Acta,* 93, 603, 1964.

Serebrovskaia, K. B. et al., Study of ribonuclease activity in lipoprotein coacervate (in Russian), *Biokhimiia,* 29, 910, 1964.

Girard, M. L. et al., Extent of the proteolytic action of *Streptomyces griseus* protease on serum lipoproteins (in French), *C. R. Acad. Sci.,* 260, 4377, 1965.

Cucuianu, M., Beta-lipoproteins and euglobulin lysis time, *Thromb. Diath. Haemorrh.,* 16, 687, 1966.

Innerfield, I. et al., Lipoprotein responses to orally given enzymes, *Exp. Med. Surg.,* 25, 169, 1967.

van Hung, N., et al., New serum protein polymorphisms by enzyme treatment of human serum. II. On the polymorphism of human X-1 and β-lipoproteins (in German), *Acta Biol. Med. Ger.,* 21, 679, 1968.

Philippot, J., Reconstitution of the activity of transport ATPase after solubilization with sodium deoxycholate. The role of Na^+ and K^+ ions in the structure of the enzyme-lipoprotein complex (in French), *Bull. Soc. Chim. Biol.,* 50, 1481, 1968.

Sharnes, R. C., In vivo interaction of endotoxin with a plasma lipoprotein having esterase activity, *J. Bacteriol.,* 95, 2031, 1968.

Sun, F. F. et al., Proteolipids. V. The activity of lipid cytochrome C, *Biochim. Biophys. Acta,* 172, 417, 1969.

Grafins, M. A. et al., Acetylcholinesterase interaction with a lipoprotein matrix, *Eur. J. Biochem.,* 22, 382, 1971.

Greten, H., Enzymatic degradation of plasma lipoproteins, *Expo. Annu. Biochim. Med.,* 31, 160, 1972.

Kutty, K. M., Possible link between cholinesterase activity and lipoprotein, *Am. J. Clin. Pathol.,* 58, 599, 1972.

Kutty, K. M. et al., Interrelationship between serum-lipoprotein and cholinesterase, *Can. J. Biochem.,* 51, 883, 1973.

Lambrecht, J. et al., A simple method for the determination of enzyme activities in turbid plasma of patients with hyperlipoproteinemia, *Z. Klin. Chem. Klin. Biochem.,* 12(4), 154, 1974.

Waite, M. et al., Utilization of serum lipoprotein lipids by the monoacylglycerol acyltransferase, *Biochim. Biophys. Acta,* 450(3), 301, 1976.

Lipoproteins and Their Immunological Properties

Berg, G., Immunological studies on serum lipoproteins and their behavior in fat transport (in German), *Verh. Dtsch. Ges. Inn. Med.,* 70, 971, 1964.

Berg, K., Comparative studies on the Lp and Ag serum type systems, *Acta Pathol. Microbiol. Scand.,* 62, 276, 1964.

Berg, K. et al., The Lp-system. Preparation of the antiserum, testing methods, results, *Humangenetik,* 1, 24, 1964.

Berg, K., Studies on the reaction between Lp(a +) human sera and anti-Lp(a) from rabbits, *Acta Pathol. Microbiol. Scand.,* 62, 613, 1964.

Blumberg, B. S. et al., γ-globulin, group specific, and lipoprotein groups in a U.S. white and negro population, *Nature (London),* 202, 561, 1964.

Hirschfeld, J. et al., Relationship of human anti-lipoprotein allotypic sera, *Nature (London),* 202, 706, 1964.

Berg, K. et al., Studies on the Lp-system. Type frequency and comparison of antisera (in German), *Humangenetik,* 1, 319, 1965.

Bundschuh, G., Studies on the thermostability of the genetically controlled lipoprotein components Lp(a) and Lp(x) (in German), *Acta Biol. Med. Ger.,* 15, 126, 1965.

Jarosch, K., A "new" hereditary characteristic of human serum Lp(a), and its phenotypic frequency in Upper Austria (in German), *Wien. Med. Wochenschr.,* 115, 839, 1965.

Kahlich-Koenner, D. M. et al., Lp-type system and β-lipoprotein concentration, *Humangenetik,* 1, 388, 1965.

Kellen, J. et al., Immunochemical findings in epidemic hepatitis. II. Immunology of β-lipoproteins (in German), *Acta Hepatosplen.,* 12, 228, 1965.

Renninger, W. et al., Contribution to problems of Lp-system (in German), *Humangenetik,* 1, 658, 1965.

Speiser, P. et al., The hereditary serum β-lipoprotein-system Lp (a,x) (in German), *Ann. Paediatr.,* 205, 193, 1965.

Yamakawa, T. et al., The tentative application of artificial lipoprotein antigens to histochemistry (in Japanese), *Adv. Neurol. Sci.,* (Tokyo), 9, 653, 1965.

Berg, K., Further studies on the Lp system, *Vox Sang.,* 11, 419, 1966.

Berg, K., Lack of linkage between the Lp and Ld serum systems, *Experientia,* 22, 600, 1966.

Jaegermann, K., Studies of β-lipoprotein (Lp) group system (in Polish), *Med. Dosw. Mikrobiol.,* 18, 287, 1966.

Seidl, S. et al., The frequency of the Lp(a)-factor in Frankfurt/Main, *Vox Sang.,* 11, 730, 1966.

Thomas, K. et al., Studies on the frequency of the Lp system in Dresden and environment, *Blut,* 13, 143, 1966.

Vamosi, M. et al., On the frequency of Lp in the Halle area (in German), *Z. Aerztl. Fortbild.,* 60, 827, 1966.

Vierucci, A. et al., Post-transfusion iso-immunization for the genetic groups of human serum β-lipoproteins (in Italian), *Riv. Emoter. Immunoematol.,* 13, 1, 1966.

Windblad, S., Studies on non-specific antistreptolysin O titre. I. The influence of serum β-lipoproteins on the non-specific antistreptolysin O titre, *Acta Pathol. Microbiol. Scand.,* 66, 93, 1966.

Berg, K., Lack of linkage between the Lp and Ag serum systems, *Vox Sang.,* 12, 71, 1967.

Chung, J. et al., Dissociation of low density lipoproteins — antibody precipitates at alkaline pH, *J. Lipid Res.,* 8, 631, 1967.

Heugel, M. et al., Use of Oudin's technic in the determination of serum β-lipoproteins, *Presse Med.,* 75, 748, 1967.

Klein, H. et al., Serum group LP: on determination of the type and gene frequencies (in German), *Dtsch. Z. Gesamte Gerichtl. Med.,* 60, 90, 1967.

Levene, C. et al., Incidence of antibodies against β-lipoproteins (Ag system), and the factors influencing isoimmunisation in transfused patients in the U.S.A. and Italy, *Lancet,* 2, 582, 1967.

Morganti, G. et al., Contribution to the genetics of the serum β-lipoproteins in man. 1. Frequency, transmission and penetrance of factors Ag(x) and Ag(y), *Humangenetik,* 4, 262, 1967.

Okochi, K., Serum lipoprotein allotypes Ag(x) and Ag(y) (in Japanese), *Vox Sang.,* 13, 319, 1967.

Vierucci, A. et al., Study of the Ag(z) factor, *Nature (London),* 216, 1231, 1967.

Gallango, M. L. et al., Lp and Ag systems in Venezuelan populations, *Bibl. Haematol.* (Pavia), 29, 299, 1968.

Morganti, G. et al., Contribution to the genetics of the serum β-lipoproteins in man. 3. Lack of association and linkage between the Ag(x,y,a¹) factors and some blood and serum protein antigens, *Humangenetik,* 6, 275, 1968.

Morganti, G. et al., Contribution to the genetics of serum β-lipoproteins in man. 2. Frequency, transmission and penetrance of the Ag(a-1) factor and its linkage with the Ag(x) and Ag(y) factors, *Humangenetik*, 5, 98, 1968.

Rittner, H., To the genetics of the Lp system. A proof of hereditary quantitative characteristics as well as prenatal selection (in German), *Humangenetik*, 5, 170, 1968.

Vierucci, A. et al., Isoantibodies to inherited types of β-lipoproteins (Ag) and immunoglobulins (Gm and Inv) serological and clinical aspects, *J. Pediatr.*, 72, 776, 1968.

Wiegandt, H. et al., Identification of a lipoprotein with antigenic activity in the Lp-system (in German), *Hoppe Seylers Z. Physiol. Chem.*, 349, 489, 1968.

Fredrickson, D. S., The regulation of plasma lipoprotein concentrations as affected in human mutants, *Proc. Natl. Acad. Sci. USA*, 64, 1138, 1969.

Graff, R. J. et al., An attempt to produce low-zone tolerance to tissue alloantigens, *Transplantation*, 8, 162, 1969.

Kahan, J. et al., Immunochemical determination of β-lipoproteins, *Scand. J. Clin. Lab. Invest.*, 24, 61, 1969.

Ütler, R., Antigen structure of lipoproteins and mechanism of isoantibody formation against plasma proteins (in German), *Bibl. Haematol.* (Pavia), 32, 34, 1969.

Utermann, G. et al., Isolation and characterization of a lipoprotein with antigenic activity in the Lp system (in German), *Humangenetik*, 8, 39, 1969.

Wood, C., Genetically determined variant in human β-lipoprotein, *Arch. Dis. Child.*, 44, 544, 1969.

Harvie, N. R. et al., Studies of Lp-lipoprotein as a quantitative genetic trail, *Proc. Natl. Acad. Sci. USA*, 66, 99, 1970.

Jaegermann, K., Studies on the application of the lipoprotein (Lp) group system in forensic medical practice (in Polish), *Folia Med. Cracov.*, 12, 157, 1970.

Kostner, G. et al., Isolation of human serum low-density lipoproteins with the aid of an immune specific adsorber, *Lipids*, 5, 501, 1970.

Simons, K. et al., Characterization of the Lp(a) lipoprotein in human plasma, *Acta Pathol. Microbiol. Scand. Sect. B*, 78, 459, 1970.

Solaas, M. H., Frequency of the Ag(x) antigen in a Norwegian population sample, *Hum. Hered.*, 20, 290, 1970.

Berg, K. et al., Genetic marker system in Arctic populations. 1. Lp and Ag data on the Greenland Eskimos, *Hum. Hered.*, 21, 129, 1971.

Boman, H., Studies on inherited antigenic variation of human serum — lipoprotein by passive hemagglutination. V. Studies on a non-precipitating antiserum, *Hum. Hered.*, 21, 614, 1971.

Boman, H. et al., Studies on inherited antigenic variation of human serum — lipoprotein by passive hemagglutination, *Int. J. Protein Res.*, 3, 175, 1971.

Budiakov, O. S., Frequency of the serum lipoprotein factor Lp(a) (in Russian), *Sud. Med. Ekspert.*, 14, 31, 1971.

Ehnholm, C. et al., Purification and quantitation of the human plasma lipoprotein carrying the Lp(a) antigen, *Biochim. Biophys. Acta*, 236, 431, 1971.

Lloyd, J. K., Inherited disorders of serum lipoproteins, *Proc. R. Soc. Med.*, 64, 899, 1971.

Utermann, G. et al., Comparative studies of the Lp(a)-lipoprotein and low density lipoproteins of human serum, *Hoppe Seylers Z. Physiol. Chem.*, 352, 938, 1971.

Boman, H., Studies on inherited antigenic variation of human serum-lipoprotein by passive hemagglutination. IV. Studies on the antibodies in two non-precipitating sera, *Int. J. Protein. Res.*, 4, 141, 1972.

Ehnholm, C. et al., Protein and carbohydrate composition of Lp(x) lipoprotein from human plasma, *Biochemistry*, 11, 3229, 1972.

Kusumi, K. et al., Immunological determination of serum beta-lipoprotein, *Jpn. J. Clin. Pathol.*, 20, 509, 1972.

Morganti, G. et al., Contribution to the genetics of serum β-lipoprotein in man. V. The linkage of Ag x-y, Ag a1-d and Ag c-g loci, *Humangenetik*, 15, 274, 1972.

Morganti, G. et al., Contribution to the genetics of serum beta-lipoprotein in man. VI. Evidence for the existence of the Ag t-z locus, closely linked to the Ag x-y, Ag a1-d and Ag c-g loci, *Humangenetik*, 16, 307, 1972.

Noseda, G. et al., Hypo-lipoproteinaemia associated with auto-antibodies against lipoproteins, *Eur. J. Clin. Invest.*, 2, 342, 1972.

Sato, J. et al., Anti-cholesterol activity in antisera against human serum lipoproteins, *Immunochemistry*, 9, 585, 1972.

Schultz, J. S. et al., Studies on the serum fraction containing soluble inhibitors of anti-HL-A sera, *Transplantation*, 13, 186, 1972.

Simons, L. et al., Autoantibodies to serum lipoproteins, *Br. Med. J.*, 4, 380, 1972.

Taylor, P. E. et al., Relationship of Milan antigen to abnormal serum lipoprotein, *Am. J. Dis. Child.*, 123, 329, 1972.

Torsvik, H. et al., An alternative procedure for production of anti-Lp(x) serum, *Acta Pathol. Microbiol. Scand. Sect. B*, 80, 270, 1972.

Utermann, G. et al., Studies on the Lp(x)-lipoprotein of human serum. IV. The disaggregation of the Lp(x)-lipoprotein, *Humangenetik*, 14, 142, 1972.

Berg, K. et al., Genetic marker system in arctic populations. V. The inherited Ag(x) serum lipoprotein antigen in Finnish Lapps, *Hum. Hered.*, 23, 241, 1973.

Ehnholm, C. et al., The occurrence of Ag determinants in different lipoproteins, *Vox Sang.*, 25, 281, 1973.

Harvie, N. R. et al., Studies on the heterogenecity of human serum Lp lipoproteins and on the occurrence of double Lp lipoprotein variants, *Biochem. Genet.*, 9, 235, 1973.

Hirschfeld, J. et al., Ag (x, y, a1, z, t) antigens in a Japanese population, *Hum. Hered.*, 23, 154, 1973.

Vogelberg, K. H. et al., Clinical findings for differentiation of Lp(a) lipoprotein, *Verh. Dtsch. Ges. Inn. Med.*, 79, 1278, 1973.

Albers, J. J. et al., Quantitative genetic studies of the human plasma Lp(a) lipoprotein, *Biochem. Genet.*, 11(6), 475, 1974.

Albers, J. J. et al., Immunochemical quantification of human plasma Lp(a) lipoprotein, *Lipids*, 9, 15, 1974.

Dahlén, G. et al., Lp(a) lipoprotein and pre-beta$_1$-lipoprotein in young adults, *Acta Med. Scand.*, 196(4), 327, 1974.

Heiberg, A. et al., On the relationship between Lp(a) lipoprotein, "sinking pre-beta-lipoprotein" and inherited hyper-beta-lipoproteinaemia, *Clin. Genet.*, 5, 144, 1974.

Müller, U. et al., Studies on the geographic distribution of the human serum beta-lipoprotein antigen Ag(x), *Hum. Hered.*, 24(5—6), 458, 1974.

Schultz, J. S. et al., The genetics of the Lp antigen. I. Its quantitation and distribution in a sample population, *Ann. Hum. Genet.*, 38(1), 39, 1974.

Sing, C. F. et al., The genetics of the Lp antigen. II. A family study and proposed models of genetic control, *Ann. Hum. Genet.*, 38(1), 47, 1974.

DiPerri, T., Hyperlipoproteinemia and immunity (in Italian), *Minerva Med.*, 66(65), 3397, 1975.

Morganti, G. et al., Contribution to the genetics of serum β-lipoproteins in man. VIII. Linkage of the Ag h/i locus with the Ag x/y, Ag a1/d, Ag c/g, and Ag t/z loci, *Humangenetik*, 30(4), 341, 1975.

Walton, K. W., The immunology of serum lipoproteins, *Adv. Exp. Med. Biol.*, 63, 85, 1975.

Berg, K. et al., Genetic lipoprotein variation and lipid levels in man, *Clin. Genet.*, 10, 97, 1976.

Brown, M. S. et al., New directions in human biochemical genetics: understanding the manifestations of receptor deficiency states, *Prog. Med. Genet.*, 1, 103, 1976.

Brown, M. S. et al., Familial hypercholesterolemia: a genetic defect in the low-density lipoprotein receptor, *N. Engl. J. Med.*, 294(25), 1386, 1976.

Dahlén, G. et al., Further evidence for the existence of genetically determined metabolic differences between Lp(a+) and Lp(a−) individuals, *Clin. Genet.*, 9(3), 357, 1976.

Kobierska-Szczepánska, A. et al., Familial hypercholesterolemia: clinical findings and immunochemical studies of lipoproteins (in Polish with English abstract), *Pediatr. Pol.*, 51(12), 1435, 1976.

Sato, H., Immunohistochemical demonstration of β-lipoproteins in the psoriatic skin by peroxidase antibody technique (author's transl.) (in Japanese), *Jpn. J. Dermatol. Ser. B*, 86(9), 515, 1976.

Weinberg, R. et al., Estimates of the heritability of serum lipoprotein and lipid concentrations, *Clin. Genet.*, 9(6), 588, 1976.

Curtis, L. K. et al., In vivo suppression of the primary immune response by a species of low density serum lipoprotein, *J. Immunol.*, 118(2), 648, 1977.

Lipoproteins and Vitamins

Baraud, J. et al., Presence, in corn kernels, of carotenolipoprotein (in French), *C. R. Acad. Sci.*, 260, 7045, 1965.

Golikov, A. P., Effect of vitamin B$_6$ on the distribution of labeled cholesterol in β-lipoproteins of blood serum under experimental hypercholesterolemia (in Russian), *Vopr. Med. Khim.*, 13, 3, 1967.

Davies, T. et al., Interrelation of serum lipoprotein and tocopherol levels, *Clin. Chim. Acta*, 24, 431, 1969.

Brubacher, G. et al., Relation between beta-lipoprotein content in serum and plasma vitamin E content. A contribution to the question of the estimation of the vitamin E status in man (English abstr.), *Int. J. Vitam. Nutr. Res.*, 44(4), 521, 1974.

Lipoproteins and Hormones

Vallance-Owen, J., Insulin antagonists (in Italian), *Clin. Ter.*, 30, 308, 1964.

Janni, A. et al., Action of HCG on the plasmatic lipoprotein pattern after a fat meal (in Italian), *Boll. Soc. Ital. Biol. Sper.*, 41, 20, 1965.

Tomarelli, R. M. et al., Bioassay of hypocholesterolemic steroids, *J. Pharm. Sci.*, 55, 1392, 1966.

Friedman, M. et al., Response of hyperlipemic subjects for carbohydrates, pancreatic hormones and prolonged fasting, *J. Clin. Endocrinol.*, 28, 1773, 1968.

Dahlén, G. et al., Pre-β_1-lipoprotein and Lp(α) antigen in relation to triglyceride levels and insulin release following an oral glucose load in middle-aged males, *Acta Med. Scand.*, 199(5), 413, 1976.

Eaton, R. P., Glucagon and lipoproteins, *Metabolism*, 25(11), Suppl. 1, 1415, 1976.

Lipoproteins and Clearing Factor

Visconti, E. et al., Hematochemical findings in young workers exposed to carbon disulfide. II. The behavior of lipoproteins and the "clearing factor," (in Italian), *Med. Lav.*, 57, 677, 1966.

Lipoproteins and Different Blood Factors

Yoshioka, M. et al., Characteristics of endotoxin altering fractions derived from normal serum. 1. Separation of β-lipoproteins and lipid from human serum fractions, *Kitasato Arch. Exp. Med.*, 36, 27, 1963.

Holobut, W. et al., Studies on the interrelationship between the quick prothrombin time, cholesterol level and lipoprotein fractions in the blood serum (in Polish), *Postepy Hig. Med. Dosw.*, 18, 657, 1964.

Howell, R. M. et al., The role of lipoproteins in the production of hypercoagulability: a new concept, *Br. J. Exp. Pathol.*, 45, 618, 1964.

Kazal, L. A., Interactions of phospholipids with lipoproteins, with serum and its proteins, and with proteolytic and nonproteolytic enzymes in blood clotting, *Trans. N.Y. Acad. Sci.*, 27, 613, 1965.

Taylor, F. B. et al., Effect of surface active lipoprotein on clotting and fibrinolysis, and of fibrinogen on surface tension of surface active lipoprotein with a hypothesis on the pathogenesis of pulmonary atelectasis and hyaline membrane in respiratory distress syndrome of the newborn, *Am. J. Med.*, 40, 346, 1966.

Zola-Sleczek, E. et al., Studies on the interrelations between blood clotting in the thromboelestographic picture and the total lipid, β-lipoprotein and cholesterol level and the degree of lipemic turbidity of the serum (in Polish), *Pol. Tyg. Lek.*, 22, 321, 1967.

Musiatowicz, J. et al., Investigations on the relation between β-lipoprotein and plasma euglobulin fibrinolysis, *Experientia*, 23, 274, 1967.

Skrzydlewski, F. et al., Influence of plasma lipoproteins on the fibrinolytic activity, *Thromb. Diath. Haemorrh.*, 17, 482, 1967.

Audram, R. et al., Study of the anticomplement power of human serum lipoproteins in vitro (in French), *C. R. Acad. Sci. Ser. D*, 268, 628, 1969.

Cultrera, G. et al., A study on lipidic metabolism by means of radioactive triolein and on fibrinolysis in aged subjects treated for six months with a duodenal sulfated polysaccharide, *Arzneim. Forsch.*, 19, 372, 1969.

Skrzydlewski, Z., The role of lipoproteins in the process of fibrinolysis (in Polish), *Wiad. Lek.*, 23, 1017, 1970.

Abe, K. et al., Lipoprotein and blood coagulation in hyperlipemia (in Japanese), *Jpn. J. Clin. Pathol.*, Suppl. 19, 177, 1971.

Zöller, H. et al., Plasmatic coagulation factors in hvperlipoproteinemias, *Muench. Med. Wochenschr.*, 116, 201, 1974.

Rymaszewski, Z., Fibrinolylic blood activity in primary hyperlipoproteinemia (English abstr.), *Pol. Arch. Med. Wewn.*, 53(2), 137, 1975.

Bajaj, S. P. et al., Human plasma lipoproteins as accelerators of prothrombin activation, *J. Biol. Chem.*, 251(17), 5233, 1976.

Carvalho, A. C., Intravascular coagulation in hyperlipiJemia, *Thromb. Res.*, 8(6), 843, 1976.

Sanberg, H. et al., Studies of the thromboplastic effect of human plasma lipoproteins, *Thromb. Haemostas.*, 35(1), 178, 1976.

Scherer, R. et al., The significance of plasma lipoproteins on erythrocyte aggregation and sedimentation., *Br. J. Haematol.*, 32(2), 235, 1976.

Lipoproteins and Pigments

Shibata, H. et al., Function of lipoproteins in pigment transport in the serum (in Japanese), *Jpn. J. Clin. Pathol.*, Suppl. 19, 179, 1971.

Lipoproteins and Ions

Brown, A. D., Hydrogen ion filtrations of intact and dissolved lipoproteins membranes, *J. Mol. Biol.*, 12, 491, 1965.

Rabinovitz, M. et al., The oxidation of ferrous ions by serum proteins demonstrated by immunoelectrophoresis, *Clin. Chim. Acta*, 14, 270, 1966.

Maibach, E., The modification of total cholesterol, β-lipoproteins and total lipid of serum by oral and parenteral calcium administration (in German), *Schweiz. Med. Wochenschr.*, 97, 418, 1967.

Burstein, M. et al., Precipitation of α- and β-serum lipoproteins by sulfate polysaccharides in the presence of manganese chloride (in French), *Nouv. Rev. Fr. Hematol.*, 9, 231, 1969.

Mores, S. G. et al., Selenium determination in human serum lipoprotein fractions by neutron activation analysis, *Anal. Biochem.,* 49, 598, 1972.

ELECTROPHORETIC METHODS FOR SEPARATION OF LIPOPROTEINS

Paper Electrophoresis

Chen, J. S., Studies on serum lipoproteins by prestaining paper electrophoresis, *J. Formosa Med. Assoc.,* 63, 53, 1964.

Wolff, R. et al., Fractionation of serum cholesterol by zone electrophoresis. Application to the study of human serum (in French), *Ann. Biol. Clin. (Paris),* 22, 1139, 1964.

Saperov, V. N., On the effect of various factors on the results of electrophoretic determination of serum lipoproteins (in Russian), *Lab. Delo,* 4, 216, 1965.

Berg, G. et al., Studies on the fat transport in serum by means of lipids electrophoresis on membrane foils (in German), *Klin. Wochenschr.,* 43, 1109, 1965.

Narayan, K. A. et al., A comparison of dyes used for staining electrophoretically separated lipoprotein components, *Clin. Chim. Acta,* 13, 532, 1966.

Simon, K. H., Determination of serum lipoproteins (in German), *Med. Wochenschr.,* 20, 234, 1966.

Strasser, H., Clinical importance of lipid- and lipoproteins determination in the serum, *Helv. Med. Acta Suppl.,* 46, 150, 1966.

Stajner, A., Rapid and simple determination of phospholipids in lipoprotein fractions directly on paper, *Vnitr. Lek.,* 12, 707, 1966.

Baker, K. J., Binding of sulfobromophthalein (BSP) sodium and indocyanine green (ICG) by plasma α_1-lipoproteins, *Proc. Soc. Exp. Biol. Med.,* 122, 957, 1966.

Kahlke, W. et al., Pathologic patterns of serum lipoproteins demonstrated with a new method of lipoprotein electrophoresis (in German), *Verh. Dtsch. Ges. Inn. Med.,* 73, 828, 1967.

Bundschuh, G. et al., Precipitation electrophoresis, a semiquantitative method for the determination of Lp-phenotypes (in German), *Z. Aerztl. Fortbild,* 61, 780, 1967.

Bundschuh, G. et al., Precipitation electrophoresis, a semiquantitative method for the determination of Lp-phenotypes (in German), *Z. Aerztl. Fortbild,* 61, 780, 1967.

Burstein, M. et al., Detergents and electrophoretic mobility of serum lipoproteins (in French), *Nouv. Rev. Fr. Hematol.,* 8, 793, 1968.

Wada, M., Clinicopathological observation of serum lipoproteins by paper electrophoresis with special reference to methodology and pre-β lipoproteins (in Japanese), *Jpn. J. Clin. Pathol.,* 17, 169, 1969.

Batsakis, J. G. et al., The electrophoresis of lipoproteins, *Mich. Med.,* 68, 223, 1969.

Moinuddin, M. et al., Serum lipoproteins: a paper electrophoresis method without albumin in the buffer, *Lipids,* 4, 186, 1969.

Lane, R. F., Serum lipoprotein patterns by electrophoresis, *J. Med. Lab. Technol.,* 26, 212, 1969.

Winkelman, J. et al., Studies on the phenotyping of hyperlipoproteinemias. Evaluation of paper electrophoresis techniques, *Clin. Chim. Acta,* 26, 25, 1969.

Oncley, J. L. et al., Lipoproteins — a current perspective of methods and concepts, *Proc. Natl. Acad. Sci. U.S.A.,* 64, 1107, 1969.

Forman, D. T., Lipoprotein electrophoresis and atherosclerosis, *Proc. Inst. Med. Chicago,* 28, 78, 1970.

Sobotka, J. et al., Determination of lipoproteins using paper electrophoresis in an albumin buffer, *Vnitr. Lek.,* 16, 397, 1970.

Buckley, G. C. et al., Variations in paper electrophoretic serum lipoprotein patterns in healthy subjects, *Can. Med. Assoc. J.,* 102(9), 943, 1970.

Hatch, F. T. et al., Semi-quantitative paper electrophoresis of serum lipoproteins, *Clin. Biochem.,* 3, 115, 1970.

Kindler, U., Lipoprotein electrophoresis in the diagnosis of hyperlipoproteinemias (in German), *Dtsch. Med. Wochenschr.,* 95, 1942, 1970.

Dyberg, J. et al., Lipoproteinelektrophorese — technic and quantitation (in Danish), *Nord. Med.,* 84, 1538, 1970.

Peyer, C. et al., Technical and nomenclatural problems in lipoprotein electrophoresis and normal ranges of glyco- and lipoproteins in the serum (in German), *Clin. Chim. Acta,* 30, 295, 1970.

Arima, T. et al., Technic of serum lipoprotein analysis by paper electrophoresis and its clinical significance (in Japanese), *Jpn. J. Clin. Pathol.,* 18, 856, 1970.

Sobotka, J. et al., Clinical importance of determination of the pre-β lipoprotein fraction with the aid of paper electrophoresis (in Czech), *Vnitr. Lek.,* 17, 36, 1971.

Kucerova, L. et al., Use of electrophoresis of serum lipoproteins in the diagnosis of primary hyperlipoproteinemias (in Czech), *Sb. Lek.,* 73, 135, 1971.

Greten, H. et al., Lipoprotein electrophoresis in the diagnosis of the hyperlipoproteinaemias, *Ger. Med. Mon.*, 1, 29, 1971.

Wilson, W. B., Evaluation of lipid disorders by lipoprotein electrophoresis, *J. Miss. State Med. Assoc.*, 12, 287-96, 1971.

Pyrovolakis, J. et al., Modification of the electrophoretic separation of lipoproteins on paper, *J. Clin. Pathol.*, 24, 368, 1971.

Dangerfield, W. G. et al., Electrophoresis of plasma lipoproteins, *Proc. R. Soc. Med.*, 64, 897, 1971.

Hayashi, T. et al., Evaluation of lipoprotein fractionation methods, *Jpn. J. Clin. Pathol.*, Suppl. 19, 163, 1971.

Tanaka, K. et al., Methodological evaluation and clinical significance of lipoprotein fractionation by filter paper electrophoresis, with special reference to correlation between lipoprotein patterns and arteriosclerosis acceleration factors (in Japanese), *Jpn. J. Clin. Pathol.*, Suppl. 19, 165, 1971.

Yokota, Y. et al., Clinical application of lipoprotein analysis by paper electrophoresis, *Jpn. J. Clin. Pathol.*, Suppl. 20, 18, 1972.

Gartzke, J., Decoloration of paper lipoprotein-pherograms, *Z. Med. Labortech.*, 13, 329, 1972.

Hayes, T. M. et al., Hyperlipoproteinaemia classification: the optimum routine electrophoretic system and its relevance to treatment, *Lancet*, 1, 696, 1972.

Kröning, G. et al., Electrophoretic separation of serum lipoproteins using ERI 10 for the assessment of the pherograms (in German with English abstract), *Z. Gesamte Inn. Med.*, 27, 211, 1972.

Lehmann, H. et al., Hyperlipoproteinaemia classification: the optimum routine electrophoretic system and its relevance to treatment, *Lancet*, 1, 557, 1972.

Yamazaki, S. et al., Lipoprotein analysis using paper electrophoresis (in Japanese), *Saishin Igaku*, 27, 429, 1972.

May, P. et al., Lipoprotein X simulating paraproteinemia in paper electrophoresis, *Clin. Chim. Acta*, 38, 255, 1972.

Petek, W. et al., Immunochemical characterization of the "pre" band in lipoprotein electrophoresis, *Clin. Chim. Acta*, 38, 460, 1972.

Oikawa, H., Ratio and fluctation of serum lipoprotein levels in patients with influenza — analysis by paper electrophoresis, *Jpn. J. Thorac. Dis.*, 10, 379, 1972.

Reed, A. H., Misleading correlations in clinical applications, *Clin. Chim. Acta*, 40, 266, 1972.

Michaux, A. et al., Characterization of pre-beta hyperlipoproteinemias using 3 methods of electrophoresis, *Ann. Biol. Clinique. (Paris)*, 31, 95, 1973.

Naito, H. K. et al., Rapid lipid-staining procedure for paper electrophoretograms, *Clin. Chem.*, 19, 106, 1973.

Yokota, Y. et al., Clinical application of electrophoresis. 6. Clinical application of lipoprotein analysis by paper electrophoresis, *Jpn. J. Clin. Pathol.*, 21, 227, 1973.

Vasilescu, F. D., Optimal separation of fractional lipoproteins in an electric field, *Microbiol. Parazitol. Epidemiol.*, 18, 271, 1973.

Revutskii, E. L. et al., Phenotyping of lipoproteinemias in patients with arteriosclerosis by means of paper electrophoresis, *Vrach. Delo*, 6, 15, 1973.

Segal, A. W. et al., Nitroblue tetrazolium — a new lipoprotein stain, *Atherosclerosis*, 18, 499, 1973.

Frajola, W. J., Letter: lipoprotein electrophoresis, *Clin. Chem.*, 19, 1414, 1973.

Lopes-Virella, L. et al., Comparative analysis of various technics for the separation of serum lipoproteins, *Rev. Clin. Esp.*, 133, 195, 1974.

Guisard, D. et al., Comparative study of lipoproteinograms obtained by three different methods (in French with English abstr.), *Clin. Chim. Acta*, 53, 79, 1974.

Phuaphairoj, S. et al., An improved and rapid method for quantitative lipoprotein electrophoresis, *J. Med. Assoc. Thai*, 57(9), 480, 1974.

Toyota, C. et al., Comparison of serum lipoprotein electrophoreses, *Jpn. J. Clin. Pathol.*, 22(10), 137, 1974.

Iammarino, R. M., Whether lipoprotein electrophoresis?, *Hum. Pathol.*, 5(6), 626, 1974.

Revutskii, E. L. et al., Determination of the lipoprotein fraction by use of horizontal electrophoresis on paper and an albumin-containing buffer solution, *Lab. Delo*, 2, 96, 1975.

Iammarino, R. M., Lipoprotein electrophoresis should be discontinued as a routine procedure, *Clin. Chem.*, 21(3), 300, 1975.

Sage, G. W., Letter: a stable lipoprotein electrophoresis control, *Clin. Chem.*, 21(3), 452, 1975.

Carmena, R. et al., Value of serum lipids and lipoprotein electrophoresis obtained from a clinical normal population (in Spanish), *Rev. Clin. Esp.*, 138(2), 125, 1975.

Carlson, K. et al., Comparison of behaviour of VLDL of type III hyperlipoproteinemia on electrophoresis on paper and on agarose gel with a note on a late (slow) pre-β-VLDL lipoprotein, *Scand. J. Clin. Lab. Invest.*, 35(7), 655, 1975.

Schwandt, P., Letter: shift in the lipoprotein distribution pattern (in German), *Med. Klin.*, 71(1), 34, 1976.

Green, J., Lipoprotein pre-staining: comparison of nitroblue tetrazolium and Sudan Black B, *Clin. Chim. Acta,* 66(3), 295, 1976.

Cellulose Acetate Electrophoresis

Charman, R. C. et al., Separation of human plasma lipoprotein by electrophoresis on cellulose acetate, *Anal. Biochem.,* 19, 177, 1967.

Berg, G. et al., Immunoelectrophoretic investigations using acetate strips (in German), *Clin. Chim. Acta,* 17, 265, 1967.

Colfs, B. et al., Electrophoresis and sudan black staining of lipoproteins on gelatinised cellulose acetate, *Clin. Chim. Acta,* 18, 325, 1967.

Ogawa, Y., Alpha-2-lipoprotein in human serum detectable by cellulose acetate electrophoresis (in Japanese), *Med. Biol.,* 76, 118, 1968.

Ogawa, Y., Staining procedure for fractionation of serum lipoproteins in cellulose acetate electrophoresis (in Japanese), *Med. Biol.,* 76, 237, 1968.

Chin, H. P. et al., Separation and quantitative analysis of serum lipoproteins by means of electrophoresis on cellulose acetate, *Clin. Chim. Acta,* 20, 305, 1968.

Crosato, M. et al., Determination by electrophoresis on cellulose acetate of the serum proteins, lipoproteins and glycoproteins in infantile mucoviscidosis (in Italian), *Minerva Pediatr.,* 20, 2326, 1968.

Chin, H. P. et al., On the precision of lipoprotein electrophoresis on cellulose acetate and its use in the diagnosis of hyperlipoproteinemia, *Clin. Chim. Acta,* 23, 239, 1969.

Farber, E. R. et al., Lipoprotein electrophoresis. A comparison of cellulose acetate and paper technics, *Tech. Bull. Regist. Med. Technol.,* 39, 55, 1969.

Kanno, T. et al., Electrophoresis of lipoproteins using a cellulose acetate membrane (in Japanese), *Jpn. J. Clin. Pathol.,* 17, 385, 1969.

Nikkari, T. et al., Anomalous mobilities of lipoproteins in cellulose acetate microelectrophoresis, *Clin. Chim. Acta,* 24, 473, 1969.

Charman, R. C. et al., Human α_2-lipoprotein bands visualized by cellulose acetate electrophoresis, *Lipids,* 4, 397, 1969.

Ross, D. L. et al., Lipoprotein fractionation by electrophoresis on cellulose acetate, *Am. J. Med. Technol.,* 35, 540, 1969.

Winkelman, J. et al., Studies on the phenotyping of hyperlipoproteinemias. Evaluation of cellulose acetate technique and comparison with paper electrophoresis, *Clin. Chim. Acta,* 26, 33, 1969.

Bertrand, F. et al., Lipoproteinograms on gel cellulose acetate in the micromethod, *Ann. Biol. Clin. (Paris),* 27, 735, 1969.

Beckering, R. E., Jr. et al., A rapid method for lipoprotein electrophoresis using cellulose acetate as support medium, *Am. J. Clin. Pathol.,* 53, 84, 1970.

Klemens, U. H. et al., Quantitative comparison of lipoprotein electrophoresis on cellulose acetate membranes with fractionation of lipoproteins in the preparative ultracentrifuge. Lipoprotein electrophoresis on cellulose acetate membranes, *Z. Klin. Chem.,* 8, 166, 1970.

Klemens, U. H. et al., Qualitative comparison of lipoprotein electrophoresis on cellulose acetate membranes with fractionation of lipoproteins in the preparative ultracentrifuge. Lipoprotein electrophoresis on cellulose acetate membranes, *Z. Klin. Chem.,* 8, 162, 1970.

Ishitoya, Y. et al., Clinical application of electrophoresis of serum lipoprotein on a cellulose acetate membrane, *J. Clin. Pathol.,* 18, 299, 1970.

Fletcher, M. J. et al., A simple method for separating serum lipoproteins by electrophoresis on cellulose acetate, *Clin. Chem.,* 16, 362, 1970.

Grabner, W. et al., Lipoprotein microzone electrophoresis on cellulose acetate strips, *Clin. Chim. Acta,* 28, 299, 1970.

Rosati, L. A. et al., Identification of type 3 hyperlipoproteinemia by electrophoresis on cellulose acetate and indirect (Schiff's) staining, *Clin. Biochem.,* 3, 171, 1970.

Sakurabayashi, I. et al., Fractionation of serum lipoproteins by cellulose acetate electrophoresis with special reference to clarification of cello-gel membrane (in Japanese), *Med. Biol.,* 81, 47, 1970.

Klemens, U. H. et al., Lipoprotein electrophoresis on cellulose acetate membranes. An analytical and semi-quantitative method suitable for the clinical routine (in German), *Z. Klin. Chem. Klin. Biochem.,* 7, 540, 1970.

Rho, G. L. et al., Electrophoresis of plasma lipoproteins on gelatinized cellulose acetate, *Clin. Chem.,* 17, 551, 1971.

Knüchel, F., Microzone-lipoprotein electrophoresis using thin cellulose acetate leafs (in German), *Med. Welt.,* 49, 1939, 1971.

Beckering, R. E., Jr. et al., Variations in cellulose acetate membranes for lipoprotein electrophoresis, *Ann. J. Clin. Pathol.,* 56, 765, 1971.

Magnani, H. N. et al., A quantitative method for blood lipoproteins using cellulose acetate electrophoresis, *J. Clin. Pathol.*, 24, 837, 1971.

Sugano, T. et al., Symposium on clinical application of electrophoresis. Clinical application of lipoprotein fractionation by the acetate membrane, *Jpn. J. Clin. Pathol.*, Suppl. 20, 19, 1972.

Kanno, T. et al., Electrophoresis analysis of lipoproteins on cellulose acetate (in Japanese), *Saishin Igaku*, 27, 434, 1972.

Schlag, B. et al., A simple, easily reproducible process for lipoprotein electrophoresis on cellulose acetate foils, *Dtsch. Gesundheitswes.*, 27, 1868, 1972.

van Kampen, E. J. et al., Lipoprotein electrophoresis on gelatinized cellulose acetate, *Clin. Chim. Acta*, 40, 485, 1972.

Bolletti Censi, M. et al., Rapid electrophoresis of lipoproteins on cellulose acetate, *Quad. Sclavo Diagn. Clin. Lab.*, 8, 1036, 1972.

Wolff, J., Electrophoresis of serum lipoproteins on cellulose acetate foils, *Z. Med. Labortech.*, 14, 328, 1973.

Kanno, T. et al., Clinical application of electrophoresis. 7. Clinical application of lipoprotein electrophoresis by acetate membrane, *Jpn. J. Clin. Pathol.*, 21, 232, 1973.

Charman, R. C. et al., An evaluation of cellulose acetate electrophoresis for the determination of human plasma lipoprotein patterns, *Atherosclerosis*, 17, 483, 1973.

Babamova-Vilarova, A., Microelectrophoretic method for determination of the serum lipoproteins (in Serbian), *God. Zb. Med. Fak. Skopje*, 20, 105, 1974.

Luthold, W. W. et al., Characterization of hyperlipemic states. Analysis of aspects of the cellulose acetate gel electrophoresis lipoproteinogram (with English abstract). *Rev. Hosp. Clin. Fac. Med. Univ. Sao Paulo*, 29(4), 180, 1974.

Palumbo, E. et al., Staining of lipoproteins for determination of the lipoelectropherogram on cellulose acetate: technical innovations, *Boll. Soc. Ital. Biol. Sper.*, 50(18), 1509, 1974.

Owada, K. et al., Serum lipoprotein levels in a rural population, separated by means of cellulose acetate electrophoresis (with English abstract), *Jpn. J. Hyg.*, 29(4), 438, 1974.

Itsumi, K. et al., Analysis of lipoprotein in the cerebrospinal fluid by cellulose electrophoresis, *Jpn. J. Clin. Pathol.*, 22(10), 143, 1974.

Sober, D. et al., A routine method of lipoprotein electrophoresis using acetate films (in German), *Clin. Chim. Acta*, 63(1), 99, 1975.

Ericson, C. et al., Some comments on the Kohn staining technigue of lipoproteins after electrophoresis on cellulose acetate, *Scand. J. Clin. Lab. Invest.*, 35(5), 479, 1975.

Uldall, A., Improvements and simplification of the ozonolysis-Schiff staining method for plasma lipoproteins on cellulose acetate, *J. Clin. Chem. Biochem.*, 14(1), 23, 1976.

Cellogel Electrophoresis
Baets, J. et al., Improved method for lipoprotein electrophoresis on cellogel, *Clin. Chim. Acta*, 32, 142, 1971.

Messerschmidt, H. J. et al., Lipid screening and lipoprotein electrophoresis by cellogel, *Clin. Chim. Acta*, 36, 51, 1972.

Berends, G. T. et al., A study of serum lipoprotein electrophoresis and cellogel: normal values, comparison with SML profiling and ultracentrifugation, *Clin. Chim. Acta*, 41, 187, 1972.

Oriente, P., Lipoprotein electrophoresis on cellogel: a practical method for screening hyperlipoproteinemias, *Adv. Exp. Med. Biol.*, 38, 247, 1973.

Ilinov, P. et al., Experience with the electrophoretic separation of lipoproteins on celogel (in Bulgarian), *Vutr. Boles.*, 14(3), 137, 1975.

Starch-Gel Electrophoresis
Lewis, L. A., Screening for serum lipoprotein abnormalities comparison of ultracentrifugal, paper and thin-layer starch-gel electrophoresis techniques, *Lipids*, 4, 60, 1969.

Aragonillo, C. et al., Nigrosine staining of wheat endosperm proteolipid patterns in starch gels, *Anal. Biochem.*, 63(2), 603, 1975.

Sepharose-Gel Electrophoresis
Borinskii, Tu. N. et al., Electrophoretic method of separation of lipoproteins of the thin-layer Sephadex G-200 (in Russian), *Lab. Delo*, 1, 34, 1976.

Agar-Gel Electrophoresis
Teichmann, B. et al., Nonspecific lipoprotein precipitations in agar gel (in German), *Acta Biol. Med. Ger.*, 12, 591, 1964.

Houtsmüller, A. J. et al., The application of Reinagar for the quantitative separation of α- and β-lipoproteins, *Clin. Chim. Acta*, 9, 497, 1964.

Churkin, E. A., Determination of serum proteins and lipoproteins by agar electrophoresis (in Russian), *Lab. Delo*, 4, 220, 1965.

Banerjee, D. et al., Serum lipoproteins in disease. A study by agar electrophoresis, *J. Assoc. Physicians India*, 13, 935, 1965.

Götz, H. et al., Animal serum proteins. Agar electrophoretic studies for characterisation of serum proteins of the goat, dog, cat, duck and hen (in German), *Zentralbl. Veterinaermed Reihe A*, 14, 385, 1967.

Burstein, M. et al., Serum lipoprotein mobility in agar gel (in French), *Nouv. Rev. Fr. Hematol.*, 8, 809, 1968.

Iammarino, R. M. et al., Agar gel lipoprotein electrophoresis: a correlated study with ultracentrifugation, *Clin. Chem.*, 15, 1218, 1969.

Ruzicka, J. et al., Electrophoresis of serum lipoproteins in psoriatic patients on agar-agar gel (in Czech), *Cesk. Dermatol.*, 46, 142, 1971.

de Gennes, J. L. et al., Detection and distribution of slow pre-β-lipoprotein in agar gel electrophoresis of total serum in 204 cases of mixed hyperlipidemia (in French), *Ann. Med. Interne*, 126(1), 19, 1975.

Agarose-Gel Electrophoresis

Rapp, W. et al., Agarose gel electrophoresis of lipoproteins (in German), *Clin. Chim. Acta*, 19, 493, 1968.

Kahlke, W. et al., A comparison of paper electrophoresis and agarose gel electrophoresis for the differentiation of hyperlipoproteinemias (in German), *Klin. Wochenschr.*, 46, 330, 1968.

Noble, R. P., Electrophoretic separation of plasma lipoproteins in agarose gel, *J. Lipid Res.*, 9, 693, 1968.

Mc Glashan, D. A. et al., A method for lipoprotein electrophoresis using agarose gel, *Clin. Chim. Acta*, 22, 646, 1968.

Noble, R. P. et al., Comparison of lipoprotein analysis by agarose gel and paper electrophoresis with analytical ultracentrifugation, *Lipids*, 4, 55, 1969.

Papadopoulos, N. M. et al., Determination of human serum lipoprotein patterns by agarose gel electrophoresis, *Anal. Biochem.*, 30, 421, 1969.

Dyerberg, J. et al., Quantitative plasma lipoprotein estimation by agarose gel electrophoresis, *Clin. Chim. Acta*, 28, 203, 1970.

Hollmann, M. et al., Gel filtration of serum lipoproteins with agarose (in German), *Klin. Wochenschr.*, 48, 493, 1970.

Zöllner, N. et al., Electrophoretic separation of serum lipoproteins in agarose gel with the addition of albumin (in German), *Z. Klin. Chem. Klin. Biochem.*, 7, 525, 1970.

Wille, L. E., Pre-β-lipoproteins in healthy persons. A study of 224 subjects with agarose gel electrophoresis, *Clin. Genet.*, 2, 228, 1971.

Sirtori, C. et al., Phenotyping of type II and IV hyperlipoproteinemias by a simple, quantitative agarose gel lipoprotein electrophoresis, *Clin. Chim. Acta*, 31, 305, 1971.

Elphick, M. C., Microscope slide electrophoresis of serum lipoproteins in agarose gel, *J. Clin. Pathol.*, 24, 83, 1971.

Kucerova, L., et al., Determination of serum lipoproteins with paper and agarose electrophoresis (in Czech), *Cas. Lek. Cesk.*, 110, 258, 1971.

Kindler, U., Lipoprotein electrophoresis in agarose gel (in German), *Med. Klin.*, 66, 639, 1971.

Papadopoulos, N. M. et al., Varieties of human serum lipoprotein pattern: evaluation by agarose gel electrophoresis, *Clin. Chem.*, 17, 427, 1971.

Johnson, E. A., Electrophoretic separation of plasma lipoproteins using agarose gel on a side, *Am. J. Med. Technol.*, 37, 233, 1971.

Dyerberg, J. et al., Quantitation of the lipoprotein complex by agarose gel electrophoresis, *Clin. Chim. Acta*, 33, 458, 1971.

Hulley, S. B. et al., Quantitation of serum lipoproteins by electrophoresis on agarose gel: standardization in lipoprotein concentration units (mg-100 ml) by comparison with analytical ultracentrifugation, *J. Lipid Res.*, 12, 420, 1971.

Nakagawa, M., Comparison between agarose gel electrophoresis and disk electrophoresis in serum lipoprotein analysis and its fluctuation in various disease conditions, *Jpn. J. Clin. Pathol.*, Suppl. 19, 167, 1971.

Kostner, G. et al., Studies on the two α-lipoprotein bands appearing in agarose gel electrophoresis, *Hoppe Seylers Z. Physiol. Chem.*, 352, 1440, 1971.

Okishio, T. et al., Analysis of lipoproteins using agarose gel and disc electrophoresis (in Japanese), *Saishin Igaku*, 27, 441, 1972.

Dabels, J. et al., Differentiation and follow-up study of hyperlipoproteinemias using the agarose gel electrophoresis (in German with English abstract), *Z. Gezamte Inn. Med.*, 27, 507, 1972.

Ahlers, I. et al., Electrophoresis of human serum lipoproteins in modified agar and agarose (in Slovenian with English abstract), *Cas. Lek. Cesk.*, 111, 592, 1972.

Tanemura, K. et al., Evaluation of an immunological assay of serum beta-lipoproteins using the agarose plate containing specific antiserum, *J. Clin. Pathol.*, 20, 658, 1972.

Kindler, U., Lipoprotein electrophoresis in agarose gel, *Med. Welt.*, 23, 1423, 1972.

Ghosh, S. et al., Agarose gel electrophoresis of serum lipoproteins: determination of true mobility, isoelectric point, and molecular size, *Anal. Biochem.*, 50, 592, 1972.

Fellin, R. et al., Electrophoresis of human serum lipoproteins on agarose-gel, *Adv. Exp. Med. Biol.*, 38, 259, 1973.

Dyerberg, J. et al., Quantitation of the lipoproteins complex by agarose gel electrophoresis, *Clin. Chim. Acta*, 43, 283, 1973.

Gros, M. et al., Electrophoretical separation of pre-stained serum lipoproteins on cellulose acetate, agarose gel and polyacrylamide, *Clin. Chim. Acta*, 45, 165, 1973.

Hatch, F. T. et al., Quantitative agarose gel electrophoresis of plasma lipoproteins: a simple technique and two methods for standardization, *J. Lab. Clin. Med.*, 81, 946, 1973.

Vogleberg, K. H. et al., The differentiation of Lp (a) -lipoprotein with the aid of agarose gel electrophoresis, *Z. Klin. Chem.Klin. Biochem.*, 11, 291, 1973.

Phillips, G. B. et al., The phospholipid composition of human serum lipoprotein fractions separated by electrophoresis on agarose gel. Demonstration of a fraction with high lysolecithin content, *Clin. Chim. Acta*, 49, 153, 1973.

Lôzsa, A., Uranyl acetate as an excellent fixative for lipoproteins after electrophoresis on agarose gel, *Clin. Chim. Acta*, 53, 43, 1974.

van Melsen, A. et al., A modified method of phenotyping of hyperlipoproteinemia on agarose electrophoresis, *Clin. Chim. Acta*, 55(2), 225, 1974.

Okuma, S. et al., Pre-alpha band of lipoproteins in agarose gel electrophoresis, *Jpn. J. Clin. Pathol.*, 22(10), 139, 1974.

Garcia-Merlo, S. et al., Serum lipoproteins: study by electrophoresis in agarose gel (in Spanish with English abstract), *Rev. Clin. Esp.*, 135(3), 225, 1974.

Menegozzo, A. et al., Simple rapid and economical procedure for agarose electrophoretic separation of serum lipoprotein fractions (with English abstract), *Quad. Sclavo Diagn. Clin. Lab.*, 11(1), 147, 1975.

Dyerberg, J., Comments on the quantitation of lipoproteins by agarose-gel electrophoresis, *Clin. Chim. Acta*, 61(1), 103, 1975.

Blaton, V., Comments on the quantitation of lipoproteins by agarose gel electrophoresis, *Clin. Chim. Acta*, 61(1), 105, 1975.

Maguire, G. F. et al., Agarose gel electrophoresis of plasma lipoproteins using the Durrum cell, *Clin. Biochem.*, 8(3), 161, 1975.

Phatak, A. G. et al., Simultaneous electrophoretic separations of serum lipoproteins and proteins on agarose, *Indian J. Med. Sci.*, 29(6-7), 167, 1975.

Badzio, T. et al., Agarose gel electrophoresis of serum lipoproteins (in Polish), *Przegl. Lek.*, 33(3), 357, 1976.

Ballantyne, D. et al., Exclusion of pre-β-lipoproteins from agarose gel electrophoresis in the presence of chylomicrons, *Scand. J. Clin. Lab. Invest.*, 36(6), 561, 1976.

Hermann, V. W. et al., The densitometric evaluation of lipoprotein electrophoresis on acetate foil and agarose gel (in German with English abstract), *Z. Gesamte Inn. Med.*, 31(23), 988, 1976.

Polyacrylamide-Gel Electrophoresis

Raymond, S. et al., Lipoprotein patterns in acrylamide gel electrophoresis, *Science*, 151, 346, 1966.

Delcourt, R., An apparatus for the kinetic study of electrophoresis of colored substances on acrylamide gel (in French), *Ann. Biol. Clin. (Paris)*, 25, 1261, 1967.

Pratt, J. J. et al., Polyacrylamide gels of increasing concentration gradient for the electrophoresis of lipoproteins, *Clin. Chim. Acta*, 23, 189, 1969.

Delcourt, R., Vertical electrophoresis of lipoproteins on acrylic gel (preliminary note), *Ann. Biol. Clin. (Paris)*, 27, 189, 1969.

Prat, J. P. et al., Staining of lipoproteins after electrophoresis in polyacrylamide gel, *Bull. Soc. Chim. Biol.*, 51, 1367, 1969.

Dangerfield, W. G. et al., An investigation of plasma lipoproteins by polyacrylamide electrophoresis, *Clin. Chim. Acta*, 30, 273, 1970.

Frings, C. S. et al., Electrophoretic separation of serum lipoproteins in polyacrylamide gel, *Clin. Chem.*, 17, 111, 1971.

Komai, Y. et al., Further studies on polyacrylamide gel electrophoresis of water-soluble proteolipid-protein from bovine brain white matter, *Experientia*, 27, 881, 1971.

Maurer, H. R. et al., Useful buffer and gel systems for polyacrylamide gel electrophoresis, *Z. Klin. Chem. Klin. Biochem.*, 10(5), 220, 1972.

Hall, F. F. et al., Serum lipoprotein electrophoresis: an improved polyacrylamide procedure, *Biochem. Med.*, 6, 464, 1972.

Wada, M. et al., Polyacrylamide-gel block-electrophoresis of plasma lipoproteins, *Clin. Chem.*, 19, 235, 1973.

Lipp, K. et al., Studies on the B-protein of human serum β-lipoprotein using SDS-polyacrylamide gel electrophoresis, *Hoppe Seylers Z. Physiol. Chem.*, 354(3), 262, 1973.

Masket, B. H. et al., The use of polyacrylamide gel electrophoresis in differentiating type 3 hyperlipoproteinemia, *J. Lab. Clin. Med.*, 81, 794, 1973.

Feliste, R. et al., Separation of human plasma lipoproteins by polyacrylamide mixed gel electrophoresis, *Clin. Chim. Acta*, 47, 329, 1973.

Mead, M. G. et al., The investigation of "mid-band" lipoproteins using polyacrylamide gel electrophoresis, *Clin. Chim. Acta*, 51, 173, 1974.

Minamisono, T. et al., Polyacrylamide gel block electrophoresis of serum lipoproteins. I. Evaluation of its clinical applicability, *Jpn. J. Clin. Pathol.*, 22(6), 433, 1974.

Minamisono, T. et al., Polyacrylamide gel block electrophoresis of human serum lipoproteins, II. Immunological study (in Japanese), *Jpn. J. Clin. Pathol.*, 22(7), 519, 1974.

Godolphin, W. J. et al., Isoelectrofocusing of human plasma lipoproteins in polyacrylamide gels: diagnosis of type III hyperlipoproteinemia ("broad beta" disease), *Clin. Chim. Acta*, 56(1), 97, 1974.

Fruchart, J. C. et al., Clinical value of the lipidogram on polyacrylamide gel in gradient density in dyslopemia tests, *Lille Med.*, 19(9), 927, 1974.

Dubarry, M. et al., Lp (a) blood types. Comparison of immunological testing with results of polyacrylamide gel electrophoresis, *Biomed. Express (Paris)*, 23(1), 28, 1975.

Akamatsu, A. et al., A clinical investigation of renogenic dyslipoproteinemia with polyacrylamide gel block electrophoresis, *Jpn. J. Clin. Pathol.*, 23(6), 453, 1975.

Sezille, G. et al., Improved serum lipoprotein electrophoresis procedure in polyacrylamide gradient gel, *Biomed. Express (Paris)*, 23(8), 315, 1975.

Desreumaux, C. et al., Analysis of lipoproteins using polyacrylamide gel electrophoresis and fractionated precipitation with polyanions and detergents. Use in the classification of hyperlipoproteinemias (in French with English abstract), *Ann. Biol. Clin. (Paris)*, 34(5), 309, 1976.

Aslanian, N. L. et al., Determination of lipoprotein fractions (classes) by the electrophoresis in polyacrylamide gel (in Russian), *Zh. Eksp. Klin. Med.*, 16(1), 47, 1976.

Desreumaux, C. et al., Fractionation of serum lipoproteins by preparative electrophoresis in polyacrylamide gel, *J. Chromatogr.*, 130, 336, 1977.

Disc-Electrophoresis

Naryan, K. A. et al., Disc electrophoresis of human serum lipoproteins, *Nature (London)*, 205, 246, 1965.

Wollenweber, J. et al., Disc electrophoresis of human serum lipoproteins (in German), *Verh. Dtsch. Ges. Inn. Med.*, 74, 254, 1968.

Garoff, H. et al., Demonstration by disc electrophoresis of the lipoprotein carrying the Lp (a) antigen in human sera, *Acta Pathol. Microbiol. Scand. Sect. B*, 78, 253, 1970.

Utermann, G. et al., Determination of the Lp (a)-protein by disc-electrophoresis of lipoprotein-fractions (in German), *Humangenetik*, 11, 66, 1970.

Koppikar, S. V. et al., Sephadex sandwich disc electrophoresis, *Anal. Biochem.*, 33, 366, 1970.

Wolfman, L. et al., Disc electrophoresis of lipoproteins: a reinterpretation of the chylomicron band, *Clin. Chem.*, 16, 620, 1970.

Wollenweber, J. et al., Comparative serum lipoprotein analysis by polyacrylamide disc gel and gel electrophoresis (in German), *Clin. Chim. Acta*, 29, 411, 1970.

Rittner, C., Disc electrophoretic variation of human serum lipoproteins (LDL). Further family data on the inheritance of the EI (C) system, *Vox Sang.*, 20, 526, 1971.

Wada, M., Disc electrophoresis of serum lipoproteins, *Jpn. J. Clin. Pathol.*, 19, 465, 1971.

Creno, R. J., Laboratory diagnosis of hyperlipoproteinemia by disc electrophoresis, *Am. J. Med. Technol.*, 37, 292, 1971.

Huismans, B. D. et al., Periodic acid-Schiff staining of unsaturated serum lipoproteins following disc electrophoresis, *Biochim. Biophys. Acta*, 248, 330, 1971.

Wada, M. et al., Disc electrophoresis of serum lipoprotein — a new method derived from Ornstein-Davis method (in Japanese), *Jpn. J. Clin. Pathol.*, 19, 849, 1971.

Wada, M. et al., Identification of serum lipoproteins by disc electrophoresis — comparative study with paper electrophoresis (in Japanese), *Jpn. J. Clin. Pathol.*, 20, 65, 1972.

Wada, M. et al., An investigation of human serum lipoprotein disc electrophoresis. 1. Behavior of serum lipoproteins in disc type acrylamide gel electrophoresis, *Jpn. Circ. J.*, 36, 121, 1972.

Utermann, G., Disc-electrophoretic patterns of human serum high density lipoproteins, *Clin. Chim. Acta*, 36, 521, 1972.

Moran, R. F. et al., Quantitation of β-lipoprotein (LDL) cholesterol by densitometric evaluation of disc electropherograms, *Clin. Chem.*, 18, 217, 1972.

Wada, M. et al., An investigation of serum lipoprotein disc electrophoresis. II. A simple method of human serum lipoprotein disc electrophoresis and identification of the resolved lipoproteins, *Jpn. Circ. J.*, 36, 335, 1972.

Magracheva, E. J., Separation of lipoproteins from blood serum by polyacrylamide disc electrophoresis, *Vopr. Med. Khim.*, 19(6), 652, 1973.

Naito, H. K. et al., Polyacrylamide gel disc electrophoresis as a screening procedure for serum lipoprotein abnormalities, *Clin. Chem.*, 19, 228, 1973.

Segal, A. W. et al., The use of nitroblue tetrasolium prestaining of serum lipoproteins on polyacrylamide disc electrophoresis, *Clin. Chim. Acta*, 53, 361, 1974.

Oya, F. et al., Serum lipoprotein estimation by polyacrylamide gel disc electrophoresis, *Nagoya J. Med. Sci.*, 37(1-2), 23, 1974.

Poliakov, L. M. et al., Method of quantitative determination of blood serum lipoproteins by means of elution sectional disc-electrophoresis with triton X-100 (in Russian), *Lab. Delo* 2, 113, 1975.

Maddock, C. et al., Separation and relative quantitation of lipoproteins by polyacrylamide gel disc electrophoresis, *Ann. Clin. Biochem.*, 13(1), 313, 1976.

Gradient Gel Electrophoresis

Melish, J. S. et al., Concentration gradient electrophoresis of plasma from patients with hyper-β-lipoproteinemia, *J. Lipid Res.*, 13, 193, 1972.

Bautovich, G. J. et al., Gradient gel electrophoresis of human plasma lipoproteins, *Clin. Chem.*, 19, 415, 1973.

Electrophoresis in Different Media

Reissell, P. K. et al., Thin-layer electrophoresis of serum lipoproteins, *J. Lipid Res.*, 7, 551, 1966.

Tippetts, R. D. et al., Stable-flow free boundary (Staflo) electrophoresis: three dimensional fluid flow properties and applications to lipoprotein studies, U. S. Atomic Energy Commission, University of California Radiation Laboratory, 16, 1967.

Vogelberg, K. H., Documentation of gel lipid electrophoresis on celophane foil (in German), *Klin. Wochenschr.*, 47, 722, 1969.

Winkelman, J., Phenotyping of hyperlipoproteinemias. Effect on electrophoretic pattern of serum storage at ambient, refrigerator, or freezing temperatures, *Clin. Chem.*, 16, 507, 1970.

Cornelissen, P. J. et al., Electrophoretic separation of lipoproteins, *Clin. Chim. Acta*, 29, 344, 1970.

Lopez, A. et al., Detection of subtle abnormalities of serum β- and pre-β-lipoproteins in "normal" individuals by turbidimetric and electrophoretic methods, *Clin. Chim. Acta*, 31, 123, 1971.

Arima, T. et al., Electrophoretic lipoprotein patterns and hyperlipemia (in Japanese), *Jpn. J. Clin. Pathol.*, 20, 44, 1972.

Valmikinathan, K. et al., Lipoprotein electrophoresis in sucrose medium, *Indian J. Biochem. Biophys.*, 9, 126, 1972.

Angela, G. C., Synopsis and recent studies of the evaluation and interpretation of electrophoretic and immunoelectrophoretic tracings of biological human fluids under normal conditions and in various morbid situations. II. Lipoproteins and glycoproteins in the blood, *Minerva Med.*, 64, 119, 1973.

Noble, R. P., Nonstainable polyester film, Cronar, for electrophoresis of plasma lipoproteins, *J. Lipid Res.*, 14, 255, 1973.

Cabezas Cerrato, J., Editorial: Current classification of hyperlipoproteinemia (electrophoretic and biochemical), *Rev. Esp. Cardiol.*, 26, 251, 1973.

Hazzard, E. R. et al., The spectrum of electrophoretic mobility of very low density lipoproteins: role of slower migrating species in endogenous hypertriglyceridemia (type IV hyperlipoproteinemia) and broad β-disease (type III). *J. Lab. Clin. Med.*, 86(2), 239, 1975.

Mc Taggart, W. G. et al., Improved method for densitometry of electrophoretic lipoprotein fraction, *Clin. Chem.*, 21(2), 183, 1975.

Kitagawa, M. et al., A new lipoprotein electrophoresis by nitroblue tetrazolium, *Jpn. J. Geriatr.*, 12(5), 315, 1975.

Badzio, T. et al., Staining of electrophoretic patterns suited for lipoprotein quantification (in Polish), *Przegl. Lek.*, 33(4), 436, 1976.

Cham, B. E. et al., Changes in electrophoretic mobilities of α- and β-lipoproteins as a result of plasma delipidation, *Clin. Chem.*, 22(3), 305, 1976.

Zadak, Z., Clinical use of various electrophoretic methods for the determination of lipoproteins (in Czech), *Cas. Lek. Cesk.*, 115(11), 326, 1976.

Wada, M., A study of the midband lipoproteins-their electrohoretic appearances and occurrences (in Japanese), *Jpn. J. Clin. Pathol.*, 24(5), 406, 1976.

Wada, M. et al., Repeatability of the electrophoretic profiles of serum lipoproteins in hyperlipidemic subjects (in Japanese), *Jpn. J. Clin. Pathol.*, 24(7), 581, 1976.

Crossed Electrophoresis
Wilkinson, P. A. et al., The analysis of serum lipoproteins by crossed electrophoresis, *Biochem. J.*, 126, 28P, 1972.

IMMUNOELECTROPHORESIS OF LIPOPROTEINS

Rabinovitz, M. et al., The use of zinctetracycline as a fluorescent stain for the serum lipoproteins in immunoelectrophoresis, *Clin. Chim. Acta*, 12, 474, 1965.
Lippi, N. et al., Immunoelectrophoretic and electrophoretic distribution of serum lipoproteins demonstrated with a new staining technic (in Italian), *Boll. Soc. Ital. Biol. Sper.*, 43, 165, 1967.
Burstein, M., Serum β-lipoproteins in reversed immunoelectrophoresis (in French), *Rev. Fr. Etud. Clin. Biol.*, 14, 918, 1969.
Balta, N. et al., Immunoelectrophoresis of serum lipoproteins in patients with chronic hepatopathies, *Fiziol. Norm. Patol.*, 17, 533, 1971.
Tomita, S., Immunological analysis of lipoproteins (in Japanese), *Saishin Igaku.*, 27, 447, 1972.
Badin, J. et al., Elimination of lipoproteins, α_1-antitrypsin and complement activity of human serum after agitation with iso-amyl alcohol. A study with special reference to bidimensional immunoelectrophoresis, *Ann. Biol. Clin. (Paris)*, 32(6), 533, 1974.
Kawaide, M. et al., Analysis of β-lipoproteins by rocket immunoelectrophoresis, *Jpn. J. Clin. Pathol.*, 22(10), 140, 1974.
Ito, N. et al., Automatic analysis of serum proteins by immunological reactions. 1. Determination of β-lipoproteins, *Jpn. J. Clin. Pathol.*, 22(10), 141, 1974.
Mekhtiev, M. A. et al., Immunoelectrophoretic picture of the serum lipoproteins under the influence of sodium selenite of various dosages, *Selen v Biologii*, Gasanov, G. G., ed., Elm, 1974, 81.
Salmon, S. et al., Two-dimensional immunoelectrophoresis of spontaneous dissociation of low density lipoproteins (in French), *Biochemie*, 57(10), 1155, 1975.

ELECTROFOCUSING OF LIPOPROTEINS

Kostner, G. et al., Analytical isoelectric focusing of human serum lipoproteins, *Hoppe Seylers Z. Physiol. Chem.*, 350, 1347, 1969.
Rittner, Ch. et al., Variations in human serum lipoprotein detected by isoelectric focusing, *Z. Klin. Chem. Klin. Biochem.*, 9(6), 503, 1971.
Scanu, A. M. et al., Application of the technique or isoelectric focusing to the study of human serum lipoproteins and their apoproteins, *Ann. N.Y. Acad. Sci.*, 209, 311, 1973.
Karmanskii, I. M., Isoelectric focusing of low-density serum lipoproteins, *Vopr. Med. Khim.*, 21(1), 60, 1975.
Stinson, R. A., Protein staining and pH gradient determination on the same gel in isoelectric focusing, *Anal. Biochem.*, 69(1), 278, 1975.
Emes, A. V. et al., The separation of plasma lipoproteins using gel electrofocusing and polyacrylamide gradient gel electrophoresis, *Clin. Chim. Acta*, 71(2), 293, 1976.

LIPOPROTEINEMIA

Newborns
Javierre, M. Q. et al., Experience at the Instituto de Puericultura de Universidad do Brasil with serum proteins and lipoproteins in infancy. II. Electrophoretic study of serum proteins and lipoproteins of the umbilical cord of premature newborns and their respective mothers (in Portugese), *Bol. Inst. Pueric. Pediatr.*, (Rio de J.), 21, 99, 1964.
Reis, S. I. et al., Experience at the Instituto de Puericultura de Universidad do Brasil with serum proteins and lipoproteins in infancy. III. Electrophoretic profile of serum proteins and lipoproteins in the umbilical cord of the normal newborn at term and the premature infant (in Portugese), *Bol. Inst. Pueric. Pediatr.*, (Rio de J.), 21, 119, 1964.
Dettori, M. et al., Metabolism of β-lipoproteins in the newborn: preliminary results obtained with the use of genetic markers (in Italian), *Atti. Accad. Fisiocrit. Siena Sez. Med. Fis.*, 14, 609, 1965.
Ginocchi, G. et al., The lipoprotein picture in the newborn (in Italian), *Osp. Magg.*, 60, 1013, 1965.

Ludyniány, K. et al., Serum lipids in the atrophic infant. I. Changes in circulating total lipids and lipoproteins during the period of weight increase (in German), *Z. Kinderheilk.*, 92, 299, 1965.

Thomas, K. et al., Studies on the hereditary β-lipoprotein system in umbilical cord blood sera (in German), *Klin. Wochenschr.*, 44, 1147, 1966.

Garlej, T., Patterns of reactions and protein fractions in the serum of mothers and newborns (in Polish), *Wiad. Lek.*, 20, 2201, 1967.

Clausen, J. et al., Changes in the liquid pattern of human sera in the neonatal period, *Acta Paediatr. Scand. Suppl.*, 206, 36, 1970.

Fris-Hansen, B. et al., Studies on serum lipoprotein in the neonatal period, *Z. Ernaehrungswiss.*, 10, 253, 1971.

Schlag, B. et al., Low density lipoproteins (LDL) in the human umbilical cord blood, *Acta Biol. Med. Ger.*, 29, 527, 1972.

Glueck, C. J. et al., Cord-blood low-density lipoprotein cholesterol: estimation versus measurement with the preparative ultracentrifuge, *J. Lab. Clin. Med.*, 82, 467, 1973.

Schettler, G., Neonatal familial type II hyperlipoproteinemia: cord blood cholesterol and follow up study in 1,323 births, *Singapore Med. J.*, 14, 334, 1973.

Ainadzhian, E. K., Protein-lipid-glycoprotein content of the blood in newborn infants and their mothers in normal and premature states and in some pathological conditions (in Russian), *Zh. Eksp. Klin. Med.*, 14(3), 55, 1974.

Csákó, G. et al., Lioprotein fractions in maternal, cord and newborn serum, *Acta Paediatr. Acad. Sci. Hung.*, 15(2), 101, 1974.

Dyerberg, J. et al., Reference values for cord blood lipid and lipoprotein concentrations, *Acta Paediatr. Scand.*, 63, 431, 1974.

Greten, H. et al., Early diagnosis and incidence of familial type II hyperlipoproteinaemia: analysis of umbilical-cord blood from 1,323 newborns (English abstr.), *Dtsch. Med. Wochenschr.*, 99(50), 2553, 1974.

Ose, L., Letter: L.D.L. and total cholesterol in cord-blood screening for familial hypercholesterolaemia, *Lancet*, 2(7935), 615, 1975.

Andersen, G. E. et al., Neonatal screening for hyperlipoproteinemia. Methods for direct estimation of cord review VLDL + LDL, *Clin. Chim. Acta*, 66(1), 29, 1976.

Andersen, G. E. et al., Neonatal diagnosis of familial type II hyperlipoproteinemia, *Pediatrics*, 57(2), 214, 1976.

Bartels, H. et al., Relations between lipoprotein X and γ-glutamyl-transferase in serum and D-saccharic acid in urine of mature, healthy newborn infants (in German), *Monatsschr. Kinderheilkd.* 124(5), 489, 1976.

Bobok, T. et al., Serum pre-β lipoprotein in the energy metabolism of newborn infants, (in German) *Kinderaerztl. Prax.*, 44(9), 395, 1976.

Brock, D. J., Protein measurements in the early prenatal diagnosis of spina bifida, *Hum. Hered.*, 26(6), 401, 1976.

de Tejada, A. et al., Hyperlipoproteinemia in children. Correlation between changes in the parents and newborn infant (in Spanish with English abstr.), *Prensa Med. Mex.*, 41(7—8), 226, 1976.

Glueck, C. J. et al., Neonatal familial hypo-β-lipoproteinemia, *Metabolism*, 25(6), 611, 1976.

Grundt, J. et al., Cord blood cholesterol, triglyceride, and lipoprotein pattern from two districts in Norway, *Scand. J. Clin. Lab. Invest.*, 36(3), 261, 1976.

Hernández, A. et al., Hyperlipidemias and abnormalities in the electrophoresis of lipoproteins. Incidence in newborn infants (in Spanish with English abstr.), *Prensa Med. Mex.*, 41(5—6), 163, 1976.

Lerdo de Tejada, A. et al., Serum lipids and electrophoresis of lipoproteins in the normal newborn infant (in Spanish with English abstr.), *Prensa Med. Mex.*, 41(3—4), 95, 1976.

Witt, I. et al., Lp-X in newborns: increased incidence of positive tests without cholestasis (author's transl.) (in German), *J. Clin. Chem. Clin Biochem.*, 14(4), 197, 1976.

Ginsburg, B. E. et al., High density lipoprotein concentrations in newborn infants, *Acta Paediatr. Scand.*, 66(1), 39, 1977.

Children

Opplt, J. J. and Blehova, B., On the results of metabolical changes in gargoylism, *Cas. Lek. Cesk.*, 97(31/32), 987, 1958.

Vierucci, A. et al., Anti-β-lipoprotein antibodies genetically determined in 2 children having received multiple transfusions in Cooley's disease (in Italian), *Boll. Soc. Ital. Biol. Sper.*, 41, 124, 1965.

Gavalov, S. M. et al., On the status of the protein fractions, glycoproteins and lipoproteins of the blood serum in chronic nonspecific pneumonia in children and their modification by sanatorium-health resort therapy (in Russian), *Zh. Eksp. Klin. Med.*, 6, 19, 1966.

Ledvina, M., The relation of body weight to the serum β-lipoprotein content in children (in German), *Sb. Ved. Pr. Lek. Fak. Karlovy Univ.*, 9, 329, 1966.

Mikolaeva, T. G., Total protein, protein, glycoprotein and lipoprotein blood levels in children with sepsis (in Russian), *Vopr. Okhr. Materin. Det.*, 11, 28, 1966.

Nigro, N. et al., Lipoproteins and diet in the healthy infant and the infant with infectious diseases (in Italian), *Minerva Pediatr.*, 18, 749, 1966.

Vierucci, A. et al., New anti-β-lipoprotein sera in transfused children with thalassemia, *Vox Sang.*, 11, 427, 1966.

Crosato, M., Physiopathology of serum protein conjugates in children (in Italian), *Minerva Pediatr.*, 19, 1321, 1967.

Lloyd, J. K., Primary disorders of lipoprotein metabolism in childhood, *Postgrad. Med. J.*, 43, 691, 1967.

Prolp'eva, G. M., Clinico-biochemical parallels in children with chronic nonspecific pneumonia (in Russian), *Pediatriia*, 46, 50, 1967.

Snezhkova, M. N., The clinical significance of determining individual serum lipoprotein fractions in blood serum of children with dysentery (in Russian), *Pediatriia*, 46, 14, 1967.

Wolff, O. H., Primary disorders of the serum lipoproteins in childhood, *Pediatrics*, 40, 1, 1967.

Jezerniczky, J. et al., Serum lipoid level in diabetic children, *Acta Paediatr. Acad. Sci. Hung.*, 9, 219, 1968.

Kniazevskaia, E. G. et al., Levels of cholesterol, phospholipids and lipoproteins in the blood of children with diabetes mellitus (in Russian), *Pediatriia*, 47, 18, 1968.

Alferov, V. P., Clinical significance of changes in lipoproteins and glucoproteins of the blood serum in kidney diseases in children (in Russian), *Pediatriia*, 48, 73, 1969.

Marcondes, E. et al., Behavior of blood lipoproteins in undernourished children and during their recovery (in Portugese), *Rev. Hosp. Clin. Fac. Med. Univ. Sao Paulo*, 24, 99, 1969.

Rudobielska, M. et al., Total lipids, lipoproteins, phospholipoproteins, total protein and its fractions in the serum of healthy children in various age groups (in Polish), *Pediatr. Pol.*, 45, 791, 1970.

Mardomingo Varela, P. et al., Primary infantile familial hyperlipemia (in Spanish), *Rev. Clin. Esp.*, 122, 353, 1971.

Pietraszun, L. et al., Behavior of whole fats, cholesterol and lipoprotein in blood serum in children with pigmentary retinal degeneration of different types (in Polish with English abstr.), *Klin. Oczna*, 41, 383, 1971.

Pietraszun, L. et al., Serum total lipids, cholesterol and lipoprotein level in children with various types of retinal pigmentary degeneration, *Pol. Med. J.*, 10, 1236, 1971.

Tamir, J., Primary disorders of serum lipoproteins in children (in Hebrew), *Harefuah*, 80, 507, 1971.

Vierucci, A. et al., Australia antigen, α-fetoprotein and lipoprotein X in some diseases in children, *Quad. Sclavo Diagn. Clin. Lab.*, 7, 829, 1971.

Bagnall, T. F., Composition of low density lipoprotein in children with familial hyperbetalipoproteinaemia and the effect of treatment, *Clin. Chim. Acta*, 42, 229, 1972.

Brown, R. E. et al., Lipid and lipoprotein studies in Reye's syndrome, *Va. Med. Mon.*, 99, 622, 1972.

Lloyd, J. K., Hyperlipoproteinaemia in childhood, *Aust. Paediatr. J.*, 8, 264, 1972.

Tamir, I. et al., Serum lipids and lipoproteins in children from families with early coronary heart disease, *Arch. Dis. Child.*, 47, 808, 1972.

Awwaad, S. et al., Changes in the blood volume, serum proteins and lipoproteins in common parasitic infestations in Egyptian children, *J. Trop. Med. Hyg.*, 76, 163, 1973.

Dyerberg, J. et al., Plasma lipid and lipoprotein levels in childhood and adolescence, *Scand. J. Clin. Lab. Invest.*, 31, 473, 1973.

Fredrickson, D. S. et al., Primary hyperlipoproteinemia in infants, *Annu. Rev. Med.*, 24, 315, 1973.

Glueck, C. J. et. al., Pediatric familial type II hyperlipoproteinemia: therapy with diet and cholestyramine resin, *Pediatrics*, 52, 669, 1973.

Adriaenssens, K., Letter: lipoprotein values in neonates, *J. Pediatr.*, 85(4), 587, 1974.

Berenson, G. S. et al., Studies of serum lipoprotein concentrations in children: preliminary report, *Clin. Chim. Acta*, 56(1), 65, 1974.

Cress, H. R. et al., Lipoproteins in neonates, *J. Pediatr.*, 84, 585, 1974.

Donde, S. M., Lipoprotein content in the bile in chronic diseases of the biliary tracts in children (in Ukranian), *Pediatr. Akush. Ginekol.*, 2, 18, 1974.

Gaburro, D. et al., Hyperlipoproteinemias in childhood, *Minerva Pediatr.*, 26(14), 743, 1974.

Glueck, C. J. et al., Plasma vitamin A and E levels in children with familial type II hyperlipoproteinemia during therapy with diet and cholestyramine resin, *Pediatrics*, 54, 51, 1974.

Mellies, M. et al., Familial and acquired hyperlipoproteinemias in children and adolescents, *Postgrad. Med.*, 56(6), 94, 1974.

Müller, W. D. et al., Influence of fat or carbohydrate rich diet on the serum lipoprotein pattern of a child with primary hypertriglyceridemia (type V hyperlipoproteinemia according to Frederickson) (in German with English abstr.), *Wien. Klin. Wochenschr.*, 86(24), 757, 1974.

Stein, E. A. et al., Familial type II hyperlipoproteinaemia. Clinical features and results of treatment in children and young adults, *S. Afr. Med. J.*, 48, 135, 1974.

Zaverbnii, M. I. et al., Blood lipid and beta-liprotein shifts in acute digestive disorders in young children, *Pediatr. Akush. Ginekol.*, 4, 19, 1974.

Bagnall, T. F. et al., Composition of low-density lipoprotein in children with hyperlipoproteinaemia, *Clin. Chim. Acta*, 59(3), 271, 1975.

Court, J. M. et al., Plasma lipid values and lipoprotein patterns during adolescence in boys, *J. Pediatr.*, 86(3), 453, 1975.

Srinivasan, S. R. et al., Serum lipid and lipoprotein profile in school children from a rural community, *Clin. Chim. Acta*, 60(3), 293, 1975.

Benakappa, D. G. et al., Low density lipoprotein levels in children with nephrotic syndrome, *Indian Pediatr.*, 13(4), 287, 1976.

Gordon, V. H. et al., Relationships of foreign protein injections (hyposensitization) in atopic children to serum lipids and lipoproteins, *Ann. Allergy*, 36(1), 1, 1976.

James, F. W. et al., Maximal exercise stress testing in normal and hyperlipidemic children, *Atherosclerosis*, 25(1), 85, 1976.

Leonard, J. V. et al., Screening for familial hyper-β-lipoproteinemia in children in hospital, *Arch. Dis. Child.*, 51(11), 842, 1976.

Myant, N. B., Changes of plasma lipoproteins in pediatrics (in Italian with English abstr.), *Recenti Prog. Med.*, 62(1), 77, 1976.

Scrinivasan, S. R. et al., Serum lipoprotein profile in children from a biracial community, the Bogalusa heart study, *Circulation*, 54(2), 309, 1976.

Tamir, I. et al., Early detection of lipoprotein abnormalities in children, *Paroi Arterielle*, 3(3), 137, 1976.

Adults

Kirkeby, K., Blood lipids, lipoproteins, and proteins in vegetarians, *Acta Med. Scand. Suppl.*, 443, 1, 1966.

Allen, M. F. et al., Patterns of Change in Serum Lipid and Lipoprotein Levels in a Selected Military Population over a 14-year Period, SAM-TR-67-101, U.S. Air Force School of Aerospace Medicine, 1967, 1—12.

Nguyen van Hung, Lp frequency in Vietnam (in German), *Z. Aerztl. Fortbild.*, 61, 784, 1967.

Das, B. C., Age — related trends of the amylase, glycoprotein, lipoprotein and serum protein in human blood, *Exp. Gerontol.*, 3, 159, 1968.

Ogawa, Y. et al., Normal values of serum lipoprotein fractions on separax in Japanese adults (in Japanese), *Med. Biol.*, 76, 283, 1968.

Sgambato, S. et al., Some metabolic aspects of senescence (in Italian), *G. Gerontol.*, 16, 837, 1968.

Herrmann, G. R., Hyperlipoproteinemias in the prematurely aged. Theories of pathogenesis, diagnosis, classification and treatment, *Geriatrics*, 25, 103, 1970.

Atanasov, K., Cholesterol, β-lipoprotein and blood protein levels in the aged (in Russian with English abstr.), *Folia Med.*, (Plovdiv), 13, 158, 1971.

Bertolini, A. M., Changes in plasma lipids in relation to age and sex in normal subjects. I. Total lipids, triglycerides, phospholipids, single NEFA (C_{16} C_{18}) α-β lipoproteic cholesterol in young adults (in Italian with English abstr.), *G. Gerontol.*, 19, 558, 1971.

Bertolini, A. M. et al., Changes of plasma lipids in normal subjects in relation to age and sex. II. Total lipids, triglycerides, phospholipids, α and β lipoproteic cholesterol and single NEFA (C_{16}-C_{18}) in aged persons, *G. Gerontol.*, 19, 657, 1971.

Bertolini, A. M. et al., Changes in plasma lipids in normal subjects in relation to age and sex. III. Total lipids, triglycerides, phospholipids, α and β lipoproteic cholesterol and single NEFA (C_{16}-C_{18}) in the presenile subjects (in Italian with English abstr.), *G. Gerontol.*, 20, 217, 1972.

Jubb, J. S. et al., Lipoprotein levels in apparently healthy men and women in the west of Scotland, *Br. Heart J.*, 38(11), 1189, 1976.

Walton, K. W. et al., A study of the Age factors in a British (West Midland) population., *Vox Sang.*, 31(4), 358, 1976.

Wood, P. D. et al., The distribution of plasma lipoproteins in middle-aged male runners, *Metabolism*, 25(11), 1249, 1976.

HYPERLIPOPROTEINEMIA

Hyperlipemia, Hyperlipidemia

Cattaneo, C. et al., Blood lipid pattern with gas chromatographic analysis of fatty acids in subjects with familial constitutional hyperlipemia (in Italian), *Boll. Soc. Ital. Biol. Sper.*, 40, 666, 1964.

Sen, S. K. et al., Studies on lipaemia., *J. Indian Med. Assoc.*, 43, 522, 1964.

Beaumont, J. L., Hyperlipidemia caused by anti-β-lipoprotein autoantibodies. A new pathological entity (in French), *C. R. Acad. Sci.*, 261, 4563, 1965.

Furman, R. H. et al., Effects of medium chain length triglyceride (MCT) on serum lipid and lipoproteins in familial hyperchylomicronemia (dietary fat-induced lipemia) and dietary carbohydrate-accentuated lipemia, *J. Lab. Clin. Med.,* 66, 912, 1965.

Lees, R. S. et al., The differentiation of exogenous and endogenous hyperlipemia by paper electrophoresis, *J. Clin. Invest.,* 44, 1968, 1965.

Laudat, P. et al., Demonstration of the heterogeneity of "hyperlipemia" by identification of specific lipid particles (in French), *Presse Med.,* 74, 1361, 1966.

Norum, K. R., Hyperlipidemia. A review on lipoprotein physiology and pathophysiology (in Norwegian), *J. Norsk. Laegeforen,* 87, 1042, 1967.

Dyerberg, J., Lipoprotein pattern in hyperlipidemia (in Danish), *Ugeskr. Laeg.,* 130, 359, 1968.

Strisower, E. H. et al., Treatment of hyperlipidemias, *Am. J. Med.,* 45, 488, 1968.

Scheiffarth, F. et al., On the problem of qualitative changes in lipoproteins in experimental hyperlipemia (in German), *Z. Klin. Chem.,* 6, 89, 1968.

Buckley, G. C. et al., A comparison of laboratory screening procedures for hyperlipidemia, *Clin. Chim. Acta,* 23, 53, 1969.

Harlan, W. R., Jr., A nomogram for determining types of hyperlipidemia., *Arch. Intern, Med.,* 124, 64, 1969.

Levy, R. I. et al., Mechanism involved in hyperlipidemia., *Mod. Treat.,* 6, 1313, 1969.

Märki, H. H., Hyperlipidemia (in Spanish), *Folia Clin. Int.,* 19, 439, 1969.

Nosek, J., Lipemia in the blood of miners from uranium mines observed by laboratory screening (in Czech), *Cas. Lek. Cesk.,* 108, 275, 1969.

Dyerberg, J. et al., Treatment of essential hyperlipidaemia, *Lancet,* 1, 422, 1970.

Knüchel, F., Diagnosis of hyperlipidemia in the general practice by simplified-lipoprotein determination, *Med. Welt,* 18, 849, 1970.

Loeper, J. et al., Mixed hyperlipemia with a distinct accumulation of β- and pre-β-lipoprotein (in French with English abstr.), *Arch. Mal Coeur,* 12(Suppl. 1), 3, 1970—1971.

Beaumont, J. L., The phenomenon of agglutination of lipid particles in hyperlipidemia by autoantibodies. Test of agglutination of activated lipid amulsions (AALE) (in French), *C. R. Acad. Sci. D,* 272, 2404, 1971.

Bélanger, M. et al., Biochemical diagnosis of hyperlipidemia: new method of serum lipoprotein fractionation and results obtained in agreement with Fredrickson's classification (in French), *Laval Med.,* 42, 249, 1971.

Fredrickson, D. S., An international classification of hyperlipidemias and hyperlipoproteinemias, *Ann. Intern. Med.,* 75, 471, 1971.

Jain, R. C., Serum cholesterol, lipoproteins, coagulability and fibrinolytic activity of blood in alimentary lipemia, *Indian J. Med. Sci.,* 25, 236, 1971.

LeMaire, A., Classification of hyperlipemia: critical study (in French), *Presse Med.,* 79, 789, 1971.

Loeper, J. et al., On mixed hyperlipemia with distinct accumulation of β-and pre-β-lipoproteins. Apropos of 78 cases (in Italian), *G. Clin. Med.* (Bologna), 52, 277, 1971.

Loeper, J. et al., Study of mixed hyperlipemia with a distinct excess of β and pre-β-lipoproteins (78 cases) (in French), *Presse Med.,* 79, 797, 1971.

Loeper, J. et al., Mixed hyperlipemia (apropos of 231 cases) (in French), *Ann. Med. Interne,* 122, 551, 1971.

Batko, B., 3 cases of idiopathic hyperlipidemia (in Polish with English abstr.), *Wiad. Lek.,* 25, 1055, 1972.

Berenson, G. S. et al., Simplified primary screening procedure for detection of hyperlipidemias in healthy individuals, *Clin. Chem.,* 18, 1463, 1972.

Belanger, M. et al., Biochemical diagnosis of hyperlipidemia. Serum lipoprotein separation and classification according to Fredrickson's classification, *Med. Chir. Dig.,* 2, 149, 1973.

Halpern, M. J. et al., Letter: 2 little-known forms of hyperlipidemia, *Nouv. Presse Med.,* 2, 3121, 1973.

Lewis, B., Hyperlipemia in a general hospital — a philosophy of classification, *Adv. Exp. Med. Biol.,* 38, 105, 1973.

Welle, H. F. et al., The importance of typing for the treatment of hyperlipidemia (in Dutch with English abstr.), *Ned. Tijdschr. Geneeskd.,* 117, 291, 1973.

Kuzuya, F. et al., Symposium on hyperlipidemia — epidemiological, clinical and experimental studies. Epidemiology of hyperlipoproteinemia in Japanese (in Japanese), *J. Jpn. Soc. Intern. Med.,* 63(10), 1139, 1974.

Lubetzki, J. et al., Hyperlipemias, *Cah. Med.,* 15(4), 173, 1974.

Sekimoto, H. et al., Symposium on hyperlipidemia — epidemiological, clinical and experimental studies. On primary familial combined hyperlipoproteinemia., *J. Jpn. Soc. Intern. Med.,* 63(10), 1145, 1974.

Yoshida, E., Lipoprotein fractions in hyperlipemia, *Jpn. J. Clin. Pathol.,* 22(10), 144, 1974.

Chong, Y. H. et al., Serum lipid levels and the prevalence of hyperlipidaemia in Malaysia, *Clin. Chim. Acta,* 65(1), 143, 1975.

Olsson, A. G., Studies in asymptomatic primary hyperlipidaemia. Clinical, biochemical and physiological investigations., *Acta Med. Scand. Suppl.,* 581, 1, 1975.

Olsson, A. G. et al., Studies in asymptomatic primary hyperlipidaemia. I. Types of hyperlipoproteinemias and serum lipoprotein concentrations, compositions and interrelations, *Acta Med. Scand. Suppl.*, 580, 1, 1975.

Olsson, A. G. et al., Studies in asymptomatic primary hyperlipidaemia. IV. ECG at rest and during exercise and its relation to various lipoprotein classes, *Acta Med. Scand.*, 198(1—2), 55, 1975.

Olsson, A. G. et al., Studies in asymptomatic primary hyperlipidaemia. V. Peripheral circulation, *Acta Med. Scand.*, 198(3), 197, 1975.

Hyperlipoproteinemia

Fredrickson, D. S. et al., A system for phenotyping hyperlipoproteinemia, *Circulation*, 31, 321, 1965.

Kahlke, W., Differentiation of essential familial hyperlipoproteinemias (hyperlipemias) (in German), *Dtsch. Med. Wochenschr.*, 91, 26, 1966.

Opplt, J. J., Lipoproteide in Blutplasma and die Semiologie ihrer Fractionen, Lipid Research. Plzeň, Czechoslovakia, 1965; published in *Lipid Res.*, 16, 194, 1966.

Sanbar, S. S. et al., Familial hyperlipoproteinemic syndromes, *Univ. Mich. Med. Cent. J.*, 32, 277, 1966.

Baes, H. et al., Essential hyperlipoproteinemia (in French), *Arch. Belg. Dermatol. Syphilig.*, 23, 163, 1967.

Ciświcka-Sznajderman, M., Studies on primary hyperlipoproteinemias, I. Clinical symptoms and serum lipid level (in Polish), *Pol. Arch. Med. Wewn.*, 39, 267, 1967.

Ciświcka-Sznajderman, M., Investigations on primary hyperlipoproteinemias I. Clinical symptoms and serum lipid level, *Pol. Med. J.*, 7, 524, 1968.

Fallon, H. J. et al., Hyperlipoproteinemia — a difference of opinion, *JAMA*, 206, 376, 1968.

Levy, R. I. et al., Diagnosis and management of hyperlipoproteinemia, *Am. J. Cardiol.*, 22, 576, 1968.

Norum, K. R., Carbohydrates and hyperlipoproteinemia (in Norwegian), *Nord. Med.*, 79, 466, 1968.

Pries, C. et al., Primary hyperlipoproteinemia, the clinico-chemical classification of the most common types, *Clin. Chim. Acta*, 19, 181, 1968.

Casdorph, H. R., Hyperlipoproteinemia, *JAMA*, 207, 151, 1969.

Cross, D. F., Recurrent pancreatitis and fat-induced hyperlipoproteinemia, *JAMA*, 208, 1494, 1969.

Fleischmajer, R., Familial hyperlipoproteinemia (in Portugese), *An. Bras. Dermatol.*, 44, 261, 1969.

Galazka, A. et al., Primary hyperlipoproteinemia (in Polish), *Przegl. Lek.*, 25, 430, 1969.

Greten, H., Diagnosis and differentiation of hyperlipoproteinemias, *Klin. Wochenschr.*, 47, 893, 1969.

Havel, R. J., Diagnosis of hyperlipoproteinemias, *Calif. Med.*, 110, 519, 1969.

Loeper, J. et al., Significance of the lipidogram (Lees-Hatch technic) associated with the determination of serum lipid fractions for the detection and classification of hyperlipoproteinemias (in French), *Presse Med.*, 77, 171, 1969.

Loeper, J. et al., Essential hyperlipoproteinemia and ischemic manifestations in young patients, *Ann. Med. Interne*, 120, 253, 1969.

Parsons, W. B., Jr., Hyperlipoproteinemia, *JAMA*, 209, 113, 1969.

Wessler, S. et al., Classification and management of familial hyperlipoproteinemia., *JAMA*, 207, 929, 1969.

Antar, M. A. et al., Interrelationship between the kinds of dietary carbohydrate and fat in hyperlipoproteinemic patients. 3. Synergistic effect of sucrose and animal fat on serum lipids, *Atherosclerosis*, 11, 191, 1970.

Beaumont, J. L. et al., Classification of hyperlipidaemias and hyperlipoproteinaemias, *Bull. WHO*, 43, 891, 1970.

Christensen, N. C. et al., Lipoprotein electrophoresis pattern in two families with hyperlipoproteinemia, *Nord. Med.*, 84, 998, 1970.

Ciświcka-Sznajderman, M., Primary hyperlipoproteinemia (in Polish), *Kardiol. Pol.*, 13, 173, 1970.

Huth, K., Classification of the hyperlipoproteinemias (in German), *Med. Klin.*, 65, 637, 1970.

Levene, D. L., The hyperlipoproteinaemias, *Postgrad. Med. J.*, 46, 713, 1970.

Magnenat, G., Lipids and hyperlipoproteinemia (in French), *Ther. Umsch.*, 27, 566, 1970.

Popović, S. S. et al., Clinical aspect and therapy of primary and secondary hyperlipoproteinemias (in Croatian), *Med. Glas.*, 24, 315, 1970.

Rifkind, B. M., The familial hyperlipoproteinaemias, *Scott. Med. J.*, 15, 223, 1970.

Scanu, A. M., Practical diagnostic aids in the study of dyslipoproteinemias, *Med. Clin. North Am.*, 54, 153, 1970.

Shanoff, H. M., Parotid enlargement and hyperlipoproteinemia, *JAMA*, 211, 2016, 1970.

Sobra, J., Inborn errors of lipid metabolism. 18. Differential diagnosis of familial hyperlipoproteinemias, *Cas. Lek. Cesk.*, 109, 69, 1970.

Isaacson, N. H. et al., Primary hyperlipoproteinemia — a medical and surgical problem. Report of a case, *Med. Ann. D. C.*, 39, 143, 1970.

Zelis, R. et al., Effects of hyperlipoproteinemias and their treatment on the peripheral circulation, *J. Clin. Invest.*, 49, 1007, 1970.

Zelis, R. et al., The hyperlipoproteinemias. A simplified classification and approach to therapy, *Calif. Med.,* 112, 32, 1970.

Billimoria, J. D. et al., The use of the esterified fatty acid index in the classification and quantitation of hyperlipoproteinaemias, *Atherosclerosis,* 14, 359, 1971.

Carter, C. O. et al., Genetics of hyperlipoproteinaemias, *Lancet,* 1, 400, 1971.

Ciświcka-Sznajderman, M., Studies on primary hyperlipoproteinemia. II. Diagnosis and classification (in Polish), *Pol. Arch. Med. Wewn.,* 46, 571, 1971.

Ciświcka-Sznajderman, M., Studies on primary hyperlipoproteinemia. 3. Disturbances of carbohydrate metabolism (in Polish), *Pol. Arch. Med. Wewn.,* 46, 577, 1971.

Frederickson, D. S., An international classification of hyperlipidemias and hyperlipoproteinemias, *Ann. Intern. Med.,* 75, 471, 1971.

Herrmann, G. R., Hyperlipoproteinemias, recognition, classification, clinical features and treatment., *Rev. Esp. Cardiol.,* 24, 153, 1971.

Kinalska, J. et al., Case of idiopathic familial hyperlipoproteinemia (in Polish), *Pol. Arch. Med. Wewn.,* 47, 693, 1971.

Levy, R. I., Hyperlipoproteinaemias, *Clin. Sci.,* 40, 15P, 1971.

Levy, R. I., Classification and etiology of hyperlipoproteinemias, *Fed. Proc.,* 30, 829, 1971.

Patterson, D. et al., Hyperlipoproteinaemias, *Lancet,* 2, 806, 1971.

Popescu, P. N. et al., Blood lipid disorder according to modern laboratory and clinical concepts (hyperlipoproteinemia) (in Rumanian), *Med. Interna,* 23, 293, 1971.

Roggenbach, H. J. et al., Correction of hyperlipoproteinaemia, *Lancet,* 1, 276, 1971.

Schwandt, P., Hyperlipoproteinemias: significance — diagnosis — therapy, *Internist,* 12, 481, 1971.

Stone, M. C. et al., Diagnosis and classification of abnormal lipoprotein patterns, *Clin. Chim. Acta,* 31, 333, 1971.

Beaumont, J. L. et al., Classification of hyperlipidemias and hyperlipoproteinemias (in Portugese), *Arq. Bras. Cardiol.,* 25, 97, 1972.

Beaumont, J. L. et al., Classification and hyperlipidemias and hyperlipoproteinemias. II (in Portugese with English abstr.), *Arq. Bras. Cardiol.,* 25, 189, 1972.

Betro, M. G., The hyperlipoproteinaemias: a review, *N. Z. Med. J.,* 75, 131, 1972.

Bron, A. J. et al., Hereditary crystalline stromal dystrophy of Schnyder. 1. Clinical features of a family with hyperlipoproteinaemia, *Br. J. Ophthalmol.,* 56, 383, 1972.

Cancio, M. et al., Hyperlipoproteinemic types among Puerto Ricans: a progress report, *Bol. Asoc. Med. P. R.,* 64, 111, 1972.

Carruthers, M. E., Hyperlipoproteinaemia classification, *Lancet,* 1, 797, 1972.

Casani, C., Primary hyperlipoproteinemias, *G. Clin. Med.,* 53, 553, 1972.

Ciświcka-Sznajderman, M., Studies of primary hyperlipoproteinemias. II. Diagnosis and classification., *Pol. Med. J.,* 11, 238, 1972.

Ciświcka-Sznajderman, M., Studies of primary hyperlipoproteinemias. III. Disturbances in the carbohydrate metabolism, *Pol. Med. J.,* 11, 244, 1972.

Galazka, A. et al., Idiopathic hyperlipoproteinemia (in Polish with English abstr.), *Wiad. Lek.,* 25, 1177, 1972.

Havel, R. J., Mechanisms of hyperlipoproteinemia, *Adv. Exp. Med. Biol.,* 26(0), 57, 1972.

Huth, K. et al., Incidence and distribution of hyperlipoproteinemias (in German), *Verh. Dtsch. Ges. Inn. Med.,* 78, 1337, 1972.

Huth, K. et al., Differential diagnosis of hyperlipoproteinemias (in German), *Z. Alternsforsch.,* 25, 239, 1972.

Klör, U. et al., Incidence of hyperlipoproteinemias in patients of the Ulm University Hospital., *Verh. Dtsch. Ges. Inn. Med.,* 78, 1349, 1972.

Kovacs, K. et al., Ultrastructural changes of hepatocytes in hyperlipoproteinaemia, *Lancet,* 1, 752, 1972.

La Rosa, J. C., Hyperlipoproteinemia. I. Diagnosis and chemical significance, *Postgrad. Med.,* 51, 62, 1972.

Madejska, M. et al., Idiopathic familial hyperlipoproteinemia (in Polish with English abstr.), *Wiad. Lek.,* 25, 913, 1972.

Mincu, J. et al., Epidemiological study of disorders of blood lipid equilibrium in the urban population (study of 7085 persons) (in Rumanian with English abstr.), *Med. Interna,* 24, 583, 1972.

Mombelloni, P. et al., Classification of hyperlipoproteinemias in hospitalized patients, *G. Clin. Med.* (Bologna), 53, 121, 1972.

Murakami, N. et al., Familial hyperlipoproteinemia (in Japanese), *Saishin Igaku,* 27, 498, 1972.

Scheig, R., Clues to the diagnosis of hyperlipoproteinemia, *Am. Fam. Physician,* 6, 82, 1972.

Schlierf, G. et al., Incidence and type distribution of hyperlipoproteinemias in hospitalized patients of a university hospital (in German with English abstr.), *Dtsch. Med. Wochenschr.,* 97, 1371, 1972.

Searcy. R. et al., A screening procedure for detecting and characterizing hyperlipoproteinemia, *Clin. Chim. Acta*, 38, 291, 1972.

Steinber, D., Some observations on hyperlipoproteinemias and their classification, *Circulation*, 45, 247, 1972.

Warembourg, H. et al., Hyperlipoproteinemia. I. Application, apropos of 133 cases, of clinical and bio-chemical criteria of classification of hyperproteinemias, *Lille Med.*, 17, 701, 1972.

Wollenweber, J. et al., Frequency of primary and secondary hyperlipoproteinemias, *Verh. Dtsch. Ges. Inn. Med.*, 78, 1334, 1972.

Zambrowicz, K. et al., Criteria for classification of hyperlipoproteinemia (in Polish with English abstr.), *Pol. Tyg. Lek.*, 27, 299, 1972.

Anjilvel, L., Familial hyperlipoproteinemia in an isolated part of Newfoundland, *Can. Med. Assoc. J.*, 108, 60, 1973.

Anjilvel, L., Familial hyperlipoproteinemia in an isolated part of Newfoundland, *Can. Med. Assoc. J.*, 109, 894, 1973.

Bassett, D. R. et al., Upper gastrointestinal radiologic findings in hyperlipoproteinemic subjects, *Am. J. Clin. Nutr.*, 26, 1269, 1973.

Brown, D. F. et al., Hyperlipoproteinemia. Prevalence in a free-living population in Albany, New York, *Circulation*, 47, 558, 1973.

Ciświcka-Sznajderman, M. et al., Capillaroscopic changes in patients with primary hyperlipoproteinemia, *Pol. Tyg. Lek.*, 28, 809, 1973.

Dioguardi, N. et al., Introduction on human hyperlipoproteinemias, *Adv. Exp. Med. Biol.*, 38, 3, 1973.

Fabien, H. D. et al., Plasma cholesterol esterification in normal and hyperlipoproteinemic subjects, *Can. J. Biochem.*, 51, 550, 1973.

Glueck, C. J. et al., Familial combined hyperlipoproteinemia: studies in 91 adults and 95 children from 33 kindreds, *Metabolism*, 23, 1403, 1973.

González-Ramos, M. et al., Genetic aspects of hyperlipoproteinemias, *Gac. Med. Mex.*, 106, 321, 1973.

Gustafson, A. et al., Plasma lipoproteins and hyperlipoproteinemia, *Acta Med. Scand.*, 193, 369, 1973.

Heiberg, A., A comparative study of different electrophoretic techniques for classification of hereditary hyperlipoproteinaemias, *Clin. Genet.*, 4, 450, 1973.

Jones, R. J., The hyperlipoproteinemias. Detection, diagnosis, and management, *Med. Clin. North Am.*, 57, 47, 1973.

Lees, R. S. et al., The familial dyslipoproteinemias, *Prog. Med. Genet.*, 9, 237, 1973.

Levy, R. I., Hyperlipoproteinemia. Some basic concepts on diagnosis and management., *JAMA*, 226, 648, 1973.

Lopez, A. et al., Automated primary screening for hyperlipoproteinemias, *Clin. Biochem.*, 6, 117, 1973.

Myant, N. B. et al., Turnover of cholesteryl esters in plasma low-density and high-density lipoproteins in familial hyperbetalipoproteinemia, *Clin. Sci. Mol. Med.*, 45, 551, 1973.

Pribor, H. C., Hyperlipoproteinemia. Survey., *J. Med. Soc. N. J.*, 70, 399, 1973.

Rifkind, B. M., Lipoproteins and hyperlipoproteinaemia, *Clin. Endocrinol. Metabol.*, 2, 3, 1973.

Wollenweber, J. et al., Hyperlipoproteinemia in out-patients. Occurrence, type distribution and correlation to clinical findings, *Dtsch. Med. Wochenschr.*, 98, 463, 1973.

Bassett, D. R., Letter: abdominal pain in hyperlipoproteinemia, *N. Engl. J. Med.*, 290, 748, 1974.

Bybee, D. E. et al., The primary hyperlipoproteinemias, *J. Ky. Med. Assoc.*, 72(12), 658, 1974.

Castoldi, G. et al., Sea-blue histiocytosis in hyperlipoproteinemia (cytomorphological, clinical and patho-genic aspects), *Haematologica*, 59(1), 68, 1974.

Cobet, L., Classification of hyperlipoproteinemias, *Rev. Med. Chir., Soc. Med. Nat. Iasi*, 78, 297, 1974.

Lorimer, A. R. et al., Prevalence of hyperlipoproteinaemia in apparently healthy men, *Br. Heart J.*, 36, 192, 1974.

Müller, G., Pathogenesis of hyperlipoproteinemias, *Z. Gesamte Inn. Med. Ihre Grenzgeb.*, 29, 469, 1974.

Plamieniak, Z., Detection and differentiation of primary hyperlipoproteinemias, *Mater Med. Pol.*, 6(3), 227, 1974.

Polchronopoulou-Trichopo, A. et al., Letter: abdominal pain in hyperlipoproteinemia, *N. Engl. J. Med.*, 290, 229, 1974.

Schlierf, G., Letter: hyperlipoproteinemia, *Dtsch. Med. Wochenschr.*, 99, 373, 1974.

Schräpler, P. et al., Frequency of abnormal lipid values and of hyperlipoproteinaemia in-patients of a gen-eral hospital, *Dtsch. Med. Wochenschr.*, 99, 1682, 1974.

Zabel, R., Classification of hyperlipoproteinemias, *Dermatol. Monatsschr.*, 160(2), 134, 1974.

Zöller, H. et al., Hyperlipoproteinemia and fibrin stabilization using blood factor 13., *Med. Welt*, 33(34), 1303, 1974.

Cucuianu, M. et al., Serum pseudocholinesterase and ceruloplasmin in various types of hyperlipoproteine-mia, *Clin. Chim. Acta*, 59(1), 19, 1975.

Gustafson, A. et al., Biochemical typing of hyperlipoproteinemia. The development of a nomogram, *Clin. Chim. Acta*, 63(1), 91, 1975.

Lowry, L. D., Hyperlipoproteinemia: cause of sensorineural hearing loss? *Trans. Pa. Acad. Ophthalmol. Otolaryngol.*, 28(1), 56, 1975.

Rajagopal, G. et al., Familial hyperlipoproteinemia in a South Indian family, *J. Assoc. Physicians India*, 23(1), 9, 1975.

Seidel, D., Normal values and pattern of findings in primary and secondary hyperlipoproteinemias (in German), *Verh. Dtsch. Ges. Inn. Med.*, 81, 534, 1975.

Seidel, D., Proceedings: normal ranges and pattern of findings in primary and secondary hyperlipoproteinemia, *Münch. Med. Wochenschr.*, 117(20), 847, 1975.

Fisher, W. R. et al., The common hyperlipoproteinemias: an understanding of disease mechanisms and their control, *Ann. Intern. Med.*, 85(4), 497, 1975.

Titov, V. N. et al., Lipid composition of blood plasma lipoproteins in primary and secondary hyperlipoproteinemia (in Russian with English abstr.), *Ter. Arkh.*, 48(9), 66, 1976.

Waddel, C. C. et al., Inhibition of lymphoproliferation by hyperlipoproteinemic plasma, *J. Clin. Invest.*, 58(4), 950, 1976.

TYPES OF DYSLIPOPROTEINEMIAS

Type I

Maldonado, N. et al., Type I hyperlipoproteinemia, *Bol. Asoc. Med. P. R.*, 62, 301, 1970.

Weber, K. et al., Hyperlipoproteinemias I (in German), *Hautarzt*, 22, 517, 1971.

Ditschuneit, H. et al., Family studies in hyperlipoproteinemia type I, *Verh. Dtsch. Ges. Inn. Med.*, 78, 1339, 1972.

Prandota, A. et al., Primary type I hyperlipoproteinemia in an 8-year-old girl, *Pediatr. Pol.*, 48, 1011, 1973.

Düchting, M. et al., Essential hyperlipoproteinemia type I Grütz-Bürger diagnosed in a 4-week old infant, *Z. Kinderheilkd.*, 116, 213, 1974.

Ireton, C. L. et al., Case report: diagnosis and management of type I hyperlipoproteinemia, *J. Am. Diet. Assoc.*, 66(1), 42, 1975.

Type IIA

Miyamoto, H., The research in the diagnosis for hyper-beta-lipoproteinemia (in Japanese), *J. Jpn. Soc. Intern. Med.*, 54, 608, 1965.

Lee, G. B. et al., Type II hyperlipoproteinemia in mother and twins, *Circulation*, 39, 183, 1969.

Lemmens, H. et al., *Urticaria pigmentosa* and familial hyper-β-lipoproteinemia in uniovular triplets (in French), *Arch. Belg. Dermatol. Syphiligr.*, 26, 409, 1970.

Raphael, S. S. et al., Deaf mutism and type-II hyperlipoproteinaemia, *Lancet*, 1, 892, 1970.

Sobra, J. et al., Inborn errors of metabolism. XXI. Familial hyperlipoproteinemia type II — serum lipids in 113 patients (in Czech), *Cas. Lek. Cesk.*, 109, 988, 1970.

Sobra, J. et al., Inborn errors of lipid metabolism. XXII. Familial hyperlipoproteinemia type II — electrophoresis of blood proteins in 84 patients (in Czech), *Cas. Lek. Cesk.*, 109, 1173, 1970.

Barbolini, G. et al., Anatomo-clinical, histopathological and histochemical study of a case of type II hyperlipoproteinemia (familial hypercholesterolemic xanthomatosis), *Arch. De Vecchi Anat. Patol. Med. Clin.*, 57, 263, 1971.

Glueck, C. J. et al., Neonatal familial type II hyperlipoproteinemia: cord blood cholesterol in 1800 births, *Metabolism*, 20, 597, 1971.

Moutafis, C. D. et al., The effect of prolonged high cholesterol intake on cholesterol metabolism in a man with type II hyperlipoproteinemia, *Clin. Sci.*, 40, 19P, 1971.

Myant, N. B., Treatment of familial hyperlipoproteinemia type II, *Clin. Sci.*, 40, 16P, 1971.

Riopérez Carmena, E. et al., Disorders of lipid metabolism (presentation of 3 cases of hyperlipemia). Frederickson's Type II (hyper-β-lipoproteinemia) (in Spanish), *Rev. Clin. Esp.*, 122, 347, 1971.

Greten, H. et al., Early diagnosis of familial hyperlipoproteinemia type II (umbilical cord blood beta cholesterol), *Verh. Dtsch. Ges. Inn. Med.*, 78, 1316, 1972.

Newall, R. G. et al., An association of very low density lipoproteins with type II hyperlipoproteinaemia, *Clin. Chim. Acta*, 38, 307—11, 1972.

Patterson, D. et al., Familial hyper-β-lipoproteinaemia, *Lancet*, 1, 1171, 1972.

Weber, K. et al., Hyperlipoproteinemias II (in German), *Hautarzt*, 23, 8, 1972.

Agostini, B., Electron microscopy of low-density plasma-lipoproteins in patients with type IIa hyperlipoproteinemia, *Naturwissenchaften*, 60, 111, 1973.

Briones, E. R. et al., Nutrition, metabolism, and blood lipids in humans with type IIA hyperlipoproteinemia, *Am. J. Clin. Nutr.*, 26, 259, 1973.

Greten, H. et al., Early diagnosis of familial type II hyperlipoproteinemia, *Nutr. Metab.*, 15, 128, 1973.

Moutafis, C. D. et al., Cholesterol metabolism in a patient with familial hyperbetalipoproteinaemia during periods of high and low cholesterol intake, *Atherosclerosis*, 17, 305, 1973.

Myant, N. B. et al., Type II hyperlipoproteinaemia, *Clin. Endocrinol. Metabol.*, 2, 81, 1973.

Merchant, R. H. et al., Hyper beta lipoproteinemia (familial hypercholesterolemia). A case report, *Indian Pediatr.*, 11 (7), 511, 1974.

Morganroth, J. et al., Pseudohomozygous type II hyperlipoproteinemia, *J. Pediatr.*, 85(5), 639, 1974.

Ballantyne, D. et al., The inheritance of mild Type II hyperlipoproteinaemia, *Scott. Med. J.*, 20(1), 23, 1975.

Gartner, U. et al., Studies with 14C-cholesterol in homozygotic hyperlipemia type IIa (in German), *Verh. Dtsch. Ges. Inn. Med.*, 81, 1434, 1975.

Haacke, H. et al., Lipid constellation in LDLs of primary type II hyperlipoproteinemia (author's transl.) (in German with English abstr.), *Klin. Wochenschr.*, 52(23), 1121, 1975.

Milewicz, A. et al., Marked inhibition of fibrinolysis in puerperium in a woman with primary hyperlipoproteinemia type IIa (in Polish), *Ginekol. Pol.*, 46(8), 881, 1975.

Newton, R. S. et al., Effect of diet on fatty acids in the lipoprotein cholesteryl esters of type IIa and normal individuals, *Lipids*, 10(12), 858, 1975.

Taketomi, T. et al, Lipids and lipoproteins of serum and organ tissues in a case of familial hyperlipoproteinemia type II, *Jpn. J. Exp. Med.*, 45(2), 55, 1975.

Cywes, S. et al., Portacaval shunt in two patients with homozygous type II hyperlipoproteinaemia, *S. Afr. Med. J.*, 50(8), 239, 1976.

Farriaux, J. P. et al., Treatment of type II familial hypercholesterolemia through portacaval anastomosis (in French with English abstr.), *Arch. Fr. Pediatr.*, 33(8), 745, 1976.

Kuo, P. T. et al., Types IIa and IIb hyperlipoproteinemia, *Circulation*, 53(2), 338, 1976.

Lee, P. et al., Isolation and carbohydrate composition of glycopeptides of human apo low-density lipoprotein from normal and type II hyperlipoproteinemic subjects, *Can. J. Biochem.*, 54(9), 829, 1976.

Type IIB

Carlson, L. A., Letter: IIB or not IIB? Classification of hyperlipoproteinaemias, *Atherosclerosis*, 19, 349, 1974.

Nikkila, E. A. et al., Inheritance of endogenous hypertriglyceridaemia type II B or IV, *Postgrad. Med. J.*, 51(8), 32, 1975.

Watermeyer, G. S. et al., The effects of carbohydrate, fat and total energy adjustment of Fredrickson's type IIB hyperlipoproteinaemia, *S. Afr. Med. J.*, 51(3), 71, 1977.

Type III

Matthews, R. J., Type 3 and IV familial hyperlipoproteinemia evidence that these two syndromes are different phenotypic expressions of the same mutant gene(s), *Am. J. Med.*, 44, 188, 1968.

Borrie, P., Type 3 hyperlipoproteinaemia, *Br. Med. J.*, 2, 665, 1969.

Bazex, A. et al., Primary, type 3-hyperlipoproteinemia, probably familial with abnormal biologic and clinical manifestations (in French), *Arch. Belg. Dermatol. Syphiligr.*, 26, 409, 1970.

Seidel, D., et al., Hyperlipoproteinemia type 3. The possibility of an immunological diagnosis (in German), *Clin. Chim. Acta*, 30, 31, 1970.

Quarfordt, S. et al., On the lipoprotein abnormality in type 3 hyperlipoproteinemia, *J. Clin. Invest.*, 50, 754, 1971.

Reimers, H. J. et al., Clinical picture of the type-3-hyperlipoproteinemia, *Verh. Dtsch. Ges. Inn. Med.*, 77, 596, 1971.

Sobra, J. et al., Inborn defects of lipid metabolism. 23. Coincidence of type 3-hyperlipoproteinemia and albinism (in Czech), *Cas. Lek. Cesk.*, 110, 451, 1971.

Hazzard, W. R. et al., Aggravation of broad-disease (type 3 hyperlipoproteinemia) by hypothyroidism, *Arch. Intern. Med.*, 130, 822, 1972.

Type IV

Gjone, E. et al., Type-IV hyperlipoproteinaemia, *Lancet*, 2, 819, 1968.

Schalm, L. et al., Hyper-prebeta-lipoproteinemia (Frederickson type IV) with steatosic infiltration of the liver and abnormal hepatic function tests, *Actualities Hepatogastroent.*, 5, 1339, 1969.

Schreibman, P. H. et al., Familial type IV hyperlipoproteinemia, *N. Engl. J. Med.*, 281, 981, 1969.

Christensen, N. C. et al., Type IV hyperlipoproteinaemia. A survey based on 23 cases, *Dan. Med. Bull.*, 17, 54, 1970.

Kunz, F. et al. Plasma phospholipids in type IV hyperlipoproteinemia, *Atherosclerosis*, 11, 265, 1970.

Hazzard, W. R. et al., Abnormal lipid composition of very low density lipoproteins in diagnosis of broad beta disease (type 3 hyperlipoproteinemia), *Metabolism,* 21, 1009, 1972.

Godolphin, W. J. et al., Type 3 hyperlipoproteinaemia in a child, *Lancet,* 1, 209, 1972.

Sobra, J. et al., Inborn errors and lipid metabolism. XXV. Familial hyperlipoproteinemia of the type 3-detection of heterozygotes (English abstr), *Cas. Lek. Cesk.,* 111, 192, 1972.

Stern, M. P. et al., Acquired type 3 hyperlipoproteinemia. Report of three cases associated with systemic types erythematosus and diabetic ketoacidosis, *Arch. Intern. Med.,* 130, 817, 1972.

Ballantyne, D. et al., Type 3 hyperlipoproteinaemia and sinking prebeta lipoprotein, *J. Clin. Pathol.,* 26, 552, 1973.

Ballantyne, D. et al., Study of the pedigree of a patient with type 3 hyperlipoproteinemia and sinking prebeta lipoprotein, *J. Clin. Pathol.,* 26, 163, 1973.

Quarfordt, S. H. et al., The kinetic properties of very low density lipoprotein triglyceride in type 3 hyperlipoproteinemia, *Biochim. Biophys. Acta,* 296, 572, 1973.

Weber, K. et al., Hyperlipoproteinemia type 3. Report on 5 cases, *Hautarzt,* 24, 179, 1973.

Weiland, H. et al., Improved techniques for assessment of serum lipoprotein patterns. II. Rapid method for diagnosis of type 3 hyperlipoproteinemia without ultracentrifugation, *Clin. Chem.,* 19, 1139, 1973.

Amatruda, J. M. et al., Type 3 hyperlipoproteinemia with mesangial foam cells in renal glomeruli, *Arch. Pathol.,* 98, 51, 1974.

Infortuna, M. et al., Broad beta band disease. Description of a case (English abstr.), *Minerva Pediatr.,* 26(12), 615, 1974.

Patsch, J. R. et al., Type 3 hyperlipoproteinaemia (broad-beta disease): diagnosis and quantitative isolation of the typical lipoprotein, *Klin. Wochenschr.,* 52(16), 792, 15, 1974.

Schneider, J. et al., Frequency of type III hyperlipoproteinemia in "broad beta" pattern of lipoproteinelectrophoresis (English abstr.), *Klin. Wochenschr.,* 52(19), 941, 1974.

Fredrickson, D. S. et al., Type III hyperlipoproteinemia: an analysis of two contemporary definitions, *Ann. Intern. Med.* 82(2), 150, 1975.

Havel, R. J., Hyperlipoproteinemia: problems in diagnosis and challenges posed by the "type III" disorder, *Ann. Intern. Med.* 82(2), 273, 1975.

Hazzard, W. R. et al., Broad-beta disease versus endogenous hypertriglyceridemia, levels and lipid composition of chylomicrons and very low density lipoproteins during fat-free feeding and alimentary lipemia, *Metabolism* 24(7), 817, 1975.

Patsch, J. R., Lipoprotein of the density 1.006—1.020 in the plasma of patients with type III hyperlipoproteinaemia in the postabsorptive state, *Eur. J. Clin. Invest.,* 5(1), 45, 1975.

Morganroth, J. et al., The biochemical, clinical, and genetic features of type III hyperlipoproteinemia, *Ann. Intern. Med.,* 82(2), 158, 1975.

Shepherd, J. et al., Diagnosis of type III hyperlipoproteinemia by chromatography of plasma lipoproteins on columns containing agarose, *Clin. Chem.,* 21(13), 1887, 1975.

Vessby, B. et al., Conversion of type III hyperlipoproteinaemia to type IV hyperlipoproteinaemia by a fat-free, carbohydrate rich diet, *Eur. J., Clin. Invest.,* 5(4), 359, 1975.

Braun-Falco O., Structure and morphogenesis of xanthomas in hyperlipoproteinemia type III. A morphologic histochemical and electron microscopy study (in German), *Hautarzt* 27(3), 122, 1976.

Glueck, C. J. et al., Pediatric familial type III hyperlipoproteinemia, *Metabolism,* 25(11), 1269, 1976.

Hazzard, W. R. et al., Delayed clearance of chylomicron remnants following vitamin-A-containing oral fat loads in broad-β disease (type III hyperlipoproteinemia), *Metabolism,* 25(7), 777, 1976.

Patsch, J. R. et al., Electron microscopic characterization of lipoproteins from patients with familial type III hyperlipoproteinaemia, *Eur. J. Clin. Invest.,* 6(4), 307, 1976.

Patsch, J. R. et al., Familial hyperlipoproteinemia type III (in German), *Dtsch. Med. Wochenschr.,* 101(44), 1612, 1976.

Reman, F. C. et al., Zero electrophoretic mobility of very low density lipoproteins in type III hyperlipidemia patients during treatment, *Clin. Chim. Acta,* 71(2), 261, 1976.

Vessby, B. et al., Lipoprotein composition and lipoprotein interrelations in 50-year-old men with hyperlipoproteinaemia III, *Upsala J. Med. Sci.,* 81(2), 71, 1976.

Vessby, B., Studies on the serum lipoprotein composition in 50-year-old men. A suggestion of chemical criteria for diagnosis of hyperlipoproteinaemia type III (broad-β disease), *Clin. Chim. Acta,* 69(1), 29, 1976.

Rerabek, J. E., Two-dimensional immunoelectrophoretic pattern of low-and very-low-density lipoproteins, with particular reference to Fredrickson's type III, *Clin. Chem.,* 23, 186, 1977.

Vessby, B. et al., Inheritance of type-III hyperlipoproteinemia. Lipoprotein patterns in first-degree relatives, *Metabolism,* 26(3), 225, 1977.

Quarfordt, S. H. et al., Very low density lipoprotein triglyceride transport in type IV hyperlipoproteinemia and the effects of carbohydrate-rich diets, *J. Clin. Invest.*, 49, 2281, 1970.

Rose, H. G., Familial hyper-pre-betalipoproteinemia with observation on the metabolism of endogenous particles, *J. Lab. Clin. Med.*, 76, 92, 1970.

Schlierf, G. et al., Diurnal patterns of plasma triglycerides and free fatty acids in normal subjects and in patients with endogenous (type IV) hyperlipoproteinemia, *Nutr. Metab.*, 13, 80, 1971.

Amidi, M., Type IV hyperlipoproteinemia in a consanguinous family, *Circulation*, 45, 988, 1972.

Glueck, C. J. et al., Pediatric familial type IV hyperlipoproteinemia, *Trans. Assoc. Am. Physicians*, 85, 139, 1972.

Goldman, J. A. et al., Musculoskeletal disorders associated with type IV hyperlipoproteinaemia, *Lancet*, 2, 449, 1972.

Kwiatkowski, S. et al., Case of type IV hyperlipoproteinemia (Fredrickson's classification), *Wiad. Lek.*, 25, 1591, 1972.

Pinals, R. S. et al., Type-IV hyperlipoproteinemia and transient osteoporosis, *Lancet*, 2, 929, 1972.

Wille, L. E., Hyperlipoproteinemia type IV and pre-beta-lipoprotein. Some new aspects of its physiology, pathogenesis and diagnosis, *J. Oslo City Hosp.*, 22, 169, 1972.

Wybenga, D. R. et al., The indirect confirmation of hyperlipoproteinemia phenotypes II, III and IV, *Clin. Chim. Acta*, 40, 121, 1972.

Mancini, M. et al., Studies of the mechanisms of carbohydrate-induced lipaemia in normal man, *Atherosclerosis*, 17, 445, 1973.

Schonfeld, G. et al., Type IV hyperlipoproteinemia. A critical appraisal, *Arch. Intern. Med.*, 132, 55, 1973.

Borrie, P. et al., A clinical syndrome characteristic of primary type IV-V hyperlipoproteinaemia, *Br. J. Dermatol.*, 90, 245, 1974.

Dyerberg, J., Familial hyperlipoproteinemia type IV, *Ugeskr. Laeg.*, 136, 769, 1974.

Freiberg, R. A. et al., Multiple intraosseous lipomas with type-IV hyperlipoproteinemia. A case report, *J. Bone Jt. Surg. Am.*, 56(8), 1729, 1974.

Halpern, M. J. et al., Letter: "Latent type IV" hyperlipoproteinemia, *Acta Diabetol. Lat.*, 11(2), 173, 1974.

Opplt, J. J., Glavaski, B. S., and Bahler, R. C., Molecular distribution of plasma lipoproteins in type IV hyperlipoproteinemia, *Clin. Chem.*, 21, 990, 1975.

Oster, P. et al., Diurnal changes in lipoproteins in patients with hyperlipoproteinemia type IV (in German), *Verh. Dtsch. Ges. Inn. Med.*, 81, 1442, 1975.

Nielsen, P. M. et al., Musculoskeletal symptoms and hyperlipoproteinemia type IV (in Danish with English abstr.), *Ugeskr. Laeg.*, 138(4), 223, 1976.

Schlierf, G. et al., Diurnal patterns of plasma lipids and lipoproteins in primary endogenous hypertriglyceridemia (type IV-hyperlipoproteinemia) (author's transl.), (in German with English abstr.). *Klin. Wochenschr.*, 55(4), 161, 1977.

Type V

Glueck, C. J. et al., Norethindrone acetate, postheparin lipolytic activity, and plasma triglycerides in familial types I, 3, IV, and V hyperlipoproteinemia. Studies in 26 patients and 5 normal persons, *Ann. Intern. Med.*, 75, 345, 1971.

Vittori, F. et al., Type V hyperlipoproteinemia. Apropos of a case, *Lyon Med.*, 226, 429, 1971.

Lossow, W. J. et al., A study of a type V hyperlipoproteinemic patient, *Clin. Chim. Acta* 36, 33, 1972.

Jacobsen, B. B., Type V hyperlipoproteinaemia, Metabolic changes in hyperlipaemic and previously hyperlipaemic ("susceptible") male patients, *Atherosclerosis*, 17, 471, 1973.

Rebollar Mesa, J., Type V hyperlipoproteinemia (in Spanish), *Rev. Clin. Esp.*, 128, 271, 1973.

Rebollar Mesa, J. et al., Type V familial hyperlipoproteinemia: study of 2 families (in Spanish with English abstr.), *Rev. Clin. Esp.*, 128, 299, 1973.

Desai, M. et al., Hyperlipoproteinemia type V (a case report), *Indian Pediatr.*, 11(10), 695, 1974.

Hüla, M. et al., Neuritis hyperplastica xanthomatosa in familial hyperlipoproteinemia type V. Case report with genetic study, *Cesk. Dermatol.*, 49, 163, 1974.

Kalofontis, A. et al., Hyperliproproteinemia type V: biochemical observations in two cases, *Clin. Chim. Acta*, 52, 361, 1974.

Kucerova, L. et al., Primary hyperlipoproteinemia of type V, *Vnitr. Lek.*, 20(9), 869, 1974.

Heckers, H. et al., Type V-hyperlipoproteinemia, *Med. Welt*, 26(39), 1766, 1975.

Raleigh, J. et al., Type V hyperlipoproteineaemia: a family study, *N.Z. Med. J.*, 82(551), 300, 1975.

Simons, L. A. et al., The biochemical composition and metabolism of lipoproteins in type V hyperlipoproteinaemia, *Clin. Chim. Acta*, 61(3), 341, 1975.

Simons, L. A. et al., Type V hyperlipoproteinaemia re-visited, findings in a Sydney population, *Aust. N. Z. J. Med.*, 5(3), 210, 1975.

Other Types

Allison, A. C. et al., Serum lipoprotein allotypes in man, *Prog. Med. Genet.*, 4, 176, 1965.

Glueck, C. J. et al., Immunoreactive insulin, glucose tolerance, and carbohydrate inducibility in types II, 3, IV and V hyperlipoproteinemia, *Diabetes*, 18, 139, 1969.

Havel, R. J., Typing of hyperlipoproteinemias. *Atherosclerosis*, 11, 3, 1970.

Herrmann, G. R., Recent practical methods for typing hyperlipoproteinemia; advances in the study of blood lipids, *Tex. Med.*, 67, 81, 1971.

Lipo, J. F. et al., Lipoprotein phenotyping, *Crit. Rev. Clin. Lab. Sci.*, 2, 461, 1971.

Roberts, W. C. et al., Hyperlipoproteinemia. A review of the five types with first report of necropsy findings in type 3., *Arch Pathol.*, 90, 46, 1970.

Winkelman, J. et al., Quantitation of lipoprotein components in the phenotyping of hyperlipoproteinemias, *Clin. Chim. Acta*, 27, 181, 1970.

Kubašta, M. et al., Classification of hyperlipoproteinemias according to the NIH system (typing according to Frederickson and collaborators) (English abstr.), *Vnitr. Lek.*, 18, 18, 1972.

Noël, C. et al., Plasma phopholipids in the different types of primary hyperlipoproteinemia, *J. Lab. Clin. Med.*, 79, 611, 1972.

Brown, H. B. et al., Mixed hyperlipemia, a sixth type of hyperlipoproteinemia, *Atherosclerosis*, 17, 181, 1973.

Kučerová, L. et al., Hepatobiliary disorders in primary hyperlipoproteinemias type IV and V, *Sb. Lek.*, 75, 237, 1973.

Rose, H. G. et al., Inheritance of combined hyperlipoproteinemia: evidence for a new lipoprotein phenotype, *Am. J. Med.*, 54, 148, 1973.

Searcy, R. L. et al., A new approach to hyperlipoproteinemia phenotyping, *Am. J. Med. Technol.*, 39, 183, 1973.

Rose, H. G. et al., Combined hyperlipoproteinemia. Evidence for a new lipoprotein phenotype, *Atherosclerosis*, 20(1), 51, 1974.

Carlson, L. A., Serum lipoprotein composition in different type of hyperlipoproteinemia, *Adv. Exp. Med. Biol.*, 63, 185, 1975.

Van Der Bijl, P. et al., Lipid and protein contents of low density lipoproteins (LDL) from patients with various types of hyperlipoproteinemia, *Clin. Chim. Acta*, 63(1), 95, 1975.

Carlson, L. A., Lipid composition of the major human serum lipoprotein density classes in different types of hyperlipoproteinemia, in *Lipoprotein Metabolism*, Greten, H., ed., Springer Verlag, Berlin, 1976.

ABNORMAL DYSLIPOPROTEINEMIAS

Hyper-α-Lipoproteinemia

Roggenbach, H. J. et al., Abnormal -α,-lipoprotein with impaired finding capacity for lipid, *Lancet*, 1, 1189, 1971.

Glueck, C. J. et al., Familial hyperalphalipoproteinemia, *Arch. Intern. Med.*, 135(8), 1025, 1975.

Glueck, C. J. et al., Familial hyper-alpha-lipoproteinemia: studies in eighteen kindreds, *Metabolism*, 24(11), 1243, 1975.

Naito, H. K. et al., Letter: Hyper-alpha-lipoproteinemia: a case study, *Clin. Chem.*, 21(4), 639, 1975.

Glueck, C. J. et al., Longevity syndromes: familial hypo-β and familial hyper-α-lipoproteinemia, *J. Lab. Clin. Med.*, 88(6), 941, 1976.

Mendoza, S. et al., Composition of HDL-2 and HDL-3 in familial hyper-α-lipoproteinemia, *Atherosclerosis*, 25(1), 131, 1976.

Hypo-α,-lipoproteinemia (See Syndromes: Tangier's Disease)

Hypo-β-lipoproteinemia

Osmond, D. H. et al., Renin "preinhibitor" in blood of anephric patients, in a case of hypobetalipoproteinemia, and evidence of its major association with plasma alpha lipoproteins, *J. Lab. Clin. Med.*, 73, 809, 1969.

Frood, J. D., Relationship between pattern of infection and development of hypoalbuminaemia and hypo-β-lipoproteinaemia in rural Ugandan children, *Lancet*, 2, 1047, 1971.

Fosbrooke, A. S. et al., Familial hypo-β-lipoproteinemia, *Arch. Dis. Child.*, 47, 671, 1972.

Rerabek, J. et al., A contribution to the knowledge of the hypo-beta-lipoproteinemia, *Acta Med. Scand.*, 194, 379, 1973.

Fosbrooke, A. et al., Familial hypo-beta-lipoproteinaemia, *Arch. Dis. Child*, 48, 729, 1973.

Aggerbeck, L. P. et al., Hypobetalipoproteinemia, clinical and biochemical description of a new kindred with 'Friedreich's ataxia,' *Neurology*, 11(24), 1051, 1974.

Brown, B. J. et al., Familial hypobetalipoproteinemia: report of a case with psychomotor retardation, *Pediatrics*, 54, 111, 1974.

Cottrill, C. et al., Familial homozygous hypobetalipoproteinemia, *Metabolism*, 23, 779, 1974.

Schüler, A., Concept and systematization of the hypolipemic syndromes, *Rev. Clin. Esp.*, 132, 399, 1974.

Wallis, K. et al., Acrodermatitis enteropathica associated with low density lipoproteins deficiency, *Commun. Behav. Biol.*, 13(9), 749, 1974.

Hypo-lipoproteinemia

Mancini, M. et al. The distribution and composition of serum lipoproteins in hypocholesterolaemic men, *Clin. Chim. Acta*, 17, 163, 1967.

Aggerbeck, L. P. et al., A study of the polypeptides of serum lipoproteins in hypo-lipoproteinemia, *J. Lab. Clin. Med.*, 78, 985, 1971.

Lloyd, J. K., Hypolipoproteinaemia, *J. Clin. Pathol.*, 5(Suppl.), 53, 1973.

Lloyd, J. K., Lipoprotein deficiency disorders, Clin. *Endocrinol. Metab.*, 2, 127, 1973.

Noseda, G., Hypolipidemias, *Schweiz. Med. Wochenschr.*, 105(39), 1233, 1975.

LIPOPROTEINS IN BIOLOGICAL FLUIDS

Cerebro-spinal fluid

Someda, K. et al., Lipoproteins in the cerebrospinal fluid: immunoelectrophoretic analysis (in German), *Klin. Wochenschr.*, 43, 230, 1965.

Clansen, J., The β-lipoprotein of serum and cerebrospinal fluid, *Acta Neurol. Scand.*, 42, 153, 1966.

de Zanche, L., Immunoelectrophoresis of the cerebrospinal fluid: some practical uses (in Italian), *G. Psichiatr. Neuropatol.*, 95, 338, 1967.

Matiar-Vahar, H., Lumbar cerebrospinal fluid obstruction in partial and complete block (in German), *Fortschr. Neurol. Psychiatr.*, 36, 247, 1968.

Messeri, E. et al., Total lipids and lipoproteins of the cerebrospinal fluid in meningitis (in Italian), *Arch. Ital. Sci. Med. Trop.*, 49, 287, 1968.

Synovial Fluid

Small, D. M. et al., Lipoproteins of symovial fluid as studied by analytical ultracentrifugation, *J. Clin. Invest.*, 43, 2070, 1964.

Lymph

Zubatsorv, D. M. et al., Coagulability and protein structure of the lymph and blood in acute hemorrhage (in Russian), *Biull. Eksp. Biol. Med.*, 57, 28, 1964.

Casley-Smith, J. R., Endothelia permeability. The passage of particles into and out of diaphragmatic lymphatics, *Q. J. Exp. Physiol.*, 49, 365, 1964.

Ockner, R. K. et al., Very low density lipoproteins in intestinal lymph: role in triglyceride and cholesterol transport during fat absorption, *J. Clin. Invest.*, 48, 2367, 1969.

Windmueller, H. G. et al., Fat transport and lymph and plasma lipoprotein biosynthesis by isolated intestine, *J. Lipid Res.*, 13, 92, 1972.

Kostner, G., Studies on the composition of the lipoproteins in human lymph, *Hoppe Seylers Z. Physiol. Chem.*, 353, 1863, 1972.

Reichl, D. et al., The lipids and lipoproteins of human peripheral lymph, with observation on the transport of cholesterol from plasma and tissues into lymph, *Clin. Sci. Mol. Med.*, 45, 313, 1973.

Bile

Clarke, G. B. et al., Effect of bile duct cannulation on plasma lipoproteins, *Proc. Soc. Exp. Biol. Med.*, 117, 355, 1964.

Nesterin, M. F. et al., Lipoprotein bile complex in the duodenal content of patients with chronic cholecystitis (in Russian), *Ter. Arkh.*, 37, 71, 1965.

Nesterin, M. F. et al., Hepatic elimination of lipoprotein complex in the bile (in Russian), *Fiziol. Zh. SSSR*, 51, 1487, 1965.

Sokolova, N. V., Electrophoretic studies of lipoproteins of gall-bladder bile in cholecystitis (in Russian), *Ter. Arkh.*, 38, 60, 1966.

Lavion, D. et al., Lack of mixed micelles bile salt-lecithin-cholesterol in bile and presence of a lipoproteic complex, *Biochemie*, 54, 529, 1972.

Picard, J. et al., Abnormal lipoproteins in human and experimental biliary obstruction, *Expo. Annu. Biochim. Med.,* 31, 27, 1972.

Einarsson, K. et al., The formation of bile acids in patients with three types of hyperlipoproteinemia, *Eur. J. Clin. Invest.,* 2, 225, 1972.

Quarfordt, S. H. et al., Liquid crystalline lipid in the plasma of humans with biliary obstruction, *J. Clin. Invest.,* 51, 1979, 1972.

Kaplun, B. L., Lipoprotein complex of bile in the evaluation of external secretory function of the liver, *Lab. Delo.,* 12, 751, 1973.

Nalbone, G. et al., Ultramicroscopic study of the bile lipoprotein complex, *Biochimie,* 55, 1503, 1973.

Fruchart, J. C. et al., Diagnosis of jaundice caused by obstruction of the extra-hepatic bile ducts. Study of serum lipids and lipoproteins, *Nouv. Presse Med.,* 2, 1505, 1973.

Einarsson, K. et al., Effect of cholic acid feeding on bile acid kinetics and neutral fecal steroid excretion in hyperlipoproteinemia (typed II and IV), *Metabolism,* 23, 863, 1974.

Milk (human)

Rapheal, B. C. et al., The serum lipoproteins as a source of milk cholesterol, *FEBS Lett,* 58(1), 47, 1975.

Parotid Saliva

Gifford, G. T et al., Protein patterns in human parotid saliva, *J. Chromatogr.,* 20, 150, 1965.

LIPOPROTEINS IN DIFFERENT ORGANS

Arterial Wall

Lawrie, T. D., The origin and distribution of lipids in aortic atherosclerosis, *Biochem. Soc. Sympos.,* 24, 101, 1963.

Lawrie, T. D. et al., A comparison of serum lipoprotein and aortic plaque fatty acids, *Clin. Sci.,* 27, 89, 1964.

Prerovska, T. et al., Effect of lead on biochemical changes in the serum and changes in the arterial wall with regard to atherosclerosis, *Int. Arch. Gewerbepathol.,)* 21, 265, 1965.

Loviagina, T. N. et al., On the β-lipoprotein content in the blood serum and aortic wall in experimental atherosclerosis (in Russian), *Vopr. Med. Khim.,* 11, 17, 1965.

Woolf, N. et al., The immunohistochemical demonstration of lipoproteins in vessel walls, *J. Pathol. Bacteriol.,* 90, 459, 1965.

Groover, M. E., Jr. et al., Electrophoretic activity of arterial plasma and experimental thrombi, *Angiology,* 16, 651, 1965.

Cottet, J. et al., The lipoproteins of the aortic wall (in French), *Rev. Atheroscler.,* 8, 34, 1966.

Walton, K. W., Lipoproteins in the vessel wall, *Rev. Atheroscler.,* 8, 41, 1966.

Woolf, N. et al., The occurrence of lipoprotein in thrombi, *J. Pathol. Bacteriol.,* 91, 383, 1966.

Campeanu, S. et al., Comparative studies on the transcapillary filtration of human protein and lipoprotein fractions in the health subject (in German), *Z. Ges. Inn. Med.,* 21, 404, 1966.

Dayton, S. et al., Movement of labeled cholesterol between plasma lipoprotein and normal arterial wall across the intimal surface, *Circ. Res.,* 19, 1041, 1966.

Fasoli, A., Lipoproteins in the blood, atherosclerosis and thrombosis (in French), *Rev. Atheroscler.,* 9, 23, 1967.

Hollander, W., Recent advances in experimental and molecular pathology; influx, synthesis, and transport of arterial lipoproteins in atherosclerosis, *Exp. Molec. Pathol.,* 7, 248, 1967.

Györkey, F. et al., The fine structure of ceroid in human atheroma, *J. Histochem. Cytochem.,* 15, 732, 1967.

Schwartzkopff, W. et al., Studies on the permeability to lipids and lipoproteins of inflamed skin capillaries, *Bibl. Anat.,* 10, 195, 1969.

Klimov, A. N. et al., Penetration of lipoproteins into the aortic wall (in Russian), *Vestn. Akad. Med. Nauk, SSSR,* 24, 43, 1969.

Scott, P. J. et al., Incorporation of radioiodinated serum albumin and low-density lipoprotein into human thrombi in vivo, *J. Pathol.,* 97, 603, 1969.

Chmielewski, H. et al., Level of lipoprotein fractions and total serum cholesterol in the course of treatment with radon-containing waters in patients with arterial disorders in the lower limbs (in Polish), *Pol. Tyg. Lek.,* 24, 635, 1969.

Schmidt, M. et al., The role of glucosaminoglucans (mucopolysaccharides) of connective tissue in the development of pathological deposits in the vascular wall. Participation of plasma globulins, β-lipoproteins and glucosaminoglucans of the arterial wall in the development of fibrinoid deposits in vitro (in Polish), *Pol. Tyg. Lek.,* 24, 881, 1969.

Scott, P. J., et al., Low-density lipoprotein accumulation in aortic and coronary artery walls, *Israel J. Med. Sci.,* 5, 1969.

Lempert, B. L. et al., Experiment on the effect of blood serum of patient with atherosclerosis on the lipolytic activity of the aortic wall, *Biull. Eksp. Biol. Med.,* 68, 41, 1969.

Scott, P. J. et al., The distribution of radio-iodinated serum albumin and low-density lipoprotein in tissues and the arterial wall, *Atherosclerosis,* 11, 77, 1970.

Virag, Sh. et al., Relation of the acid mucopolysaccharide content of the aortic wall to the degree of absorption of β-lipoproteins J-125 (in Russian), *Biull. Eksp. Biol. Med.,* 70, 37, 1970.

Hoff, H. F., Studies on the pathogenesis of atherosclerosis with experimental model systems. IV. Ultrastructural and lipid-histochemical changes in the lipoprotein-filled doubly ligated carotid artery, *Virchows Arch. (Pathol. Anat.)* 352, 99, 1971.

Zubzhitskii, Iu. N., et al., Laboratory production of antisera for the detection of β-lipoproteins in the arterial wall in atherosclerosis, *Lab. Delo,* 5, 282, 1971.

Dousset, J. C. et al., Composition of the lipid fractions of aortic lipoproteins of the human and the calf (in French with English abstract), *Ann. Pharm. Fr.,* 30, 277, 1972.

Klimov, A. H. et al., Preparation of lipoproteins with a double radioactive label and a study of the penetration into vessel walls, *Vopr. Med. Khim.,* 18, 434, 1972.

Srinivasan, S. R. et al., Isolation of lipoprotein-acid mucopolysaccharide complexes from fatty streaks of human aortas, *Atherosclerosis,* 16, 95, 1972.

Smith, E. B. et al., Lipids and lipoproteins in aging aortic intima, *Proc. R. Soc. Med.,* 65, 675, 1972.

Gerö, S., et al., Studies in the functional properties of vascular mucopolysaccharides, *Adv. Metab. Disord. Suppl.,* 2, 103, 1973.

Kramsch, D. M. et al., The interaction of serum and arterial lipoproteins with elastin of the arterial intima and its role in the lipid accumulation in atherosclerotic plaques, *J. Clin. Invest.,* 52, 236, 1973.

Smith, E. B. et al., Quantitative studies on fibrinogen and low-density lipoprotein in human aortic intima, *Atherosclerosis,* 18, 479, 1973.

Fisher-Dzoga, K. et al., Effects of serum lipoproteins on the morphology, growth, and metabolism of arterial smooth muscle cells, *Adv. Exp. Med. Biol.,* 43, 299, 1974.

Berenson, G. S. et al., Mucopolysaccharide-lipoprotein complexes in atherosclerotic aorta, *Adv. Exp. Med. Biol.,* 43, 141, 1974.

Hoff, H. F. et al., Localization of low-density lipoproteins in atherosclerotic lesions from human normolipemics employing a purified fluorescent-labeled antibody, *Biochim. Biophys. Acta,* 351, 407, 1974.

Popov, A. V. et al., Histoautoradiographic and biochemical study of the transport of atherogenetic lipoproteins in the vascular wall (in Russian), *Arkh. Patol.,* 37(10), 25, 1975.

Shimamoto, T. et al., Immunofluorescent demonstration of plasma protein entry into arterial wall by cholesterol, epinephrine, norepinephrine and angiotensin II, *Acta Pathol. Jpn.,* 25(1), 51, 1975.

Camejo, G. et al., The participation of aortic proteins in the formation of complexes between low density lipoproteins and intima-media extracts, *Atherosclerosis,* 21(1), 77, 1975.

Srinivasan, S. R. et al., Lipoprotein-acid mucopolysaccharide complexes of human atherosclerotic lesions, *Biochim. Biophys. Acta,* 388(1), 58, 1975.

Vekhoff, D. et al, Blood lipoprotein transfers in the normal arterial wall. Study auto-historadiography in optic microscopy (in French with English abstract), *Paroi Arterielle,* 2(4) 241, 1975.

Hoff, H. F. et al., Lipoproteins in atherosclerotic lesions. Localization by immunofluorescence of apo-low density lipoproteins in human atherosclerotic arteries from normal and hyperlipoproteinemias, *Arch. Pathol.,* 99(5), 253, 1975.

Denisenko, A. D. et al., Comparative characteristics of the fatty acid composition of lipoproteins in human blood plasma and aortic wall, *Biull. Eksp. Biol. Med.,* 79(6), 44, 1975.

Hoff, H. F. et al., Localization patterns of plasma apolipoproteins in human atherosclerotic lesions, *Circ. Res.,* 37(1), 72, 1975.

Popov, A. V. et al., A method for quantitative isolation of lipoproteins from blood vessel walls in an electric field, (in Russian with English abstract), *Vopr. Med. Khim.,* 21(6), 655, 1975.

Hoff, H. F. et al., Ultrastructural localization of plasma lipoproteins in human intracranial arteries, *Virchows Arch. A.,* 369(2), 111, 1975.

Walton, K. W., Factors affecting lipoprotein deposition in the arterial wall, in *Blood and Arterial Wall in Atherogenesis and Arterial Thrombosis,* Hautvast, J. G. et al., Eds., Leiden, Brill, 1975, 79.

Seidel, D., Structure and metabolism of plasma lipoproteins: impact upon deposition in the arterial wall, in *Blood and Arterial Wall in Atherogenesis and Arterial Thrombosis,* Hautvast, J. G. et al., Eds. Leiden, Brill, 1975, 87.

Camejo, G. et al., Differences in the structure of plasma low-density lipoproteins and their relationship to the extent of interaction with arterial wall-components, *Ann. N. Y. Acad. Sci.,* 275, 153, 1976.

Ghosh, S. et al., Evaluation of the permeability parameters (influx, efflux and volume of distribution) of arterial wall for LDL and other proteins, *Adv. Exp. Med. Biol.,* 67(00), 191, 1976.

Hoff, H. F. et al., Apo-lipoprotein localization in human atherosclerotic arteries, *Adv. Exp. Med. Biol.*, 67(00), 109, 1976.

Vinogradov, A. G. et al., Role of foam cells in lipoprotein transformation within the vascular wall (with English abstract), *Kardiologiia*, 16(2), 37, 1976.

Klimov, A. N. et al., Mechanism of lipoprotein and cholesterol transport into the vascular wall (with English abstract), *Kardiologiia*, 16(2), 30, 1976.

Onitiri, A. C. et al., Lipoprotein concentrations in serum and in biopsy samples of arterial intima: a quantitative comparison, *Atherosclerosis*, 23(3), 513, 1976.

Denisenko, A. D. et al., Comparative characteristics of the lipoproteins of the human vascular wall and blood plasma (in Russian with English abstract), *Kardiologiia*, 16(7), 64, 1976.

Stoff, H. F., Apolipoprotein localization in human cranial arteries, coronary arteries, and the aorta, *Stroke*, 7(4), 390, 1976.

Hollander, W., Unified concept on the role of acid mucopolysaccharides and connective tissue proteins in the accumulation of lipids, lipoproteins, and calcium in the atherosclerotic plaque, *Exp. Mol. Pathol.*, 25(1), 106, 1976.

Smith, E. B. et al. The release of an immobilized lipoprotein fraction from atherosclerotic lesion by incubation with plasmin, *Atherosclerosis*, 25(1), 71, 1976.

Day, C. E., Control of the interaction of cholesterol ester-rich lipoproteins with arterial receptors, *Atherosclerosis*, 25(2—3), 199, 1976.

Different Tissues

Deutsch, E. et al., Studies on tissue thromboplastin. 1. Purification chemical characteristics and separation into protein and lipid components (in German), *Thromb. Diath. Haemorrh.*, 12, 12, 1964.

la Torre, J. L. et al., Isolation of a cholinergic proteolipid receptor from tissue, *Proc. Natl. Acad. Sci., U.S.A.*, 65, 716, 1970.

Nemerson, Y. et al., Purification and characterization of the protein component of tissue factor, *Biochemistry*, 9, 5100, 1970.

Pitlick, F. A. et al., Binding of the protein component of tissue factor to phospholipides, *Biochemistry*, 9, 5105, 1970.

Popp, J. et al., Studies on the assimilation of lipoproteins by isolated adipose tissue, *Verh. Dtsch. Ges. Int. Med.*, 77, 616, 1971.

de Robertis, E. et al., Multiple binding sites for acetylcholine in a proteolipid from electric tissue, *Mol. Pharmacol.*, 7, 97, 1971.

Maca, R. D. et al., The role of β-lipoprotein, cholesterol, and various sera in tissue culture intracellular lipidosis, *Proc. Soc. Exp. Biol. Med.*, 136, 457, 1971.

Wigglesworth, V. B., Bound lipid in the tissues of mammal and insect: a new histochemical method, *J. Cell Sci.*, 8, 709, 1971.

Smith, E. B. et al., An immuno-electrophoretic assay of β-lipoprotein and albumin in human aortic intima by direct electrophoresis from the tissue sample into an antibody-containing gel, *Biochem. J.*, 123, 39P, 1971.

Mezesova, V. et al., Lipoproteins of the cell nuclei of the central nervous system, *Bratisl. Lek. Listy*, 58, 174, 1972.

Smith, E. B. et al., Relationship between plasma lipids and arterial tissue lipids, *Nutr. Metab.*, 15, 17, 1973.

Sodhi, H. S. et al., Correlating metabolism of plasma and tissue cholesterol with that of plasma-lipoproteins, *Lancet*, 1, 513, 1973.

Beran, M., Serum inhibitors of in vitro colony formation: relation to haemopoietic tissue in vivo, *Exp. Hematol.*, 2(2), 58, 1974.

Klimov, A. N. et al., Preparation of tissue fluid of the vessel wall and determination of its lipoproteins, *Atherosclerosis*, 19, 243, 1974.

Pitlick, F. A., Concanavalin A inhibits tissue factor coagulant activity, *J. Clin. Invest.*, 55(1), 175, 1975.

Hruzova-Kukurova, K. et al., The effect of concentrating of the tissue extract by dialysis on the structure of lipoproteins in CNS (in Slovenian), *Bratisl. Lek. Listy*, 64(1), 35, 1975.

Adamczyk, A., Lipoprotein balance and its physiological and clinical aspects. Metabolism of plasma adipose bodies and their participation in lipid balance (in Polish), *Przegl. Lek.*, 33(7), 689, 1976.

Bowness, J. M., Lipoproteins versus structural glycoproteins: atherosclerosis as an analogue and competitor of normal connective tissue interactions, *Med. Hypotheses*, 2(5), 200, 1976.

Heart

Holczabek, W., On the chromotropic lipoids (lipoproteins) of the cardiac muscle fibers (in German), *Z. Rechtsmed.*, 67, 99, 1970.

Morozova, L. et al., Study of serum protein, lipoprotein, glycoproteins and hemodynamics in mitral defects of the heart (in Russian with English abstract), *Vrach. Delo*, 1, 32, 1976.

Aonuma, S. et al., Studies on heart. XVI. Biologically active substances in thymus-isolation and properties of heart function promoting substances from calf thymus (in Japanese with English abstract), *J. Pharm. Soc. Jpn.*, 96(9), 1057, 1976.

Muscle (skeletal)
Knieriem, H. J. et al., Actomyosin and myosin and the deposition of lipids and serum lipoproteins, *Arch. Pathol.*, 84, 118, 1967.

Siegelman, S. S. et al., Hyperlipoproteinemia with skeletal lesions, *Clin. Orthop.*, 87, 228, 1972.

Liver
Opplt, J. J. and Syllaba, J., On the so-called "Hepatic Diabetes", *Cas. Lek. Cesk.*, 95, 44/55, 1247, 1956.

Opplt, J. J. and Syllaba, J., Clinical and laboratory findings in hepatical chronic inflammatory processes and hepatic cirrhosis, *Cas. Lek. Cesk.*, 96, 40/41, 1303, 1957.

Opplt, J. J. and Syllaba, J., The value of analyses of blood lipoproteins for the diagnosis and prognosis of hepatic diseases, *Cas. Lek. Cesk.*, 16, 449, 1960.

Takahashi, T. et al., Congenital abnormalities of liver metabolism (in Japanese), *Naika*, 15, 455-60, 1965.

Dudarova, S. I., Role of so-called secretory function of the liver in the pathogenesis of acute toxic fatty dystrophy (in Russian), *Vopr. Med. Khim.*, 71, 68, 1965.

Nesterin, M. F. et al., Secretion of lipoprotein complex in the liver bile (in Russian), *Biull. Eksp. Biol. Med.* 60, 56, 1965.

Nesterin, M. F. et al., Lipoprotein complex secretion in the liver bile with deficient lipotropic substances in the diet (in Russian), *Vopr. Pitan.*, 24, 56, 1965.

Leites, S. M. et al., The influence of chloropropamide on certain biochemical indices of the liver under normal conditions and in experimental toxic hepatitis (in Russian), *Biull. Eksp. Biol. Med.*, 61, 55, 1966.

Hirooka, T. et al., Selenium metabolism. 3. Serum proteins, lipoproteins and liver injury, *Biochim. Biophys. Acta*, 130, 321, 1966.

Koga, S., Lipoprotein biosynthesis in the liver in various conditions (in Japanese), *J. Jpn. Biochem. Soc.*, 39, 221, 1967.

Heimberg, M. et al., Hepatic lipid metabolism in experimental diabetes, II. Incorporation of (1-14C) palmitate into lipids of the liver and of the d less than 1.020 perfusate lipoproteins, *Biochim. Biophys. Acta*, 137, 435, 1967.

Buckley, J. T. et al., The relationship of protein synthesis to the secretion of the lipid mobility of low density lipoprotein by the liver, *Can. J. Biochem.*, 46, 341, 1968.

Koga, S. et al., Disturbed release of lipoprotein from ethanol-induced fatty liver, *Experientia*, 24, 438, 1968.

Feyrter, F. et al., The chromotropic lipoid-like (lipoproteid-like) granulate of liver cells and the chromotropic granulated liver cells (in German), *Zentralbl. Allg. Pathol. Anat.*, 112, 433, 1969.

Mookerjea, S., Studies on the plasma glycolipoprotein synthesis by the isolated perfused liver: effect of early choline deficiency, *Can. J. Biochem.*, 47, 125, 1969.

Feyrter, F. et al., The chromatropic lipoid (lipoprotein) granulation of the liver cells of man (in German), *Wien. Med. Wochenschr.*, 119, 368, 1969.

Spagna, C. et al., Liver function and experimental amylacetate poisoning: changes in the lipoproteic picture (in Italian), *Folia. Med. (Naples)*, 52, 579, 1969.

Infante, R. et al., Biosynthesis of plasma lipoproteins by the regenerating liver (in French), *Biochim. Biophys. Acta*, 187, 335, 1969.

Judah, J. D. et al., Role of liver-cell potassium ions in secretion of serum albumin and lipoproteins, *Biochem. J.*, 116, 663, 1970.

Mahley, R. W. et al., Identity of very low density lipoprotein apoproteins of plasma and liver Golgi apparatus, *Science*, 168, 380, 1970.

Topping, D. L. et al., Direct stimulation by insulin and fructose of very-low-density lipoprotein secretion by the perfused liver, *Biochem. J.*, 119, 48P, 1970.

Stein, O. et al., Lipoproteins and the liver, *Biochem. J.*, 119, 48P, 1970.

Stein, O. et al., Lipoproteins and the liver, *Prog. Liver Dis.*, 4, 45, 1972.

Tomas, J., Beta-lipoprotein synthesis in the liver after partial hepatectomy and its influencing by adrenal glucocorticoids, *Sb. Ved. Pr. Lek. Fak. Karlovy Univ.*, Suppl. 15, 433, 1972.

Topping, D. L. et al., The immediate effects of insulin and fructose on the metabolism of the perfused liver. Changes in lipoprotein secretion, fatty acid oxidation and esterification, lipogenesis and carbohydrate metabolism, *Biochem. J.*, 126, 295, 1972.

Manilow, M. R. et al., Muscular activity and the degradation of cholesterol by the liver, *Atherosclerosis*, 15, 153, 1972.

Renger, F. et al., Liver findings in primary hyperlipoproteinemia (HLP), *Dtsch. Z. Verdau. Stoffwechselkr.*, 33, 199, 1973.

Práce, K., Liver changes in hyperlipoproteinemia (type IV—Fredrickson's) (with English abstract), *Cesk. Gastroenterol. Vyz.,* 27(1), 1, 1973.

Orci, L. et al., Letter: role of microtubules in lipoprotein secretion by the liver, *Nature (London),* 244, 30, 1973.

Stein, Y. et al., Serum lipoproteins and the liver, synthesis and catabolism, *Horm. Metab. Res.,* Suppl. 4, 16, 1974.

Reaven, E. P. et al., Ultrastructural and physiological evidence for corticosteroid-induced alterations in hepatic production of very low density lipoprotein particles, *J. Lipid Res.,* 15, 74, 1974.

Sniderman, A. D. et al., Paradoxical increase in rate of catabolism of low-density lipoproteins after hepatectomy, *Science,* 183, 526, 1974.

Stein, Y. et al., Serum lipoproteins and the liver, synthesis and catabolism, *Ann. Otolaryngol. Chir. Cervicofac.,* 91(4-5), 16, 1974.

Wilcox, H. G. et al., The effect of sex on certain properties of the very low-density lipoprotein secreted by the liver, *Biochem. Biophys. Res. Commun.,* 58, 919, 1974.

Kirsten, E. S. et al., Regulation of 3-hydroxy-3-methyl-glutaryl coenzyme A reductose in hepatoma tissue culture cells by serum lipoproteins, *J. Biol. Chem.,* 249(19), 6104, 1974.

Starzl, T. E. et al., Letter: portacaval shunt in hyperlipidaemia, *Lancet,* 2(7891) 1263, 1974.

Grajewski, O. et al., Proceedings: serum lipoproteins in experimental liver damage, *Naunyn Schmiedebergs Arch. Pharmacol.,* Suppl. 287, R85, 1975.

Korman, M. G. et al., Fasting serum bile acid levels in the primary hyperlipoproteinemias, *Mayo Clin. Proc.,* 50(2), 76, 1975.

Nervi, F. O. et al., The kinetic characteristics of inhibition of hepatic cholesterogenesis by lipoprotein of intestinal origin, *J. Biol. Chem.,* 250(11), 4145, 1975.

Goodman, Z. D. et al., Transfer of esterified cholesterol from serum lipoproteins to the liver, *Biochim. Biophys. Acta,* 398(2), 325, 1975.

Negishi, I. et al., Effect of phospholipid on the release of beta-lipoprotein in orotic acid-induced fatty liver, *Chem. Pharm. Bull.,* 23(9), 1928, 1975.

Starzl, T. E. et al., Portal diversion. Treatment for glycogen storage disease and hyperlipemia, 233(9), 955, 1975.

Montaguti, U. et al., The liver and abnormal lipoproteins, *Minerva Med.,* 66(65), 3428, 1975.

Nervi, F. O. et al., Ability of six different lipoprotein fractions to regulate the rate of hepatic cholesterogenesis in vivo, *J. Biol. Chem.,* 250(22), 8704, 1975.

Husakova, A. et al., Effect of adrenal glucocorticoids on β-lipoprotein production by resected liver, *Physiol. Bohemoslov.,* 25(2), 147, 1976.

Soler-Argilaga, C. et al., The effect of sex on the quantity and properties of the very low density lipoprotein secreted by the liver in vitro, *J. Lipid Res.,* 17(2), 139, 1976.

Nervi, F. O. et al., Dissociation of β-hydroxy-β-methylglutaryl-CoA reductase activity from the overall rate of cholesterol synthesis in the liver following the intervenous administration of lipid, *J. Biol. Chem.,* 251(12), 3831, 1976.

Wilcox, H. G. et al., Effects of a mixture of a saturated with an unsaturated fatty acid on secretion of the very low density lipoprotein by the liver, *Biochem. Biophys. Res. Commun.,* 73(3), 733, 1976.

Spleen

Ferrans, V. J. et al., The spleen in type 1 hyperlipoproteinemia. Histochemical, biochemical, microfluorometric and electron microscopic observations, *Am. J. Pathol.,* 64, 67, 1971.

Rywlin, A. M. et al., Ceroid histiocytosis of the spleen in hyperlipemia: relationship to the syndrome of the sea-blue histiocyte, *Am. J. Clin. Pathol.,* 56, 572, 1971.

Kidney

Zühlke, V., The reaction of serum lipoproteins and serum immunoglobulins following hemologous kidney transplantation in humans (in German), *Langenbecks Arch. Chir.,* 322, 542, 1968.

Faulk, W. P. et al., Glomerular beta-lipoprotein in childhood renal diseases, *Scand. J. Immunol.,* 3(5), 665, 1974.

Wada, M. et al., Studies on the effects of hemodialysis in plasma lipoproteins, *Trans. Am. Soc. Artif. Intern. Organs* 21, 464, 1975.

Samar, R. E. et al., Lipoprotein binding and hypertriglyceridemia in chronic uremia, *Trans. Am. Soc. Artif. Intern. Organs,* 21, 455, 1975.

Brain

Lees, M. B. et al., Purification of bovine brain white matter proteolipids by dialysis in organic solvents, *Biochim. Biophys. Acta,* 84, 464, 1964.

Ninfo, V. et al., Studies on complex lipids in human tissue. VI. Complex lipids of the human adenohypophysis in normal and pathological conditions (in Italian), *Riv. Anat. Pat. Oncol.,* 26, 473, 1964.

Mezes, M. et al.,Apropos of lipoprotein metabolism by the brain (in Czech), *Bratisl. Lek. Listy,* 45, 33, 1965.

Tsumita, T. et al., Chemical composition of isolated nuclei from bovine brain cortex and white matter, *Jpn. J. Exp. Med.,* 35, 11, 1965.

Kies, M. W. et al. , The relationship of myelin proteins to experimental allergic encephalomyelitis, *Ann. N. Y. Acad. Sci.,* 122, 148, 1965.

Wajda, I. J. et al., Transglutaminase and experimental allergic encephalomyelitis, *Life Sci.,* 4, 1853, 1965.

Komnatnaia, L. I., Lipoproteins of the human brain (in Ukrainian), *Ukr. Biokhim. Zh.,* 37, 243, 1965.

Miul'borg, A. A. et al., Electrophoretic analysis of water-insoluble proteins and lipoproteins of brain tissue (in Russian), *Ukr. Biokhim. Zh.,* 38, 328, 1966.

Robinson, N., Proteolipids and proteins in subcellular particles of human brain, *Clin. Chim. Acta,* 13, 541, 1966.

Curri, S. et al., Identification of free and bound aminoacids in proteolipids extracted from the hypothalamus. Further observation in the chemical nature of corticotrophin releasing factors, *Ric. Sci.,* 36, 753, 1966.

Lees, M. B., Influence of sucrose on the extraction of proteolipids from brain and other tissue, *J. Neurochem.,* 13, 1407, 1966.

Schneider, G. et al., Qualitative and quantitative studies on loss of substance during formol and Carnoy fixation of human brain tissue (in German), *Acta Histochem.,* 28, 227, 1967.

Prensky, A. L. et al., Changes in the amino acid composition of proteolipids of white matter during maturation of the human nervous system, *J. Neurochem.,* 14, 117, 1967.

Lees, M. B. et al., Tryptic hydrolysis of brain proteolipid, *Biochem. Biophys. Res. Commun.,* 28, 185, 1967.

Zand, R., Solution properties and structure of brain proteolipids, *Biopolymers,* 6, 939, 1968.

Bass, N. H., Pathogenesis of myelin lesions in experimental cyanide encephalopathy. A microchemical study, *Neurology,* 18, 167, 1968.

Matveeva, T. S., The pathomorphology of human cerebral cortex neurons in the presence of changes in their lipoprotein structure (in Russian), *Zh. Nevropatol. Psikhiatr.,* 70, 1819, 1970.

Vásques, C. et al., Electron microscopy of proteolipid macromolecules from cerebral cortex, *J. Mol. Biol.,* 52, 221, 1970.

Brodskaia, N. J. et al., The effect of brain tumor removal on several biochemical and hematologic indices (in Russian), *Vopr. Neirokhir.,* 35, 22, 1971.

Lunt, G. G. et al., Association of the acetylcholine-phosphatidyl inositol effect with a "receptor" proteolipid from cerebral cortex, *Nature New Biol.,* 230, 187, 1971.

Stoffyn, P. et al., On the type of linkage binding fatty acids present in brain white matter proteolipid apoprotein, *Biochem. Biophys. Res. Commun.,* 44, 157, 1971.

Craven, P. A. et al., Properties of the glucose 6-phosphate-solubilized brain hexokinase. Evidence for a lipoprotein complex, *Biochim. Biophys. Acta,* 255, 620, 1972.

Barrantes, F. J. et al., Studies on proteolipid proteins from cerebral cortex. 1. Preparation and some properties, *Biochim. Biophys. Acta,* 263, 368, 1972.

Volkova, T. N. et al., Clinico-biochemical studies of patients with hypothalamic pathology in the practice of occupation medicine, *Zh. Nevropatol. Psikhiatr.,* 73, 1808, 1973.

Lu, C. Y. et al., Isolation of acidic lipoproteins from brain chromatin. Their relation to the acidic nonhistone proteins, *FEBS Lett.,* 34, 48, 1973.

Carnegie, P. R. et al., Phosphorylation of selected serine and threonine residues in myelin basic protein by endogenous and exogenous protein kinases, *Nature (London),* 249, 147, 1974.

Wallgren, H. et al., Ethanol-induced changes in cation-stimulated endosine triphosphatase activity and lipid-proteolipid labeling of brain microsomes, *Adv. Exp. Med. Biol.,* 59, 23, 1975.

Tandler, C. J. et al., The use of tetrahydrofuran for delipidation and water solubilization of brain proteolipid proteins, *Life Sci.,* 17(9), 1407, 1975.

Mezesova, V. et al., Phospholipid metabolism in soluble and unsoluble lipoproteins in the brain (in Slovenian), *Bratisl. Lek. Listy,* 65(1), 23, 1976.

Ovary

Guraya, S. S., Histochemical analysis of the interstitial gland tissue in the human ovary at the end of pregnancy, *Am. J. Obstet. Gynecol.,* 96, 907, 1966.

Placenta

Paluszak, J. et al., Studies on the lipolytic system of the human placenta, *Acta Med. Pol.,* 8, 179, 1967.

Salivary Gland

Kurata, Y. et al., Immunological studies of insoluble lipoproteins. II. On the salivary gland-characteristic antigens, *Int. Arch. Allerg.*, 35, 392, 1969.

Lung (surfactant)

Abrams, M. E., Isolation and quantitative estimation of pulmonary surface-active lipoprotein, *J. Appl. Physiol.*, 21, 718, 1966.

Takruri, H. et al., Interactions of surfactants with lipoproteins, *J. Pharm. Sci.*, 55, 979, 1966.

Wilhite, J. L. et al., Lung surfactant: a review, *J. Tenn. Med. Assoc.*, 60, 524, 1967.

Heinemann, H. O., Surfactant of the lung, *Adv. Intern. Med.*, 14, 83, 1968.

Garbagni, R. et al., Effects of lipid loading and fasting on pulmonary surfactant, *Respiration*, 25, 458, 1968.

Rüfer, R., The influence of surface active substances on alveolar mechanics in the respiratory distress syndrome, *Respiration*, 25, 441, 1968.

Permutt, S., Pulmonary surfactant: its depletion and regeneration in excised lungs, *Aspen Emphysema Conf.*, 9, 43, 1968.

Hackney, J. D. et al., Organotypic culture of mammalian lung—studies on morphology ultra-structure, and surfactant, *Aspen Emphysema Conf.*, 9, 13, 1968.

Henry, J. N., The effect of shock on pulmonary alveolar surfactant. Its role in refractory respiratory insufficiency of the critically ill or severely injured patient, *J. Trauma*, 8, 756, 1968.

Greenfield, L. J. et al., The role of surfactant in the pulmonary response to trauma, *J. Trauma*, 8, 735, 1968.

Zelkowitz, P. S. et al., Effects of ether and halothane inhalation on pulmonary surfactant, *Am. Rev. Resp. Dis.*, 98, 795, 1968.

Reynolds, E. O. et al., Hyaline membrane disease, respiratory distress, and surfactant deficiency, *Pediatrics*, 42, 758, 1968.

Klein, R. M. et al., Purification of pulmonary surfactant by ultracentrifugation, *J. Appl. Physiol.*, 25, 654, 1968.

Schoedel, W. et al., Time-dependent changes in the surface pressure of alveolar surface layers in the Langmuir trough (in German), *Pflueger Arch.*, 306, 20, 1969.

Schoedel, W. et al., Determination of surface characteristics of lung alveolar surfactant by a bubble method (in German), *Pflueger Arch.*, 307, R15, 1969.

Modell, J. H. et al., The effects of wetting and antifoaming agents on pulmonary surfactant, *Anesthesiology*, 30, 164, 1969.

Galdston, M. et al., Isolation and characterization of a lung lipoprotein surfactant, *J. Colloid Interface Sci.*, 29, 319, 1969.

Rhodes, M. L., Pulmonary surfactant, *N. Engl. J. Med.*, 280, 331, 1969.

Pasquier, C. et al., Fixation of lanthanum on lung intra-alveolar lipoproteins (in French), *C. R. Acad. Sci. Ser. D*, 268, 1129, 1969.

Redding, R. A. et al., New activity on lung surfactant, *N. Engl. J. Med.*, 280, 1298, 1969.

Jiménez, J. M., Studies on the pulmonary surface-active lipoprotein. 1. Isolation and chemical characterization of its constituents, *Rev. Biol. Trop.*, 17, 125, 1969.

Rosenberg, E., Analysis of the properties of pulmonary surfactant using modified Wilhelmy balances, *Resp. Physiol.*, 7, 72, 1969.

Mcclenahan, J. B. et al., Effect of ethanol on surfactant of ventilated lungs, *J. Appl. Physiol.*, 27, 90, 1969.

Lloyd, J. K., The pulmonary surfactant, *Can. Med. Assoc. J.*, 101, 109, 1969.

Pariente, R., Pulmonary surfactant. Current physiological and physiopathological concepts, *Presse Med.*, 77, 1871, 1969.

Benzer, H. et al., Significance of surface tension in cystic pulmonary diseases (in German), *Pneumonologie*, 143, 127, 1970.

Pasquier, C. et al., Analysis of fatty acids in the purified lipids of pulmonary surfactant (in French), *Ann. Biol. Clin. (Paris)*, 28, 343, 1970.

Pasquier, C. et al., Biochemical aspect of alveolar retention preceding liquid aerosols of rare earths (in French), *Inhaled Part Vap.*, 3, 301, 1970.

Iwainsky, H. et al., Determination and marking of surface-active lung lipoproteins (in German), *Acta Biol. Med. Ger.*, 24, 47, 1970.

Corrin, B., Phagocytic potential of pulmonary alveolar epithelium with particular reference to surfactant metabolism, *Thorax*, 25, 110, 1970.

Weisser, K., Idiopathic respiratory distress syndrome in premature infants (in German), *Anaesthesist*, 19, 101, 1970.

Dermer, G. B., The fixation of pulmonary surfactant for electron microscopy. II. Transport of surfactant through the air-blood barrier, *J. Ultrastruct. Res.,* 31, 229, 1970.

Onellet, Y., Pulmonary surfactant: anti-atelectasis factor (in French), *Laval Med.,* 41, 627, 1970.

Clements, J. A., Pulmonary surfactant, *Am. Rev. Resp. Dis.,* 101, 984, 1970.

Thomas, P. A. et al., Pulmonary surfactant in lungs of patients with desquamative interstitial pneumonia, *Am. Rev. Resp. Dis.,* 101, 967, 1970.

Scarpelli, E. M. et al., A search for the surface-active pulmonary lipoprotein, *Am. Rev. Resp. Dis.,* 102, 285, 1970.

Clements, J. A. et al., Pulmonary surfactant and evolution of the lungs, *Science,* 169, 603, 1970.

Ellison, L. T. et al., Alteration in pulmonary surfactant associated with cardiopulmonary bypass. Further observations and conclusions, *Ann. Thorac. Surg.,* 10, 258, 1970.

Dermer, G. B., The pulmonary surfactant content of the inclusion bodies found within type II alveolar cells, *J. Ultrastruct. Res.,* 33, 306, 1970.

Nozaki, M. et al., Relationship between surface properties of lung washings and ventilatory functions of patients with pulmonary tuberculosis before and after the operation (in Japanese), *Iryo,* 24, 915, 1970.

Scarpelli, E. M., The pulmonary surfactant system, *Clin. Notes Resp. Dis.,* 9, 3, 1970.

Sláma. H. et al, Determination of surface active properties of lung alveolar surfactants by a bubble method (in German), *Pfluegers Arch.,* 322, 355, 1971.

Schoedel, W. et al., Time-dependent behaviour of lung alveolar surfactant films (in German), *Pfluegers Arch.,* 322, 336, 1971.

Levitsky, S. et al., Depletion of alveolar surface active material by transbronchial plasma irrigation of the lung, *Ann. Surg.,* 173, 107, 1971.

Thomas, P. A., A comparative study of lung compliance and pulmonary surfactant activity in human subjects, *Ann. Thorac Surg.,* 11, 133, 1971.

Spitzer, H. L. et al., The biosynthesis and turnover of surfactant lecithin and protein, *Arch. Intern. Med. Med.,* 127, 429, 1971.

Morgan, T. E., Biosynthesis of pulmonary surface-active lipid, *Arch. Intern. Med.,* 127, 401, 1971.

Ramirez, J. et al., Biochemical comparison of human pulmonary washings, *Arch. Intern. Med.,* 127, 395, 1971.

Clemens, J. A., Comparative lipid chemistry of lungs, *Arch. Intern. Med.,* 127, 387, 1971.

Cherng, M. J. et al., The characterization of pulmonary surfactant lipoprotein, *Chest,* Suppl. 59, 65, 1971.

vonWichert, P., The pulmonary surface active system (in German), *Med. Welt.,* 43, 1694, 1971.

Scarpelli, E. M. et al., Protein and lipid-protein fractions of lung washings: immunological characterization, *J. Appl. Physiol.,* 34, 750, 1973.

Groniowski, J. et al., Electron microscopic observations on pulmonary alveolar lipoproteinosis, *Ann. Med. Sect. Pol. Acad. Sci.,* 19(2), 109, 1974.

Aonuma, S. et al., Studies on biologically active substances formed by bacilli. VI. Inhibitory effect of the serum calcium decreasing active lipoprotein alpha-I to cytolytic activity of surfactin (in Japanese with English abstract), *J. Pharm. Soc. Jpn.,* 94(5), 587, 1974.

Sawada, H. et al., Reassembly in vitro of lung surfactant lipoprotein, *Biochem. Biophys. Res. Commun.,* 74(3), 1263, 1977.

LIPOPROTEINS IN CELLS

Membranes

Ways, P. et. al., The role of serum in acanthocyte autohemolysis and membrane lipid composition, *J. Clin. Invest.,* 43, 1322, 1964.

Vandenheuvel, F. A., Structural studies of biological membranes, the structure of myelin, *Ann. N. Y. Acad. Sci.,* 122, 57, 1965.

Joo, F. et al., Lipoprotein substances in the post-synaptic membrane of the myoneural junction, *Acta Histochem.,* 20, 280, 1965.

Razin, S. et al., Membrane subunits of Mycoplasma laidlawii and their assembly to membranelike structures, *Proc. Natl. Acad. Sci. U.S.A.,* 54, 219, 1965.

Harwalkar, V. R. et al., Effect of dissociating agents on physical properties of fat globule membrane fractions, *J. Dairy Sci.,* 48, 1139, 1965.

Deborin, G. A. et al., Study of the surface films of the lipoprotein complex from the M. lysodeikticus membrane (in Russian), *Dokl. Akad. Nauk SSSR,* 166, 231, 1966.

Pollak, J. K. et al., The formation of a membranous reticulum by the interaction of reticulosomes and micellar phospholipids, *J. Mol. Biol.,* 16, 564, 1966.

Benson, A. A., On the orientation of lipids in chloroplast and cell membranes, *J. Am. Oil Chem. Soc.,* 43, 265, 1966.

Vandenheuvel, F. A. Lipid-protein interactions and cohesional forces in the lipoproteins system of membranes, *J. Am. Oil Chem. Soc.,* 43, 258, 1966.

Blazheevich, N. V. et al., Effect of vitamin D on the adenosine triphosphatase activity and the stability of erythrocyte membranes (in Russian), *Vopr. Med. Khim.,* 12, 424, 1966.

Brown, A. D. et al., Sedimentation, viscosity and partial specific volumes of membrane proteins and lipoproteins, *Biochem. J.,* 103, 24, 1967.

Graham, J. M. et al., The binding of sterols in cellular membranes, *Biochem. J.,* 103, 16C, 1967.

Terry, T. M. et al., Characterization of the plasma membrane of Mycoplasma laidlawii. II. Modes of aggregation of solubilized membrane components, *Biochim. Biophys. Acta,* 135, 391, 1967.

Dreher, K. D. et al., The stability and structure of mixed lipid monolayers and bilayers. 1. Properties of lipid and lipoprotein monolayers on OsO_4 solutions and the role of cholesterol, retinol, and tocopherol in stabilizing lecithin monolayers, *J. Ultrastruct. Res.,* 19, 586, 1967.

de Robertis, D. et al., Cholinergic binding capacity of proteolipids from isolated nerve-ending membranes, *Science,* 158, 928, 1967.

Bruckdorfer, K. R., et al., The incorporation of steroid molecules into lecithin sols, β-lipoproteins and cellular membranes, *Eur. J. Biochem.,* 4, 512, 1968.

Ji, T. H. et al, Association of lipids and proteins in chloroplast lamellar membrane, *Biochim. Biophys. Acta,* 150, 686, 1968.

Hatch, F. T. et al., Amino-acid composition of soluble and membranous lipoproteins, *Nature (London),* 218, 1166, 1968.

Prezbindowski, K. S. et al., Membrane structure: binary membranes of mitochondrial cristae, *Exp. Cell. Res.* 57, 385, 1969.

Lievremont, M. et al., Cholinergic receptor in the neuromuscular junction. Intervention of a lipoprotein (in French), *Bull. Soc. Chim. Biol.,* 52, 23, 1970.

Tiffany, J. M. et al., The interaction of fetuin with phosphatidylcholine monolayers. Characterization of a lipoprotein membrane system suitable for the attachment of myxoviruses, *Biochem. J.,* 117, 377, 1970.

Hauser, H. et al., Physical studies of the interactions of acetylcholine chloride with membrane constituents, *Biochem. J.,* 120, 329, 1970.

de Robertis, E. et al., Acetylcholinesterase and acetylcholine proteolipid receptor: two different components of electroplax membranes, *Biochim. Biophys. Acta,* 219, 388, 1970.

Claude, A., Growth and differentiation of cytoplasmic membranes in the course of lipoprotein granule synthesis in the hepatic cell. 1. Elaboration of elements of the Golgi complex, *J. Cell Biol.,* 47, 745, 1970.

Shapot, V. S. et al., Liporibonucleoprotein as an integral part of animal cell membranes, *Progr. Nucleic Acid Res. Mol. Biol.,* 11, 81, 1971.

Ochva, E. et al., The effect of noradrenaline on artificial lipidic membranes containing a proteolipid with adrenoceptor properties, *J. Pharm. Pharmacol.,* 24, 75, 1972.

Folch-Pi, J. et al., Proteolipids from membrane systems, *Ann. N. Y. Acad. Sci.,* 195, 86, 1972.

Wooding, F. B., Milk microsomes, viruses, and the milk fat globule membrane, *Experientia,* 28, 1077, 1972.

Nemerson, Y. et al., Activation of a proteolytic system by a membrane lipoprotein: mechanism of action of tissue factor, *Proc. Natl. Acad. Sci. U.S.A.,* 70, 310, 1973.

Okoda, S., Studies on tissue-specific antigens and cellular membranes, *Acta Pathol. Jpn.,* 23, 249, 1973.

Bell, F. P., Transfer of cholesterol between serum lipoproteins, isolated membranes, and intact tissue, *Exp. Mol. Pathol.,* 19, 293, 1973.

Kurebe, M., Proceedings: mechanism of intestinal mucosal barrier absorption. 3. Property of aggregation by divalent cations of separated lipoprotein from microvillus membrane, *Jpn. J. Pharmacol.,* Suppl. 24, 118, 1974.

Moran, A. et al., The effect of prymnesin on the electric conductivity of thin lipid membranes, *J. Membr. Biol.,* 16, 237, 1974.

Copley, A. L. et al., The reducing action of highly purified gamma globulin and beta lipoprotein on the viscous resistance of surface layers of fibrinogen, *Thromb. Res.,* 4, 193, 1974.

Schubert, D. et al., Proceedings: protein-lipid and protein-protein interaction in recombined lipoprotein membranes, *Hoppe Seylers Z. Physiol. Chem.,* 355(10), 1253, 1974.

Sogor, B. V. et al., Studies of a serum albumin-liposome complex as a model lipoprotein membrane, *Biochim. Biophys. Acta,* 375(3), 363, 1975.

Moshkov, D. A. et al., IR-spectroscopic study of cytochrome C-phospholipid lipoprotein and proteolipid model membranes (with English abstract), *Biofizika,* 20(2), 233, 1975.

Shapot, V. S. et al., Fractionation of liporibonucleoprotein complexes from animal cell plasma membranes, *Dokl. Akad. Nauk. SSSR,* 222(1), 240, 1975.

Laggner, P. et al., The interaction of a proteolipid from sacroplasmic reticulum membranes with phospholipids. A spin label study, *Arch. Biochem. Biophys.,* 170(1), 92, 1975.

Kondrashin, A. A. et al., Role of phospholipids in the generation of membrane potentials by proteoliposomes (in Russian with English abstract), *Biokhimia,* 40(5), 1071, 1975.

Reader, T. A. et al. The incorporation of hydrophobic protein receptors and artificial lipid membranes, *Rec. Adv. Stud. Cardiac Struct. Metab.,* 9, 149, 1976.

Tall, A. R. et al., Salubilisation of phospholipid membranes by human plasma high density lipoproteins, *Nature (London)*, 265(5590), 163, 1977.

Lee, N. et al., Optical properties of an outer membrane lipoprotein from Escherichia coli, *Biochim. Biophys. Acta*, 465(3), 650, 1977.

Nerve Cells

Wolfgram, F., A new proteolipid fraction of the nervous system. 1. Isolation and amino acid analysis, *J. Neurochem.*, 13, 461, 1966.

Rumsby, M. G. et al., The action of organic solvents on the myelin sheath of peripheral nerve tissue. II. Short-chain aliphatic alcohols, *J. Neurochem.*, 13, 1509, 1966.

Landolt, R. et al., Regional distribution of some chemical structural components of the human nervous system. II. Cerebrosides, proteolipid proteins and residue proteins, *J. Neurochem.* 13, 1453, 1966.

Lapetina, E. G. et al., Action of Triton X-100 on lipids and proteolipids of nerve-ending membranes, *Life Sci.*, 7, 203, 1968.

Porcellati, G., Protein synthesis in peripheral nerve fibers, with some considerations on the synthesis in the central nervous system (in Italian), *Acta Neurol. (Naples)*, 23, 474, 1968.

Barnola, F. V. et al., Ionic channels and nerve membrane lipoproteins: DDT-nerve membrane interaction, *Int. J. Neurosci.*, 1, 309, 1971.

Mann, D. M. et al., Lipoprotein pigments—their relationship to ageing in the human nervous system. I. The lipofuscin content of nerve cells, *Brain*, 97(3), 481, 1974.

Myelin

Wolfgram, F., Macromolecular constituents of myelin, *Ann. N. Y. Acad. Sci.*, 122, 104, 1965.

Thompson, E. B. et al., Current studies on the lipids and proteins of myelin, *Ann. N. Y. Acad. Sci.*, 122, 129, 1965.

Klee, C. B. et al., Amino acid incorporation into proteolipid of myelin in vitro, *Proc. Natl. Acad. Sci., U.S.A.*, 53, 1014, 1965.

Murdock, D. D. et al., Preparation of myelin using the L-4 zonal ultracentrifuge, *Can. J. Biochem.*, 47, 818, 1969.

Wolman, M. et al., Phase transitions in myelin from lamellar into water in oil and oil in water patterns, *Proc. Soc. Exp. Biol. Med.*, 131, 1460, 1969.

Soller, M. et al., Lipoprotein subunit of myelin: isolation, characterization, in vitro formation of myelin membranes, *Trans. Am. Neurol. Assoc.*, 95, 309, 1970.

Folch-Pi, J., Distribution of constituents in myelin, *Neurosci. Res. Prog. Bull.*, 9, 554, 1971.

Palo, J. et al., Biochemical studies on human peripheral nerve myelin, *Acta Neurol. Scand. Suppl.*, 51, 405, 1972.

Banik, N. L. et al., Lipid and basic protein interaction of myelin, *Biochem. J.*, 143(1), 39, 1974.

Feinstein, M. B. et al., Reactions of fluorescent probes with normal and chemically modified myelin basic protein and proteolipid. Comparisons with myelin, *Biochemistry*, 14(14), 3049, 1975.

Erythrocytes

Schiebel, W. et al., The mechanism of erythrocyte sedimentation. X. Studies on the inhibitor of accelerated erythrocyte sedimentation (in German), *Hoppe Seylers Z. Physiol. Chem.*, 338, 198, 1964.

Vulpis, G. et al., Red cell agglutination by S protein, a lipoprotein from arythrocyte stroma, *Experientia*, 20, 211, 1964.

Basford, J. M. et al., Exchange of cholesterol between human β-lipoproteins and erythrocytes, *Biochim. Biophys. Acta.*, 84, 764, 1964.

Douste-Blazy L. et al., Separation of erythrocyte lipoproteins on dextran gels (in French), *Ann. Biol. Clin. (Paris)*, 23, 265, 1965.

Sakagami, T. et al., Behavior of plasma lipoproteins during exchange of phospholipids between plasma and erythrocytes, *Biochim. Biophys. Acta*, 98, 111, 1965.

Nicoli, J. , Erythrocyte receptors for arboviruses. Cell lipoproteins (in French), *Ann. Inst. Pasteur (Paris)*, 109, 472, 1965.

Morgan, T. E. et al., Solubilization and characterization of a lipoprotein from erythrocyte stroma, *Biochemistry*, 5, 1099, 1966.

Khachadurian, A. K. et al., Persistent elevation of the erythrocyte sedimentation rate (ESR) in familial hypercholesterolemia. With a preliminary report on the effect of plasma β-lipoproteins on ESR, *J. Med. Liban*, 20, 31, 1967.

Kamat, V. B. et al., Proton magnetic resonance (PMR) spectra of erythrocyte membrane lipoproteins and apoproteins, *Chem. Phys. Lipids*, 4, 323, 1970.

Quarfordt, S. H. et al., Quantitation of the in vitro free cholesterol exchange of human red cells and lipoproteins, *J. Lipid Res.*, 11, 528, 1970.

Zahler, P. et al., Reconstitution of membranes by recombining proteins and lipids derived from erythrocyte stroma, *Biochim. Biophys. Acta,* 219, 320, 1970.

Chevallier, F. et al., Exchange of free cholesterol between plasma lipoproteins and erythrocytes, and globular origin of plasma esterified cholesterol in vitro (in French), *C. R. Acad. Sci. (D),* 272, 1028, 1971.

Kang, K. W. et al., Genetic variability of human plasma and erythrocyte lipids, *Lipids,* 6, 595, 1971.

Cooper, R. A. et al., The relationship between serum lipoproteins and red cell membranes in abetalipoproteinemia: deficiency of lecithin: cholesterol acyltransferase, *J. Lab. Clin. Med.,* 78, 323, 1971.

Juliano, R. L. et al., Properties of an erythrocyte membrane lipoprotein fraction, *Biochim. Biophys. Acta,* 249, 227, 1971.

Jacob, H. S., Concomitant abnormalities in red cell membrane lipoprotein conformation and rheology in hereditary spherocytosis and acanthocytosis, *Biorheology,* 8, 109, 1971.

Hollander, F. et al., In vitro exchange of cholesterol between-and-lipoproteins in rat plasma and between the lipoprotein and erythrocytes (in French with English abstract), *Biochim. Biophys. Acta,* 260, 110, 1972.

Hamnström, B. et al., Influence of species of origin of red blood cells on in vitro inhibition of streptolysis O by serum lipoproteins, *Int. Arch. Allergy Appl. Immunol.,* 42, 590, 1972.

Schubert, D. et al., Association of protein fractions and lipids from human erythrocyte membranes. I. Studies on a strongly bound protein fraction, *Hoppe Seylers Z. Physiol. Chem.,* 353, 1034, 1972.

Bell, F. P. et al., Membrane-lipid exchange: exchange of cholesterol between porcine serum lipoproteins and erythrocytes, *Pathology,* 4, 205, 1972.

Redman, C. M., Proteolipid involvement in human erythrocyte membrane function, *Biochim. Biophys. Acta,* 282, 123, 1972.

Casu, A. et al., Analysis of the lipo-glycoprotein components of sheep erythrocyte membranes using disc electrophoresis, *Haematologica,* 58, 689, 1973.

Schubert, D., Association of protein fractions and lipids from human erythrocyte membranes. II. Studies on a loosely bound protein fraction, *Hoppe Seylers Z. Physiol. Chem.,* 354(7), 781, 1973.

Springer, G. F. et al., Functional aspects and nature of the lipopolysaccharide-receptor of human erythrocytes, *J. Infect. Dis.,* Suppl. 128, 202, 1973.

Cho, K. S. et al., Interactions of acyl carnitines and other lysins with erythrocytes and reconstituted erythrocyte lipoproteins, *Biochim. Biophys. Acta,* 318, 50, 1973.

Wehrli, E. et al., Recombination of human erythrocyte apoprotein and lipid. Visualization of apoprotein-lipid bilayer complex by freeze-etching, *J. Supramol. Struct.,* 2, 71, 1974.

Morse, P. D., 2nd, Recombination of human erythrocyte apoprotein and lipid. Interaction of apoprotein and lipid at the air-water interface, *J. Supramol. Struct.,* 2, 60, 1974.

Longdon, R. G., Serum lipoprotein apoproteins as major protein constituents of the human erythrocyte membrane, *Biochim. Biophys. Acta,* 342, 213, 1974.

Carey, C., et al., A critical evaluation of the proposal that serum apolipoproteins are the major constituents of the human erythrocyte membrane, *Biochim. Biophys. Acta,* 401(1), 6, 1975.

Kirchhof, B. et al., J blood group active lipoproteins extracted from bovine erythrocytes, *Anim. Blood Groups Biochem. Genet.,* 7(1),51, 1976.

Beaumont, J. L. et al., A new serum lipoprotein associated erythrocyte antigen which reacts with a monoclonal IgM. The stored human red blood cell SHRBC antigen, *Vox Sang.,* 30(1), 36, 1976.

Wehrli, E. et al., Effects of pH during recombination of human erythrocyte membrane apoprotein and lipid, *Biochim. Biophys. Acta,* 426(2), 271, 1976.

Berger, K. V. et al., Reversible transformation of precipitated and nonprecipitated lipoproteins recombined from patients and lipids of erythrocyte membranes, *Z. Naturforsche C, 31(3—4), 174, 1976.*

Leukocytes

Thumb, N. et al., Study on the relations of blood basophils and single serum lipid fractions (in German), *Wien. Z. Inn. Med.,* 47, 26, 1966.

Lynn, M. et al., In vivo distribution of Pseudomonas aeruginosa slime glycolipoprotein: association with leukocytes, *Infect. Immun.,* 15(1), 109, 1977.

Lymphocytes

Rothblat, G. H., The effect of serum components on sterol biosynthesis in L cells, *J. Cell Physiol.,* 74, 163, 1969.

Chisari, F. V. et al., Lymphocyte E rosette inhibitory factor: a regulatory serum lipoprotein, *J. Exp. Med.,* 142(5), 1092, 1975.

Reichl, D. et al., Uptake and catabolism of low density lipoprotein by human lymphocytes, *Nature (London),* 260(5552), 634, 1976.

Curtiss, L. K. et al. Regulatory serum lipoproteins: regulation of lymphocyte stimulation by a species of low density lipoprotein, *J. Immunol.,* 116(5), 1452, 1976.

Ho, Y. K. et al., Regulation of low density lipoprotein receptor activity in freshly insolated human lymphocytes, *J. Clin. Invest.,* 58(6), 1465, 1976.

THROMBOCYTES

Moga, A. et al., Study of adhesiveness of thrombocytes in different phases of development of arteriosclerosis: Correlations with blood lipid disorders, *Stud. Cercet. Med. Intern.*, 7, 589—595, 1966 (in Rumanian).

Martynov, Sm. et al., Role of β-lipoproteinemia in the development of nonspecific of β-lipoproteinemia in the development of nonspecific agglutination (aggregation) of thrombocytes, *Prob. Gemat.*, 14, 24—28, 1969 (in Russian).

Storozhev, A. L. et al., Adrenaline-induced release from thrombocytes of lipoprotein, alkaline phosphates and 5'-nucleotidase, *Biull Eksp. Biol. Med.*, 81 (5), 545—547, 1976 (in Russian with English abstract).

Fibroblasts

Brown, M. S. et al., Restoration of a regulatory response to low density lipoprotein in acid lipase-deficient human fibroblasts, *J. Biol. Chem.*, 251(11), 3277, 1976.

Anderson, R. G. et al., Localization of low density lipoprotein receptors on plasma membrane of normal human fibroblasts and their absence in cells from a familial hypercholesterolemia homozygote, *Proc. Natl. Acad. Sci., U.S.A.*, 73(7), 2434, 1976.

Brown, M. S. et al., Analysis of a mutant strain of human fibroblasts with a defect in the internalization of receptor-bound low density lipoprotein, *Cell, 9(4PT2) 663, 1976.*

Different Other Cells

Green, C., Structral lipoproteins of the mucosal cell, *Biochem. J.*, 90, 543, 1964.

Day, A. J. et al., The incorporation of ^{14}C-labeled particulate cholesterol into lipoprotein by macrophages, *J. Atheroscler. Res.*, 4, 497, 1964.

Pollak, J. K. et al., The isolation and characterization of a new cycloplasmic component, the 'reticulosome', *Biochem. J.*, 93, 36C, 1964.

Rejnek, J. et al., The effect of human serum fractions on cell growth and its relation to the state of serum χ-lipoprotein, *Exp. Cell Res.*, 37, 65, 1965.

Clegg, J. A., Secretion of lipoprotein by Mehlis' gland in Fasciola hepatica, *Ann. N. Y. Acad. Sci.*, 118, 969, 1965.

Casley-Smith, J. R. et al., The uptake of lipid and lipoprotein by macrophages in vitro: an electron microscopical study, *Q. J. Exp. Physiol.*, 51, 1, 1966.

Tsumita, T. et al., Systemic fractionation of calf thymus cell nucleus and characterization of water insoluble lipoproteins, *Jpn. J. Exp. Med.*, 39, 519, 1969.

Tomonaga, M. et al., The ultrastructure of protagon (pi) granules (in German), *Acta Neuropathol.*, 15, 56, 1970.

Spector, A. A. et al., Utilization of free fatty acids complexed to human plasma lipoproteins by mammalian cell suspension, *J. Lipid Res.*, 12, 545, 1971.

Stein, O. et al., The removal of cholesterol from Landschutz ascites cells by high-density apolipoproteins, *Biochim. Biophys. Acta*, 326, 232, 1973.

Saito, M. et al., Chemical conversion of lysine residue of chromatin proteins and other proteins to alpha-aminoadipic acid residue in the presence of chlorine. Reexamination on the occurrence of alpha-aminoadipic acid in a lipoprotein of cell nucleus, *Jpn. J. Exp. Med.*, 43, 523, 1973.

Koenig, H., The soluble acidic lipoproteins (SALPS) of storage granules. Matrix constituents which may bind stored molecules, *Adv. Cytopharmacol.*, 2, 273, 1974.

Hardwicke, P. M. et al., The effect of delipidation on the adenosine triphosphatase of sarcoplasmic reticulum. Electron microscopy and physical properties, *Eur. J. Biochem.*, 42, 183, 1974.

Zotikov, A. A., A cytofluorimetric approach to analysis of lipoprotein complexes in the cell (with English abstract), *Tsitologiia*, 16(4), 476, 1974.

Brenneman, D. E. et al., Utilization of ascites plasma very low density lipoprotein triglycerides by Ehrlich cells, *J. Lipid Res.*, 15, 309, 1974.

Carvalho, A. C. et al., Platelet function in hyperlipoproteinemia, *N. Engl. J. Med.*, 290, 434, 1974.

Anghileri, L. J., The metabolism of calcium and phosphorus and the cellular lipoproteins of the Ehrilich's ascites tumor cell, *Int. J. Clin. Pharmacol.*, 10, 23, 1974.

Bates, S. R. et al., Regulation of cellular sterol flux and synthesis by human serum lipoproteins, *Biochim. Biophys. Acta*, 360, 38, 1974.

Kichev, G., New protein-glycolipoprotein complex formation in diluted semen, in *Physical and Chemical Bases of Biological Information Transfer,* Vassileva-Popova, J. G., Ed., Plenum Press, New York, 1975.

Stein, Y. et al., Role of serum lipoproteins in the transport of cellular cholesterol, Greten H., Ed., *Lipoprotein Metabolism,* Springer-Verlag, Berlin, 1976, 99.

Gassmann, A. E., et al., Estimation of lipoprotein phagocytosis by intraoperative acquired tissue macrophages in humans (in German), *Helv. Chir. Acta*, 43(1-2), 161, 1976.

Kovanen, P. T. et al., Cholesterol exchange between fat cells, chylomicrons and plasma lipoproteins, *Biochim. Biophys. Acta*, 441(3), 357, 1976.

Glickman, R. M. et al., Localization of apolipoprotein B in intestinal epithelial cells, *Science,* 193(4259), 1254, 1976.

Arbogast, L. Y. et al., Cellular cholesterol ester accumulation induced by free cholesterol-rich lipid dispersions, *Proc. Natl. Acad. Sci. U.S.A.,* 73(10), 3680, 1976.

McGee, R. et al., Regulation of fatty acid biosynthesis in Ehrlich cells by ascites tumor plasma lipoproteins, *Lipids,* 12(1), 66, 1977.

Cell Organells

Neifakh, S. A. et al.,   Kinasine   — a glycolysis stimulating protein of heart mitochondria, *Biochem. Biophys. Res. Commun.,* 14, 86, 1964.

Repin, V. S., On the site of action of the mitochondrial protein enhancing factor in the glycolysis reaction system (in Russian), *Biokhimiia,* 29, 255, 1964.

Kazakova, T. B., Formation of a molecular complex between the mitochondrial proteins actomyosin and kinazine (in Russian), *Vop. Med. Khim.,* 10, 324, 1964.

Sjöstrand, F. S. et al., Myelin-like figures formed from mitochondrial material, *Nature (London),* 202, 1075, 1964.

Neifakh, S. A. et al., Glycolysis-stimulating protein factor in cardiac mitochondria, *Fed. Proc.,* 23, 677, 1964.

Manson, L. A. et al., Microsomal lipoproteins as transplantation antigens, *Ann. N. Y. Acad. Sci.,* 120, 251, 1964.

Wolman, M., Study of the nature of lysosomes and of their acid phosphotase, *Z. Zellforsch. Mikrosk. Anat.,* 65, 1, 1965.

Neifakh, S. A. et al., Mechanism of the controlling function of mitochondria, *Biochim. Biophys. Acta,* 100, 329, 1965.

Day, A. J. et al., Removal of double-labelled lipid mixtures and double-labelled lipoprotein preparations by reticulo-endothelial cells, *J. Atheroscler. Res.,* 5, 466, 1965.

Napier, E. A., Jr. et al., Cellular lipoproteins. 1. The isolation of lipoprotein fractions from cellular mitochondria and microsomes, *J. Biol. Chem.,* 240, 4244, 1965.

Feyrter, F. et al., On the relation between chromotropic lipoids (lipoproteins) and mitochondria *(in German), Zentralbl. Allg. Pathol. Anat.,* 109, 40, 1966.

Byington, K. H. et al., On the fragmentation of mitochondria by diethylstilesterol. 1. Conditions for maximizing fragmentation, *Arch. Biochem.,* 128, 762, 1968.

Coleman, R. et al., Structural and fractional modifications induced in muscle microsomes by trypsin, *Biochim. Biophys. Acta,* 173, 51, 1969.

Cattell, K. J. et al., The isolation of dicyclohexylcarbodi-imide-binding proteins from mitochondrial membranes, *Biochem. J.,* 117, 1011, 1970.

Goldstone, A. et al., Isolation and characterization of acidic lipoprotein in renal and hepatic lysosomes, *Life Sci.,* 9, (11), 607, 1970.

Stone, W. L., Hydrophobic interaction of alkanes with liposomes and lipoproteins, *J. Biol. Chem.,* 250(11), 4368, 1975.

Stoffel, W., Carbon 13 NMR- spectroscopic studies on liposomes and human high density lipoproteins, *Lipoprotein Metabolism,* Greten, H., Ed., Berlin, Springer-Verlag, Berlin, 1976, 132.

Platelets

Greenwalt, T. J. et al., Characterization of autoimmune antiplatelet activity, *Bibl. Haematol. (Basel),* 23, 7, 1965.

Spaet, T. H. et al., Studies on platelet factor-3 availability, *Br. J. Haematol.,* 11, 269, 1965.

White, J. G. et al., The ultrastructural localization and release of platelet lipids, *Blood,* 27, 167, 1966.

Moga, A. et al., Study of adhesiveness of thrombocytes in different phases of development of arteriosclerosis: Correlations with blood lipid disorders (in Rumanian), *Stud. Cercet. Med. Intern.,* 7, 589, 1966.

Bolton, C. H. et al., Nature of the transferable factor which causes abnormal platelet behaviour in vascular disease, *Lancet,* 2, 1101, 1967.

Farbiszewski, R. et al., Enhancement of platelet aggregation and adhesiveness by β-lipoprotein, *J. Atheroscler. Res.,* 8, 988, 1968.

Rodriguez-Erdmann, F., Production of the Sanarelli-Schwartzman phenomenon by platelet-factor 3, *Thromb. Diath. Haemorrh.,* Suppl. 36, 36, 63, 1969.

Farbiszewski, R. et al., The effect of lipoprotein fractions on adhesiveness and aggregation of blood platelets, *Thromb. Diath. Haemorrh.,* 21, 89, 1969.

Farbiszewski, R. et al., The effect of modified β-lipoproteins on adhesiveness and on aggregation of blood platelets, *J. Atheroscler. Res.,* 9, 339, 1969.

Martynov, Sm. et al., Role of β-lipoproteinemia in the development of nonspecific agglutination (aggregation) of thrombocytes (in Russian), *Probl. Gemat.,* 14, 24, 1969.

Kommerell, B. et al., Relationship between platelet factor 3 and adenosine diphosphate and the product from contact activation of the blood coagulation system, *Blut,* 20, 31, 1970.

Sixma, J. J. et al., Characteristics of platelet factor 3 release during ADP induced aggregation. Comparison with 5 hydroxytryptamine release, *Thromb. Diath. Haemorrh.,* 24, 206, 1970.

Renaud, S. et al., Hypercoagulability induced by hyperlipemia in rat, rabbit and man. Role of platelet factor 3, *Circ. Res.,* 27, 1003, 1970.

MacKenzie, R. D. et al., A modified Stypven test for the determination of platelet factor 3, *Am. J. Clin. Pathol.,* 55, 551, 1971.

Reimers, H. J. et al., Blood coagulation with special reference to blood platelet turnover in hyperlipoproteinemias (in German with English abstract), *Klin. Wochenschr.,* 50, 12, 1972.

Zöller, H. et al., Thrombocyte function analysis in hyperlipoproteinemias, *Blut,* 28, 24, 1974.

Miettinen, T. A., Hyperlipoproteinemia-relation to platelet lipido, platelet function and tendency to thrombosis, *Thromb. Res.,* Suppl. 4, 41, 1974.

Lopes-Virella, M. F. et al., Platelet lipoproteins. A comparative study with serum lipoproteins, *Biochim. Biophys. Acta,* 439(2), 339, 1976.

Storozhev, A. L. et al., Adrenaline-induced release from thrombocytes of lipoprotein, alkaline phosphatase and 5′-nucleotidase (in Russian with English abstract), *Biull. Eksp. Biol. Med.,* 85(5), 545, 1976.

Shattil, S. J. et al., Abnormalities of cholesterol-phospholipid composition in platelets and low-density lipoproteins of human hyper-β-lipoproteinemia, *J. Lab. Clin. Med.,* 89(2), 341, 1977.

LIPOPROTEINS IN CELLS IN CULTURES

Milcon, S. M. et al., Action of various isolated thymus extracts on the proliferation of cells cultivated "in vitro" (in French), *Acta Biol. Med. German,* 16, 606, 1966.

Robertson, A. L., Jr., Transport of plasma lipoproteins and ultrastructure of human arterial intimacytes in culture, *Wistar Inst. Symp. Monogr.,* 6, 115, 1967.

Rothblat, G. H. et al., Cholesterol uptake by L5178Y tissue culture cells: studies with delipidized serum, *Biochim. Biophys. Acta,* 164, 327, 1968.

Illingworth, D. R. et al., The exchange of phospholipids between plasma lipoproteins and rapidly dividing human cells grown in tissue culture, *Biochim. Biophys. Acta,* 306, 422, 1973.

Brown, M. S. et al., Regulation of 3-hydroxy-3-methylglutaryl coenzyme A reductase activity in human fibroblasts by lipoproteins, *Proc. Natl. Acad. Sci. U.S.A.,* 70, 2162, 1973.

Bailey, J. M. et al., Cholesterol uptake from doubly labeled alpha-lipoproteins by cells in tissue culture, *Arch. Biochem. Biophys.,* 159, 580, 1973.

Bierman, E. L. et al., Very low density lipoprotein "remnant" particles: uptake by aortic smooth muscle cells in culture, *Biochim. Biophys. Acta,* 329, 163, 1973.

Brown, M. S. et al., Familial hypercholesterolemia: defective binding of lipoproteins to cultured fibroblasts associated with impaired regulation of 3-hydroxy-3-methylglutaryl coenzyme A reductase activity, *Proc. Natl. Acad. Sci. U.S.A.* 71, 788, 1974.

Goldstein, J. L. et al., Binding and degradation of low density lipoproteins by cultured human fibroblasts. Comparison of cells from a normal subject and from a patient with homozygous familial hypercholesterolemia, *J. Biol. Chem.,* 249, 5153, 1974.

Goldstein, J. L. et al., Esterification of low density lipoprotein cholesterol in human fibroblasts and its absence in homozygous familial hypercholesterolemia, *Proc. Natl. Acad. Sci. U.S.A.,* 71(11), 4288, 1974.

Goldstein, J. L. et al., Steroid requirements for suppression of HMG CoA reductase activity in cultured human fibroblasts, *Adv. Exp. Med. Biol.,* 63, 77, 1975.

Goldstein, J. L. et al., Genetic heterogeneity in familial hypercholesterolemia: evidence for two different mutations affecting functions of low density lipoprotein receptor, *Proc. Natl. Acad. Sci. U.S.A.,* 72(3), 1092, 1975.

Brown, M. S. et al., Role of the low density lipoprotein receptor in regulating the content of free and esterified cholesterol in human fibroblasts, *J. Clin. Invest.,* 55(4), 783, 1975.

Maynard, J. R. et al., Association of tissue factor activity with the surface of cultured cells, *J. Clin. Invest.,* 55(4) 814, 1975.

Bierman, E. L. et al., Lipoprotein uptake by cultured human arterial smooth muscle cells, *Biochim. Biophys. Acta,* 388(2), 198, 1975.

Offner, H. et al., Morphological changes of astrocyte-like cells induced by serum β-lipoprotein in brain cell culture, *Neurobiology,* 5(3), 192, 1975.

Brown, M. S. et al., Lipoprotein receptors and genetic control of cholesterol metabolism in cultured human cells, *Naturwissenschaften,* 62(8), 385, 1975.

Stein, O. et al., Surface binding and interiorization of homologous and heterologous serum lipoproteins by rat aortic smooth muscle cells in culture, *Biochim. Biophys. Acta,* 398(3), 377, 1975.

Wenzel, D. G. et al. Cholesterol and β-lipoprotein on lipid inclusions and lysosomal and mitochondrial permeability of cultured heart muscle and endothelioid cells, *Res. Commun. Chem. Pathol. Pharmacol.*, 12(4), 789, 1975.

Goldstein, J. L. et al., The LDL pathway in human fibroblasts: a receptor-mediated mechanism for the regulation of cholesterol metabolism, *Curr. Top. Cell Regul.*, 11, 147, 1976.

Brown, M. S. et al., The low-density lipoprotein pathway in human fibroblasts: relation between cell surface receptor binding and endocytosis of low-density lipoprotein, *Ann. N. Y. Acad. Sci.*, 275, 244, 1976.

Bierman, E. L. et al., Lipoprotein uptake and degradation by cultured human arterial smooth muscle cells, *Adv. Exp. Med. Biol.*, 67(00), 437, 1976.

Biermann, E. L. et al., Lipoprotein uptake and degradation of human arterial smooth muscle cells in tissue culture, *Ann. N. Y. Acad. Sci.*, 275, 199, 1976.

Stein, O. et al., Binding, internalization, and degradation of low density lipoprotein by normal human fibroblasts and by fibroblasts from a case of homozygous familial hypercholesterolemia, *Proc. Natl. Acad. Sci. U.S.A.*, 73(1), 14, 1976.

Albers, J. J. et al., The effect of hypoxia on uptake and degradation of low density lipoproteins by cultured human arterial smooth muscle cells, *Biochim. Biophys. Acta*, 424(3), 422, 1976.

Stein, O. et al., High density lipoproteins reduce the uptake of low density lipoproteins by human endothelial cells in culture, *Biochim. Biophys. Acta*, 431(2), 363, 1976.

Howard, B. V. et al., Triglyceride accumulation in cultured human fibroblasts: the effects of hypertriglyceridemic serum, *Atherosclerosis*, 23(3), 521, 1976.

Ho, Y. K. et al., Binding, internalization, and hydrolysis of low density lipoprotein in long-term lymphoid cell lines from a normal subject and a patient with homozygous familial hypercholesterolemia, *J. Exp. Med.*, 144(2), 444, 1976.

Stein, Y. et al., Lipoprotein interaction and cholesterol metabolism in vascular cells in culture, *Paroi Arterielle*, 3(3), 135, 1976.

Steinberg, D. et al., Binding, uptake, and catabolism of low density (LDL) and high density lipoproteins (HDL) by cultured smooth muscle cells, *Lipoprotein Metabolism, Greten, H., Ed.*, Springer-Verlag, Berlin, 1976, 90.

LIPOPROTEINS IN DIFFERENT SYNDROMES

Acanthocytosis

Levine, T. M. et al., Hereditary neurological disease with acanthocytosis. A new symdrome, *Arch. Neurol.*, (Chicago), 19, 403, 1968.

Betts, J. J. et al., Acanthocytosis with normolipoproteinaemia: biophysical aspects, *Postgrad. Med. J.*, 46, 702, 1970.

Critchley, E. M. et al., Acanthocytosis normolipoproteinaemia and multiple tics, *Postgrad. Med. J.*, 46, 698, 1970.

Alcoholism

Cavalieri, U. et al., Influence of ethyl alcohol on blood lipid modifications induced by a fat meal in elderly subjects, *G. Gerontol.*, 13, 1079, 1965.

Johansson, B. G. et al., Disorders of serum α-lipoproteins after alcoholic intoxication, *Scand. J. Clin. Lab. Invest.*, 23, 231, 1969.

Madsen, N. P., Reduced serum very low-density lipoprotein levels after acute ethanol administration, *Biochem. Pharmacol.*, 18, 261, 1969.

Greten, H., Alcoholic hyperlipidemia, *Nutr. Rev.*, 29, 140, 1971.

Papenberg, J. et al., Ethanol-inducable hyperlipoproteinemias, *Verh. Dtsch. Ges. Inn. Med.*, 77, 599, 1971.

Brun, D. et al., Alcohol-induced chylomicronemia in type IV hyperlipoproteinemia, *Union Med. Can.*, 103, 1710, 1973.

Mendelson, J. H. et al. Alcohol-induced hyperlipidemia and beta lipoproteins, *Science*, 180, 1372, 1973.

Johansson, B. G. et al., Increase in plasma alpha-lipoproteins in chronic alcoholics after acute abuse, *Acta Med. Scand.*, 195, 272, 1974.

Mishkel, M. A., Letter: alcohol and alpha-lipoprotein cholesterol, *Ann. Intern. Med.*, 81(4), 564, 1974.

Avogaro, P. et al., Changes in the composition and physico-chemical characteristics of serum lipoproteins during ethanol-induced lipaemia in alcoholic subjects, *Metabolism*, 24(11), 1231, 1975.

Allergy

Nikolaeva, T. G., Total protein content, protein fractions, glycoproteins and lipoproteins in the blood serum of children with bronchial asthma (in Russian), *Vopr Okhr. Materin. Det.*, 9, 32, 1964.

Fisherman, E. W. et al., Serum triglyceride and cholesterol levels and lipid electrophoretic patterns in intrinsic and extrinsic allergic states, *Ann. Allergy*, 38(1), 46, 1977.

Amyloidosis
Gombert, J., Behavior of serum proteins of patients suffering from amyloidosis as demonstrated with Congo red and thioflavine T, *Rev. Fr. Etud. Clin. Biol.*, 14, 921, 1969.

Anemia
Reutova, R. L. et al., Protein and lipoprotein serum fractions and pneumonias with and without anemia in children. (Preliminary communication) (in Russian), *Sov. Med.*, 28, 68, 1964.

Martine-Maldonado, M., Role of lipoproteins in the formation of spur cell anaemia, *J. Clin. Pathol.*, 21, 620, 1968.

Anesthesia
Cavagna, R., et al., Behavior of lipoproteins in relation to penthrane anesthesia in humans (in Italian), *Minerva Anestesiol.*, 32, 768, 1966.

Arthritis
Deliamure, L. L. et al., Diagnostic role of the determination of lipoproteins in the blood serum in rheumatism (in Russian), *Vopr. Revm.*, 5, 40, 1965.

Barwik-Schramm, A. et al., Fat conversion disorders in degenerative rheumatism (in German), *Z. Gesamte Inn. Med. Ihre Grenzgeb.*, Suppl. 21, 230, 1966.

Bernacka, K. et al., Rheumatoid arthritis as an antiatheromatous syndrome, *Pol. Med. J.*, 6, 346, 1967.

Morozova, L. I., Serum lipoproteins in patients with rheumatism (in Russian), *Vrach. Delo*, 6, 69, 1967.

Milewski, B. et al., Behavior of plasma lipids in rheumatoid arthitis (in Polish), *Reumatologia*, 7, 139, 1969.

Glinski-Urban, D. et al., Comparative serological studies of the serum β-lipoprotein level in patients with psoriatic arthritis and rheumatoid arthritis (in Polish), *Reumatologia*, 8, 283, 1970.

Bluestone, R., Hyperlipoproteinanemia and arthritis (two cases), *Proc. R. Soc. Med.*, 64, 669, 1971.

Van Kerckhove, H., Oligoarthritis and hyperlipoproteinemia (type 3: [4 cases]) (in French), *J. Belge Rhumatol. Med. Phys.*, 26, 45, 1971.

Frank, O. et al., The prevalence of lipid metabolism disorders in rheumatological diseases, *Wien. Med. Wochenschr.*, 124 (37), 527, 1974.

Atherosclerosis
Bernasconi, C. et al., Effect of coenzyme A administration on lipid and lipoprotein alterations of the blood in arteriosclerotic patients (in Italian), *Clin. Ter.*, 31, 522, 1964.

Bode, G., New possibilities in the diagnosis and measures for drug control of degenerative vascular diseases (in German), *Med. Welt*, 42, 2251, 1964.

Cloetens, W. et al., Action of β-sitosterol on cholesterolemia and the lipidogram in arteriosclerosis. Long term results (in French), *Bruxelles Med.*, 44, 1391, 1964.

Delfino, H., Electrophoretic study of arteriosclerosis (in Spanish), *Rev. Esp. Tuberc. Arch. Nac. Enferm.*, 33, 259, 1964.

Romero, R. L. et al., Alpha-2-lipo-haptoglobulin and siderophilin in atherosclerosis (in Spanish), *Rev. Clin. Esp.*, 95, 363, 1964.

Shanoff, H. M. et al., Studies of male survivors of myocardial infarction due to "essential" atherosclerosis. 3. Corneal arcus: incidence and relation to serum lipids and lipoproteins, *Can. Med. Assoc. J.*, 91, 835, 1964.

Derepa, K. P., Electrophoretic study of the blood serum proteins and lipoproteins in scleroma patients (in Russian), *Zh. Ushn. Nos. Gorl. Bolezn;* 25, 60, 1965.

Kao, V. C. et al., A study of immunohistochemical localization of serum lipoproteins and other plasma proteins in human atherosclerotic lesions, *Exp. Mol. Pathol.*, 4, 465, 1965.

Krčilková, M. et al., Plasma lipids in 6—7 year old children from two different regions: a paediatrician's contribution to epidemiological research on atherosclerosis, *J. Atheroscler. Res.*, 5, 342, 1965.

Levymate, M., On atherosclerosis (in French), *Gaz. Med. Fr.*, 72, 3171, 1965.

Nagasawa, S., Studies on lipid metabolism disorders in arteriosclerosis, with special reference to lipoproteins in the vascular wall and its fatty acid composition, *Saishin Igaku*, 20, 1929, 1965.

Ilinski, B. V., The early diagnosis of atherosclerosis (in Rumanian), *Med. Interna*, 17, 393, 1965.

Anisimov, V. E., The value of estimating the blood serum cholesterol extraction factor in atherosclerosis (in Russian), *Kardiologiia*, 6, 73, 1966.

Belizhenko, V. D., Metabolism of nucleic acids, phosphoproteins and phospholipids of the liver in experimental atherosclerosis (in Russian), *Biull. Eksp. Biol. Med.*, 62, 44, 1966.

Kawana, H., Fatty acid composition of serum lipoprotein-lipids in normal and arteriosclerotic subjects. I. Fatty acid composition of lipoprotein normal and arteriosclerotic subjects. II. Effects of ethyl linoleate on the fatty acid composition of serum lipoprotein (in Japanese), *Jpn. Circ. J.*, 30, 1417, 1966.

Opplt. J. J., Evaluation of analysis of serum lipoproteins in arteriosclerosis (in Czech), *Acta Univ. Carol. Med.*, 12, 425, 1966.

Solez, C., Premature arteriosclerosis: metabolic findings, *J. Am. Geriatr. Soc.*, 14, 1185, 1966.

Albertini, E. et al., Comparative serological study of the evolutive changes of biochemical and enzymatic components in humans and swine in relation to the pathogenesis of atherosclerosis (in Italian), *Arch. Sci. Med.*, 123, 1, 1967.

Dallocchio, M. et al., Attempted prevention of experimental atherosclerosis by immunization with human β-lipoproteins from atherosclerotic subjects (in French), *Rev. Atheroscler.*, 9 (Suppl. 1), 220, 1967.

Enachescu, G. et al., Research on certain humoral aspects of the preclinical and clinical stage of atherosclerosis in relation to age and sex (in French), *G. Gerontol.*, 15, 695, 1967.

Etienne, Ayrault-Jarrier, M., et al., Effects of the nature of the lipoproteins in various clinical types of atherosclerosis on the extractibility of cholesterol, *Presse Med.*, 75, 1397, 1967.

Gero, S., Some data on the influence of cholesterol atherosclerosis by immunological means, *Rev. Atheroscler.*, 9 (Suppl. 1), 194, 1967.

Gola, A., Interdependence of fat and carbohydrates metabolism disorders in arteriosclerosis (in Polish), *Postepy Hig. Med. Dosw.*, 21, 119, 1967.

Kipshidze, N. N. et al., An immunological study of β-lipoproteins in healthy humans and in patients with atherosclerosis (in Russian), *Kardiologiia*, 7, 16, 1967

Mazzei, E. S. et al., Effect of sodium d-thyroxine on the hyperlipidemia, hypercholesteremia, triglycerides and serum lipoproteins in patients with arteriosclerosis (in Spanish), *Presse Med. Argent.*, 54, 199, 1967.

Martinelli, M., Blood lipids and atherosclerosis (in Italian), *Bull. Sci. Med.*, 139, 454, 1967.

Musiatowicz, J. et al., Relationship between β-lipoprotein concentration and the activity of the fibrinolytic system in arteriosclerosis and in postpartum (in Polish), *Pol. Tyg. Lek.*, 22, 1760, 1967.

Yasugi, T. et al., Atherosclerosis and lipoprotein (in Japanese), *Saishin Igaku*, 22, 1719, 1967.

Brauner, R. et al., Metabolic interrelations in atherosclerosis (in Rumanian), *Med. Interna*, 20, 1447, 1968.

Khomulo, P. S., Nervous system, emotions and atherosclerosis (in Russian), *Patol. Fiziol. Eksp. Ter.*, 12, 3, 1968.

Nikol'skaia, O. N. et al., Method and clinical significance of determination of lipoproteins by turbidimetric method in atherosclerosis (in Russian), *Lab. Delo*, 10, 579, 1968.

Rouffy, J. et al., Hereditary hyperlipoproteinaemia and atherosclerosis (in French) *Ann. Cardiol. Angeiol.*, 17, 135, 1968.

Szajbel, W. et al., Effect of clinical exacerbations and ACTH therapy on the level of lipids in blood serum of patients with sclerosis (in Polish), *Neurol. Neurochir. Pol.*, 2, 691, 1968.

Watts, H. F., The role of lipoproteins in the formation of the atherosclerotic plague, *Ann. N.Y. Acad. Sci.*, 149, 725, 1968.

Amanasov, K., Diagnostic value of cholesterin , β-lipoproteins and proteinogramms in atherosclerosis and hypertension (in Russian), *Folia Med.*, (Plovdiv), 11, 19, 1969.

Hoff, H. F. et al., Studies on the pathogenesis of atherosclerosis with experimental model systems. II. An electron microscopy study on the uptake of egg lipoproteins by endothelial and smooth muscle cells of the doubly-ligated rabbit carotid artery, *Virchows Arch.*, 348, 77, 1969.

Morganti, G. et al., Blood lipoproteins may be key to understanding atherosclerosis, *Il. Med. J.*, 135, 399, 1969.

Page, I. H. et al., A long-time study of the blood lipids of two students of atherosclerosis, *Circulation*, 40, 915, 1969.

Podobedova, N. S., The effect of Matsesta baths in the general complex of health resort therapy on the content of heparin and β-lipoproteins in the blood serum of patients with atherosclerosis (in Russian), *Vopr. Kurortol. Fizioter. Lech. Fiz. Kult.*, 34, 255, 1969.

Shakhuazarov, A. B. et al., Stability of serum lipoproteins and their significance in the pathogenesis and diagnosis of atherosclerosis (in Russian), *Vrach. Delo*, 2, 8, 1969.

Zhelezovskaia, I. B., Concentration of vitamin A in blood and its relation to the content of proteins, cholesterin and β-lipoproteids in the blood of patients with atherosclerosis (in Russian), *Vopr. Med. Khim.*, 15, 506, 1969.

Samuel, P. et al., Long-term decay of serum cholesterol radio-activity: body cholesterol metabolism in normals and in patients with hyperlipoproteinemia and atherosclerosis, *J. Clin. Invest.*, 49, 346, 1970.

Fasoli, A., Classification of hyperlipoproteinemias and the problem of "atherogenicity," *G. Ital. Cardiol.*, 1, 287 1971.

Ježková, Z. et al., β-lipoprotein screening test in relation to biochemical findings in the epidemiology of atherosclerosis, *Vnitr. Lek.*, 17, 908, 1971.

Malcolm, J. et al., Recent concepts on arteriosclerosis seen in special correlation to lipoprotein particles, *Bruxelles Med.*, 51, 43, 1971.

Vecchi, G. P. et al., Thyroid hormones, lipoproteins and arteriosclerosis (in Italian with English abstr.), *Acta Gerontol.*, 21, 81, 1971.

Boyle, E., Jr. et al., Evaluation of rapid screening methods for the detection of abnormal lipid states related to atherosclerosis, *Ann. Clin. Lab. Sci.,* 2, 393, 1972.

Danish, G. I., Lipid metabolism disorders and atherosclerotic changes in endomyocarditis, *Vrach. Delo,* 10, 16, 1972.

Drash, A. et al., The identification of risk factors in normal children in the development of atherosclerosis, *Ann. Clin. Lab. Sci.,* 2, 348, 1972.

Jaillard, J. et al., Fatty acid composition of the steroids of the main classes of β-lipoproteins isolated by ultracentrifugation in normal and hyperglyceridemic subjects with or without associated atheroma (in French with English abstr.), *Pathol. Biol.,* 20, 51, 1972.

Kajiyama, G. et al., Study on the change of the serum lipids and lipoproteins in patients with arteriosclerosis due to aging, (in Japanese with English abstr.), *Jpn. Arch. Intern. Med.,* 19, 71, 1972.

Khomulo, P. S. et al., Role of endocrine mechanisms in the development of atherosclerosis (in Russian), *Biull. Eksp. Biol. Med.,* 73, 17, 1972.

Wardle, E. N. et al., A study of patients with atherosclerosis and renal disease with respect to lipoprotein types, fibrinolysis, and carbohydrate tolerance, *Q. J. Med.,* 41, 15, 1972.

Waters, L. L., Insoluble lipoproteins in the pathogenesis of experimentally induced atherosclerosis, *Arch. Pathol.,* 93, 525, 1972.

Furaeva, N. V., Examination of beta-lipoproteins and ether test in patients with arteriosclerosis and hypertensive disease, *Lab. Delo,* 12, 737, 1973.

Hoff, H. F. et al., Plasma lipoproteins and atherosclerosis in man: an immunohistochemical study, *Cardiovasc. Res. Cent. Bull. (Houston), 12(2), 29, 1973.*

Khaletskiĭ, M. E. et al., Beta-lipoproteins of the blood and results of the prednisolone-glucose tolerance test in patients with arteriosclerosis, *Vrach. Delo,* 3, 43, 1973.

Kučerová, L. et al., Arteriosclerosis in primary hyperlipoproteinemia, *Sb. Lek.,* 75, 161, 1973.

Malcolm, J. et al., Recent concepts of arteriosclerosis, studied in correlation with the classification of hyperlipoproteinemia and the indicated laboratory methods, *Bruxelles Med.,* 53, 179, 1973.

Ballantyne, D. et al., Relationship of plasma lipids and lipoprotein concentration to cerebral atherosclerosis and electrocardiographic findings, *J. Neurol. Sci.,* 23(2), 323, 1974.

Bavina, M. V. et al., Spectra of blood serum lipoproteins in arteriosclerosis, *Ter. Arkh.,* 46, 28, 1974.

Fridliand, L. M. et al., Certain biochemical manifestations of atherosclerosis, *Vrach. Delo,* 3, 24, 1974.

Gerasimova, E. N. et al., The effect of cortisol on lipoprotein transformation in plasma in experimental atherosclerosis (English abstr.), *Vopr. Med. Khim.,* 20(4), 418, 1974.

Kajiyama, G. et al., The lowered serum phospholipids in X-lipoprotein in patients with atherosclerosis, *Hiroshima J. Med. Sci.,* 23(4), 229, 1974.

Micheli, H. et al., Cholesterol and triglycerides of serum lipoproteins isolated by means of ultracentrifugation correlation with arteriosclerosis, *Schweiz. Med. Wochenscher.,* 104(49), 1794, 1974.

Smedile, G. et al., Relations between carbohydrate loading and arteriosclerosis. II. Behavior of blood cholesterol and serum lipoprotein fractions in normal subjects and patients with cardiovascular diseases after intravenous administration of glucose, *Boll. Soc. Ital. Biol. Sper.,* 50(13), 936, 1974.

Smith, E. B., The relationship between plasma and tissue lipids in human atherosclerosis, *Adv. Lipid Res.,* 12(0), 1, 1974.

Walton, K. W. et al., A study of methods of identification and estimation of Lp(a) lipoprotein and of its significance in health, hyperlipidaemia and atherosclerosis, *Atherosclerosis,* 20(2), 323, 1974.

Goldstein, J. L. et al., Lipoprotein receptors, cholesterol metabolism, and atherosclerosis, *Arch. Pathol.,* 99(4), 181, 1975.

Schettler, G., Plasma lipoproteins and atherosclerosis, *Folia Clin. Int.,* 25(5), 262, 1975.

Dahlén, G. et al., Lpαlipoprotein/pre-β-lipoprotein, serum lipids and atherosclerotic disease, *Clin. Genet.,* 9(6), 558, 1976.

Fellin, R., Structure and metabolism of plasma lipoproteins: effect of atherosclerosis (in Italian), *G. Ital. Cardiol.,* 6(5), 961, 1976.

Ferlito, S. et al., The blood pattern in healthy and arteriosclerotic aged subjects, (in Italian with English abstr.), *Minerva Cardioangiol.,* 24(3), 179, 1976.

Torkhovskaia, T. I. et al., Various changes in the composition of blood serum phospholipids and lipoproteins in arteriosclerosis (in Russian with English abstr.), *Kardiologiia,* 16(2), 46, 1976.

Avogaro, P. et al., Chemical composition of ultracentrifugal fractions in different patterns of human atherosclerosis, *Atherosclerosis,* 26(2), 163, 1977.

Atherosclerosis, Cerebral

Gandini, S. et al., Statistical observations on variations of electrophoretic lipoprotein fractions in a group of 20 patients with cerebral arteriosclerotic vascular diseases in blood serum of the carotid, internal jugular and the crease of the elbow, *Rass. Studi Psichiatr.,* 53, 541, 1964.

Komai, Y. et al., Biochemical study of cerebral edema: with special reference to dynamic metabolism of proteolipid protein (in Japanese), *Brain Nerve,* 23, 251, 1971.

Markiewicz, M., Serum lipoprotein fractions in patients with apoplexy (in Polish with English abstr.), *Neurol. Neurochir. Pol.,* 5, 801, 1971.

Szulc-Kuberska, J. et al., Hereditary hypo-β-lipoproteinemia with cerebellar atrophy syndrome (in Polish with English abstr.), *Neurol. Neurochir. Pol.* 5, 811, 1971.

Liepelt, F. et al., Lipo-protein patterns in cerebrovascular disorders, *Acta Neurol. Scand. Suppl.,* 51, 455, 1972.

Schmidt, R. C. et al., Frequency of hyperlipoproteinemias in patients with ischemic cerebral infarction, *Verh. Dtsch. Ges. Inn. Med.,* 78, 668, 1972.

Pilgeram, L. O. et al., Evidence for abnormalitics in clotting and thrombolysis as a risk factor for stroke, *Stroke,* 4, 643, 1973.

Hollanders, F. D. et al., Serum lipid changes following the completed stroke syndrome, *Postgrad. Med. J.,* 51(596), 386, 1975.

Mathew, N. T. et al., Hyperlipoproteinemia in occlusive cerebro-vascular disease, *JAMA, 232(3), 262, 1975.*

Kukhtevich, I.I. et al., Lipoprotein metabolism disorders in hypertension with : cerebral syndrome and their diagnostic value for medico-occupational expertise, (in Russian with English abstr.,) *Zh. Nevropatol. Psikhiatr.,* 76(8), 1153, 1976.

Molchanov, V. V. et al., Changes in the stability of cholesterol binding to serum proteins and vascular immunologic reactivity in cerebral atherosclerosis (in Russian with English abstr.), *Zh. Nevropatol. Psikhiatr.,* 16(1), 45, 1976.

Takamatsu, S. et al., Symposium on the cerebral stroke. 4. Studies on influences of serum lipoprotein and hypertension in cerebrovascular disorders (in Japanese), *J. Jpn. Soc. Intern. Med.,* 65(12), 1367, 1976.

Ionescu, M. et al., Serum lipoproteins as a risk factor in cerebral arteriosclerosis (in Rumanian), *Rev. Roum. Med.,* 21(1), 1, 1976.

Atherosclerosis, Coronary

Bandyopadhyay, A. et al., Plasma lipids in some cardiovascular disorders, *Am. J. Med. Sci.,* 248, 203, 1964.

Baratta, P. F. et al., Action of pancreatic lipase on the electrophoretic pattern of serum lipoproteins after a fat meal in coronary sclerosis patients (in Italian), *Minerva Dietol.,* 4, 64, 1964.

Krasno, L. R. et al., Low-density serum lipoproteins in survivors and nonsurvivors of myocardial infarction, *Postgrad. Med.,* 36, 602, 1964.

Kroman, H. et al., Lipids in normals and patients with coronary artery disease, *Am. J. Med. Sci.,* 248, 571, 1964.

Braunsteiner, H. et al., Lipid values in normal persons and patients with myocardial infarct. Statistical evaluation by means of an electronic computer (in German), *Wien. Klin. Wochenschr.,* 77, 859, 1965.

Erdélyi, G. et al., Comparison of the diagnostic value of a modified 131-J triolein test with that of the tests for total serum lipid, cholesterol and β-lipoprotein in coronary heart disease, *J. Atheroscler. Res.,* 5, 255, 1965.

Groover, M. E., Jr., Heparin and the electrophoretic mobility of β-lipoprotein in coronary disease, *Am. J. Cardiol.,* 15, 13, 1965.

Nestel, P. J., Metabolism of linoleate and palmitate in patients with hypertriglyceridemia and heart disease, *Metabolism,* 14, 1, 1965.

Rifkind, B. M., The incidence of arcus senilis in ischaemic heart-disease, its relation to serum-lipid levels, *Lancet,* 1, 312, 1965.

Stazka, Z., Behavior of serum proteins, glycoproteins and lipoproteins in myocardial infarction, *Pol. Tyg. Lek.,* 20, 511, 1965.

Szczelik, E. et al., Behavior of lipoproteins in the blood serum and heparinocytes in the blood in myocardial infarction (in Polish), *Kardiol. Pol.,* 8, 215, 1965.

Weitzman, D., Blood fats and the coronary arteries, *Geriatrics,* 20, 172, 1965.

Grover, M. E., Jr. et al., Physicochemical properties of blood in postmyocardial infarction patients and controls, *Angiology,* 17, 85, 1966.

Jenkins, C. D., Components of the coronary-prone behavior pattern. Their relation to silent myocardial infarction and blood lipids, *J. Chronic Dis.,* 19, 599, 1966.

Mills, G. L. et al., Plasma lipid levels and the diagnosis of coronary arteriosclerosis in England, *Br. Heart J.,* 28, 638, 1966.

Nestel, P. J., Triglyceride turnover in coronary heart disease and the effect of dietary carbohydrate, *Clin. Sci.,* 31, 31, 1966.

Rosenman, R. H. et al., The prediction of immunity to coronary heart disease, *JAMA,* 198, 1159, 1966.

Szczeklik, E. et al., Relations between heparinocytes and blood lipoproteins in myocardial infarction, *Cor Vasa,* 8, 34, 1966.

Opplt, J. J., The evaluation of serum lipoprotein analyses in atherosclerosis, *Acta Univ. Carol. Medica*, 12, 617, 443, 1966.

Synchuk, A. N., On the interrelationship between the blood coagulation system and lipoid-protein metabolism in patients with myocardial infarction (in Russian), *Vrach. Delo*, 6, 48, 1966.

Brunner, D. et al., Physical activity, lipoproteins and ischemic heart disease, *Pathol. Microbiol.*, 30, 648, 1967.

Demerdash, H. et al., The rate of utilization of glucose and the lipoprotein pattern in patients with coronary atherosclerosis, *J. Egypt. Med. Assoc*, 50, 466, 1967.

Koike, S. et al., Serum lipids and lipoprotein in hypertension and ischemic heart disease (in Japanese), *Jpn. J. Hyg.*, 22, 451, 1967.

Petrova, T. R., Comparative study of basophils (heparinocytes), eosinophils and β-lipoproteins of the blood in myocardial infarct and stenocardia, *Ter. Arkh.*, 39, 22, 1967.

Blankenhorn, D. H. et al., Ischemic heart disease in young adults. Metabolic and angiographic diagnosis and the prevalence of type IV hyperlipoproteinemia, *Ann. Intern. Med.*, 69, 21, 1968.

Fomina, L. F. et al., Use of the complex method of Il'k, Dodik and Iovanovich for diagnosis of humoral lipid disorders in coronary atherosclerosis (in Russian), *Ter. Arkh.*, 40, 61, 1968.

Naimi, S., Unsaturated fats and coronary heart disease, *Lancet*, 2, 1141, 1968.

Vastesaeger, E. et al. Dyslipidemias or dyslipoproteinemias? How can one detect the etiology of lipid metabolism disorders in subjects disposed to coronary accidents?(in French), *Bruxelles Med.*, 48, 655, 1968.

Babei, E. et al., Influence of heparin on plasmatic lipoproteins in patients with coronary arteriosclerosis (in Rumanian), *Stud. Cercet. Med. Interna*, 10, 537, 1969.

Bassett, D. R. et al., Blood lipids and lipoproteins, glucose tolerance and plasma insulin response in Chinese men with and without coronary heart disease in Hawaii, *Isr. J. Med. Sci.*, 5, 666, 1969.

Chahud, A. et al., The study of lipoprotein in myocardial infarction, *Prensa Med. Argent.*, 56, 536, 1969.

Heinle, R. A. et al., Lipid and carbohydrate abnormalities in patients with angiographically documented coronary artery disease, *Am. J. Cardiol.*, 24, 178, 1969.

Mills, G. L. et al., Analysis of the distribution of lipoprotein patterns in healthy men and in patients with myocardial infarction, *Clin. Chim. Acta*, 26, 67, 1969.

Slack, J., Risks of ischemic heart-disease in familial hyperlipoproteinaemic states, *Lancet*, 2, 1380, 1969.

Balgma, K. et al., Blood serum lipids in ischemic disease of the heart (in Russian), *Sov. Med.*, 33, 134, 1970.

Bang, H. O. et al., Lipoprotein pattern in patients with coronary occlusion, *Nord. Med.*, 84, 998, 1970.

Bosch, V. et al., Serum lipids and lipoproteins in normal persons and in patients with ischemic heart disease in Venezuela (in Spanish), *Acta Cient. Venez.*, 21, 94, 1970.

Dyerberg, J. et al., Plasma lipids and lipoproteins in patients with myocardial infarction and in a control material, *Acta Med. Scand.*, 187, 353, 1970.

Fomina, L. G. et al., Relationship between humoral lipid disorders and changes in ECG in patients with coronary atherosclerosis, *Ter. Arkh.*, 42, 61, 1970.

Shanoff, H. M. et al., Studies of male survivors of myocardial infarction. XII. Relation of serum lipids and lipoproteins to survival over a 10-year period, *Can. Med. Assoc. J.*, 103, 927, 1970.

Avogaro, P. et al., Lipid and lipoprotein changes during acute myocardial infarct and during the following post-infarction heart diseases, *Cardiol. Prat.*, 22, 129, 1971.

Chopra, J. S. et al., Congenital hyperlipoproteinaemias and ischaemic heart-disease, *Lancet*, 1, 117, 1971.

Dalderup, L. M. Mutants, hyperlipoproteinaemia, and coronary artery disease, *Br. Med. J.*, 2, 771, 1971.

Fredrickson, D. S., Mutants, hyperlipoproteinaemia, and coronary artery disease, *Br. Med. J.*, 2, 187, 1971.

Kannel, W. B. et al., Serum cholesterol, lipoproteins, and the risk of coronary heart disease, The Framingham study, *Ann. Intern. Med.*, 74, 1, 1971.

Masarei, J. R. et al., Lipoprotein electrophoretic patterns, serum lipids, and coronary heart disease, *Br. Med. J.*, 1, 78, 1971.

Steiner, G. et al., Early coronary atherosclerosis in primary type V hyperlipoproteinemia, *Can. Med. Assoc. J.*, 105, 1172, 1971.

Tzagournis, M. et al., Lipoproteins and coronary disease, *Ann. Intern. Med.*, 74, 796, 1971.

Zahler, P., Congenital hyperlipoproteinemias and ischaemic heart-disease, *Lancet*, 1, 117, 1971.

Birchwood, B., Serum lipids and lipoprotein following acute myocardial infarction in women, *Nutr. Metab.*, 14, 38, 1972.

Bordia, A. et al., Serum magnesium, serum cholesterol and serum lipoproteins in cases of coronary artery disease, *Indian Heart J.*, 24, 277, 1972.

Ciswicka-Sznajderman, M. et al., Incidence of hyperlipoproteinemia in patients with past myocardial infarct, *Pol. Tyg. Lek.*, 27, 1673, 1972.

Diehl, H. J. et al., Incidence of hyperlipoproteinemia and other risk factors in chronic and acute phases of patients with myocardial infarct, *Arzneim. Forsch.*, 22, 1815, 1972.

Golod, I. S., Beta-lipoprotein reaction with blood serum in cardiovascular diseases, *Sov. Med.*, 35, 22, 1972.

Gustafson, A. et al., Serum lipids and lipoproteins in men after myocardial infarction compared with representative population sample, *Circulation*, 46, 709, 1972.

Klemens, U. H. et al., Primary hyperlipoproteinemias in patients with coronary disease and peripheral arterial circulatory disorders, *Verh. Dtsch. Ges. Inn. Med.*, 78, 1342, 1972.

Klemens, U. H. et al., Hyperlipoproteinemia and coronary disease. Occurrence, distribution of types, age and sex factors (in German), *Klin. Wochenschr.*, 50, 139, 1972.

Klemens, U. H. et al., Frequency of hyperlipoproteinemia with type distribution in patients with myocardial infarct and angina factors (in German), *Z. Ernaehrungswiss. Suppl.* 12, 61, 1972.

Mihail, A. et al., Problems in cardiology. Correlations between the types of hyperlipoproteinemias. Clinically manifested atherosclerosis and risk factors. Comparative functional study of 5 criteria of classification, *Med. Interna*, 24, 1195, 1972.

Mundy, G. R. et al., The relationship between serum lipid abnormalities and other major risk factors in myocardial infarction, *Aust. N.Z. J. Med.*, 2, 8, 1972.

Nikkilä, E. A., Serum lipids, lipoproteins, and coronary disease, *Duodecim*, 88, 1271, 1972.

Pagnan, A. et al., Prevalence of hyperlipoproteinemia during the course of recent myocardial infarct (in Italian with English), *Cardiol. Prat.*, 23, 237, 1972.

Passowicz, L. et al., Fat tolerance test in patients with past myocardial infarct undergoing long-term physical exercise treatment, *Pol. Tyg. Lek.*, 27, 1306, 1972.

Patterson, D. et al., Lipid abnormalities in male and female survivors of myocardial infarction and their first-degree relatives, *Lancet*, 1, 393, 1972.

Ritland, S. et al., Changes in the lipoprotein pattern during two years following myocardial infarction, *Acta Med. Scand.*, 191, 447, 1972.

Schwartzkoff, W. et al., Hyperlipoproteinemia. A risk factor in peripheral-arterial and coronary occlusive disease, *Arzneim. Forsch.*, 22, 1811, 1972.

Stone, N. J., et al., The hyperlipidemias and coronary artery disease, 1972. *Dis. Mon.*, 1, 1972.

Stone, N. J. et al., Hyperlipoproteinemia and coronary heart disease, *Prog. Cardiovasc. Dis.*, 14, 341, 1972.

Sullivan, J. M. et al., Studies of platelet adhesiveness, glucose tolerance and serum lipoprotein patterns in patients with coronary artery disease, *Am. J. Med. Sci.*, 264, 475, 1972.

von Löwis, P. et al., Coincidence of disorders in carbohydrate metabolism and types of hyperlipoproteinemias in patients with myocardial infarct, *Arzneim. Forsch.*, 22, 1819, 1972.

Aro, A., Serum lipids and lipoprotein in first degree relatives of young survivors of myocardial infarction, *Acta Med. Scand.*, 1, 1973.

Babei, E., et al., Blood cholesterol and its distribution in the lypoproteins of patients with coronary atherosclerosis, *Rev. Roum. Med. Interne*, 10, 493, 1973.

Barats, S. S. et al., Androgen excretion in men in angina pectoris and cardialgia and changes in lipoprotein metabolism and in the blood coagulation system in androgen deficiency, *Kardiologiia*, 13, 74, 1973.

Hazzard, W. R. et al., Hyperlipidemia in coronary heart disease. 3. Evaluation of lipoprotein phenotypes of 156 genetically defined survivors of myocardial infarction, *J. Clin. Invest.*, 52, 1569, 1973.

Kaniak, J. et al., Disseminated occlusions of coronary arteries during primary hyperlipoproteinemia type II, *Wiad. Lek.*, 26, 2259, 1973.

Koren, E. et al., Significance of increased serum pre-beta lipoproteins in myocardial infarction, *Angiology*, 24, 708, 1973.

Nikilä, E. A. et al., Family study of serum lipids and lipoproteins in coronary heart-disease, *Lancet*, 1, 954, 1973.

Noya, M. et al., Incidence of hyperlipoproteinemia in coronary heart disease, *Rev. Clin. Esp.*, 130, 237, 1973.

Papadopoulos, N. M. et al., Serum lipoprotein patterns in patients with coronary atherosclerosis, *Clin. Chim. Acta*, 44, 153, 1973.

Rosing, D. R. et al., Impairment of the diurnal fibrinolytic response in man. Effects of aging, type IV hyperlipoproteinemia, and coronary artery disease, *Circ. Res.*, 32, 752, 1973.

Schoonmaker, F. W., et al., Hyperlipidemias in predicting coronary atherosclerosis by arteriography, *Rocky Mount. Med. J.*, 70, 32, 1973.

Snyder, S. et al., Serum glycoproteins in hyperlipoproteinemic patients with and without clinical vascular disease, *Clin. Chim. Acta*, 46, 191, 1973.

Stone, M. C. et al., Prevalence of hyperlipoproteinaemias in a random sample of men and in patients with ischaemic heart disease, *Br. Heart J.*, 35, 954, 1973.

Válek, J. et al., Relationship of the increased level of lipoproteins in blood and arteriosclerosis of the coronary arteries detected by angiography, *Vnitr. Lek.*, 19, 649, 1973.

Voslarova, Z. et al., Changes in serum cholesterol and prebeta-lipoproteins in patients with ischemic heart diseases, *Sb. Lek.*, 75, 251, 1973.

Berenson, G. S. et al., Serum lipoproteins and coronary heart disease, *Am. J. Cardiol.*, 34(5), 588, 1974.

Berg, K. et al., Lp(a) lipoprotein and pre-beta 1-lipoprotein in patients with coronary heart disease, *Clin. Genet.*, 6(3), 230, 1974.

Clark, D. A., Allen M. F., and Wilson, F. H., Jr., USAFAM cardiovascular disease followup study: comparisons of serum lipid and lipoprotein levels, *Aerosp. Med.*, 45(10), 1167, 1974.

Dahlén, G., The pre-β-lipoprotein phenomenon in relation to serum cholesterol and triglyceride levels, the L-α lipoprotein and coronary heart disease, *Acta Med. Scand. Suppl.*, 570, 1, 1974.

Dahlén, G. et al., Pre-beta-1-lipoprotein and early detection of risk factors for coronary heart disease, *Acta Med. Scand.*, 195, 341, 1974.

Frick, M. H. et al., Serum pre-beta-1 lipoprotein fraction in coronary atherosclerosis, *Acta Med. Scand.*, 195, 337, 1974.

Gorbachev, V. V. et al., Typing of hyperlipoproteinemia in patients with coronary atherosclerosis, *Kardiologiia*, 14(11), 113, 1974.

Jonker, J. J. et al., Platelet survival time in angina pectoris and hyperlipoproteinaemia, *Thromb. Res.*, 4 (Suppl.), 65, 1974.

Lewis, B. et al., Serum lipoprotein abnormalities in patients with ischaemic heart disease: comparisons with a control population, *Br. Med. J.*, 3(5929), 489, 1974.

Lewis, B. et al., Proceedings: serum lipoprotein abnormalities in ischaemic heart disease and peripheral vascular disease, *Clin. Sci. Mol. Med.*, 46, 12P, 1974.

Lewis, B. et al., Frequency of risk factors for ischaemic heart-disease in a healthy British population. With particular reference to serum-lipoprotein levels, *Lancet*, 1, 141, 1974.

Olsson, A. G., Hyperlipidemia, lipoproteins and coronary disease, *Acta Cardiol.*, Suppl. 20, 37, 1974.

Pagnan, A. et al., Coronary disease and the "accessory" pre-beta band: a causal relationship?, *Minerva Med.*, 65(60), 3123, 1974.

Rymar, B., Blood cholesterol, beta lipoproteins and free fatty acids in myocardial infarct, *Pol. Tyg. Lek.*, 29(52), 2249, 1974.

Salel, A. F. et al., The importance of type IV hyperlipoproteinemia as a predisposing factor in coronary artery disease, *Am. J. Med.*, 57(6), 897, 1974.

Válek, J. et al., Analysis of lipid disturbances in patients with angiographically confirmed coronary artery disease, *Nutr. Metab.*, 16, 193, 1974.

Atukorale, D. P., S. C. Paul Oration — 1975: lipoprotein patterns after myocardial infarction, *Ceylon Med. J.*, 20(3), 136, 1975.

Berenson, G. S. et al., A study of serum lipoproteins and angiographic evidence of coronary artery disease, *South. Med. J.*, 68(12), 1513, 1975.

Carlson, L. A. et al., Quantitative and qualitative serum lipoprotein analysis. Part 2. Studies in male survivors of myocardial infarction, *Atherosclerosis*, 21(3), 435, 1975.

Dahlén, G. et al., Lp(a) lipoprotein/pre-beta 1-lipoprotein in Swedish middle-aged males and in patients with coronary heart disease, *Clin. Genet.*, 7(4), 334, 1975.

Dahlén, G. et al., Further studies of Lp(a) lipoprotein/pre-beta 1-lipoprotein in patients with coronary heart disease, *Clin. Genet.*, 8 (3), 183, 1975.

Gajewska-Lipka, J., Hyperlipoproteinemia as a risk factor in coronary disease in males, *Wiad. Lek.* 28(3), 173, 1975.

Goldstein, J. L. et al., Hyerlipidemia in coronary heart disease: a biochemical genetic approach, *J Lab. Clin. Med.*, 85(1), 15, 1975.

Lewis, B., et al., Serum lipoprotein abnormalities in ischaemic heart disease, *Postgrad. Med., J.*, 51(8), 37, 1975.

Neverov, I. V., Glycosaminoglycans-lipoproteins in coronary atherosclerosis, *Sov. Med.*, 4, 16, 1975.

Perova, N. V. et al., Electrophoretic mobility and antigenic properties of lipoproteins in patients with coronary arteriosclerosis, *Ter. Arkh.*, 47(12), 25, 1975.

Reiniš, Z. et al., Serum lipoproteins in the epidemiology of ischaemic heart disease, *Cas. Lek. Cesk.*, 114(34—35) 1067, 1975.

Slack, J., The genetic contribution to coronary heart disease through lipoprotein concentrations, *Postgrad. Med. J.*, 51(8) 27, 1975.

Wada, M. et al., Ischemic heart disease and hyperlipidemia, *Jpn. Circ. J.*, 39(3), 325, 1975.

Yamanaka, T. et al., Characteristics of serum lipid and lipoprotein patterns in patients with coronary sclerosis and with the slightest hyperlipidemia, *Jpn. Circ. J.*, 39(3), 331, 1975.

Belćhenko, D. I. et al., Phospholipid composition of blood serum lipoproteins in healthy persons and in patients with arteriosclerotic cardiosclerosis, *Kardiologiia*, 16(2), 141, 1976.

Berg, K. et al., Serum-high-density-lipoprotein and atherosclerotic heart-disease, *Lancet*, 1(7958), 499, 1976.

Berg, K. et al., Letter: serum-H.D.L. in atherosclerotic heart-disease, *Lancet*, 2(7975), 40, 1976.

Camejo, G. et al., The affinity of low density lipoproteins for an arterial macromolecular complex. A study in ischemic heart disease controls, *Atherosclerosis*, 24(3), 341, 1976.

Miller, G. J. et al., Inverse relationship in Jamaica between plasma high-density lipoprotein cholesterol concentration and coronary-disease risk as predicted by multiple risk-factor status, *Clin. Sci. Mol. Med.*, 51(5), 475, 1976.

Neto, A. F. et al., Fourteen year cardiological follow-up of patients with initially known levels of low density lipoproteins and early death from myocardial infarction *Arg. Bras. Cardiol.*, 29(6), 447, 1976.

Nikkilä, E. A., Letter: serum high-density-lipoprotein and coronary heart-disease, *Lancet*, 2(7980), 320, 1976.

Ose, L. et al., Serum β-lipoprotein subfractions in polyacrylamide gel electrophoresis associated with coronary heart disease, *Scand. J. Clin. Lab. Invest.*, 36(1), 75, 1976.

Rhoads, G. G. et al., Serum lipoproteins and coronary heart disease in a population study of Hawaii Japanese men, *N. Engl. J. Med.*, 294(6), 293, 1976.

Titov, V. N. et al., Lipid composition of lipoproteins and excretion of 17-ketosteroids in coronary arteriosclerosis (in Russian with English.), *Kardiologiia*, 16(2), 49, 1976.

Arterial Peripheral Disease

Greenhalgh, R. M. et al., Serum lipids and lipoproteins in peripheral vascular disease, *Lancet*, 2, 947, 1971.

Newall, R. G., A lipid and lipoprotein study of patients with peripheral arterial disease using micronephelometry, *Clin. Chim. Acta*, 32, 185, 1971.

Cotton, R. C. et al., Inter-relationships between platelet response to adenosine diphosphate, blood coagulation and serum lipids in patients with peripheral occlusive atherosclerosis, *Atherosclerosis*, 16, 337, 1972.

Ballantyne, D. et al., Prevalence of lipoprotein abnormalities in patients with peripheral vascular disease, *Ann. Biol. Clin.* (Paris), 31, 145, 1973.

Newall, R. G., et al., Lipoproteins and the relative importance of plasma cholesterol and triglycerides in peripheral arterial disease, *Angiology*, 24, 297, 1973.

Barndt, R., Jr. et al., Prevalence of asymptomatic femoral artery atheromas in hyperlipoproteinemic patients, *Atherosclerosis*, 20(2), 253, 1974.

Farid, N. R. et al., Hyperlipoproteinaemia in peripheral arterial disease in the north of England, *J. Cardiovasc. Surg.*, 15, 366, 1974.

Fellin, R., et al., Lipoprotein profile estimated by nephelometry in peripheral arterial disease, *G. Ital. Cardiol.*, 4(5), 613, 1974.

Greenhalgh, R. M. et al., Proceedings: a comparison of fasting serum lipid concentrations and lipoprotein patterns in patients with stenosing and dilating forms of peripheral arterial disease, *Br. J. Surg.*, 61, 327, 1974.

Jipp, P., et al., Lipid up-take of lipoproteins in patients with arterial occlusion of extremities, *Verh. Dtsch. Ges. Inn. Med.*, 80, 954, 1974.

Greenhalgh, R. M. et al., A comparison of fasting serum lipid concentrations and lipoprotein patterns in patients with stenosing and dilating forms of peripheral arterial disease, *J. Cardiovasc. Surg.*, 16(2), 150, 1975.

Ballantyne, D. et al., Relationship of plasma uric acid to plasma lipids and lipoproteins in subjects with peripheral vascular disease, *Clin. Chim. Acta*, 70(2), 323, 1976.

Botkin's Disease

Stepina, N. G. et al., The blood serum β-lipoproteins in Botkin's disease, *Vrach. Delo*, 5, 110, 1966.

Burns

Birke, G., et al., Lipid metabolism and trauma. 3. Plasma lipids and lipoproteins in burns, *Acta Med. Scand.*, 178, 337, 1965.

Borisova, T. A. et al., Changes in lipoproteins of blood plasma in burns, *Eksp. Khir. Anesteziol.*, 11, 43, 1966.

Harlan, W. R. et al., Echinocytes and acquired deficiency of plasma lipoproteins in burned patients, *Arch. Intern. Med.*, 136(1), 71, 1976.

Bypass

Strisower, E. H. et al., Effect of ileal bypass on serum lipoproteins in essential hypercholesterolemia, *J. Atheroscler. Res.*, 8, 525, 1968.

Coli, R. D. et al., Partial ileal bypass in the management of familial type II hyperlipoproteinemia, *R. I. Med. J.*, 55, 377, 1972.

Starzl, T. E. et al., Portacaval shunt in hyperlipoproteinaemia, *Lancet*, 2, 940, 1973.

Balfour, J. F. et al., Homozygous type II hyperlipoproteinemia treatment. Partial ileal bypass in two children, *JAMA*, 227, 1145, 1974.

Celiac Syndrome

Tolentino, P. et al., Celiac syndrome, retinal dystrophy, acanthocytosis, without defect of β-lipoprotein, *Ann. Paediatr.* (Basel), 203, 178, 1964.

Collagenoses

Salamatina, V. V., The effect of hormone therapy on the cholesterin and lipoprotein levels in rheumatism and other collagenoses (in Russian), *Vopr. Revm.*, 6, 39, 1966.

Cholecystectomy

Sil'chenko, K., Biochemical disorders following cholecystectomy (in Russian), *Vrach. Delo*, 2, 4, 1967.

Cholecystitis

Khamidova, M. K., Serum lipoproteins in chronic cholecystitis patients (in Russian), *Ter. Arkh.*, 36, 90, 1964.

Karaman, N. V., The lipoproteins in the blood serum in cholecystitis (in Russian), *Vrach. Delo*, 3, 17, 1966.

Cholestasis

Etienne, G. et al., The abnormal proteins of cholestasis, the extractibility of their phospholipids (in French), *Rev. Int. Hepatol.*, 16, 169, 1966.

Picard, J. et al., Separation of abnormal serum lipoproteins in cholestasis (in French), *C. R. Acad. Sci. Ser. D*, 270, 1845, 1970.

Picard, J. et al., Abnormal lipoproteins in cholestasis (in French), *Presse Med.*, 78, 2319, 1970.

Picard, J. et al., Abnormal serum lipoprotein in cholestasis: identification and isolation, *Clin. Chim. Acta*, 30, 149, 1970.

Hamilton, R. L. et al., Cholestasis: lamellar structure of the abnormal human serum lipoprotein, *Science*, 172, 475, 1971.

Havel, R. J., The abnormal lipoprotein of cholestasis, *N. Engl. J. Med.* 285, 578, 1971.

Infante, R., Plasma lipoproteins in cholestasis (in French), *Rev. Med. Chir. Mal. Foie*, 46, 1, 1971

Diabetes

Opplt, J. J. and Syllaba, J., Des resultats de laboratoire clinique dans 68 sujets avec diabete sucre complique par nephropathie diabetique, Diabetes mellitus, (Comptes rendus III. Kongress der International Diabetes Federation, Düsseldorf, 1958), Georg Thieme Verlag, Stuttgart, 1959, 200

Opplt, J. J. and Syllaba, J., Classification du syndrome Kimmelstiel-Wilson base sur des experiences cliniques et des recherches biochimiques, Comptes rendus 4e Congres de la Federation internationale du Diabete, Geneve, 1961, Ed. Medicine et Hygiene, Geneve, 1962, 513.

Lapteva, N. N., Dynamics of protein fractions, lipo- and glycoproteins in the blood in alloxan diabetes (in Russian), *Patol. Fiziol. Eksp. Ter.*, 8, 69, 1964.

Sanwald, R. et al., On the action of lipoproteins in diabetics under treatment with biguanides, *Schweiz. Med. Wochenschr.*, 94, 1459, 1964.

Horváth, E., et al., Atherosclerosis and lipemia in diabetics (in German), *Dtsch. Z. Verdau. Stoffwechselkr*, 24, 273, 1965.

Davies, M. J. et al., Immunohistochemical studies in diabetic glomerulosclerosis, *J. Pathol. Bacteriol.*, 92, 441, 1966.

Kobierska-Szczepanska, A., Proteinogram, lipoproteinogram and glycoproteinogram of serum protein in diabetic children (in Polish), *Pediatr. Pol.*, 41, 1151, 1966.

Opplt, J. J., Serum proteins in diabetic angiopathies, *Cas. Lek. Cesk.*, 15, 799, 1966.

Simms, H. S. et al., The lipfanogen-antilipfanogen system in relation to lipoproteins, diabetes and species, *Angiology*, 17, 872, 1966.

Kumar, D., et al., Serum cholesterol phospholipids and β-lipoproteins in untreated diabetics, *J. Assoc. Physicians India*, 15, 357, 1967.

Shestakova, S. A., Effect of "diabetic" serum on respiration and carbohydrate metabolism in leukocytes of rabbit exudate in experimental diabetes (in Russian), *Vopr. Med. Khim.*, 13, 461, 1967.

Bacon, G. E. et al., Serum lipids and lipoproteins in diabetic children, *Univ. Mich. Med. Cent. J.*, 34, 84, 1968.

Sil'nitskii, P. A., Dynamics of blood serum lipid and lipoprotein indices in diabetes mellitus, treated with chlorpropamide and butamide (in Russian), *Probl. Endokrinol.*, 14, 9, 1968.

Bhu, N. et al., Serum lipoproteins and glycoproteins in diabetic subjects, *J. Assoc. Physicians India*, 17, 573, 1969.

Chance, G. W. et al., Serum lipids and lipoproteins in untreated diabetic children, *Lancet*, 1, 1126, 1969.

Chance, G. W. et al., Lipids and lipoproteins in untreated diabetes, *Lancet*, 2, 544, 1969.

Haller, H. et al., Flotation analysis studies of lipoproteins in correlation with the lipid pattern in essential hyperlipemia and diabetic metabolism status, *Schweiz. Med. Wochenschr.*, 99, 813, 1969.

Mazovetskii, A. G. et al., Blood serum lipoproteins in patients with diabetic angiopathy and obliterating atherosclerosis (in Russian), *Ter. Arkh.*, 41, 63, 1969.

Meduri, D. et al., Changes of the blood lipid pattern in diabetics under treatment with chlorpropamide and phenethyl-1-biguanide (in Italian), *Minerva Med.*, 60, 2439, 1969.

Murthy, D. Y. et al., β-lipoproteins in children with diabetes mellitus. A study, *Mo. Med.*, 66, 273, 1969.

Roe, R. L. et al., Lipids and lipoproteins in untreated diabetes, *Lancet*, 2, 496, 1969.

Skopichenko, N. F., Study of blood proteins, lipids and lipoproteins in patients with diabetic glomerulosclerosis (in Russian), *Vrach. Delo*, 3, 22, 1969.

Wille, L. E., Lipoproteins and diabetes mellitus, *Tidsskr. Nor. Laegeforen.*, 90, 313, 1970.

Wille, L. E., A slow moving pre-β-lipoprotein band in serum. Report of a diabetic serum subjected to lipoprotein electrophoresis, *Acta Med. Scand.*, 188, 241, 1970.

Bazex, A. et al., Type IV hyperlipoproteinemia in a female diabetic (in French), *Bull. Soc. Fr. Dermatol. Syphiligr.*, 78, 74, 1971.

Bhan, C. K. et al., Studies on neutral fat, lipoproteins and lipoprotein lipase in relation to vascular disease in young Indian diabetics, *Acta Diabetol. Lat.*, 8, 638, 1971.

Hart, A. et al., Lipoprotein and fibrinogen studies in diabetes, *Postgrad. Med. J. Suppl.*, 47, 435, 1971.

Jonsson, A. et al., Lipoprotein electrophoresis in diabetic patients with various degrees of vascular insufficiency, *Nord. Med.*, 86, 1284, 1971.

Nakano, E. et al., Serum lipoprotein fractions in diabetes mellitus, *Jpn. J. Clin. Pathol.*, Suppl. 19, 171, 1971.

Petrides, P. et al., Behavior of lipoproteins in diabetic coma, *Verh. Dtsch. Ges. Inn. Med.*, 77, 603, 1971.

Siedek, H. et al., Influence of caffein on the daily profile of metabolites from lipid and carbohydrate metabolism as well as on insulin secretion, *Verh. Dtsch. Ges. Inn. Med.*, 77, 606, 1971.

Tishenina, R. S. et al., Influence of incubating the serum of donors and diabetes mellitus patients with insulin on its effect with regard to several lipid metabolism indices (in Russian with English.), *Probl. Endokrinol.*, 17, 19, 1971.

Zorilla, E. et al., Diabetes and hyperlipoproteinemia in coronary atherosclerosis, *Acta Diabetol. Lat.*, 8, 629, 1971.

Birkbeck, J. A., Dyslipoproteinemias and diabetes mellitus, *Acta Diabetol. Lat.*, Suppl. 9, 79, 1972.

Bricker, L. A. et al., The hyperlipoproteinemias: mechanisms, managements and implications, *Acta Diabetol. Lat.*, Suppl. 9, 53, 1972.

Brown, D. F., The dyslipoproteinemias, *Acta Diabetol. Lat.*, Suppl. 9, 11, 1972.

Hase, M. et al., Abnormalities of serum lipids and lipoproteins in diabetes mellitus, *Saishin Igaku*, 27, 510, 1972.

Kremer, G. J. et al., Distribution of various hyperlipoproteinemia patterns in patients with patent diabetes mellitus or peripheral circulation disorders, *Verh. Dtsch. Ges. Inn. Med.*, 78, 1346, 1972.

Steiner, G., Biosynthesis, physiologic role and normal patterns of lipoproteins in normal human plasma, *Acta Diabetol. Lat.*, Suppl. 9, 3, 1972.

Středa, M. et al., Serum lipoproteins in diabetes mellitus, *Cas Lek. Cesk.*, 111, 1026, 1972.

Righetti, A. et al., Clinical study of the relationship between diabetes, hyperlipoproteinemia and atheromatosis, *Schweiz. Med. Wochenschr.*, 103, 668, 1973.

Schwandt, P., Hyperlipoproteinemia and diabetes mellitus, *Hippokrates*, 44, 457, 1973.

Středa, M. et al., Serum lipoproteins in diabetes mellitus, *Rev. Czech. Med.*, 19, 72, 1973.

Vogelberg, K. A. et al., Clinical picture and treatment of insulin resistance in primary hyperlipoproteinaemia, *Dtsch. Med. Wochenschr.*, 98, 1751, 1973.

Wille, L. E et al., Demonstration of hyper-alpha-lipoproteinemia in three diabetic patients, *Clin. Genet.*, 4, 281, 1973.

Davidsen, O., Immunoelectrophoretic determination of serum globulins in newborn infants of diabetic mothers, *Acta Paediatr. Scand.*, 63(6), 833, 1974.

Gligore, V. et al., Significance of hyperlipoproteinemia in diabetes mellitus, *Oeff. Gesundheitswes.*, 36, 1259, 1974.

Howard, C. F., Jr., Correlations of serum triglyceride and prebetalipoprotein levels to the severity of spontaneous diabetes in *Macaca nigra*, *J. Clin. Endocrinol. Metab.*, 38, 856, 1974.

Naruszewicz, M. et al., Phenotyping of secondary hyperlipoproteinemia in diabetics, *Pol. Tyg. Lek.*, 29, 653, 1974.

Schonfeld, G. et al., Apolipoprotein B levels and altered lipoprotein composition in diabetes, *Diabetes*, 23(10), 827, 1974.

Strat, C. et al., Study of serum lipoproteins in diabetes mellitus. Clinicometabolic significance (English abstr.), *Rev. Med. Chir. Soc. Med. Nat. Iasi*, 78(3), 565, 1974.

Telner, A. et al., In vitro synthesis of aortic glycosaminoglycans. Effect of diet, lipoproteins and diabetes, *Atherosclerosis*, 20(1), 81, 1974.

Tiengo, A. et al., Insulin secretion in hyperlipoproteinemias, *Acta Diabetol. Lat.*, 11(2), 148, 1974.

Wahl, P. et al., Diabetes and hyperlipoproteinaemias, *Dtsch. Med. Wochenschr.*, 99(43), 2158, 2163, 1974.

Wille, L. E. et al., Lipoprotein and lipids in diabetics with and without coronary heart disease, *J. Oslo City Hosp.*, 24(9), 113, 1974.

Sachlová, M. et al., Increased prebeta lipoproteins in the serum as an indicator of latent diabetes, *Vnitr. Lek.*, 21(10), 974, 1975.

Syllaba, J., Lipids and lipoproteins in diabetes mellitus (in Czech with English), *J. Vnitr., Lek.*, 21(11), 1078, 1975.

Billimoria, J. D. et al., A lipid and lipoprotein profile of treated and untreated diabetics, *Ann. Clin. Biochem.*, 13(1), 315, 1976.

Chase, H. P. et al., Juvenile diabetes mellitus and serum lipids and lipoprotein levels, *Am. J. Dis. Child.*, 130(10), 113, 1976.

Kissebah, A. H. et al., The metabolic fate of plasma lipoproteins in normal subjects and in patients with insulin resistance and endogenous hypertriglyceridaemia, *Diabetologia*, 12(5), 501, 1976.

Krejsová, Z. et al., Hyperlipoproteinaemia and diabetes mellitus, *Cas. Lek. Cesk.*, 115(24), 721, 1976.

Pometta, D., Hyperlipemia and diabetes (in French with English), *Schweiz. Med. Wochenschr.*, 106(31), 1054, 1976.

Digestive Tract Disorders

Kirkeby, K., Lipids, lipoproteins and proteins in serum following partial gastrectomy, *Acta Med. Scand.*, 178, 433, 1965.

Rousanov, E. et al., Lipoproteins from glandular mucoprotein of gastric juice, *Dokl. Bolg. Akad. Nau.*, 18, 871, 1965.

Bihari-Varga, M. et al., Role of intimal mucoid substances in the pathogenesis of atherosclerosis. Investigations on the interacting components in the mucopolysaccharide-β-lipoprotein complex formation in vitro, *Acta Physiol. Acad. Sci. Hung.*, 29, 273, 1966.

Dobbins, W. O., An ultrastructural study of the intestinal mucosa in congenital β-lipoprotein deficiency with particular emphasis upon the intestinal absorptive cell, *Gastroenterology*, 50, 195, 1966.

Roheim, P. S. et al., Extrahepatic synthesis of lipoproteins of plasma and chyle: role of the intestine, *J. Clin. Invest.*, 45, 297, 1966.

Zuber, E. et al., Total lipids, cholesterol and β-lipoproteins in blood serum in certain diseases of the digestive tract (in Polish), *Pol. Tyg. Lek.*, 22, 872, 1967.

Werner, M. et al., Serum protein changes after gastrectomy as a model of acute phase reaction, *J. Lab. Clin. Med.*, 70, 302, 1967.

Adamczyk, B. et al., Studies on protein fractions in patients with ulcer disease of the stomach (in Polish), *Pol. Przegl. Chir.*, 40, 553, 1968.

Shimoyama, T. et al., Altered fatty acid pattern in plasma lipoproteins of patients with intestinal malabsorption, *Clin. Sci.*, 40, 18P, 1971.

Gangl, A., The lipids metabolism of the small intestine and its correlation to the lipid and lipoprotein metabolism of the total organism, *Acta Med. Austriaca*, 2(3), Suppl., 1, 1975.

Glickman, R. M. et al., Intestinal lipoprotein formation: effect of cholchicine, *Gastroenterology*, 70(3), 347, 1976.

Drowning

Bondoli, A. et al., Changes in pulmonary alveoli β-lipoprotein levels following fresh water drowning and cardiac resuscitation (in Italian), *Minerva Anestesiol.*, 36, 315, 1970.

Dysentery

Cherepanova, G. P., Blood serum protein fractions, glycoproteins and lipoproteins in acute dysentery (in Russian), *Klin. Med. Moscow*, 47, 104, 1969.

Dystrophy, Myotonic

Watters, G. V. et al., Early onset myotonic dystrophy. Clinical and laboratory findings in five families and a review of the literature. *Arch. Neurol.*, 17, 137, 1967.

Ear Disease

Opplt, J. J. and Hlavacek, V., Investigation of blood lipoprotein in otosclerosis, *Cas. Lek. Cesk.*, 7, 198, 1957.

Naumann, H. W., The behavior of lipoproteins in otosclerosis patients (paper electrophoresis studies) (in German), *Arch. Ohren Nasen Kehlkopfheilkd.*, 184, 143, 1964.

Spencer, J. T., Jr., Hyperlipoproteinemias in the etiology of inner ear disease, *Laryngoscope,* 83, 639, 1973.
Nims, J. C. et al., Proteins in exudates of experimental otitis media, *J. Laryngol. Otol.,* 88(9), 863, 1974.
Spencer, J. T., Jr., Hyperlipoproteinemia and inner ear disease, *W. Va. Med. J.,* 70, 215, 1974.

Eye Disease

Daniele, S. et al., Electrophoretic research on the behavior of the blood serum glycoproteins and lipoproteins in subjects with senile cataract (in French), *Ann. Ocul.,* 197, 880, 1964.
Wolff, O. H. et al., A beta-lipoproteinaemia with special reference to the visual defect, *Exp. Eye Res.,* 3, 439, 1964.
Smirnova, S. P., Results following study of the blood serum protein and lipoprotein fraction composition in patients with pigmentary degeneration of the retina (in Russian), *Vestn. Oftal.,* 78, 60, 1965.
Cardia, L. et al., Research on the protein composition of ocular tissue of humans and various animals in normal and pathological conditions with new methods of electrophoretic investigation. I. Proteins and lipoproteins of the retina (in Italian), *Boll. Ocul.,* 46, 423, 1967.
Cardia, L. et al., Study of the protein composition of eye tissues with new electrophoretic methods. IV. Corneal lipoproteins (in Italian), *Minerva Oftalmol.,* 10, 17, 1968.
Gjone, E. et al., Corneal arcus and hyperlipoproteinaemia, *Lancet,* 2, 359, 1968.
Klenka, L. et al., Are there correlations between the β/α index of the serum lipoproteins and arteriosclerotic changes of the fundus oculi? (in German), *Klin. Monatsbl. Augenheilkd.,* 152, 225, 1968.
Waters, L. L., Removal of serum lipoproteins from the cornea in vivo, *Yale J. Biol. Med.,* 41, 257, 1968.
Scuderi, G. et al., Protein composition of eye tissues by means of new methods of electrophoretic research (proteins, lipoproteins, glycoproteins of the cornea) (in French), *Rev. Bras. Oftalmol.,* 28, 175, 1969.
Hayashi, Y., Studies on the influence of retinal stimulation due to luminous stimuli on the living body. 4. Fluctuations of total cholesterol and ester, and the fraction of lipoprotein in blood (in Japanese), *Folia Ophthalmol. Jpn.,* 21, 79, 1970.
Vancea, P. et al. Contribution to the problem of lipidic metabolism troubles in children suffering from flictenular kerato-conjunctivitis (in German), *Ber. Dtsch. Ophthalmol. Ges.,* 70, 42, 1970.
Vinger, P. F. et al., Ocular manifestations of hyperlipoproteinemia, *Am. J. Ophthalmol.,* 70, 563, 1970.
Dvořák, V. et al., Eye changes in hyperlipoproteinemia (in Czech), *Cesk. Oftalmol.,* 27, 149, 1971.
Goswamy, S. et al. Lipoproteins of the crystalline lens and serum factors in senile cataract, *Indian J. Med. Res.,* 59, 1460, 1971.
Williams, H. P. et al., Hereditary crystalline corneal dystrophy with an associated blood lipid disorder, *Trans. Ophthalmol. Soc. UK.,* 91, 531, 1971.
Graner, L. E. et al., Studies on atherogenesis: the reaction of the corneal model to repeated injections of lipoprotein-rich and lipoprotein-poor homologous serum, *Yale J. Biol. Med.,* 45, 93, 1972.
Mathur, R. L. et al., Proteolipids and phosphatido-peptides in senile cataract, *Indian J. Med. Res.,* 60, 641, 1972.
Muto, Y., Metabolism of fat soluble vitamins — retinol binding protein, *Protein Nucleic Acid Enzyme* 19(6), 411, 1974.
Schmut, O. et al., Immunological determination of aqueous humor lipoproteins, *Albrecht von Graefes Arch. Klin. Exp. Ophthalmol.,* 191(1), 19, 1974.
Bron, A. J., Dyslipoproteinemias and their ocular manifestations, *Birth Defects,* 12(3), 257, 1976.
Heller, J. et al., Transport of retinol from the blood to the retina: involvement of high molecular weight lipoproteins as intracellular carriers, *Exp. Eye Res.,* 22(5), 403, 1976.
Kurs, G. H. et al., The retina in type 5 hyperlipoproteinemia, *Am. J. Ophthalmol.,* 82(1), 32, 1976.
Miyata, M. et al., Biochemical analysis of peripheral blood in patients with retinal pigmentary degeneration, abnormal value of lipo- and glycoprotein on electrophoresis (author's transl. in Japanese with English abstr.), *Acta Soc. Ophthalmol. Jpn.,* 80(10), 1264, 1976.
Yee, R. D. et al., A typical retinitis pigmentosa in familial hypo-β-lipoproteinemia, *Am. J. Ophthalmol.,* 82(1), 64, 1976.

Endocrine Disorders

Koppers, L. E. et al., Lipid disturbances in endocrine disorders, *Med. Clin. North Am.,* 56, 1013, 1972.
Szczeklik, A., Differential diagnosis of hyperlipidemias and hypolipidemias in endocrinologic diseases, *Przegl. Lek.,* Suppl., 31(0), 79, 1974.

Fabry's Disease

Clarke, J. T. et al., Neutral glycosphingolipids of serum lipoproteins in Fabry's disease, *Biochim. Biophys. Acta,* 431(2), 317, 1976.

Farber's Disease

Rampini, S. et al., Farber's disease (disseminated lipogranulomatosis). Clinical picture and summary of the chemical findings (in German), *Helv. Paediatr. Acta*, 22, 500, 1967.

Gout

Bluestone, R. et al., Hyperlipoproteinaemia in gout, *Ann. Rheum. Dis.*, 30, 134, 1971.

Mertz, D. P. et al., Classification of hyperlipoproteinaemia in primary gout (in German with English abstr.), *Dtsch. Med. Wochenschr.*, 97, 600, 1972.

Wiedemann, E. et al., Plasma lipoproteins, glucose tolerance and insulin response in primary gout, *Am. J. Med.*, 53, 299, 1972.

Wollenweber, J. et al., Classification of hyperlipoproteinemia in primary gout (in German), *Dtsch. Med. Wochenschr.*, 97, 924, 1972.

Mertz, D. P. et al., Dyslipoproteinaemia in primary gout, *Dtsch. Med. Wochenschr.*, 98, 1457, 1973.

Mielants, H. et al., Gout and its relation to lipid metabolism. II. Correlations between uric acid, lipid, and lipoprotein levels in gout, *Ann. Rheum. Dis.*, 32, 506, 1973.

Mielants, H. et al., Gout and its relation to lipid metabolism. I. Serum uric acid, lipid, and lipoprotein levels gout, *Ann. Rheum. Dis.*, 32, 501, 1973.

Marcolongo, R. et al., Changes in the lipid and lipoprotein pattern in patients with primary gout (English abstr.), *Reumatismo*, 27(1), 180, 1975.

Gynecologic Disorders and Pregnancy

Opplt, J. J. and Blekta, M., Metabolical changes in the course of late toxemia, *Acta Univ. Carol*, 5, 1—3, *Medica*, 1—512, 192, 1958.

Opplt, J. J. and Blekta, M., The relationship between the clinical picture and the biochemical and functional changes in late toxemia, *Acta Univ. Carol*, 5, 1—3; *Medica*, 1—512, 165, 1958.

Mullick J. et al., Serum lipid studies in pregnancy, *Am. J. Obstet. Gynecol.*, 89, 766, 1964.

Opplt, J. J. Novotny, A., Dvorak, V., and Schreiber, B., The changes in the spectrum of serum proteins and lipoproteins in women after gynecological operations, *Acta Univ. Carol. Med. Suppl.*, 19, 129, 1964.

Capra, E. et al., On some aspects of protein and lipoprotein fractionation at term of pregnancy and in the first days of puerperium (electrophoretic studies) (in Italian), *Boll. Ist. Sieroter. Milan.*, 44, 153, 1965.

Jörgensen, G. et al., Studies of the β-lipoprotein system according to Berg in healthy, sick and pregnant persons (in German), *Humangenetik*, 1, 476, 1965.

Aurell, M. et al., Serum lipids and lipoproteins in human pregnancy, *Clin. Chim. Acta*, 13, 278, 1966.

Opplt, J. J., Novtony, A., and Dvorak, V., Dyslipoproteinemia after chirurgical castration in women, *Cas. Lek. Cesk.*, 105, 569, 1966.

Koroleva, A. M. et al., Blood lipoproteins in normal and pathological pregnancy (in Russian), *Vopr. Okkr. Materin. Dets.*, 12, 92, 1967.

Opplt, J. J., Novotny, A., and Dvorak, V., Dyslipoproteinemia after surgical castration of women (in English), *Rev. Czech. Med.*, 13, 151, 1967.

Opplt, J. J., Novotny, A., and Dvorak, V., Dyslipoproteinemia after surgical castration of women (in Russian), *Ceskoslovackoje Med. Obozrenije.*, 13, 3, 160, 1967.

Rendina, G. M. et al., A new factor in late gestoses: dehydroepiandrosterone deficiency (in Italian), *Quad. Clin. Ostet. Ginecol.*, 22, 793, 1967.

Sotnikova, L. G., Content of proteins, lipoproteins, glycoproteides and sialic acids in healthy non-pregnant and pregnant women (in Russian), *Lab. Delo*, 8, 476, 1967.

Opplt, J. J., Dvorak, V., and Novotny, A., The changes in total proteins and electrophoretic spectrum on serum-lipoproteins, occurring after operations of gynecological malignancies, *Cas. Lek. Cesk.*, 107(3), 69, 1968.

Opplt, J. J., Dvorak, V., and Novotny, A., The changes of total cholesterin triglycerides and beta-lipoproteins after gynecological operations, *Cas. Lek. Cesk.*, 107, 34—35, 1068, 1968.

Roszkowski, T. et al., Contents of proteins and lipoproteins and blood serum of mothers and their offspring (in Polish), *Ginekol. Pol.*, 39, 303, 1968.

Roszkowski, I. et al., Protein and lipoprotein levels in serum samples from women who delivered malformed babies b)in Polish), *Ginekol. Pol.*, 39, 521, 1968.

Sotnikova, L. G., Changes in protein, lipoproteid, glycoproteid and sialic acid content during normal pregnancy and in late toxemia (in Russian), *Vopr. Okhr. Materin. Dets.*, 13, 57, 1968.

Krawczyk, B. et al., Protein and lipoprotein fractions, cholesterol levels, aminotransferases and guanosine deaminase levels in serum of women in early normal pregnancy (in Polish), *Ginekol. Pol.*, 40, 133, 1969.

Suenaga, M., Study on lipoprotein of the vascular wall of the placenta with special reference to toxemia of late pregnancy (in Japanese), *J. Kumamoto Med. Soc.*, 43, 444, 1969.

Nencioni, T. et al., Albumin, transferrin, IgA, IgG, IgM and β-lipoprotein concentration in maternal serum, amniotic fluid and umbilical cord serum, in normal pregnancy near term (in French), *Gynecol. Obstet.*, 69, 219, 1970.

Skrzydlewski, Z. et al., Concentration of β-lipoproteins and activity of the fibrinolytic system in the blood of women in last days of pregnancy and in fetal blood (in Polish with English.), *Wiad. Lek.*, 25, 321, 1973.

Jezuita, J. et al., Serum beta-lipoproteins and plasma fibrinolytic activity in women with different post-labor blood loss, *Wiad. Lek.*, 26, 719, 1973 .

Knopp, R. H. et al., Lipid metabolism in pregnancy. I. Changes in lipoprotein triglyceride and cholesterol in normal pregnancy and the effects of diabetes mellitus, *J. Reprod. Med.*, 10, 95, 1973.

Gustafson, A. et al., Human plasma high-density lipoprotein composition during the menstrual cycle, *Scand. J. Clin. Lab. Invest. Suppl.*, 33, 63, 1974.

Ylöstalo, P. et al., Serum lipids and lipoproteins in hepatosis of pregnancy and pre-eclampsia, *Ann. Chir. Gynaecol. Fenn.*, 63, 11, 1974.

Casu, A. et al., Analysis of proteins and phospholipids in human amniotic fluid, II. Study of a lipoprotein fraction characteristic of the amniotic fluid (in Italian), *Pathologica*, 67(971—972), 395, 1975.

Gehrmann, J. et al., Dynamic behavior of lipids and lipoproteins during pregnancy and puerperium with special reference to hormonal influences (in German), *Dtsch. Ges. Inn. Med.*, 81, 1438, 1975.

Hillman, L. et al., Apolipoproteins in human pregnancy, *Metabolism*, 24(8), 943, 1975.

Johnson, P., Studies in cholestasis of pregnancy. IV. Serum lipids and lipoproteins in relation to duration of symptoms and severity of the disease, and fatty acid composition of lecithin in relation to duration of symptoms, *Acta Obstet. Gynecol. Scand.*, 54(4), 307, 1975.

Johnson, P. et al., Studies in cholestasis of pregnancy, *Acta Obstet. Gynecol. Scand.*, 54(2),105, 1975.

Wankowicz, Z. et al., β-cholesterol in pregnancy cholestasis. III. Protein metabolism in women with pregnancy cholestasis and the cholesterol contents in serum β-lipoproteins (in Polish with English.), *Ginekol. Pol.*, 46(12), 1241, 1975.

Wankowicz, Z. et al., β-cholesterol in pregnancy cholestasis. I. Total cholesterol present in β-lipoprotein fractions (in Polish with English.), *Ginekol. Pol.*, 46(10), 1051, 1975.

Wankowicz, Z. et al., β-cholesterol in pregnancy cholestasis. II. Enzymatic changes in blood serum samples from women with pregnancy cholestasis and cholesterol content in β-lipoprotein fractions (in Polish with English.), *Ginekol. Pol.*, 46(11), 1147, 1975.

Wankowicz, Z. et al., Cholesterol concentration in beta and pre-beta-lipoprotein fractions in serum samples from women in normal pregnancy, labor and puerperium (English.), *Ginekol. Pol.*, 46(5), 503, 1975.

Warth, M. R. et al., Lipid metabolism in pregnancy. II. Altered lipid composition in intermediate, very low, low and high-density lipoprotein fractions, *J. Clin. Endocrinol. Metab.*, 41(4), 649, 1975.

Ferlito, S. et al., Serum lipoprotein typing in pregnant women and the fetus at them (in Italian with English.), *Minerva Ginecol.*, 28(10), 771, 1976.

Guminski, S. et al., Pattern of total lipids, β-lipoproteins, fatty acids and total cholesterol in blood serum samples in women with a normal course of gestation and in protracted pregnancy and the condition of the newborn infant. II. Protracted pregnancy and the condition of the newborn infant (in Polish with English.), *Ginekol. Pol.*, 47(11), 1251, 1975.

Novotný, A. et al., Serum β-lipoproteins following gynecologic surgery (in German with English.), *Zentralbl. Gynaekol.*, 98(5), 937, 1976.

Novotný, V. A. et al., Changes in the serum lipoprotein spectrum after gynecologic operations especially after castration (in German with English.), *Zentralbl. Gynaekol.*, 98(5), 299, 1976.

Wankowicz, Z. et al., β-cholesterol in the cholestasis of pregnancy. IV. Concentration of total bilirubin and iron in the serum and occurrence of pruritus in pregnant women with cholestasis and content of cholesterol in blood β-lipoprotein fractions (in Polish with English.), *Ginekol. Pol.*, 47(1), 1, 1976.

Hodgkin's Disease

Tremblay, E. C. et al., Subacute Hodgkin's disease with increased myeloid hyperleukocytosis — interesting palliative effect of marrow fraction FM7A, γ-globulin and β-lipoproteins (in French), *Bibl. Haematol.*, 23, 224, 1965.

Christophe, A. et al.,The effect of different fat feedings on lipids and lipoproteins in ascites fluid in a patient with Hodgkin disease, *Arch. Int. Physiol. Biochim.*, 84(3), 591, 1976.

Hypertension

Gagov, S. et al., Blood cholesterol and lipoprotein level in hypertension and their modification in decreased blood pressure by means of low-frequency impulse currents (in Bulgarian), *Izv. Inst. Fiziol.*, 8, 209, 1964.

Belousenko, E. F., Some biochemical indices of the blood in patients with neurocirculatory dystonia of the hypertension type (in Russian), *Voen. Med. Zh.*, 6, 47, 1967.

Koike, S. et al., Serum lipids and lipoprotein in hypertension and ischemic heart disease (in Japanese), *Jpn. J. Hyg.*, 22, 451, 1967.

Bavina, M. V. et al., Types of hyperlipoproteinemia in various forms of arterial hypertension, *Ten. Arkh.*, 45, 88, 1973.

Pirrelli, A. et al., Lipid and lipoprotein variations in the course of essential hypertension (English.), *Boll. Soc. Ital. Cardiol.*, 19(5), 443, 1974.

Hypotension

Kleiner, G. M., Serum lipoproteins in patients with hypotension (in Russian), *Klin. Med.*, 44, 97, 1966.

Infection

Angelopoulos, B. et al., Electrophoretic analysis of serum proteins, glucoproteins and lipoproteins in acute infectious disease in infants and children, *Med. Pharmacol. Exp.*, 14, 517, 1966.

Badin, J. et al., Non-specific inhibition of streptolysin O and β-lipoprotein cholesterol in hemolytic streptococcus infections (in French with English.), *Pathol. Biol.*, 19, 1081, 1971.

Kireeva, R. I., Blood serum protein fractions and lipoproteins in tick-borne rickettsiosis (in Russian with English.), *Klin. Med., (Moscow)*, 49, 117, 1971.

Griffiths, E. K., et al., Antibody response to enterobacterial lipoprotein of patients with varied infections due to Enterobacteriaceae, *Proc. Soc. Exp. Biol. Med.*, 154(2), 246, 1977.

Kwashiorkor

Truswell, A. S. et al., Serum lipoproteins and phospholipids in relation to fatty liver in kwashiorkor, *S. Afr. Med. J.*, 40, 887, 1966.

Truswell, A. S. et al., Relation of serum lipids and lipoproteins to fatty liver in kwashiorkor, *Am. J. Clin. Nutr.*, 22, 568, 1969.

Coward, W. A. et al., Changes in serum-lipoprotein concentration during the development of kwashiorkor and in recovery, *Br. J. Nutr.*, 27, 383, 1972.

Onitiri, A. C. et al., Serum lipids and lipoproteins in children with kwashiorkor, *Br. Med. J.*, 3(5984), 630, 1975.

Leprosy

Hariprasad, C. et al., Serum beta lipoprotein levels in leprosy, *Int. J. Lepr.*, 39, 896, 1971

Leptospirosis

Kasarov, L. B. et al., Metabolism of the lipoproteins of serum by leptospires: degradation of the triglycerides, *J. Med. Microbiol.*, 2, 165, 1969.

Leukemia

Tanaka, H. et al., Case of juvenile form of metachromatic leukodystrophy associated with low serum levels of beta-lipoproteins and vitamin E, *Clin. Neurol.* (Tokyo), 15(6), 355, 1975.

Lipomatosis

Greene, M. L. et al., Benign symmetric lipomatosis (Launois-Bensande adenolipomatosis) with gout and hyperlipoproteinemia, *Am. J. Med.*, 48, 239, 1970.

Wengeler, H. et al., Serum cholesterol esterification in liver disease. Combined determinations of lecithin, cholesterol acyltransferase and lipoprotein-X, *Eur. J. Clin. Invest.*, 2, 372, 1972.

Wollenweber, J. et al., Pool size and turnover of primary bile acids in hyperlipoproteinemias: different findings in type II and type IV hyperlipoproteinemias (in German with English abstr.), *Klin. Wochenschr.*, 50, 33, 1972.

Fassati, P. et al., Demonstration of beta-lipoproteins in patients with chronic hepatitis by beta-L test, *Cas. Lek. Cesk.*, 112, 1095, 1973.

Hartmann, L., et al., Temporary suppression of the quaternary structure of plasma alphalipoproteins (HDL) in the early stage of common viral hepatitis, *C.R. Acad. Sci. Ser. D*, 276, 2597, 1973.

Janecki, J. et al., One (of) alpha-1-lipoprotein in acute viral hepatitis, *Ann. Biol. Clin.*, 31, 100, 1973.

Johnson, P., Studies in cholestasis of pregnancy with special reference to lipids and lipoproteins, *Acta Obstet. Gynecol. Scand. Suppl.*, 27, 1, 1973.

Schmitz, J. et al., Serum lipoproteins and lipids in patients with hepatitis, *Dtsch. Med. Wochenschr.*, 98, 2436, 1973.

Seidel, D. et al., Significance of the LP-X test in differential diagnosis of jaundice, *Clin. Chem.*, 19, 86, 1973.

Stein, O. et al., Obstructive jaundice lipoprotein particles studied in ultrathin sections of livers of bile duct-ligated mice, *Lab. Invest.*, 29, 166, 1973.

Baumgarten, M. et al., Serum lipoproteins in acute viral hepatitis, *Klin. Wochenschr.*, 52, 617, 1974.

Hartmann, L. et al., Abnormality of the structure and composition of plasma lipoproteins (HDL) during different types of hepatitis, *Biomed. Express*, 21(12), 481, 1974.

Picard, J. et al., Abnormal serum lipoproteins in cholestatic viral hepatitis (in French with English abstr.), *Clin. Chim. Acta*, 51, 5, 1974.

Tokarskaia, Z. B., et al., Biochemical indices of the blood serum in various liver diseases (English abstr.), *Ter. Arkh.*, 46(4), 53, 1974.

Campbell, D. P. et al., Determination of serum lipoprotein-X for the early differentiation between neonatal hepatitis and biliary atresia, *J. Surg. Res.*, 18(4), 385, 1975.

Chernova, G. V. et al., Characteristic features of blood serum in mechanical jaundice, *Lab. Delo*, 3, 140, 1975.

Greiner, S. et al., Serum lipids and serum lipoproteins in liver diseases (English abstr.), *Med. Klin.*, 70(6), 231, 1975.

Milewski, B. et al., Usefulness of serum lipoprotein-X (LP-X) detection test in the diagnosis of cholestasis in chronic liver diseases (English abstr.), *Pol. Arch. Med. Wewn.*, 53(5), 445, 1975.

Ritland, S., Quantitative determination of the abnormal lipoprotein of cholestasis, LP-X, in liver disease, *Scand. J. Gastroenterol.*, 10(1), 5, 1975.

Ritland, S. et al., Plasma concentration of lipoprotein-X (LP-X) in experimental bile duct obstruction, *Scand. J. Gastroenterol.*, 10(1), 17, 1975.

Seidel, D., Lipoproteins in liver diseases (in German), *Med. Welt*, 26(47), 2131, 1975.

Thalassinos, N. et al., Plasma alpha-lipoprotein pattern in acute viral hepatitis, *Am. J. Dig. Dis.*, 20(2), 148, 1975.

Agorastos, J. et al., Plasma lipoproteins in liver disease, *Biochem. Soc. Trans.*, 4(4), 593, 1976.

Devi, C. S. et al., Serum lipoprotein profile in infectious hepatitis, *Jpn. J. Exp. Med.*, 46(1), 15, 1976.

Kirchmayer, S., Postgraduate teaching. Lipoprotein metabolism — its physiological and clinical aspects. Dyslipoproteinemia in the course of liver diseases (in Polish), *Przegl. Lek.*, 33(12), 996, 1976.

Koch, K. S. et al., Control hepatic proliferation: a working hypothesis involving hormones, lipoproteins, and novel nucleotides, *Metabolism*, Suppl. 25, 1419, 1976.

Kostner, G. M. et al., Investigation of the abnormal low-density lipoproteins occurring in patients with obstructive jaundice, *Biochem. J.*, 157(2), 401, 1976.

Mordasini, R. C. et al., Changes in serum lipids and lipoproteins in acute hepatitis (in German with English abstr.), *Schweiz. Med. Wochenschr.*, 106(35), 1173, 1976.

Takasugi, Y., Serum lipids in patients with liver disease (author's transl.), (in Japanese), *Jpn. J. Clin. Pathol.*, 24(3), 218, 1976.

Fehér, J. et al., Serum lipids and lipoproteins in chronic liver diseases (in Hungarian), *Orv. Hetil.*, 118(4), 194, 1977.

Liver Disease

Opplt, J. J. and Syllaba, J., On the so-called "Hepatic diabetes", *Cas. Lek. Cesk.*, 95(44/55), 1247, 1956.

Opplt, J. J. and Syllaba, J., Clinical and laboratory findings in hepatical chronic inflammatory processes and hepatic cirrhosis, *Cas. Lek. Cesk*, 96(40/41), 1303, 1957.

Opplt, J. J. and Syllaba, J., The significance of the investigation and analysis of serum-lipoproteins for the diagnosis and prognosis of hepatic diseases, *Cas. Lek. Cesk.*, 99(16), 500, 1960.

Bilousov, I. V., Lipoproteins in the blood serum of children with hepato-cholecystitis (in Rumanian), *Pediatr. Akush. Ginekol.*, 2, 8, 1965.

Kremmer, T. et al., Study of serum lipoproteins in jaundice. The role of abnormal lipoproteins in the Jirgl test (in Hungarian), *Orv. Hetil.*, 106, 405, 1965.

Crosato, M. et al., Electrophoretic behavior of serum lipoproteins in subjects with viral hepatitis before and after treatment with lysozyme (in Italian), *Epatologia*, 12, 353, 1966.

Kremmer, T. et al., Serum lipids in experimental jaundice, *Acta Med. Acad. Sci. Hung.*, 22, 219, 1966.

Burstein, M. et al., Abnormal serum lipoproteins in the course of certain jaundices by retention (in French), *Rev. Fr. Etud. Clin. Biol.*, 12, 898, 1967.

Franczak, T. et al., Correlation between cholesterol, β-lipoproteins and Quick's prothrombin time in the course of virus hepatitis in children (in Polish), *Pol. Tyg. Lek.*, 22, 459, 1967.

Switzer, S., Plasma lipoproteins in the differential diagnosis of liver disease, *Gastroenterology*, 53, 790, 1967.

Switzer, S., Plasma lipoproteins in liver disease. 1. Immunologically distinct low-density lipoproteins in patients with biliary obstruction, *J. Clin. Invest.*, 46, 1855, 1967.

Worth, G. et al., Pulmonary alveolar proteinosis combined with granulomatous hepatitis (in German), *Med. Welt*, 33, 1883, 1967.

Albano, O. et al., Hepatis synthesis of β-lipoproteins in experimental carbon tetrachloride poisoning (in French), *Pathol. Biol.* (Paris), 16, 735, 1968.

Burstein, M. et al., Flocculation by polyvinylpyrrolidone of abnormal serum lipoproteins during retention jaundice (in French), *Rev. Fr. Etud. Clin. Biol.,* 13, 404, 1968.

Burstein, M. et al., Isolation and study of abnormal serum lipoproteins during retention jaundice after flocculation by polyvinylpyrrolidone (in French), *Rev. Fr. Etud. Clin. Biol.,* 13, 387, 1968.

Monari, E. et al., Considerations on the behavior of the antistreptolysin titer and of the Burstein reaction in infantile virus hepatitis (in Italian), *G. Mal. Infett. Parassit.,* 20, 763, 1968.

Padolecchia, N. et al., Fractionation of the β-lipoproteins of the blood of subjects with hepatocellular jaundice and obstructive jaundice using gel filtration on an agarose column (in Italian), *Boll. Soc. Ital. Biol. Sper.,* 44, 1533, 1968.

Wilcox, H. G. et al., Hepatic lipid metabolism in experimental diabetes. IV. Incorporation of amino acid 14-C into lipoprotein-protein and triglyceride, *J. Biol. Chem.,* 243, 666, 1968.

Burstein, M. et al., Abnormal blood lipoproteins in hepatobiliary pathology (in French), *Rev. Med. Chir. Mal. Foie,* 44, 125, 1969.

Burstein, M. et al., β-lipoproteins in obstructive jaundice. Dosage of the 2 immunologically distinct fractions (in French), *Rev. Fr. Etud. Clin. Biol.,* 14, 68, 1969.

Charache, S. et al., Effect of plasma from patients with liver disease on resistance of red blood cells to lysis by saponin, *J. Lab. Clin. Med.,* 73, 951, 1969.

Kupershtein, A. P. et al., Phototurbidimetric determination of β-lipoproteins (Burstein-Samaille test) in the acute phase of epidemic hepatitis in children (in Russian), *Lab. Delo,* 77, 661, 1969.

Mills, G. L. et al., Ultracentrifugal characterization of a lipoprotein occurring in obstructive jaundice, *Clin. Chim. Acta,* 26, 239, 1969.

Seidel, D. et al., A lipoprotein characterizing separation and identification of lipoproteinin jaundiced subjects, *J. Clin. Invest.,* 48, 1211, 1969.

Grunevska, B. et al., Lipidogram in the diagnosis and follow-up of disease course in acute and chronic infectious hepatitis (in Croatian), *God. Zb. Med. Fak. Skopje,* 16, 197, 1970.

Kater, R. M. et al., Relationship of serum tocopherol to β-lipoprotein concentrations in liver diseases, *Am. J. Clin. Nutr.,* 23, 913, 1970.

Mincu, I. et al., Significant decrease of rapid lipoproteins in the blood serum in patients with jaundice, *Med. Interna.,* 22, 49, 1970.

Papadopoulos, N. M. et al., Serum lipoprotein patterns in liver disease, *Proc. Soc. Exp. Biol. Med.,* 134, 797, 1970.

Ross, A. et al., Occurrence of an abnormal lipoprotein in patients with liver disease, *Gut,* 11, 1035, 1970.

Seidel, D. et al., A lipoprotein characterizing obstructive jaundice. II. Isolation and partial characterization of the protein moieties of low density lipoprotein, *J. Clin. Invest.,* 49, 2396, 1970.

Seidel, D. et al., Advances in the analysis of plasma lipoproteins. A new possibility for differential diagnosis of icterus, *Klin. Wochenschr.,* 48, 1, 1970.

Klör, U. et al., Studies on changes of low-density lipoproteins (LDL) in liver disease, *Verh. Dtsch. Ges. Inn. Med.,* 77, 1388, 1971.

Leonard, R. F. et al., Serum lipoproteins in infants with liver disease, *Aust. Paediatr. J.,* 7, 3, 1971.

Lukěs, J. et al., Isolation and properties of abnormal lipoprotein of human serum in cholestasis, *Cas. Lek. Cesk.,* 110, 1195, 1971.

Picard, J. et al., Abnormal serum lipoproteins in cholestasis: identification of the components (in French), *C. R. Acad. Sci. Ser. (D),* 273, 418, 1971.

Seidel, D., The abnormal lipoprotein of cholestasis, *N. Engl. J. Med.,* 285, 1538, 1971.

Zamret, P. et al., Serum lipoproteins and intrahepatic biliary tract atresia in children (in French with English abstr.), *Arch. Fr. Pediatr.,* 28, 711, 1971.

Beresneva, A. B. et al., Activity of humoral factors on non-specific immunity in infectious hepatitis (English.), *Zh. Mikrobiol. Epidemiol. Immunobiol.,* 49, 120, 1972.

Cooper, R. A. et al., An analysis of lipoproteins, bile acids, and red cell membranes associated with target cells and spur cells in patients with liver disease, *J. Clin. Invest.,* 51, 3182—3193, 1972.

Herfort, K. et al., Contribution to the prevalence of cholecystolithiasis in patients with familial hyperlipoproteinaemia (in Czech with English.), *Sb. Lek.,* 74, 176, 1972.

Klör, U. et al., Comparative studies on lipoprotein changes after cholestasis in rats and humans, *Med. Welt,* 23, 1428, 1972.

Lukěs, J. et al., β-lipoprotein lipids of the human serum in liver diseases. IV. Composition of ultracentrifugal classes, *Cas. Lek. Cesk.,* 11, 78, 1972.

Magnani, H. N. et al., A method for the quantitative determination of the abnormal lipoprotein (LP-X) of obstructive jaundice, *Clin. Chim. Acta,* 38, 405, 1972.

Picard, J. et al., Identification of the apolipoprotein of abnormal serum lipoproteins in cholestasis (in French with English.,) *Clin. Chim. Acta,* 37, 483, 1972.

Picard, J. et al., Fatty acid composition of phospholipids in abnormal serum lipoproteins of cholestasis (in French with English.), *Clin. Chim. Acta,* 36, 247, 1972.

Seidel, D. et al., Further aspects on the characterization of high and very low density lipoproteins in patients with liver disease, *Eur. J. Clin. Invest.,* 2, 359, 1972.

Seidel, D., Plasma lipids and lipoproteins in patients with liver disease, *Scand. J. Gastroenterol.,* 7, 105, 1972.

Seidel, D. et al., Structure of an abnormal plasma lipoprotein (LP-X) characterizing obstructive jaundice, *Biochim. Biophys. Acta,* 260, 146, 1972.

Vogt, N. et al., Lipids and lipoproteins in liver diseases, *Verh. Dtsch. Ges. Inn. Med.,* 78, 1330, 1972.

Williams, P. F. et al., Plasma lipoproteins in pregnancy *Horm. Res.,* 7(2), 83, 1976.

Lung Disorders

Higa, M., Clinical and experimental studies on serum lipoproteins in pulmonary diseases, *Acta Med. Nagasaki.,* 9, 242, 1965.

Omori, Y. et al., Serum lipoprotein in lung diseases (in Japanese), *Jpn. J. Clin. Pathol.,* 13, 497, 1965.

Kungurov, T. A., The relationship between the lipoprotein metabolism of patients with chronic bronchopulmonary diseases and the stage of the process and degree of pulmonary insufficiency (in Russian), *Ter. Arkh.,* 40, 36, 1968.

Safronova, O. N. et al., Clinical evaluation of lipid metabolism indices in children with bronchial asthma (in Russian), *Pediatriia* (Moscow), 50, 72, 1971.

Malabsorption

Islam, S. S., Classifying protein-calorie malnutrition, *Lancet,* 2, 720, 1966.

Press, M. et al., Plasma lipoproteins and post-heparin lipolytic activity in patients with intestinal malabsorption, *Clin. Sci.,* 42, 17P, 1972.

Shimoyama, T. et al., Fatty acid composition of plasma lipoproteins in control subjects and in patients with malabsorption, *Gut,* 14, 716, 1973.

Thompson, G. R. et al., Plasma lipid and lipoprotein abnormalities in patients with malabsorption, *Clin. Sci. Mol. Med.,* 45, 583, 1973.

Devi, C. S. et al., Plasma lipids and lipoproteins in proteins-caloric malnutrition, *Acta Paediatr. Scand.,* 65(2), 161, 1976.

Mental Disorders

Eastham, R. D. et al., Serum cholesterol fractions, total esterified fatty acids and lipoprotein electrophoresis in mentally retarded patients, *J. Med. Sci.,* 3, 557, 1970.

Wada, M. et al., Arteriosclerosis and plasma lipoprotein-lipoprotein abnormalities and sclerotic cardiovascular diseases in patients with mental disorders (in Japanese), *Saishin Igaku,* 25, 2197, 1970.

Kruchinina, N. A. et al., Lipids and the lipoprotein spectrum of blood in men engaged in intensive mental activity (English abstr.), *Kardiologiia,* 15(4), 104, 1975.

Migraine

Leviton, A. et al., Migraine associated with hyper-pre-β-lipoproteinemia, *Neurology,* 19, 963, 1969.

Mole

Ma, L. et al., Serum lipoproteins in hydatidiform mole, *Clin. Chim. Acta,* 11, 561, 1965.

Morfan's Syndrome

Suschke, J. et al., Electrophoretic and immunologic findings in Morfan's Syndrome, (in German), *Dtsch. Med. Wochenschr.,* 94, 2289, 1969.

Multiple Sclerosis

Korin, M. M., The protein and protein-lipoid complexes in the blood serum of patients with multiple sclerosis (in Russian), *Zh. Nevropatol. Psikhiatr. im. S. S. Korsakov,* 65, 1623, 1965.

Ketelaer, C. J. et al., On the occurrence of proteolipids in plasma in multiple sclerosis, *Acta Neurol. Belg.,* 66, 270, 1966.

Csögör, S. et al., Changes of lipoproteins and congo-red binding capacity of serum albumins in patients with multiple sclerosis, (in German), *Psychiatr. Neurol.,* 154, 201, 1967.

Szabó, S. et al., Study of serum proteins in multiple sclerosis, (in Rumanian), *Stud. Cercet. Neurol.,* 12, 371, 1967.

Tichý, J. et al., Serum lipoproteins, cholesterol esters and phospholipids in multiple sclerosis, *Acta Neurol. Scand.,* 45, 32, 1969.

Rieder, H. P. et al., Sex differences in the serum lipoprotein relations of normal persons and multiple sclerosis patients, *Clin. Chim. Acta,* 30, 305, 1970.

Offner, H. et al., Precipitation of myelin basic protein by beta-lipoprotein of human serum. A study of similarities and differences of the precipitation in normal sera and sera from patients with multiple sclerosis, *Acta Neurol. Scand.,* 50, 221, 1974.

Muscular Atrophy

Quarfordt, S. H. et al., Familial adult-onset proximal spinal muscular atrophy. Report of a family with type II hyperlipoproteinemia, *Arch. Neurol.* (Chicago), 22, 541, 1970.

Mycoplasma

Sethi, K. K. et al., Enzymatic degradation of human lipoproteins by mycoplasmas, *Experientia,* 26, 804, 1970.

Myeloma

Andreeva, N. E. et al., Proteins, glycoproteins and lipoproteins of the blood in multiple myeloma and Waldenström's macroglobulinemia (in Russian), *Probl. Gematol.,* 9, 18, 1964.

Neufeld, A. H. et al., Beta-2-lipoprotein myelomatosis, *Can. J. Biochem.,* 42, 1499, 1964.

Beaumont, J. L. et al., Presence of an anti-β-lipoprotein autoantibody in a myeloma serum (in French), *C. R. Acad. Sci.,* 260, 5960, 1965.

Beaumont, J. L., A myeloma γ-A-globulin with a specific anti-lipoprotein activity. The anti-Pg autoantibody (in French), *C. R. Acad. Sci. Ser. D,* 263, 2046, 1966.

Aubert, L. et al., Clinical and cytologic study of a 2nd case of xanthomatous IgA myeloma with circulating anti-lipoprotein antibodies (in French), *Sem. Hop. Paris,* 43, 3014, 1967.

Beaumont, J. L., A common specificity in serum α- and β-lipoproteins demonstrated by a myeloma autoantibody. The Pg antigen (in French), *C.R. Acad. Sci. Ser. D,* 264, 185, 1967.

Beaumont, J. L. et al., Myeloma and hyperlipemia. IV. Nature of the specific anti-lipoprotein activity (in French), *Nouv. Rev. Fr. Hematol.,* 7, 481, 1967.

Beaumont, J. L. et al., Anti-lipoprotein autoantibodies (anti-Pg) of γ-A-myeloma with hyperlipidemia: method of isolation and purification from circulating complexes, (in French), *Ann. Biol. Clin.,* 25, 655, 1967.

Spikes, J. L., Jr. et al., The identification of a myeloma serum factor which alters serum β-lipoproteins, *Clin. Chim. Acta,* 20, 413, 1968.

Beaumont, J. L., A second type of myeloma antilipoprotein autoantibody: IgG anti-Lp-A1, (in French), *C.R. Acad. Sci. Ser. D.,* 269, 107, 1969.

Valdiguié, P. et al., Lipids of myeloma proteins. Comparison of free fatty acids, triglyceride fatty acids and serum sterol esters of gamma G myeloma and Bence-Jones protein (in French), *C.R. Acad. Sci. Ser. D,* 269, 1570, 1969.

Beaumont, J. L. et al., Myeloma anti-lipoprotein autoantibodies. Comparative study of 2 types anti-Lp P.G. IgA and anti-Lp A.S. IgG., (in French), *Ann. Biol. Clin.,* 28, 387, 1970.

Leon, M. A. et al., Specificity for phosphorylcholine of six murine myeloma proteins reactive with Pneumococcus C polysaccharide and β-lipoprotein, *Biochemistry,* 10, 1424, 1971.

Marien, K. J. et al., Generalized planar xanthomata associated with multiple myeloma and hyperlipoproteinemia, *Arch. Belg. Dermatol. Syphiligr.,* 29(4), 317, 1973.

Zittoun, R. et al., Waldenström's macroglobulinemia with preserved antierythrocyte and antilipoprotein antibody activity (in French with English.), *Sem. Hop. Paris,* 51(48), 2943, 1975.

Shulmann, G. et al., Serum β-lipoprotein and other specific protein concentrations in patients with immunocytoma, *J. Clin. Pathol.,* 29(5), 458, 1976.

Nervous System

Curri, S. B. et al., Biological activity of lipoprotein complexes in man. (Preliminary observations on the effect of phospholipid-cytochrome C complexes on dysmetabolic-hypoxic conditions of the central nervous system (in Italian), *Acta Neurol.,* 21, 781, 1966.

Critchley, E. H. et al., Acanthocytosis and neurological disorder without β-lipoproteinemia., *Arch. Neurol.* (Chicago), 18, 134, 1968.

Araki, S. et al., Serum proteins in familial amyloidotic polyneuropathy (in Japanese), *Med. Biol.* (Tokyo), 78, 153, 1969.

Fokin, A. S., Dynamics of cholesterol-lipoprotein metabolism and hormonal shifts in the body in overstrain of the central nervous system, *Patol. Fiziol. Eksp. Ter.,* 2, 52, 1975.

Niemann Pick Syndrome
Speigel-Adolf, M. et al., Hematologic studies in Niemann-Pick and Wolman's disease. (Cytology and electrophoresis.), *Confin. Neurol.,* 28, 399, 1966.

Obesity
Opplt, J. J. and Vamberova, M., The obesity in children, *Vnitr. Lek.,* VII-8, 875, 1961.

Opplt, J. J. Skamenova, B., and Misak, J., Some remarks on the metabolism of obese women, *Cas. Lek. Cesk.,* 101(16/17), 516, 1962.

Greenberg, S. R. et al., Alteration in serum lipids induced by metrical in obese patients, *Am. J. Med. Sci.,* 248, 221, 1964.

Hornet, N. N. et al., Metabolic disorders in obesity (in Rumanian), *Med. Interna,* 16, 1079, 1964.

Rifkind, B. M. et al., Relationship of plasma lipids and lipoproteins to obesity, *Proc. Roy. Soc. Med.,* 59, 1277, 1966.

Miturzynska-Stryjecka, H., Activity of lipoprotein lipase and effect of heparin on blood lipids in exogenous obesity, *Pol. Med. J.,* 6, 1459, 1967.

Kuroshima, A. et al., Relationship of plasma lipids and β-lipoproteins to obesity in children, *Jpn. J. Hyg.,* 24, 368, 1969.

Kikuchi, H. et al., Lipoprotein disorders in obesity (in Japanese), *Saishin Igaku,* 27, 531, 1972.

Jourdan, M. et al., The turnover rate of serum glycerides in the lipoproteins of fasting obese women during weight loss, *Am. J. Clin. Nutr.,* 27, 850, 1974.

Segers, M. J. et al., Anxiety associated with hyperlipoproteinemia with and without obesity (in French with English.), *J. Psychosom. Res.,* 19(13), 169, 1975.

Pancreatic Disorders
Kikuchi, S., Round table discussion. 3. Lipid metabolism in pancreatic impairments (in Japanese), *J. Jpn. Soc. Intern. Med.,* 53, 1387, 1965.

Kessler, J. I. et al., Hyperlipemia in acute pancreatitis. Metabolic studies in a patient and demonstration of abnormal lipoprotein-triglyceride complexes resistant to the action of lipoprotein lipase, *Am. J. Med.,* 42, 968, 1967.

Salen, S. et al., The development of pancreatic secretory insufficiency in a patient with recurrent pancreatis and type V hyperlipoproteinemia, *Mt. Sinai J. Med. N.Y.,* 37, 103, 1970.

Pancreatitis
Farmer, R. G. et al., Hyperlipoproteinemia and pancreatitis, *Am. J. Med.,* 54, 161, 1973.

Peritonitis
Fedorov, V. D., Changes in lipid metabolism in patients with peritonitis, *Khirurgiia,* 49, 131, 1973.

Phenylketonuria
Menkes, J. H., Cerebral proteolipids in phenylketonuria, *Neurology,* 18, 1003, 1968.

Pneumonia
Lobova, I. V., Blood serum lipids in patients with chronic pneumonia (in Russian), *Klin. Med.,* 46, 86, 1968.

Kondrat'ev, V. G., Serum lipoproteins in pneumonia (in Russian), *Ter. Arkh.,* 47(3), 81, 1975.

Poisoning
Weinstein, I. et al., Hepatic lipid metabolism in carbon tetrachloride poisoning. Incorporation of palmitate-1-^{14}C into lipids of the liver and of the d less than 1.020 serum lipoprotein, *Biochem. Pharmacol.,* 15, 851, 1966.

Lukoshkina, L. P. et al., Lipid-protein metabolism in workers exposed to divinyl and styrol, *Gig. Tr. Prof. Zabol.,* 17, 42, 1973.

Polyarthritis
Noseda, G. et al., Autoantibodies against lipoproteins and hypolipidemia in seronegative primary chronic polyarthritis (in German with English.), *Schweiz. Med. Wochenschr.,* 102, 969, 1972.

Polycythemia
Bryla, R. et al., Effect of treatment with radioactive phosphorus (^{32}P) on glycoproteins and lipoproteins in patients with polycythemia vera (in Polish with English.), *Pol. Arch. Med. Wewn.,* 56(6), 495, 1976.

Porphyria
Lees, R. S. et al., Hyperbeta-lipoproteinemia in acute intermittent porphyria; preliminary report, *N. Engl. J. Med.,* 282, 432, 1970.

Prostatic Disorders
Fabris, P. et al., Electrophoretic studies of normal prostate homogenate and adenoma, with special reference to the prostate homogenate of patients treated with Raveron (in Italian), *Arch. Ital. Urol.,* 38, 391, 1966.
Mattila, S. P. et al., An immunoelectrophoretic study of the soluble antigens of the human prostate, *J. Urol.,* 97, 117, 1967.

Pyridoxine Deficiency
Audet, A. et al., Lipoprotein metabolism in pyridoxine deficiency, *Rev. Can. Biol.,* 31, 171, 1972.

Radiation Injury
Pasynskii, A. G. et al., Effect of radiation injury of nucleoprotein and lipoprotein interfaces on rate of enzyme reactions, *Fed. Proc.,* Transl., Suppl., 24, 129, 1965.

Renal Diseases
Paul, C. et al., Serum protein changes in nephrotic syndrome, *Indian J. Med. Sci.,* 18, 513, 1964.
Bertone, E., The determination of the protein fractions, glycoproteins and lipoproteins in blood serum. Experimental contribution for the diagnosis of renal diseases (in Italian), *Arch. Sci. Med.,* 121, 248, 1966.
Daubresse, J. C. et al., Lipids and lipoproteins in chronic uraemia. A study of the influence of regular haemodialysis, *Eur. J. Clin. Invest.,* 6(2), 159, 1966.
Lewis, L. A. et al., Renal regulation of serum α-lipoproteins. Decrease of α-lipoproteins in the absence of renal function, *N. Engl. J. Med.,* 275, 1097, 1966.
Markiewicz, K., Electrophoretic picture of serum lipoproteins in the nephrotic syndrome (in Polish), *Pol. Arch. Med. Wewn.,* 36, 55, 1966.
Markiewicz, K., Serum lipoprotein pattern in the nephrotic syndrome, *Pol. Med. J.,* 5, 737, 1966.
Casciani, C., Renal function in chronic uremic patients treated with periodic depuration (in Italian), *Minerva Nefrol.,* 14, 189, 1967.
Nikiforova, N. V., On the pathochemical characteristics of experimental nephrotic syndrome (in Russian), *Vopr. Med. Khim.,* 13, 69, 1967.
Bryla, R., Urinary and serum lipoproteinogram in certain renal diseases (in Polish), *Pol. Tyg. Lek.,* 23, 7, 1968.
De Simoni, G. et al., Seroprotein and lipoprotein pattern in patients with tubercular nephropathy and in patients with other nonspecific nephropathies associated with tuberculosis in different areas of the body, mainly the lung, *Ann. Ist. Carlo Forlanini,* 28, 264, 1968.
Svistunenko, L. N., Comparative study: blood lipids, lipoproteins and proteins in nephritis in children (in Russian), *Vopr. Okhr. Materin. Dets.,* 13, 85, 1968.
Zühlke, V., The reaction of serum lipoproteins and serum immunoglobulins following homologous kidney transplantation in humans (in German), *Langenbecks Arch. Chir.,* 322, 542, 1968.
Derevianko, N. A., Changes in protein fractions and lipoproteins in azotemia (in Russian), *Lab. Delo,* 1, 37, 1969.
Jović, R. et al., Protein and lipoprotein fractions in endemic nephropathy in the village of Moravac, *Srp. Arh. Celok. Lek.,* 97, 1035, 1969.
Taskovska, D. et al., Electrophoretic study of serum lipoproteins. II. Study of serum lipoproteins in acute glomerulonephritis in children, *God. Zb. Med. Fak. Skopje,* 15, 279, 1969.
Taskovska, D. et al., Electrophoretic study of serum lipoproteins. I. Study of serum lipoproteins in nephrotic syndrome (in German), *God. Zb. Med. Fak. Skopje,* 15, 267, 1969.
Nestel, P., Low density lipoprotein peptide metabolism in the nephrotic syndrome, *Australas. Ann. Med.,* 19, 62, 1970.
Scott, P. J. et al., Low density lipoprotein peptide metabolism in nephrotic syndrome. A comparison with patterns observed in other syndrome characterized by hyperlipoproteinaemia, *Australas. Ann. Med.,* 19, 1, 1970.
Chopra, J. S. et al., Hyperlipoproteinemias in nephrotic syndrome, *Lancet,* 1, 317, 1971.
Brons, M. et al., Hyperlipoproteinemia in patients with chronic renal failure, *Acta Med. Scand.,* 192, 119, 1972.
Sorge, F. et al., Relations between insulin, growth hormone and hyperlipoproteinemia in chronic kidney failure, *Verh. Dtsch. Ges. Inn. Med.,* 78, 1472, 1972.
Yasugi, T. et al., Hyperlipoproteinemia in the nephrotic syndrome (in Japanese), *Saishin Igaku,* 27, 504, 1972.

Arora, K. K. et al., Changes in glucose tolerance, insulin, serum lipid and lipoproteins in patients with renal failure on intermittent haemodialysis, *Postgrad. Med. J.,* 49(57), 293, 1973.

Chatys-Górska, L., Blood lipids and lipoproteins in children with nephrotic syndrome, *Pediatr. Pol.,* 48, 713, 1973.

Mann, J. I., Serum lipoprotein pattern in Bantu and Indian patients with nephrotic syndrome, *S. Afr. Med. J.,* 47, 552, 1973.

Shanley, B. C. et al., Serum lipoprotein patterns in Bantu and Indian patients with nephrotic syndrome, *S. Afr. Med. J.,* 47, 187, 1973.

Faulk, W. P. et al., Glomerular beta-lipoprotein in childhood renal disease, *Scand. J. Immunol.,* 3(5), 665, 1974.

Prakash, C. et al., Hyperlipidaemias and hyperlipoproteinemias in proteinuric glomerulonephritis — an experimental study, *J. Assoc. Physicians India,* 22(9), 637, 1974.

Shafrir, E. et al., Lipoprotein synthesis in hypoproteinemia of experimental nephrotic syndrome and plasmapheresis, in *Plasma Protein Turnover,* Bianchi, R. et al., Eds., University Park Press, Baltimore, 1976, 343.

Lutz, W., Disturbances in lipid composition of serum lipoproteins in patients with chronic renal failure, *Acta Med. Pol.,* 16(2), 127, 1975.

Newmark, S. R. et al., Lipoprotein profiles in adult nephrotics, *Mayo Clin. Proc.,* 50(7), 359, 1975.

Mydlik, M. et al., Lipids and lipoproteins in acute renal failure, (English.), *Cas. Lek. Cesk.,* 114(10—11), 337, 1975.

Samar, R. E. et al., Lipoprotein binding and hypertriglyceridemia in chronic uremia, *Trans. Am. Soc. Artif. Intern. Organs,* 21, 455, 1975.

Wada, M. et al., Studies on the effects of hemodialysis on plasma lipoprotein, *Trans. Am. Soc. Artif. Intern. Organs,* 21, 464, 1975.

Way, R. C. et al., Relationship between serum cholinesterase and low density lipoproteins in children with nephrotic syndrome, *Clin. Biochem.,* 8(2), 103, 1975.

Bagdade, J. et al., Effects of chronic uremia hemodialysis, and renal transplantation of plasma lipids and lipoproteins in man, *J. Lab. Clin. Med.,* 87(1), 38, 1976.

De Mendoza, S. G. et al., High density lipoproteinuria in nephrotic syndrome, *Metabolism,* 25(10), 1143, 1976.

Handa, Y. et al., Fatty acid composition of serum lipoprotein in renal diseases (author's transl.), *Jpn. J. Clin. Pathol.,* 24(10), 847, 1976.

Norbeck, H. E. et al., Serum lipid and lipoprotein concentration in chronic uremia, *Acta Med. Scand.,* 200(6), 487, 1976.

Mydlik, M. et al., Serum lipids and lipoproteins in acute renal failure, *Proc. Eur. Dial. Transplant Assoc.,* 12, 389, 1976.

Segura, R. et al., Lipid and lipoprotein abnormalities in renal disease, *Perspect. Nephrol. Hypertens.,* 3, 159, 1976.

Sacrez Mac-Mahon Tannhauser Disease

Thomas, M. et al., Sacrez Mac-Mahon Thannhauser Disease. Apropos of a case. Clinical aspects and biologic diagnosis (in French), *Rev. Med. Chir. Mal. Foie,* 46, 39, 1971.

Sarcoidosis

Oleneva, J. N. et al., Some hematologic and biochemical changes in the blood of patients with sarcoidosis (in Russian), *Probl. Tuberk.,* 45, 33, 1967.

Scoliosis

Dezyna, M. et al., Studies on lipoproteins of the blood serum in children with idiopathic scoliosis (in Polish), *Chir. Narzadow Ruchu Ortop. Pol.,* 31, 667, 1966.

Schizophrenia

Depczynski, L., Lipids and enzymatic lipolysis in the blood serum of schizophrenic patients (in Polish), *Psychiatr. Pol.,* 1, 667, 1967.

Gottieb, J. S. et al., Schizophrenia — new concepts, *South. Med. J.,* 64, 743, 1971.

Nicol, S. et al., Serum from schizophrenic patients. Effect on cellular lactate stimulation and tryptophan uptake, *Arch. Gen. Psychiatry,* 29, 744, 1973.

Silicosis

Szabó, S. et al., Immunity and blood proteins in silicosis. 3. Serum proteins, lipoproteins, glycoproteins and cryoagglutinins in silicosis (in French), *Arch. Roum. Pathol. Exp. Microbiol.,* 28, 109, 1969.

Skin Diseases

Havemann, K., Binding of albumin-lipoprotein complexes in lupus erythematosus (in German), *Klin. Wochenschr.*, 43, 606, 1965.

Nigro, N. et al., Electrophoretic behavior of blood lipoprotein fractions in the child affected by seborrheic eczema (in Italian), *Minerva Pediatr.*, 17, 1615, 1965.

Martina, G. et al., Behavior of the lipoprotein pattern in various stages of evolution of psoriasis, *Minerva Dermatol.*, 42, 489, 1967.

Gluek, C. J. et al., Acquired type 1 hyperlipoproteinemia with systemic lupus erythematosus, dysglobulinemia and heparin resistance, *Am. J. Med.*, 47, 318, 1969.

Holasek, A., Blood lipoproteins and the skin, *Z. Haut Geschlechtskr.*, 46, 673, 1971.

Berg, K. et al., Presence of lipoprotein in serum of patients treated for acrodermatitis enteropathica (Danbolt's disease), *Clin. Genet.*, 3, 401, 1972.

Sugihara, I. et al., Fatty acids composition of very low density lipoprotein in psoriatic serum, *Jpn. J. Dermatol. Ser. B*, 82(4), 107, 1972.

Dean, F. D., Lipoproteins in relation to skin disease, *Br. J. Dermatol.*, 88, 191, 1973.

Lebedeva, N. E. et al., Lipoproteins, mucopolysaccharides and isoenzymes in the skin and mucus of the minnow *Phoxinus phoxinus*, *Zh. Evol. Biokhim. Fiziol.*, 9, 527, 1973.

Cerniková, M., Serum lipoproteins in psoriatic patients, *Cesk. Dermatol.*, 49, 30, 1974.

Ishikawa, H. et al., Demonstration of beta-lipoproteins in the psoriatic skin by immunofluorescent technique, *Arch. Dermatol. Forsch.*, 249, 191, 1974.

Kovtunenko, V. S., Study of lipoproteins in blood serum, liver and skin in eczema (in Russian), *Vrach. Delo*, 6, 123, 1974.

Zabel, R., Hyperlipoproteinemias and their skin manifestations, *Paediatr. Grenzgeb.*, 13(2—3), 99, 1974.

Kostiushov, V. V., Protein sulhydryl groups of blood serum and their relation to the indicators of lipid metabolism in patients with psoriasis (English.), *Vrach. Delo*, 1, 116, 1975.

Pototskii, I. I. et al., Stability of β-lipoprotein complexes and cholesterol concentration in the serum of psoriasis patients (in Russian with English.), *Vestn. Dermatol. Venerol.*, 7, 16, 1976.

Smoking

Gupta, N. N. et al., Tobacco smoking and serum cholesterol and lipoproteins in healthy males, *J. Indian Med. Assoc.*, 43, 372, 1964.

Modzelewski, A. et al., Patterns of various blood lipids in smokers (in Polish), *Wiad. Lek.*, 22, 229, 1969.

Howell, R. W., Smoking and vascular disease, *Br. Med. J.*, 2, 2321, 1972.

Billimoria, J. D. et al., Effect of cigarette smoking on lipids, lipoproteins, blood coagulation, fibrinolysis and cellular components of human blood, *Atherosclerosis*, 21(1), 61, 1975.

Spinocerebellar Degeneration Familial

Korula, J. et al., A case of familial spinocerebellar degeneration with hypo-β-lipoproteinemia, *Neurol. India*, 24(1), 41, 1976.

Surgery

Kaspar, F. et al., Studies on blood changes after hypothermia and extracorporeal circulation (in German), *Thoraxchirurgie*, 12, 475, 1965.

Gingiaro, A. et al., Behavior of total lipemia, cholesteremia and lipoprotein fractions in the postoperative period in children (in Italian), *Minerva Pediatr.*, 18, 262, 1966.

Morozova, L. I., Study of the level of blood serum proteins, lipoproteins and glycoproteins after surgical correction of mitral heart defects (in Russian), *Grudn. Khir.*, 6, 105, 1974.

Syphilis

Madievskaia, N. N. et al., Lipoproteins and free cholesterol in patients with syphilitic aortitis (in Russian), *Vestn. Dermatol. Venerol.*, 38, 64, 1964.

Mirakhmedov, U. M., Protein and lipoprotein fractions of the blood serum in therapy of syphilitic patients (in Russian), *Vestn. Dermatol. Venerol.*, 43, 52, 1969.

Tangier Disease

Hoffman, H. N. et al., Tangier disease (familial high density lipoprotein deficiency). Clinical and genetic features in two adults, *Am. J. Med.*, 39, 582, 1965.

Engel, W. K. et al., Neuropathy in Tangier disease. Alpha-lipoprotein deficiency manifesting as familial recurrent neuropathy and intestinal lipid storage, *Arch. Neurol.* (Chicago), 17, 1, 1967.

Kocen, R. S. et al., Familial α-lipoprotein deficiency (Tangier disease) with neurological abnormalities, *Lancet*, 1, 1341, 1967.

Kummer, H. et al., Familial analphalipoproteinemia (Tangier disease) (in German), *Schweiz. Med. Wochenschr.*, 98, 406, 1968.

Shacklady, M. M. et al., Red-cell lipids in familial α-lipoprotein deficiency (Tangier disease), *Lancet*, 2, 151, 1968.

Spiess, H. et al., Polyneuropathy in familial analphalipoproteinemia (Tangier disease), *Nervenarzt*, 40, 191, 1969.

Huth, K. et al., Tangier disease (hypo-α-lipoproteinemia) (in German), *Dtsch. Med. Wochenschr.*, 95, 2357, 1970.

Bale, P. M. et al., Pathology of Tangier Disease, *J. Clin. Pathol.*, 24, 609, 1971.

Clifton-Bligh, P. et al., Tangier disease. Report of a case and studies of lipid metabolism, *N. Engl. J. Med.*, 286, 567, 1972.

Lux, S. E. et al., Studies on the protein defect in Tangier disease. Isolation and characerization of an abnormal high density lipoprotein, *J. Clin. Invest.*, 51, 2505, 1972.

Kostner, G. et al., Immunochemical study and analytical isoelectric focusing of serum of a case of Tangier disease (in German with English.), *Clin. Chim. Acta*, 38, 155, 1972.

Tyminski, W. et al., A case of hypocholesteremia and analphalipoproteinemia (Tangier's disease?) (in Polish with English.), *Pol. Tyg. Lek.*, 27, 598, 1972.

Garcia-Merlo, S. et al., Acquired alpha-lipoprotein deficiency, *Rev. Clin. Esp.*, 129, 455, 1973.

Kocen, R. S. et al., Nerve biopsy findings in two cases of Tangier disease, *Acta Neuropathol.*, 26, 319, 1973.

Thomas, P. K. et al., Peripheral nerve involvement in Tangier disease, *Trans. Am. Neurol. Assoc.*, 98, 73, 1973.

Greten, H. et al., Lipoproteins and lipolytic plasma enzymes in a case of Tangier disease, *N. Engl. J. Med.*, 291, 548, 1974.

Haas, L. F. et al., Tangier disease, *Brain*, 97(2), 351, 1974.

Stanios, W. et al., Cholesterolosis with alpha-lipoprotein deficiency (Tangier disease), *Wiad. Lek.*, 27, 805, 1974.

Ferrans, V. J., The pathology of Tangier disease. A light and electron microscopic study, *Am. J. Pathol.*, 78(1), 101, 1975.

Golabek, W. et al., Tangier disease, *Mater Med. Pol.*, 7(3), 237, 1975.

Utermann, G. et al., Plasma lipoprotein abnormalities in a case of primary high-density lipoprotein (HDL) deficiency, *Clin. Genet.*, 8(4), 258, 1975.

Varkonyi, A. et al., Simultaneous occurrence of Tangier disease and mucoviscidosis, *Orv. Hetil.*, 116(37), 2187, 1975.

Assmann, G. et al., Structure-function relationships of lipoproteins in Tangier disease, pp106—110, in *Lipoprotein Metabolism*, Greten, H., Ed., Springer, Berlin, 1976.

Gagyi, J., Letter: clinical aspects of Tangier disease (in Hungarian), *Orv. Hetil.*, 117(3), 184, 1976.

Vivell, O. et al., Hypo-α-lipoproteinemia (Tangier disease). (in German with English.), *Klin. Paediatr.*, 188(1), 82, 1976.

Assmann, G. et al., The lipoprotein abnormality in Tangier disease: quantitation of A apoproteins, *J. Clin. Invest.*, 59(3), 565, 1977.

Thalassemia

Nigro, N. et al., Study of the serum lipoprotein fractions in the child with "thalassemia major," (in Italian), *Minerva Pediatr.*, 18, 2119, 1966.

Pantelakis, S. N. et al., Serum lipoproteins in schoolboys in relation to glucose-6-phosphate dehydrogenase deficiency and thalassaemia trait, *Arch. Dis. Child.*, 42, 328, 1967.

Chabaud, F. et al., Letter: anti-β-lipoprotein (anti-Ag) antibodies in a patient with major thalassemia (in French), *Nouv. Presse Med.*, 4(39), 2814, 1974.

Desai, M. et al., Lipids and lipoproteins in thalassaemia major (part 1), *Indian Pediatr.*, 13(9), 663, 1976.

Thymomegaly

Nigro, N. et al., Changes in the blood electrophoretic lipoproteins picture following treatment of thymomegaly (in Italian), *Minerva Pediatr.*, 17, 861, 1965.

Thyroid Disorders

Moses, C. et al., Hydrocortisone and/or desiccated thyroid in physiologic dosage. XI. Effects of thyroid hormone excesses on lipids and other blood and serum solutes, *Metabolism*, 13, 717, 1964.

Roitt, J. M. et al., The cytoplasmic auto-antigen of the human thyroid. 1. Immunological and biochemical characteristics, *Immunology*, 7, 375, 1964.

Shevel', E. I., Serum lipoproteins in thyrotoxicosis patients and their modification in therapy (in Russian), *Ter. Arkh.,* 36, 99, 1964.

Grigor'ev, P. I., Changes in the content of blood cholesterol and lipoprotein in thyrotoxicosis (in Russian), *Ter. Arkh.,* 37, 91, 1965.

Kliachko, V. R. et al., Examination of the blood lipoprotein and glucoprotein content in myxedema treated with triiodothyronine (in Russian), *Probl. Endokrinol. Gormonoter.,* 11, 8, 1965.

Sfikakis, P. et al., Serum proteins, glycoproteins and lipoproteins in thyroid disease and their relation to thyroid autoantibodies, *Z. Immunitaetsforsch,* 129, 462, 1965.

Sterling, K. et al., Isolation of lipoprotein from thyroxine-labeled human serum, *Endocrinology,* 77, 398, 1965.

Walton, K. W. et al., The significance of alterations in serum lipids in thyroid dysfunction. 1. The relation between serum lipoproteins, carotenoids and vitamin A on hypothyroidism and thyrotoxicosis, *Clin. Sci.,* 29, 199, 1965.

Kurata, Y. et al., Immunological studies of insoluble lipoproteins. 1. Antigen analysis of thyroidal lipoproteins, *Int. Arch. Allerg.,* 29, 495, 1966.

Strisower, E. H., The combined use of CP1B and thyroxine in treatment of hyperlipoproteinemias, *Circulation,* 33, 291, 1966.

Durić, D. S. et al., Lipidogram of patients with hypothyrosis, *Acta Med. Iugosl.,* 21, 166, 1967.

Kurata, Y., Immunopathology of thyroiditis, with special reference to the purification of tissue specific antigens (in Japanese), *Trans. Soc. Pathol. Jpn.,* Suppl. 56, 374, 1967.

Nishizuka, F., Variations in plasma total fatty acid fractions and serum lipoproteins fractions in thyroid diseases (in Japanese), *Clin. Endocrinol.* (Tokyo), 15, 743, 1967.

Kurata, Y., Immunopathological studies on thyroiditis: an attempt to isolate the tissue-specific lipoproteins particle, *Acta Pathol. Jpn.,* 18, 501, 1968.

Sachs, B. A. et al., Lipid and clinical response to a new thyroid hormone combination, *Am. J. Med. Sci.,* 256, 232, 1968.

Slukhai, I. F., Blood serum β-lipoprotein in thyroid gland diseases (in Russian), *Probl. Endokrinol.,* 14, 52, 1968.

Dyerberg, J., Type 3 hyperlipoproteinemia with low plasma thyroxine binding globulin, *Metabolism,* 18, 50, 1969.

Kostner, G. et al., The binding of thyroid hormones to human serum lipoprotein (in German), *Z. Klin. Chem.,* 8, 60, 1970.

Okada, S. et al., Immunological studies of insoluble lipoproteins. 3. Characterization of the lipoprotein-bound thyroid gland-specific antigen, *Int. Arch. Allerg. Appl. Immunol.,* 39, 6, 1970.

Contreras, M. C. et al., The role of the thyroid gland in the arterial lipolytic activity in rabbits on atherogenic diets, *Arch. Biol. Med. Exp.,* 8(1—3), 8, 1971.

Chopra, I. J. et al., Lats inhibition by a soluble lipoprotein from human thyroid gland, *J. Clin. Endocrinol. Metab.,* 32, 772, 1971.

Lepiavko, A. G. et al., Lipid metabolism indices in endemic goiter, *Vrach. Delo,* 2, 38, 1972.

Bommer, J. et al., Characteristic plasma lipoprotein pattern in patients suffering from hyperthyroidism, *Acta Endocrinol. Suppl.* (Copenhagen), 173, 142, 1973.

Fokin, A. S., Thyroid gland function and cholesterol-lipoprotein metabolism under prolonged functional stress of the central nervous system (English abstr.), *Patol. Fiziol. Eksp. Ter.,* 1, 37, 1974.

Rossner, S. et al., Serum lipoproteins and the intravenous fat tolerance test in hypothyroid patients before and during substitution therapy, *Atherosclerosis,* 20(2), 365, 1974.

Bommer, J. et al., Characteristic lipoprotein changes in hyperthyroidism (in German), *Verh. Dtsch. Ges. Inn. Med.,* 81, 1563, 1975.

Bommer, J. et al., Characteristic plasma lipoprotein pattern in hyperthyroid patients, *Thyroid Research,* Robbins, J. et al., Eds., Excerpta Medica, Amsterdam, 1976, 369.

Transfusion

Solaas, M. H. et al., Antibodies to serum lipoprotein antigens and Australia-SH antigen in multiply transfused patients, *Scand. J. Haematol.,* 7, 233, 1970.

Berg, K. et al., Individual antigenic differences of a human serum lipoprotein revealed by a "cold-precipitin" in the serum of a transfused patient, *Nature,* (London), 206, 312, 1975.

Tropical Diseases

Sytinskii, I. A. et al., Paper electrophoresis of serum proteins and lipoproteins of patients with tropical diseases (in Russian), *Vopr. Med. Khim.,* 14, 513, 1968.

Tuberculosis

Bert, R. et al., Behavior of the serum protein, glycoprotein and lipoprotein curves and immuno-electrophoretic pattern in recent pulmonary tuberculous processes and in chronic pulmonary tuberculosis (in Italian), *Arch. Fisiol.,* 19, 385, 1964.

Jevrić, S. et al., Studies on the lipoprotein index in tuberculous disease (in Polish), *Vojnosanit. Pregl.,* 21, 555, 1964.

Kaminskaiá, G. O., Clinical significance of the lipoproteinogram in tuberculosis (in Russian), *Probl. Tuberk.,* 42, 52, 1964.

Klent, D. et al., Lipoproteins in tuberculous patients (in Serbian), *Tuberkuloza,* 16, 372, 1964.

Teichmann, B. et al., Unspecific lipoprotein precipitation in experiments for detection of tuberculosis-specific antibodies, *Nature,* (London), 202, 916, 1964.

Jevrić, S. et al., Significance of lipoproteins in chronic tuberculosis and recurrence (in Croatian), *Tuberkuloza,* 19, 670, 1967.

Bellis, E. et al., Behavior of serum beta-lipoproteins in pulmonary emphysema, pneumothorax and severe pulmonary destructive processes of tubercular origin (in Italian), *Rass. Int. Clin. Ter.,* 48, 715, 1968.

Klent, D. et al., Ratio of total lipids and lipoproteins in the serum tuberculous of patients (in Croatian), *Plucne Boles. Tuberk.,* 22, 254, 1970.

Tumors

Barclay, M. et al., Serum lipoproteins and human neoplastic disease, *Clin. Chim. Acta,* 10, 39, 1964.

Falor, W. H. et al., Electrophoresis of chyle in bronchogenic carcinoma, *Arch. Surg.,* 91, 671, 1965.

Gatta, L. et al., Evaluation of the serum protein picture in malignant neoplasms of the female genital system. 3 Lipoproteins (in Italian), *Quad. Clin. Ostet. Ginecol.,* 20, 112, 1965.

Kristofova, H. et al., The role of lipoprotein and deoxyribonucleic acid in antigenicity of the Ehrlich ascites tumour, *Folia Biol.,* 11, 237, 1965.

Lossow, W. J. et al., Uptake, hydrolysis and synthesis of cholesterol esters by a transplantable adrenal cortical tumor, *Proc. Soc. Exp. Biol. Med.,* 119, 126, 1965.

Potop, I. et al., The influence of a lipoprotein extract isolated from the thymus gland on the development of experimental tumors (in Rumanian), *Stud. Cercet. Endocrinol.,* 16, 25, 1965.

Nakagawa, H., Studies on α-lipoprotein in serum of patients with cancer (in Japanese), *J. Jpn. Soc. Intern. Med.,* 55, 65, 1966.

Adamczyk, B. et al., Studies on globulin fractions of patients with gastric cancer by the Ouchterlony test (in Polish), *Nowotwory,* 17, 103, 1967.

Narayan, K. A., Serum lipoprotein during chemical carcinogenesis, *Biochem. J.,* 103, 672, 1967.

Baylet, R. et al., Serum inhibitors of streptolysin O in primary liver cancer (in French), *Bull. Soc. Med. Afr. Noire Lang. Fr.,* 13, 553, 1968.

Helle, K. B., The chromogranin of the adrenal medulla: a high-density lipoprotein, *Biochem. J.,* 109, 43P, 1968.

Kellen, J., The serum β-lipoproteins in different human malignant diseases, *Neoplasma,* 15, 139, 1968.

Roszkowski, I. et al., Studies on proteins and lipoproteins in patients with uterine cancer and myoma (in Polish), *Ginekol. Pol.,* 39, 307, 1968.

Sakamoto, Y., Diagnostic evaluation of the correlation between serum total cholesterol and α-lipoprotein in malignant tumors (in Japanese), *Jpn. J. Cancer Clin.,* 14, 1049, 1968.

Barclay, M. et al., Serum lipoproteins in children with cancer, *Clin. Chim. Acta,* 24, 225, 1969.

Sakamoto, Y., Changes in the correlation between serum total cholesterol and α-lipoproteins and their clinical significance in malignant cancer patients under various treatments (in Japanese), *Jpn. J. Cancer Clin.,* 15, 49, 1969.

Barclay, M. et al., Effects of cancer upon high-density and other lipoproteins, *Cancer Res.,* 30, 2420, 1970.

Yamamoto, Y. et al., Antibody formation for malignant tumor. II. Antigenicity of Ehrlich ascites tumor lipoprotein, *Acta Med. Okayama,* 24, 527, 1970.

Barclay, M. et al., Lipoproteins in cancer patients, *Ca,* 21, 202, 1971.

Feldman, E. B. et al., Circulating lipids and lipoproteins in women with metastatic breast carcinoma, *J. Clin. Endocrinol. Metab.,* 33, 8, 1971.

Skipski, V. P. et al., A new proteolipid apparently associated with cancer, *Proc. Soc. Exp. Biol. Med.,* 136, 1261, 1971.

Holczinger, L. et al., Chemical composition and electron microscopy of residual lipoproteins of sarcoma 180 and NK-lymphoma ascites tumour cells, *Neoplasma,* 19, 95, 1972.

Nydegger, V. E. et al., Serum lipoprotein levels in patients with cancer, *Cancer Res.,* 32, 1756, 1972.

Riesen, W. et al., Autoantibody-induced serum alpha lipoprotein decrease in carcinoma patients, *Verh. Dtsch. Ges. Inn. Med.,* 78, 847, 1972.

Feldman, E. B. et al., Diethylstilbestrol and circulating lipids in women with metastatic breast cancer: lack of dose response over 3 log units, *J. Clin. Endocrinol. Metab.,* 36, 381, 1973.

Kline, J. A., The relation of lipoproteins to malignant disease and Down's syndrome, *J. Am. Osteopath. Assoc.*, 73, 206, 1973.

Barclay, M. et al., Lipoproteins in relation to cancer, *Prog. Biochem. Pharmacol.*, 10, 76, 1975.

Riesen, W. et al., Autoantibodies with antilipoprotein specificity and hypolipoproteinemia in patients with cancer, *Cancer Res.*, 35(3), 535, 1975.

Sachs, H. et al., Proceedings: clinical-epidemiological analysis of patients with breast carcinoma with special reference to serum β-Lp content, *Arch. Gynaekol.*, 219(1—4), 147, 1975.

Sakamoto, Y., Plasma lipoprotein abnormalities in patients with advanced cancer, (in Japanese with English abstr.), *J. Jpn. Soc. Cancer Ther.*, 10(3), 256, 1975.

Skipski, V. P. et al., Tumor proteolipids, *Prog. Biochem. Pharmacol.*, 10, 112, 1975.

Mathur, S. N. et al., Characterization of the Ehrlich ascites tumor plasma lipoproteins, *Biochim. Biophys. Acta*, 424(1), 45, 1976.

Viral Diseases

Nicoli, J. et al., Serum inhibitors of arbovirus hemagglutination (in French), *Ann. Inst. Pasteur Paris*, 108, 22, 1965.

Bidwell, D. E. et al., Serum non-specific inhibitors of arbovirus haemagglutination, *J. Comp. Pathol.*, 78, 469, 1968.

Bninc, D., Nonspecific serum inhibitors of hemagglutination of arboviruses; biochemical data, mechanism of action, methods of separation (in Rumanian), *Stud. Cercet. Inframicrobiol.*, 19, 207, 1968.

Lees, R. S. et al., Effects of an experimental viral infection on plasma lipid and lipoprotein metabolism, *Metabolism*, 21, 825, 1972.

Granström, M., Studies of inhibitors of bone marrow colony formation in normal human sera and during a viral infection, *Exp. Cell. Res.*, 82, 426, 1973.

Blom, H. et al., Identification of non-specific serum inhibitors of rubella virus haemagglutination, *Med. Microbiol. Immunol.* (Berlin), 159, 271, 1974.

Perdue, M. L. et al., Studies of the molecular anatomy of the L-M cell strain of equine herpes virus type 1: proteins of the nucleocapsid and intact virion, *Virology*, 59, 201, 1974.

Shortridge, K. F. et al., Human serum lipoproteins as inhibitors of haemagglutination for selected togaviruses, *J. Gen. Virol.*, 23, 113, 1974.

Allen, R. et al., Simple procedure for the removal of nonspecific inhibitors of rubella virus hemagglutination, *J. Clin. Microbiol.*, 2(6), 524, 1975.

Cox, R. A. et al., Effect of simian virus 40 subcutaneous tumors on circulating lipids and lipoproteins in the Syrian hamster, *J. Natl. Cancer Inst.*, 54(2), 379, 1975.

Shortridge, K. F. et al., Studies on the inhibitory activities of human serum lipoproteins for Japanese encephalitis virus, *Southeast Asian J. Trop. Med. Public Health*, 6(4), 461, 1975.

Ho, W. K. et al., The role of lipids in the serum lipoproteins as inhibitors of hemagglutination by Japanese encephalitis virus, *Jpn. J. Med. Sci. Biol.*, 29(5), 283, 1976.

Shortridge, K. F., Comparison of the activities in inhibition of haemagglutination by different togaviruses for human serum lipoproteins and their constituents, *J. Gen. Virol.*, 33(3), 523, 1976.

Gokcen, M. et al., Virus-induced hyperlipoproteinemia (letter), *JAMA*, 237(13), 1311, 1977.

Ho, W. K., Serum lipoproteins as inhibitors of haemagglutination by rubella virus, *Lipids*, 12(1), 85, 1977.

Wolman's Disease

Eto, Y., et al., Wolman's disease with hypolipoproteinemia and acanthocytosis: clinical and biochemical observations, *J. Pediatr.*, 77, 862, 1970.

Xanthomatosis

Beaumont, J. L. et al., Chemical composition of S$_f$ 0-12 lipoproteins in familial xanthomatous hypercholesteremia (in French), *Rev. Fr. Etud. Clin. Biol.*, 10, 221, 1965.

Jacotot, B., et al., Myeloma, hyperlipidemia and xanthomatosis. II. Complementary research on the association between paraprotein and fast lipoproteins (in French), *Nouv. Rev. Fr. Hematol.*, 5, 777, 1965.

Polano, M. K., Xanthomatosis (in Dutch), *Ned. Tijdschr. Geneeskd.*, 111, 2057, 1967.

Scott, P. J. et al., Low-density lipoprotein accumulation in actively growing xanthomas, *J. Atheroscler. Res.*, 7, 207, 1967.

Baes, H. et al., Distribution of various forms of xanthomata in three types of primary hyperlipoproteinemia, *Dermatologica*, 136, 301, 1968.

Baes, H. et al., Lipid composition of various types of xanthoma, *J. Invest. Dermatol.*, 51, 286, 1968.

Polano, M. K., Cutaneous xanthomatosis in relation to the blood lipoprotein pattern, *Br. J. Dermatol.*, Suppl. 81, 2, 39, 1969.

Polano, M. K. et al., Xanthomata in primary hyperlipoproteinemia. A classification based on the lipoprotein pattern of the blood, *Arch. Dermatol.*, 100, 387, 1969.

Buligesco, L. et al., Study of a case of familial hypercholesterolemic xanthomatosis, pseudo-tumoral in form, with hyperalphalipoproteinemia (in French), *Acta Clin. Belg.*, 24, 276, 1969.

Schreiber, M. M. et al., Secondary eruptive xanthoma. Type V hyperlipoproteinemia, *Arch. Dermatol.*, 100, 601, 1969.

Beaumont, V. et al., Composition of the β-lipoproteins in normal subjects and in xanthomatous familial hypercholesterolemia, *Pathol. Biol.*, (Paris), 18, 643, 1970.

Cornelius, C. E., Xanthomata in primary hyperlipoproteinemia, *Arch. Dermatol.*, 101, 701, 1970.

Moschella, S. L., Cutaneous xanthomatoses: a review and their relationship with the current classification of the hyperlipoproteinemias, *Lahey Clin. Found. Bull.*, 19, 103, 1970.

Parker, F. et al., Xanthomatosis associated with hyperlipoproteinemia, *J. Invest. Dermatol.*, 55, 71, 1970.

Fanta, D. et al., Behavior of lipoproteins in xanthomatoses (in German), *Wien. Klin. Wochenschr.*, 83, 313, 1971.

Storey, G. O., Xanthomatosis, hyperlipoproteinaemia (type III Frederickson), gout, cardiac infarction, *Proc. R. Soc. Med.*, 64, 59, 1971.

Weintrob, L., A case of type 3 hyperlipoproteinemia with xanthomata, *Br. J. Radiol.*, 44, 215, 1971.

Kodama, H. et al., Plane xanthomatoses with antilipoprotein autoantibody, *Arch. Dermatol.*, 105, 722, 1972.

Lindeskog, G. R. et al., Serum lipoprotein deficiency in diffuse "normolipemic" plane xanthoma, *Arch. Dermatol.*, 106, 529, 1972.

Rufli, T. et al., Xanthomatosis as a symptom of hyperlipidemic metabolic disorders. Relationship of the morphological picture to the lipid of three patients, *Dermatologica*, 144, 270, 1972.

Braun-Falco, O., Origin, structure, and function of the xanthoma cell, *Nutr. Metab.*, 15, 68, 1973.

Elias, P. et al., Intertriginous xanthomata in type 2 hyperbetalipoproteinemia, *Arch. Dermatol.*, 107, 761, 1973.

Polano, M. K., Xanthoma types in relation to the type of hyperlipoproteinemia, *Nutr. Metab.*, 15, 107, 1973.

Scher, R. K., Hyperlipoproteinemias and xanthomatosis, *J. Natl. Med. Assoc.*, 65, 45, 1973.

Walton, K. W. et al., The pathogenesis of xanthomata, *J. Pathol.*, 109, 271, 1973.

Vigh, C. et al., Xanthomatosis eruptiva papulosa (hyperlipoproteinemia type V), *Z. Hautkr.*, 49(19), 819, 1974.

Bulkley, B. H. et al., Tuberous xanthoma in homozygous type II hyperlipoproteinemia. A histologic, histochemical, and electron microscopial study, *Arch. Pathol.*, 99(6), 293, 1975.

Heiberg, A., The lipoprotein and lipid pattern in xanthomatosis, *Acta Med. Scand.*, 198(3), 183, 1975.

Balarac, N. et al., A vertical study of the HL-A system in a family affected with hypercholesteremic xanthinoma tendinosum (proceedings) (in French), *Diabete Metab.*, 2(3), 148, 1976.

Heiberg, A., Inheritance of xanthomatosis and hyper-β-lipoproteinemia. A study of 7 large kindreds, *Clin. Genet.*, 9(1), 92, 1976.

Schink, W., Essential hypercholesteremic xanthomatosis (Fredrickson's hyperlipoproteinemia type II-B) (in German), *Handchirurgie*, 8(1), 37, 1976.

THERAPY OF DYSLIPOPROTEINEMIAS

Exercise

Carlson, L. A. et al., Acute effects of prolonged, heavy exercise on the concentration of plasma lipids and lipoproteins in man, *Acta Physiol. Scand.*, 62, 51, 1964.

Semenov, A. I. et al., Dynamics of the lipoprotein content in the blood serum of atherosclerosis patients during treatment in a sanatorium (in Russian), *Voen. Med. Zh.*, 10, 63, 1965.

Haralambie, G., Biochemical serum levels and the overstrain syndrome in athletes (in French), *Acta Biol. Med. German*, 17, 34, 1966.

Gehrke, J., The effect of physical therapy. Terrain therapy, medical exercise therapy for lipid metabolism in Ohlstadt-Obb (in German), *Med. Welt*, 39, 2305, 1967.

Tiavokin, V. V., The effect of limiting muscular activity (hypodynamia) on the concentration of total cholesterol and lipoproteins in the blood (in Russian), *Biull. Eksp. Biol. Med.*, 69, 34, 1970.

Shumskaya, V. I. et al., Change in the strength of protein-lipid bonds and esterification coefficient in patients with arteriosclerosis and hypertension under the effect of resort therapy at Kislovodsk (in Russian), *Sov. Med.*, 34, 107, 1971.

Todorović, B. et al., The effect of the type of physical activity on serum beta lipoproteins, *Glas Srp. Akad. Nauka. (Med.)* 24, 253, 1971.

Oscai, L. B. et al., Normalization of serum triglycerides and lipoprotein electrophoretic patterns by exercise, *Am. J. Cardiol.*, 30, 775, 1972.

Lopez, A. et al., Effect of exercise and physical fitness on serum lipids and lipoproteins, *Atherosclerosis*, 20(1), 1, 1974.

Diet

Fasoli, A., Diet and the lipoprotein picture in the aged (in Italian), *Minerva Diet,* 4, 202, 1964.

Bierman, E. L. et al., Characterization of fat particles in plasma of hyperlipemic subjects maintained on fat-free high-carbohydrate diets, *J. Clin. Invest.,* 44, 261, 1965.

Morisio-Guidetti, L., First results of evaluation of blood lipoproteins in subjects with a prevalently milk diet, in high mountains (in Italian), *Minerva Med.,* 56, 4161, 1965.

Loranskaia, T. I., The influence of diet therapy on the blood serum β-lipoprotein content in patients with atherosclerosis, obesity and chronic liver diseases (in Russian), *Vopr. Pitan,* 25, 61, 1966.

Chen, J. S., The effect of long-term vegetable diet on serum lipid and lipoprotein levels in man, *J. Formosan Med. Assoc.,* 65, 65, 1966.

Bertolami, V. et al., Action of a 2.2 percent solution of androsterone, in chlorphenisate, on the serum lipoproteins of low density (LDL) in patients resistant to diet therapy (in Portuguese), *Arq. Brasil Cardiol.,* 19, 111, 1966.

Nestel, P. J. et al., Influence of diet on the composition of plasma cholesterol esters in man, *J. Lipid Res.,* 7, 487, 1966.

Little, J. A. et al., Dietary carbohydrate and fat, serum lipoproteins and human atherosclerosis, *Am. J. Clin. Nutr.,* 20, 133, 1967.

Pavel, J., Various hematological parameters in ruminants on normal and semisynthetic diets (in German), *Acta Vet. Acad. Sci. Hung.,* 19, 379, 1969.

Spritz, N. et al., Effects of dietary fats on plasma lipids and lipoproteins: an hypothesis for the lipid lowering effect of unsaturated fatty acids, *J. Clin. Invest.,* 48, 78, 1969.

Shorokhov, I. A., Changes in the blood serum lipid and beta-lipoprotein composition from a low-protein diet (in Russian), *Vopr. Pitan,* 28, 16, 1969.

Kuo, P., Dietary treatment of hyperlipidemia, *Mod. Treat.,* 6, 1328, 1969.

Little, J. A. et al., Interrelationship between the kinds of dietary carbohydrate and fat in hyperlipoproteinemic patients. 1. Sucrose and starch with polyunsaturated fat, *Atherosclerosis,* 11, 173, 1970.

Birchwood, B. L. et al., Interrelationship between the kinds of dietary carbohydrate and fat in hyperlipoproteinemic patients. 2. Sucrose and starch with mixed saturated and polyunsaturated fats, *Atherosclerosis,* 11, 183, 1970.

Schlierf, G. et al., Change in lipoprotein composition with dietary carbohydrate, *Nutr. Rev.,* 28, 153, 1970.

Butkus, A. et al., Effects of diets rich in saturated fatty acids with or without added cholesterol on plasma lipids and lipoproteins, *Lipids,* 5, 896, 1970.

Lewis, L. A. et al., Ten years dietary treatment of primary hyperlipidemia, *Geriatrics,* 25, 64, 1970.

Collins, F. D. et al., Plasma lipids in human linoleic acid deficiency, *Nutr. Metab.,* 13, 150, 1971.

Elphick, M. C., Dietary fat and carbohydrate induction of hyperlipemia, *Nutr. Rev.,* 29, 59, 1971.

Greten, H., Dietary fat and carbohydrate induction of hyperlipemia, *Nutr. Rev.,* 29, 59, 1971.

Brunner, D. et al., Heredity, environment, serum lipoproteins and serum uric acid. A study in a community without familial eating pattern, *J. Chronic Dis.,* 23, 763, 1971.

Lewis, B., Diet and plasma lipoproteins, *Clin. Sci.,* 40, 16P, 1971.

Levy, R. I. et al., Dietary management of hyperlipoproteinemia, *J. Am. Diet Assoc.,* 58, 406, 1971.

Dryden, F. D. et al., Bovine serum lipids. II. Lipoprotein quantitative and qualitative composition as influenced by added animal fat diets, *J. Anim. Sci.,* 32, 1016, 1971.

Maha, G. E., Diet and drug therapy for hyperlipoproteinemia, *Med. Times,* 99, 49, 1971.

Wilson, W. S. et al., Serial lipid and lipoprotein responses to the American Heart Association fat-controlled diet, *Am. J. Med.,* 51, 491, 1971.

Bide, R. W., Changes in fowl plasma lipoproteins caused by starvation, *Poult. Sci.,* 51, 305, 1972.

Boquillon, M. et al., Composition, particle size and role in dietary fat transport of two different lipoproteins of the intestinal lymph, *Lipids,* 7, 409, 1972.

La Rosa, J. C., Hyperlipoproteinemia. 2. Dietary management, *Postgrad. Med.,* 52, 75, 1972.

Khegai, M. D. et al., The effect of adresone and alimentary sucrose loading on blood lipids and beta-lipoprotein composition, *Vopr. Pitan.,* 31, 63, 1972.

Levy, R. I. et al., Dietary and drug treatment of primary hyperlipoproteinemia, *Ann. Intern. Med.,* 77, 267, 1972.

Seakins, A. et al., Effect of a low-protein diet on the incorporation of amino acids into rat serum lipoproteins, *Biochem. J.,* 129, 793, 1972.

Grzeskowiak, B. et al., Diet therapy of hyperlipoproteinemias from the diabetological viewpoint, *Dtsch. Gesundheitsw.,* 27, 1868, 1972.

Kottke, B. A. et al., Sterol balance studies in patients on solid diets: comparison of two "nonabsorbable" markers, *J. Lab. Clin. Med.,* 80, 530, 1972.

Stone, M. C., The role of diet in the management of hyperlipoproteinaimias, *Proc. Nutr. Soc.,* 31, 311, 1972.

Chapman, M. J. et al., The effect of a lipid-rich diet on the properties and composition of lipoprotein particles from the Golgi apparatus of guinea-pig liver, *Biochem. J.*, 131, 177, 1973.

Chait, A. et al., Influence of saturated and unsaturated dietary fat on diurnal changes in plasma triglyceride and lipoprotein levels, *Clin. Sci.*, 44, 7P, 1973.

Khalil, M. et al., The effect of diarrhoea on low density beta lipoproteins and cholesterol in protein calorie malnutrition, *J. Trop. Med. Hyg.*, 76, 71, 1973.

Matzkies, F. et al., Dietary plans in the treatment of hyperlipoproteinemia, *Fortschr. Med.*, 91, 336, 1973.

Oriente, P. et al., Hyperlipoproteinemias and their dietetic treatment, *Minerva Med.*, 64, 1259, 1973.

Huth, K. et al., Dietetic and drug therapy of hyperlipidemias and hyperlipoproteinemia, *Med. Klin.*, 68, 1089, 1973.

Rojas-Hidalgo, E. et al., Influence of diet, sex and age on lipid and lipoprotein levels in the blood (in Spanish with English abstract), *Rev. Clin. Esp.*, 130, 195, 1973.

Wille, L. E. et al., Changes in serum pre-beta-lipoprotein following the feeding of sucrose. A preliminary trial in physically active and inactive young male students, *J. Oslo City Hosp.*, 23, 141, 1973.

Christophe, A. et. al., Differential postprandial effects of single free fatty acid or fat feedings on serum lipoproteins, *Arch. Int. Physiol. Biochim.*, 81, 961, 1973.

Sailer, V. S. et al., Modification of the plasma concentration of individual lipoprotein density classes through starvation and carbohydrate-rich diet (with English abstract), *Acta Med. Austriaca*, 2(1), 55, 1974.

Abrahamsson, H. et al., Polyunsaturated fatty acids in hyperlipoproteinemia. 1. Influences of a sucrose-rich diet on fatty acid comparison of serum lipoprotein lipids, *Nutr. Metab.*, 17(6), 329, 1974.

Sachlová, M. et al., Incidence of hyperlipoproteinemias, their classification and influencing through diet during a course of spa treatment. II. Clinical part, *Vnitr. Lek.*, 20, 15, 1974.

Heyden, S., Classification and diet therapy of hyperlipemia, *Dtsch. Med. Wochenschr.*, 99, 141, 1974.

Canzler, H., Diet therapy in hyperlipoproteinemias, *Z. Allgemeinmed.*, 50, 53, 1974.

Mann, J. I. et al., Effects of omitting dietary sucrose and isoenergetic substitution of starch in primary type IV hyperlipoproteinaemia, *Proc. Nutr. Soc.*, 33(1), 2A, 1974.

Lisch, H. J. et al., Effect of body weight changes on plasma lipids in patients with primary hyperlipoproteinemia, *Atherosclerosis*, 19, 477, 1974.

Calvert, G. D. et al., Serum lipoproteins in pigs on high-cholesterol-high-triglyceride diets, *Atherosclerosis*, 19, 485, 1974.

Teoh, P. C. et al., Primary type V hyperlipoproteinaemia (mixed hyperlipaemia): successful response to dietary restricton and simfibrate, *Singapore Med. J.*, 15(2), 149, 1974.

Glueck, C. J. et al., Plasma vitamin A and E levels in children with familial type II hyperlipoproteinemia during therapy with diet and cholestyramine resin, *Pediatrics*, 54, 51, 1974.

Christophe, A. et al., Proceedings: Effect of prolonged fasting in obese men, *Hoppe Seylers Z. Physiol. Chem.*, 355(10), 1184, 1974.

Glascock, R. F. et al., Contribution of the fatty acids of three low density serum lipoproteins to bovine milk fat, *J. Dairy Sci.*, 57(11), 1364, 1974.

Matzkies, F. et al., Studies on the influence of a fat-modified diet in healthy subjects and in patients with hyperlipoproteinemia (in German with English abstract), *Fortschr. Med.*, 92(31), 1257, 1974.

Sirtori, C. R. et al., Diet lipids and lipoproteins in patients with peripheral vascular disease, *Am. J. Med., Sci.*, 268(6), 325, 1974.

Ditschuneit, H. H. et al., Effect of a diet high in fat and protein and low in carbohydrates on the feeling of satiation, lipoproteins, uric acid and insulin in the blood of children (in German), *Verh. Dtsch. Ges. Inn. Med.*, 81, 1415, 1975.

Jaeger, H. et al., Gas chromatographic analysis of blood lipoproteins during treatment with a diet high in fat and protein (in German), *Verh. Dtsch. Ges. Inn. Med.*, 81, 1411, 1975.

Vessby, B. et al., Effects of dietary treatment on lipoprotein levels in hyperlipoproteinaemia, *Postgrad. Med. J.*, 51(8), 52, 1975.

Torsvik, H. et al., Effects of intravenous hyperalimentation of plasma-lipoproteins in severe familial hyper-cholesterolaemia, *Lancet*, 1(7907), 601, 1975.

Schellenberg, B. et al., Proceedings: Reciprocal fluctuations of lipoproteins in fasting therapy of hyperlipoproteinemia, *Munch. Med. Wochenschr.*, 117(20), 844, 1975.

Gordon, D. T. et al., Effects of diet and type IIa hyperlipoproteinemia upon structure of triacylglycerols and phosphalidyl cholines from human plasma lipoproteins, *Lipids*, 10(5), 270, 1975.

Sacks, F. M. et al., Plasma lipids and lipoproteins in vegetarians and controls, *N. Engl. J. Med.*, 292(22), 1148, 1975.

Wu, C. H. et al., Human plasma triglyceride labeling after high sucrose feeding. II. Study on triglyceride kinetics and postheparin lipolytic activity, *Metabolism*, 24(6), 755, 1975.

Bang, H. O. et al., Fat content of the blood and composition of the diet in a population group in West Greenland (with English abstract), *Ugeskr. Laeg.*, 137(29), 1641, 1975.

Kuo, P. T. et al., Combined para-aminosalicylic acid and dietary therapy in long-term control of hypercholesterolemia and hypertriglyceridemia (Types IIa and IIb hyperlipoproteinemia), *Circulation,* 53(2), 338, 1976.

Schonfeld, G. et al., Alterations in levels and interrelations of plasma apolipoproteins induced by diet, *Metabolism,* 25(3), 261, 1976.

Bahler, R. C. and Opplt, J. J., Dietary influence on molecular distribution of serum lipoproteins in subjects with proven artery disease, *Proceedings of IV International Conference on Atherosclerosis,* Plenum Press, New York, 1975; Atherosclerosis Monogr., in *Adv. Exp. Med. Biol.,* 28, 195, 1977.

Lampman, R. M. et al., Comparative effects of physical training and diet in normalizing serum lipids in men with Type IV hyperlipoproteinemia, *Circulation,* 55(4), 652, 1977.

Surgical Procedures

Thompson, G. R. et al., Ileal bypass in the treatment of hyperlipoproteinaemia, *Lancet,* 819, 35, 1973.

Wagner, A., Treatment of hyperlipoproteinemia-type IIa- using portacaval anastomosis, *Hippokrates,* 46(1), 115, 1975.

Bilheimer, D. W., Low density lipoprotein metabolism and cholesterol synthesis in familial homozygous hypercholesterolemia: influence of portacaval shunt surgery, *Lipoprotein Metabolism,* Greten, H., Ed., Springer-Verlag, Berlin, 1976, 44.

Schwartz, M. Z. et al., Treatment of heterozygous type II hyperlipidemia by partial ileal bypass in a pediatric population, *J. Pediatr. Surg.,* 11(3), 411, 1976.

THERAPEUTIC AGENTS

Fatty Acids Infusions

Schwartzkopff, W. et al., Studies on the transport and sojourn of intravenously administered fat emulsions (in German), *Dtsch. Med. Wochenschr.,* 90, 116, 1965.

Schwartzkopff, W. et al., The fate of intravenously administered fat emulsions, *Ger. Med. Monthly,* 10, 305, 1965.

Apostolakis, M. et al., The effect of orally administered lipids on the fat content and composition of plasma and lymph in man (in German), *Klin. Wochenschr.,* 43, 1094, 1965.

Kachorovskii, B. V., Changes in serum lipoproteins in parenteral feeding with a fatty emulsion (in Russian with English abstract), *Biull. Eksp. Biol. Med.,* 73, 45, 1972.

Lakshminarayana, G. et al., Preparation of ^{131}I-labeled fat emulsions and distribution of the label in serum lipoproteins and lipid classes following intravenous infusion to normal and hyperlipemic subjects, *Am. J. Clin. Nutr.,* 25, 531, 1972.

Thompson, G. R. et al., Changes in the fatty acid composition of plasma lipoproteins during administration of Intralipid to a patient with essential fatty acid deficiency, *Clin. Sci. Mol. Med.,* 47(4), 387, 1974.

Heparin

Petrova, T. R., The relationship between certain indices of lipid metabolism and heparin and heparinocytes of the blood in patients with coronary arteriosclerosis (in Russian), *Vrach. Delo.,* 3, 37, 1967.

Solaro, F. et al., Behavior of serum lipoproteins in elderly patients treated with subcutaneous injection of heparin (preliminary note) (in Italian), *G. Gerontol.,* 15, 587, 1967.

Nichols, A. V. et al., Analysis of change in ultracentrifugal lipoprotein profiles following heparin and ethyl-p-chlorophenoxyisobutyrate administration, *Clin. Chim. Acta,* 20, 277, 1968.

Skorepa, J. et al., A study on the plasmatic lipoproteins metabolism. 3. The effect of lipolytic enzymes of postheparinic serum (in Czech), *Sborn. Lek.,* 70, 287, 1968.

Burstein, M. et al., Precipitation by heparin of abnormal serum lipoproteins in obstructive icterus (in French), *Rev. Fr. Etud. Clin. Biol.,* 14, 182, 1969.

Burstein, M. et al., Effect of heparin and protamine on the electrophoretic mobility of the serum lipoproteins (in French), *Nouv. Rev. Fr. Hematol.,* 9, 365, 1969.

Burstein, M. et al., Precipitation of serum lipoproteins in the presence of anionic detergents and protamine, or cationic detergents and heparin (in French), *Rev. Fr. Transfus.,* 12, 271, 1969.

Tinterova, Z. et al., Immunoelectrodiffusional determination of β-lipoproteins in the plasma of heparin-treated patients, *Clin. Chim. Acta,* 26, 583, 1969.

Keler-Bacoka, M., The specific property of paraprotein lipids in serum concerning the heparin clearing effect, *Z. Klin. Chem.,* 8, 9, 1970.

la Rosa, J. C. et al., Changes in high-density lipoprotein protein composition after heparin-induced lipolysis, *Am. J. Physiol.,* 220, 785, 1971.

Greten, H., Post-heparin plasma phospholipases in normals and patients with hyperlipoproteinemia, *Klin. Wochenschr.,* 50, 39, 1972.

Baylin, S. B. et al., Response of plasma histaminase activity to small doses of heparin in normal subjects and patients with hyperlipoproteinemia, *J. Clin. Invest.*, 52, 1985, 1973.

Mookerjea, S., Phosphorylcholine: a specific promoter of heparin and serum-lipoprotein interaction, *Biochem. Biophys. Res. Commun.*, 53, 580, 1973.

Hara, T., Studies on the lipoprotein metabolism — changes of lipids, lipoprotein and apolipoprotein after heparin-induced lipolysis (in Japanese), *J. Jpn. Soc. Intern. Med.*, 63(6), 547, 1974.

Srinivasan, S. R. et al., Studies on the interaction of heparin with serum lipoproteins in the presence of Ca2 + , Mg2 + , and Mn2 + , *Arch. Biochem. Biophys.*, 170(1), 334, 1975.

Smith, G. D. et al., A sedimentation procedure for the detection of binding to concanavalin A and heparin to lipoproteins, *Anal. Biochem.*, 68(2), 637, 1975.

Bleyl, H. et al., Proceedings: heparin neutralizing effect of lipoproteins, *Thromb. Diath. Haemorrh.*, 34(2), 549, 1975.

Homma, Y. et al., Changes in plasma lipoprotein constituents during constant infusions of heparin, *Atherosclerosis*, 22(3), 551, 1975.

Lee, D. M. et al., Studies on the hydrolyses of triacylglycerols in chylomicrons, very-low- and low-density lipoproteins by C-inactivated lipoprotein lipase from post-heparin plasma of normal human subjects, *FEBS Lett.*, 64(1), 163, 1976.

Thyroid Hormones

Searcy, R. L. et al., Effects of dextrothyroxine on serum lipoprotein and cholesterol levels, *Curr. Ther. Res.*, 10, 177, 1968.

Schwartzkopff, W., Therapy of hyperlipoproteinemias type IIa and IIb using dextro-thyroxine or DL-alpha-methyl-thyroxine-ethylester, *Verh. Dtsch. Ges. Inn. Med.*, 78, 1621, 1972.

Koschinsky, T. et al., Decreasing effect of DL-alpha-methylthyroxine-ethylester, D-thyroxine (DT4), D-triiodothyronine (DT3) and a combination of DT4 + DT3 on serum lipids in primary hyperlipoproteinemia Type IIa and b, *Verh. Dtsch. Ges. Inn. Med.*, 80, 1262, 1974.

Klemens, U. H. et al., Treatment of primary lipoproteinemia type IIa and IIb with highly purified dextro-thyroxine (D-T4). Controlled study in ambulant conditions, *Dtsch. Med. Wochenschr.*, 99, 487, 1974.

Koschinsky, T. et al., Treatment of primary hyperlipoproteinemia type IIa and IIb with dextrothyroxine, *Dtsch. Med. Wochenschr.*, 99, 494, 1974.

Myburgh, D. P. et al., D-Thyroxine in hyperbetalipoproteinaemia, *S. Afr. Med., J.*, 49(42), 1757, 1975.

Nicotinic Acid

Kedra, M. et al., The effect of nicotinic acid on the metabolism of lipids, especially the blood cholesterol in cases of arteriosclerosis (in Polish), *Pol. Tyg. Lek.*, 19, 1397, 1964.

Ferliga, G., Favorable effect of nicotinic acid on lipoprotein normalization with temporary drug association (in Italian), *G. Gerontol.*, 15, 1381, 1967.

Carlson, L. A. et al., Effect of a simple dose of nicotinic acid on plasma lipids in patients with hyperlipoproteinemia, *Acta Med. Scand.*, 183, 457, 1968.

Carlson, L. A. et al., Effect of nicotinic acid on plasma lipids in patients with hyperlipoproteinemia during the first week of treatment, *J. Atheroscler. Res.*, 8, 739, 1968.

Carlson, L. A., Arteriosclerosis — nicotinic acid therapy. 7. Effect of nicotinic acid on lipoproteins (in Swedish), *Lakartidningen*, 68, 1763, 1971.

Gustafson, A., Arteriosclerosis — nicotinic acid therapy. 2. Alterations in serum lipoproteins. Diagnosis, therapy (in Swedish), *Lakartidningen*, 68, 1744, 1971.

Carlson, L. A. et al., Serum and tissue lipid metabolism and effect of nicotinic acid in different types of hyperlipidemia, *Adv. Exp. Med. Biol.*, 26(0), 165, 1972.

Avogaro, P. et al., Effects of the combination of nicotinic acid and propranolol in very low doses on blood lipids in man, *Atherosclerosis*, 20(2), 395, 1974.

Beaumont, J. L. et al., Binding to plasma lipoproteins of chlorophenoxyisobutyric, tibric and nicotinic acids and their esters, *Proc. R. Soc. Med.*, Suppl. 69, 2, 41, 1976.

Clofibrate (Atromid-S)

Feinberg, L. J. et al., The effects of clofibrate on the metabolism of ^{14}C-labeled tripalmitin in the human subject, *Metabolism*, 16, 618, 1967.

Gabrys, B. et al., Effect of Atromid-S on lipid metabolism in coronary insufficiency (in Polish), *Pol. Tyg. Lek.*, 23, 1608, 1968.

Scott, P. J. et al., Effect of clofibrate on low-density lipoprotein. Turnover in essential hypercholesterolaemia, *J. Atheroscler. Res.*, 9, 25, 1969.

Goodman, D. S. et al., Cholesteryl ester turnover in human plasma lipoproteins during cholestyramine and clofibrate therapy, *J. Lipid Res.*, 11, 183, 1970.

Modzelewski, A. et al., Effect of short-term use of clofibrate on the blood level of cholesterol, β-lipoproteins and free fatty acids (in Polish), *Wiad. Lek.*, 23, 1937, 1970.

Takayasu, K. et al., Human plasma fatty acid composition: the features of hyperlipoproteinemia and the influence of -linolenate and clofibrate, *Jpn. Circ. J.*, 35, 1059, 1971.

Miethinen, T. A. et al., Change of lipoprotein pattern by clofibrate in hyperglyceridaemia and mixed hyperlipidaemia, *Acta Med. Scand.*, 192, 177, 1972.

Bridgman, J. F. et al., Complications during clofibrate treatment of nephrotic-syndrome hyperlipoproteinaemia, *Lancet*, 2, 506, 1972.

Neuman, J. et al., A double-blind comparison of the hypolipidemic and hypouricemic action of halofenate and clofibrate in patients with hyperlipoproteinemia (types III, IV and V), *J. Cardiovasc. Surg.*, Spec. No., 532, 1973.

Hanefeld, M. et al., Problems of long-term treatment of hyperlipoproteinemia with clofibrinic acid, *Z. Gesamte Inn. Med.*, Suppl. 28, 10, 160c, 1973.

Beaumont, J. L. et al., Serum lipoproteins and antilipemic drugs related to clofibrate. Their in vitro interaction, *Atherosclerosis*, 17, 419, 1973.

von Löwis, P. et al., Treatment of primary hyperlipoproteinaemia types, IIa, IIb, 3 and IV with clofibrate, *Dtsch. Med. Wochenschr.*, 98, 2328, 1973.

Fenderson, R. W., Jr. et al., Effect of clofibrate on plasma glucose and serum immunoreactive insulin in patients with hyperlipoproteinemia, *Am. J. Clin. Nutr.*, 27, 22, 1974.

Plamieniak, L., Therapeutic effect on clofibrate in a patient with primary hyperlipoproteinemia type 3, *Wiad. Lek.*, 27, 811, 1974.

Carvalho, A. C. et al., Clofibrate reversal of platelet hypersensitivity in hyperbetalipoproteinemia, *Circulation*, 50(3), 570, 1974.

Beaumont, V. et al., Comparative study of several hypolipidemic agents related to clofibrate, *Atherosclerosis*, 20(2), 141, 1974.

Knüchel, F., Effect of a new lipid-reducing preparation (Etofibrat) on hyperlipoproteinemia type II, *Med. Welt*, 25(44), 1810, 1974.

Stein, E. A. et al., Colestipol, Clofibrate, Cholestyramine and combination therapy in the treatment of familial hyperbetalipoproteinaemia, *S. Afr. Med. J.*, 49(31), 1252, 1975.

Ditzel, J. et al., Clofibrate in type II hyperlipoproteinemia, *Acta Med. Scand.*, 200(1—2), 55, 1976.

Dujovne, C. A. et al., One-year trials with Halofenate, Clofibrate, and Placebo, *Clin. Pharmacol. Ther.*, 19(3), 352, 1976.

Rose, H. G. et al., Clofibrate-induced low density lipoprotein elevation. Therapeutic implications and treatment by colestipol resin, *Atherosclerosis*, 23(3), 413, 1976.

Beaumont, J. L. et al., Binding to plasma lipoproteins of chlorophenoxyisobutyric, tibric and nicotinic acids and their esters: its significance for the mechanism of lipid lowering by clofibrate and related drugs, *Atherosclerosis*, 25(2—3), 255, 1976.

Bielmann, P. et al., Hypolipidemic effect of tibric acid. A comparison with clofibrate and placebo in type IV hyperlipoproteinemia, *Int. J. Clin. Pharmacol. Biopharm.*, 15(4), 166, 1977.

Opplt, J. J. and Bahler, R. C., Lipoprotein metabolism on a molecular level during Atromid-S therapy, *Atherosclerosis IV*, Schettler, G. Goto, Y., Hata, Y., and Klose, G., Eds., Springer-Verlag, Berlin, 1977, 545.

Alufibrate
Stähelin, H. B. et al., Proceedings: effect of alufibrate (atherolip) on pyruvate and lipoprotein metabolism in patients with hyperlipidemia Type IV, *Helv. Med. Acta*, 37(5—6), 388, 1974.

Stähelin, H. B. et al., Effect of Al-chlorophenoxyisobutyrate (Alufibrate) on pyruvate metabolism and on the fate of very low density lipoprotein lipids in hyperlipidemic patients, *Clin. Chim. Acta*, 54, 115, 1974.

Cholestyramine
Jones, R. J. et al., Lipoprotein lipid alterations with cholestyramine administration, *J. Lab. Clin. Med.*, 75, 953, 1970.

Montafis, C. D. et al., Cholestyramine and nicotinic acid in the treatment of familial hyper-β-lipoproteinaemia in the homozygous form, *Atherosclerosis*, 14, 247, 1971.

Christensen, N. C. et al., Cholestyramine therapy in type II hyperlipoproteinemia, *Ngeskr. Laeger*, 133, 2431, 1971.

Glueck, C. J. et al., Colestipol and cholestyramine resin. Comparative effects in familial type II hyperlipoproteinemia, *JAMA*, 222, 676, 1972.

Schade, R. W. et al., Treatment of type II hyperlipoproteinaemia with cholestyramine, *Neth. J. Med.*, 16, 254, 1973.

Sachs, B. A. et al., Response of hyperlipoproteinemia to colestipol, *N.Y. State J. Med.*, 73, 1068, 1973.

Levy, R. I. et al., Cholestyramine in type II hyperlipoproteinemia. A double-blind trial, *Ann. Intern. Med.*, 79, 51, 1973.

Vålek, J. et al., On the treatment of hyperlipoproteinaemia type II with cholestyramine, *Cas. Lek. Cesk.*, 112, 1531, 1973.

Einarsson, K. et al., The effect of cholestyramine on the elimination of cholesterol as bile acids in patients with hyperlipoproteinaemia type II and IV, *Eur. J. Clin. Invest.*, 4(6), 405, 1974.

Clifton-Bligh, P. et al., Changes in plasma lipoprotein lipids in hypercholesterolaemic patients treated with the bile acid-sequestering resin, colestipol, *Clin. Sci. Mol. Med.*, 47(6), 547, 1974.

Miller, N. E. et al., Differences among hyperlipoproteinaemic subjects in the response of lipoprotein lipids to resin therapy, *Eur. J. Clin. Invest.*, 5(3), 241, 1975.

Lees, A. M. et al., Results of colestipol therapy in type II hyperlipoproteinemia, *Atherosclerosis*, 24(1—2), 129, 1976.

Witztum, J. L. et al., The effects of colestipol on the metabolism of very low density lipoproteins in man, *J. Lab. Clin. Med.*, 88(6), 1008, 1976.

Gemfibrozil

Olsson, A. G. et al., Effect of gemfibrozil on lipoprotein concentrations in different types of hyperlipoproteinaemia, *Proc. R. Soc. Med.*, 69 Suppl. 2, 29, 1976.

Nikkilä, E. A. et al., Gemfibrozil: effect on serum lipids, lipoproteins, postheparin plasma lipase activities and glucose tolerance in primary hypertriglyceridaemia, *Proc. R. Soc. Med.*, 69 Suppl., 2, 58, 1976.

Janus, E. D. et al., The evaluation of lipoprotein changes during gemfibrozil treatment, *Proc. R. Soc. Med.*, 69 Suppl. 2, 76, 1976.

Gemfibrozil: a new lipid lowering agent. Open forum, *Proc. R. Soc. Med.*, 69 Suppl. 2, 115, 1976.

Probucol (Lorelco)

Drake, J. W. et al., The Effect of [4,4′-(isoproplyidenedithio)bis (2,6-di-t-butylphenol)] (DH-581) on serum kipids and lipoproteins in human subjects, *Metabolism*, 18/11, 916, 1969.

Miettinen, T., Mode of action of a new hypocholesterolaemic drug (DH-581) in familial hypercholesterolemia, *Atheroschlerosis*, 15, 163, 1972.

Parsons, W. B., Effect of a new cholesterol-reducing agent (Probucol) in hyperlipoproteinemic humans, Am. Heart Assoc. Meerting, Dallas, 1972.

Salel, A. F. et al., Probucol: a new cholesterol-lowering drug effective in patients with type II hyperlipoproteinemia, *Clin. Pharmacol. Ther.*, 20(6), 690, 1976.

Murphy, B F., Probucol (LORELCO) in Treatment of Hyperlipemia, *JAMA*, 238/23, 2537, 1977.

ECPIB

Giannini, S. D. et al., Action of ethyl-chlorophenoxyisobutyrate (CPIB) on blood lipids (in Portuguese), *Rev. Brazil Med.*, 22, 702, 1965.

Keller, L. et al., The effect of ethyl-chlorophenoxyisobutyrate on the serum lipids in coronary sclerosis (in German), *Z. Ges. Inn. Med.*, 21, 169, 1966.

Sabella, G., Plasma lipid-normalizing activity of the combination of ethylchlorophenoxyisobytyrate and duodenal sulfomucopolysaccharide extract (in Italian), *Farmaco Ed. Prat.*, 22, 786, 1967.

Vastesaeger, M. M. et al., Action of albuminum bis-parachlorophenoxyisobutyrate (2CPIBA) on various hyperlipoproteinemias: initial results (in French), *Brux. Med.*, 49, 615, 1969.

Bronzini, A. et al., Effect of parachlorophenoxyisobutyric acid (CPIB acid) and of diet on blood cholesterol and on the β-α-lipoprotein ratio in aged subjects (in Italian), *G. Gerontol.*, 17, 1067, 1969.

Davignon, J. et al., Heterogeneity on familial hyperlipoproteinemia type II on the basis of fasting plasma triglyceride-cholesterol ratio and plasma cholesterol response to ethyl p-chlorophenoxyisobutyrate, *Rev. Can. Biol.*, 30, 307, 1971.

Riboflavin Tetrabutyrate

Trakenaka, T. et al., Effect on riboflavin -2′, 3′, 4′, 5′-tetrabutyrate on total serum cholesterol and β-lipoprotein contents (in Japanese with English abstract), *Iryo*, 26, 153, 1972.

Kuzuya, F., Effect of riboflavin tetrabutyrate on degradation of lipoprotein by fatty acid peroxide, *J. Vitaminol.*, 18, 65, 1972.

Estrogens

Kroman, H. S. et al., The interrelationship of blood lipids and estrogens. 2. A preliminary study in normal humans and patients with coronary artery disease, *J. Atheroscler. Res.*, 6, 247, 1966.

Balestreri, R. et al., Study of some metabolic effects of an orally active steroidyl-17-β-enolether (guinbolone) with anabolic action (in Italian), *Arch. Maragliano Pat. Clin.*, 22, 369, 1966.

Notelovitz, M. et al., Metabolic effect of conjugated oestrogens (USP) on lipids and lipoproteins, *S. Afr. Med. J.,* 48(61), 2552, 1974.

Contraceptives
Aurell, M. et al., Serum lipids and lipoproteins during long-term administration of an oral contraceptive, *Lancet,* 1, 291, 1966.

Glueck, C. J. et al., Amelioration of hyperlipoproteinaemia by progestational drugs in familial type-V hyperlipoproteinaemia, *Lancet,* 1, 1290, 1969.

Wynn, V. et al., Effect of oral contraceptives on serum lipids and lipoproteins, *Lancet,* 2, 256, 1969.

Sachs, B. A. et al., Plasma lipid and lipoprotein alterations during oral contraceptive administration, *Obstet. Gynecol.,* 34, 530, 1969.

Wynn, V. et al., Fasting serum triglyceride, cholesterol and lipoprotein levels during oral contraceptive therapy, *Lancet,* 2, 756, 1969.

Sachs, B. A. et al., Plasma lipid and lipoprotein changes during "pill-a month" contraceptive steroid administration, *Am. J. Obstet. Gynecol.,* 109, 155, 1971.

Rössner, S. et al., Effects of an oral contraceptive agent on plasma lipids, plasma lipoproteins, the intravenous fat tolerance and the post-heparin lipoprotein lipase activity, *Acta Med. Scand.,* 190, 301, 1971.

Gad-el-Mawla, N. et al., Plasma lipids and lipoproteins in bilharzial females during oral contraceptive therapy, *J. Egypt Med. Assoc.,* 55, 137, 1972.

Kuku, S. B. et al., Fasting serum lipids and serum lipoprotein distribution during oral contraceptive therapy in Nigerians, *J. Obstet. Gynaecol. Br. Commonw.,* 80, 750, 1973.

Zirm, M. et al., Increase of lipoprotein occurring simultaneously with retinal damage through intake of an oral contraceptive (in German with English abstract), *Klin. Monatsbl. Augenheilkd.,* 165(3), 470, 1974.

Fiser, R. H., Complications of oral contraceptive agents — a symposium. Contraceptives and lipid metabolism, *West. J. Med.,* 122(1), 35, 1975.

Wankowicz, Z. et al., Content of cholesterol in pre-β and β-lipoprotein fractions of blood serum and β-lipoprotein fractions of blood serum and total cholesterol concentrations in women ingesting progestagen preparations as contraceptives (with English abstract), *Ginekol. Pol.,* 46(9), 965, 1975.

Different Therapeutic Agents
Ojala, L., Experiences on deca-durabolin treatment of diabetic retinopathy, *Acta Opthal.,* 42, 519, 1964.

Navarranne, P. et al., 1st biological results of a new lipotropic formula (in French), *Clinique,* 59, 443, 1964.

Meciani, L. et al., Possibilities and limits of ambulatory treatment of the dyslipidoses in coronary disease with Triac (triiodothyroacetic acid). Use of Triac alone (in Italian), *Minerva Med.,* 55, 2865, 1964.

Knüchel, F., Behavior of serum lipids and lipoproteins with administration of Regelan (ethyl-p-chlorophenoxyisobutyrate (in German), *Med. Welt,* 47, 2530, 1964.

el-Sheikh, A. et al., Effect of sorbite on serum cholesterol and lipidogram in man, *J. Egypt Med. Assoc.,* 48, 415, 1965.

Diekmeier, L., Effect of terrain therapy on the cholesterol and β-lipoprotein content of the blood serum (in German), *Med. Klin.,* 60, 1357, 1965.

Howard, R. P. et al., Effects of triparanol administration on certain aspects of protein and steroid metabolism, serum lipids and lipoproteins, *J. Atheroscler. Res.,* 5, 580, 1965.

Abitbol, L. et al., Preliminary studies on the action of sulfinpyrazone in serum uric acid, cholesterol and lipoproteins (in Portuguese), *Rev. Brazil Med.,* 22, 748, 1965.

Modzelewski, A. et al., Effect of linodoxine treatment on cholesterol and β-lipoprotein levels in the blood serum of patients with atherosclerosis, *Pol. Med. J.,* 5, 1221, 1966.

Modzelewski, A. et al., Pattern of certain lipids in the blood serum in geriocaine therapy (in Polish), *Wiad. Led.,* 19, 1925, 1966.

Manzini, E. et al., Action of flavone-7-ethyl-hydroxyacetate on blood levels of cholesterol, lipoid phosphorus and α- and β-lipoproteins (in Italian), *Clin. Ter.,* 40, 341, 1967.

Melo, E. et al., Contribution to the study of anabolic drugs in hyperlipoproteinemias (in Portuguese), *Arq. Brazil. Eudocr.,* 16, 163, 1967.

Faloona, G. R. et al., The effects of actinomycin D on the biosynthesis of plasma lipoproteins, *Biochemistry,* 7, 720, 1968.

Guidetti, E., 15 years' clinical experience with antineoplastic therapy of an immunological type (in Italian), *Clin. Ter.,* 44, 487, 1968.

Sachs, B. A. et al., Effect of oxandrolone on plasma lipids and lipoproteins of patients with disorders of lipid metabolism, *Metabolism,* 17, 400, 1968.

Chu, Y. C. et al., Hyperlipoproteinemia. Status after one year of treatment, *N.Y. J. Med.,* 68, 1417, 1968.

Oriente, P. et al., Reduction of serum lipids in familial hyperlipidemias by a tetralin derivative, *Int. Z. Klin. Pharmakol. Ther. Toxik.,* 2, 348, 1969.

Drake, J. W. et al., The effect of (4,4′-(isopropylidenethio) bis (2, 6-di-t-butyl-phenol) (DH-581) on serum lipids and lipoproteins in human subjects, *Metabolism,* 18, 916, 1969.

Jepson, E. W. et al., Treatment of essential hyperlipidaemia, Lancet, 1, 423, 1970.

Whitehead, T. P. et al., The effect of low molecular weight dextran infusions on plasma lipids and lipoproteins, *Clin. Sci.,* 38, 233, 1970.

Kuo, P. T. et al., Study of serum insulin in atherosclerotic patients with endogenous hypertriglyceridemia (type 3 and IV hyperlipoproteinemia), *Metabolism,* 19, 372, 1970.

Sobra, J., Inborn errors of lipid metabolism. XIX. Basic principles of complex therapy of hyperlipoproteinemias, *Cas. Lek. Cesk.,* 109, 457, 1970.

Thiffault, C. et al., Treatment of no. II type essential hyperlipoproteinemia with a new therapeutic agent, celluline, *Can. Med. Assoc. J.,* 103, 165, 1970.

Schönbeck, M. et al., Therapeutic effect of Ciba 13437-Su in 40 patients with familial essential hyperlipemia, *Dtsch. Med. Wochenschr.,* 95, 1761, 1970.

Bertolotto, E., Effect of treatment with phospholipids on the serum lipoprotein pattern (in Italian with English abstract), *Pathologica,* 62, 195, 1970.

Ceruso, D. et al., Glucose metabolism, lipid fractions and blood uric acid level following administration of sodium dextro-thyroxine, *G. Gerontol.,* 18, 868, 1970.

Boberg, J. et al., Serum lipid reducing principles — possibilities for treatment. 2. Classification of hyperlipidemia (in Swedish), *Lakartidningen,* 67 Suppl., IV, 40t, 1970.

Klimov, A. N. et al., Binding of the aromatic heptaen levorin and the nonaromatic heptaen amphotericin-B with proteins and lipoproteins of the blood, *Antibiotiki,* 16, 325, 1971.

Mookerjea, S., Action of choline in lipoprotein metabolism, *Fed. Proc.,* 30, 143, 1971.

Morgan, J. P. et al., Hypolipidemic, uricosuric, and thyroxine-displacing effects of MK-185 (halofenate), *Clin. Pharmacol. Ther.,* 12, 517, 1971.

Levy, R. I. et al., Hypolipidemic drugs and hyperlipoproteinemia, *Ann. N. Y. Acad. Sci.,* 179, 475, 1971.

Greten, H., Therapy of primary hyperlipidemias (in German), *Dtsch. Med. Wochenschr.,* 96, 1449, 1971.

Cloarec, M., Treatment of hyperlipoproteinemia in relation to recent classifications, *Minerva Med.,* 62, 3452, 1971.

Fasoli, A., Results and prospects of the use of drugs acting on serum lipoproteins in atherosclerosis, *Minerva Med.,* 62, 3458, 1971.

Brusco, O. J. et al., Action of the new hypocholestermic agent DH-581 (dithiobisphenol) in humans, *Prensa Med. Argent.,* 58, 1587, 1971.

Fleischmajer, R., Diagnosis and treatment of familial lipoproteinemias, *Int. J. Dermatol.,* 10, 251, 1971.

Ciswicka-Sznajderman, M., Primary hyperlipoproteinemia. IV. Treatment (in Polish with English abstract), *Pol. Arch. Med. Wewn.,* 47, 589, 1971.

Levy, R. I., Beta lipoprotein protein turnover: an approach to the understanding of drug and disease mechanisms, *Expo. Annu. Biochim. Med.,* 31, 91, 1972.

Shuliakovskaia, T. A. et al., Effect of oat and galascorbin polyphenols on the cholesterol and β-lipoprotein level in the blood serum in the fasting, *Ukr. Biokhim. Zh.,* 44, 376, 1972.

Groh, J. et al., The effect of Brodilan on the levels of lipoproteins, cholesterol and liposoluble vitamins in the plasma, *Sb. Ved. Pr. Lek. Fak. Karlovy Univ.,* 15, 155, 1972.

Levy, R. I. et al., Hypolipidemic drugs and lipoprotein metabolism, *Adv. Exp. Med. Biol.,* 26(0), 155, 1972.

Hauser, J. et al., Diagnosis and treatment of hyperlipoproteinemias, *Schweiz. Med. Wochenschr.,* 102, 265, 1972.

Solyom, A., Effect of androgens on serum lipids and lipoproteins, *Lipids,* 7, 135, 1972.

Sirtori, C. et al., Clinical evaluation of MK-185: a new hypolipidemic drug, *Lipids,* 7, 96, 1972.

Danz, J. et al., Influencing of various serum lipid levels by bath treatments (in German), *Z. Physiother.,* 24, 101, 1972.

Klimov, A. N. et al., Types of hyperlipoproteinemia, their relationship to atherosclerosis and treatment, *Kardiologia,* 12, 133, 1972.

Menci, S. et al., Changes induced by S-adenosylmethiomine (SAM) in the cholesterol and triglyceride plasma levels and in the β-α-lipoprotein ratio in cases of hyperlipemia (in Italian with English abstract), *G. Gerontol.,* 20, 523, 1972.

Trimmer, E. J., Lipid lowering agents, *Br. Med. J.,* 3, 174, 1972.

Carlson, L. A. et al., Hyper-lipoproteinemia in men exposed to chlorinated hydrocarbon pesticides, *Acta Med. Scand.,* 192, 29, 1972.

Jepson, E. M. et al., A comparative study of a new drug MK 185 with clofibrate in the treatment of hyperlipidaemias, *Atherosclerosis,* 16, 9, 1972.

la Rosa, J. C., Hyperlipoproteinemia. 3. Drug therapy, *Postgrad. Med.,* 52, 128, 1972.

Gustafson, A. et al., Gonadal steroid effects on plasma lipoproteins and individual phospholipids, *Endocrinolology,* 35, 203, 1972.

Bogdade, J. D., Therapy of the hyperlipoproteinemias, *Acta Diabetol. Lat. Suppl.,* 9, 98, 1972.

Zorrilla, E. et al., Effect of oxandrolone on the serum lipids in type IV and IIb hyperlipidemia, *Arch. Inst. Cardiol. Mex.,* 43, 270, 1973.

Schoger, G. A., Symptoms and therapy in hyperlipoproteinemias, *Med. Monatsschr.,* 27, 118, 1973.

Berg, G. et al., Behavior of serum lipoproteins after constant infusion of xylitol, fructose and sorbitol, *Klin. Wochenschr.,* 51, 1124, 1973.

Kolmodin-Hedman, B., Changes in drug metabolism and lipoproteins in workers occupationally exposed to DDT and lindane, *Arh. Hig. Rada Toksikol,* 24, 289, 1973.

Klimova, T. A. et al., Study of active hypolipemic preparations based on mevalonic acid derivatives, *Vestn. Akad. Med. Nauk. SSSR,* 28, 45, 1973.

Faergeman, O. et al., Increase of post-heparin lipase activity by oxandrolone in familial hyperchylomicronemia, *Scand. J. Clin. Lab. Invest.,* 31, 27, 1973.

Modzelewski, A., Effect of intravenous procaine infusions on the blood level of free acids and beta lipoproteins, *Wiad. Lek.,* 26, 23, 1973.

Boechko, F. F., Effect of manganese on the dynamics of the neutral fat and lipoprotein content in the blood serum under conditions of alimentary hypercholesteremia (in Russian with English abstract), *Vopr. Pitan.,* 32, 57, 1973.

Jain, R. C. et al., Effect of onion ingestion on serum triglyceride, betalipoprotein-cholesterol and phospholipids in alimentary lipaemia, *J. Assoc. Physicians India,* 21, 357, 1973.

Fellin, R. et al., Treatment of primary hyperlipoproteinaemias with a tetralin derivative (Su-13´437), *Atherosclerosis,* 17, 383, 1973.

Knüchel, F. et al., Therapeutic results in directed and nondirected therapy of hyperlipoproteinemia, *Med. Welt,* 24, 1019, 1973.

Klemens, U. H. et al., Therapy of primary hyperlipoproteinemia types IIa, IIb and IV using beta pyridylcarbinol. Study on three phases under ambulatory conditions, *Dtsch. Med. Wochenschr.,* 98, 1197, 1973.

Peeters, H. et al., Action of "essential" phospholipids on plasma lipids and fatty acids in type II hyperlipoproteinemias, *Munch. Med. Wochenschr.,* 115, 1358, 1973.

Salvati, A., Drugs which influence plasma fatty acids and lipoproteins. Opinion of Prof. Rodolfo Paoletti, *Minerva Med.,* 64, 3256, 1973.

Barton, G. M. et al., Trial of an anabolic steroid (oxymetholone) in atherosclerosis with hyperlipoproteinaemia, *Atherosclerosis,* 18, 505, 1973.

Gerasimova, E. N. et al., Steroid hormones in arteriosclerosis with different types of hyperlipoproteinemia, *Kardiologiia,* 15, 14, 1974.

Crepaldi, G. et al., Trea tment of hyperlipoproteinemias, *G. Ital. Cardiol.,* 4, 332, 1974.

Kuo, P. T., Hyperlipidemia and coronary artery disease. Principles of diet and drug treatment, *Med. Clin. North. Am.,* 58, 351, 1974.

Olsson, A. G. et al., Effects of oxandrolone on plasma lipoproteins and the intravenous fat tolerance in man, *Atherosclerosis,* 19, 337, 1974.

Fleischmajer, R. et al., Familial hyperlipidemias. Diagnosis and treatment, *Arch. Dermatol.,* 110(1), 43, 1974.

Frommeyer, W. B., Jr., The diagnosis and treatment of the hyperlipoproteinemias: emphasis on Types IIa, IIb and IV, *Ala. J. Med. Sci.,* 2(3), 207, 1974.

Knuchel, F., Effect of a new lipid-reducing preparation in case of hyperlipoproteinemia, *Med. Welt,* 25(43), 1766, 1974.

Chait, A. et al., Reduction of serum triglyceride levels by polyunsaturated fat. Studies on the mode of action and on very low density lipoprotein composition, *Atherosclerosis,* 20(2), 347, 1974.

Barter, P. J. et al., Lowering of serum cholesterol and triglyceride by para-aminosalicylic acid in hyperlipoproteinemia. Studies in patients with types II-A and II-B, *Ann. Intern. Med.,* 81(5), 619, 1974.

Herscovici, R., Hyperlipoproteinemias: classification and treatment, *Rev. Med. Chir. Soc. Med. Nat. Iasi,* 78(4), 983, 1974.

Murphy, B. F., Management of hyperlipidemias, *JAMA,* 230 (12), 1683, 1974.

Noseda, G. et al., Clinical trial of tibric acid, a new hyperlipidemic agent (in German with English abstract), *Schweiz. Med. Wochenschr.,* 104(51), 1917, 1974.

Berg, C. et al., Serum lipoproteins and ketone bodies after constant infusion of sorbitol i.v. (with English abstract), *Klin. Wochenscher.,* 53(4), 187, 1975.

Torsvik, H. et al., Effects of intravenous hyperalimentation of plasma-lipoproteins in severe familial hypercholesterolemia, *Lancet,* 1(7907), 601, 1975.

Schwartzkopff, W. et al., Long-term treatment of hyperlipoproteinaemia, types IIa and IIb, with etiroxate (with English abstract), *Dtsch. Med. Wochenschr.,* 100(15), 815, 1975.

Schwartzkopff, W. et al., Comparative studies of the lipid-lowering activity of etiroxate hydochloride and dextrothyroxine (with English abstract), *Münch. Med. Wochenschr.,* 117(19), 827, 1975.

Cywinska, R. et al., Effect of vitamin E and geriocaine treatment on serum cholesterol and β-lipoprotein levels in elderly subjects, (in Polish with English abstract), *Wiad. Lek.,* 28(10), 819, 1975.

Bielmann, P. et al., Dose-response to tibric acid: a new hypolipidemic drug in type IV hyperlipoproteinemia, *Clin. Pharmacol. Ther.*, 17(5), 606, 1975.

Thompson, G. R. et al., Contrasting effects on plasma lipoproteins of intravenous versus oral administration of a triglyceride-phospholipid emulsion, *Eur. J. Clin. Invest.*, 5(5), 373, 1975.

Day, C. E. et al., Hypo-beta-lipoproteinemic agents. 1 Bicyclo (2.2.2) octyloxyaniline and its derivatives, *J. Med. Chem.*, 18(11), 1065, 1975.

Blaton, V. et al., The human plasma lipids and lipoproteins under influence of EPL-therapy, in *Phosphatidylcholine*, Peeters, H., Ed., Springer-Verlag, Berlin, 1976, 125.

Lees, R. S. et al., Effects of sitosterol therapy on plasma lipid and lipoprotein concentrations, in *Lipoprotein Metabolism*, Greten, H., Ed., Springer-Verlag, Berlin, 1976, 117.

Torsvik, H. et al., The effects of intravenous hyperalimentation on plasma proteins in severe hypercholesterolemia (proceedings), *Acta Chir. Scand. Suppl.*, 466, 124, 1976.

Rouffy, J. et al., Atilipidemic drugs. Part 5: Evaluation of the hypolipidemic effect of LF 178 in 191 patients affected by the atherogenic form of endogenous hyperlipoproteinemia (types II. a, II. b and IV), *Arzneim. Forsch.*, 26(5), 901, 1976.

Day, C. E. et al., Biological activity of a hypo-β-lipoproteinemic agent, *Adv. Exp. Med. Biol.*, 67(00), 231, 1976.

Britov, A. N. et al., Effect of preparations of polyunsaturated fatty acids on the level of blood serum lipoproteins in patients with ischemic heart disease (in Russian), *Sov. Med.*, 1, 118, 1976.

Weisweiler, P. et al., Treatment of primary hyperlipoproteinemias type IV with different biquanides, *Klin. Wochenschr.*, 54(6), 283, 1976.

Ikai, A., Stepwise degradation of serum low density lipoprotein by sodium dodecyl sulfate, *J. Biochem.*, 79(3), 679, 1976.

Simons, L. A. et al., Changes in plasma lipoproteins in subjects treated with the bile acid-sequestering resin polidexide (Secholex), *Aust. N. Z. J. Med.*, 6(2), 127, 1976.

Nilsen, O. G., Serum albumin and lipoproteins as the quinidine binding molecules in normal human sera, *Biochem. Pharmacol.*, 25(9), 1007, 1976.

Lavieunville, M. et al., Lipoproteins and metformin (in French), *J. Annu. Diabetol. Hotel Dieun*, 349, 1976.

Thompson, G. R. et al., Effects of intravenous phospholipid on low density lipoprotein turnover in man, *Eur. J. Clin. Invest.*, 6(3), 241, 1976.

Kissebah, A. H. et al., Transport kinetics of plasma free fatty acids, very low density lipoprotein triglycerides and apoprotein in patients with endogenous hypertriglyceridaemia: effects of 2,2-dimethyl, 5 (2,5-xylyoxy) valeric acid therapy, *Atherosclerosis*, 24(1—2), 199, 1976.

de Gennes, J. L. et al., The effect of tiadenol on type II.A hyperlipidemias (with English abstract), *Therapie*, 31(4), 455, 1976.

Fedele, D. et al., Hypolipidemic effects of metformin in hyper-pre-β-lipoproteinemia, *Diabete Metab.*, 2(3), 127, 1976.

Oster, P. et al., Sitosterol in familial hyperlipoproteinemia type II. A randomized double-blind crossover study (in German with English abstract), *Dtsch. Med. Wochenschr.*, 101(36), 1308, 1976.

Tanaka, N. et al., Effect of chronic administration of propranolol on lipoprotein composition, *Metabolism*, 25(10), 1071, 1976.

Bagnarello, A. G. et al., Unusual serum lipoprotein abnormality induced by the vehicle of miconazole, *N. Engl. J. Med.*, 296(9), 497, 1977.

Index

INDEX